MW00760110

Studies in Systems, Decision and Control

Volume 50

Series editor

Janusz Kacprzyk, Polish Academy of Sciences, Warsaw, Poland
e-mail: kacprzyk@ibspan.waw.pl

About this Series

The series "Studies in Systems, Decision and Control" (SSDC) covers both new developments and advances, as well as the state of the art, in the various areas of broadly perceived systems, decision making and control- quickly, up to date and with a high quality. The intent is to cover the theory, applications, and perspectives on the state of the art and future developments relevant to systems, decision making, control, complex processes and related areas, as embedded in the fields of engineering, computer science, physics, economics, social and life sciences, as well as the paradigms and methodologies behind them. The series contains monographs, textbooks, lecture notes and edited volumes in systems, decision making and control spanning the areas of Cyber-Physical Systems, Autonomous Systems, Sensor Networks, Control Systems, Energy Systems, Automotive Systems, Biological Systems, Vehicular Networking and Connected Vehicles, Aerospace Systems, Automation, Manufacturing, Smart Grids, Nonlinear Systems, Power Systems, Robotics, Social Systems, Economic Systems and other. Of particular value to both the contributors and the readership are the short publication timeframe and the world-wide distribution and exposure which enable both a wide and rapid dissemination of research output.

More information about this series at http://www.springer.com/series/13304

Muhammad Zeeshan Shakir
Muhammad Ali Imran · Khalid A. Qaraqe
Mohamed-Slim Alouini · Athanasios V. Vasilakos
Editors

Energy Management in Wireless Cellular and Ad-hoc Networks

Editors
Muhammad Zeeshan Shakir
Department of Systems and Computer Engineering
Carleton University
Ottawa, ON
Canada

Muhammad Ali Imran
Center for Communication Systems Research
University of Surrey
Guildford
UK

Khalid A. Qaraqe
Department of Electrical and Computer Engineering
Texas A&M University at Qatar
Doha
Qatar

Mohamed-Slim Alouini
Kind Abdallah University of Science and Technology
Thuwal
Saudi Arabia

Athanasios V. Vasilakos
Department of Computer Science, Electrical and Space Engineering
Luleå University of Technology
Skellefteå
Sweden

ISSN 2198-4182 ISSN 2198-4190 (electronic)
Studies in Systems, Decision and Control
ISBN 978-3-319-27566-6 ISBN 978-3-319-27568-0 (eBook)
DOI 10.1007/978-3-319-27568-0

Library of Congress Control Number: 2015957785

© Springer International Publishing Switzerland 2016
This work is subject to copyright. All rights are reserved by the Publisher, whether the whole or part of the material is concerned, specifically the rights of translation, reprinting, reuse of illustrations, recitation, broadcasting, reproduction on microfilms or in any other physical way, and transmission or information storage and retrieval, electronic adaptation, computer software, or by similar or dissimilar methodology now known or hereafter developed.
The use of general descriptive names, registered names, trademarks, service marks, etc. in this publication does not imply, even in the absence of a specific statement, that such names are exempt from the relevant protective laws and regulations and therefore free for general use.
The publisher, the authors and the editors are safe to assume that the advice and information in this book are believed to be true and accurate at the date of publication. Neither the publisher nor the authors or the editors give a warranty, express or implied, with respect to the material contained herein or for any errors or omissions that may have been made.

Printed on acid-free paper

This Springer imprint is published by SpringerNature
The registered company is Springer International Publishing AG Switzerland

Contents

About the Book

Along with the advent of the 5G era in telecommunications systems, new emerging applications, services and engineering are now being announced to facilitate the wide range of end user demands over cellular and ad hoc networks. Therefore, such rapid advancements in mobile and wireless communications are expected to increase the demand for higher data rates by several orders of magnitude over the next decade. The resulting customer demand for ubiquitous network access and wireless services is mainly responsible for increased energy consumption and, consequently, for the growing carbon footprint of the wireless communications industry.

Energy Management in Wireless Cellular and Ad-hoc Networks will bring together academic and industrial researchers and experts from communication and signal processing to present recent advances, identify technical challenges and forecast future trends related to a wide range of topics associated with energy efficiency in wireless cellular and ad-hoc networks. The book:

- Investigates energy management approaches for energy efficient or energy-centric system design and architecture.
- Presents end-to-end energy management in the recent heterogeneous-type wireless network medium.
- Considers energy management in wireless sensor and mesh networks by exploiting energy efficient transmission techniques and protocols.
- Explores energy management in emerging applications, services and engineering to be facilitated with 5G networks such as WBANs, VANETS and cognitive networks.
- Examines the energy management practices in emerging wireless cellular and ad-hoc networks.

Considering the broad scope of energy management in wireless cellular and ad-hoc networks, this book is organized into six sections covering a range of energy efficient systems and architectures; energy efficient transmission and techniques; energy efficient applications and services.

Muhammad Zeeshan Shakir
Muhammad Ali Imran
Khalid A. Qaraqe
Mohamed-Slim Alouini
Athanasios V. Vasilakos

Part I
Energy Management in Heterogeneous Networks

Outage Detection Framework for Energy Efficient Communication Network

Ahmed Zoha, Oluwakayode Onireti, Arsalan Saeed, Ali Imran, Muhammad Ali Imran and Adnan Abu-Dayya

Abstract In this chapter, we present a Cell Outage Detection (COD) framework for Heterogeneous Networks (HetNets) with split control and data planes. COD is a pre-requisite to trigger fully automated self-healing recovery actions following cell outages or network failures not only to ensure reliable recovery of services but also to significantly minimize wastage of energy. To cope with the idiosyncrasies of both the data and control planes, our proposed framework incorporates control COD and data COD mechanisms. The control COD leverage the relatively larger number of UEs in the control cell to gather large scale Minimize Drive Testing (MDT) reports data. These measurements are further pre-processed using multidimensional scaling method and are employed together with state-of-the art machine learning algorithms to detect and localize anomalous network behaviour. On the other hand, for data cells COD, we propose a heuristic Grey-Prediction based approach, which can work with the small number of UEs in the data cell, by exploiting the fact that the control BS manages UE-data BS connectivity, by receiving a periodic update of the Received Signal Reference Power (RSRP) statistic between the UEs and data BSs in its coverage. The detection accuracy of the heuristic data COD algorithm is further

A. Zoha (✉) · A. Abu-Dayya
QMIC, Qatar Science and Technology Park, Doha, Qatar 210531
e-mail: ahmedz@qmic.com

O. Onireti · A. Saeed · M.A. Imran
Institute of Communication Systems (ICS), University of Surrey, Guildford, GU2 7XH, UK
e-mail: o.s.onireti@surrey.ac.uk

A. Saeed
e-mail: a.saeed@surrey.ac.uk

M.A. Imran
e-mail: m.imran@surrey.ac.uk

A. Imran
University of Oklahoma, Tulsa 71435, USA
e-mail: ali.imran@ou.edu

A. Abu-Dayya
e-mail: adnan@qmic.com

© Springer International Publishing Switzerland 2016

M.Z. Shakir et al. (eds.), *Energy Management in Wireless Cellular and Ad-hoc Networks*, Studies in Systems, Decision and Control 50,
DOI 10.1007/978-3-319-27568-0_1

improved by exploiting the fourier series of residual error that is inherent to grey prediction model. We validate and demonstrate the effectiveness of our proposed solution for detecting cell outages in both data and control planes via performing network simulations under various operational settings.

1 Introduction

The increased demands of high throughput, coverage and end user quality of service (QoS) requirements, driven by ever increasing mobile usage, incur additional challenges for the network operators. Fueled by the mounting pressure to reduce capital and operational expenditures (CAPEX & OPEX) and improve efficiency in legacy networks, the Self-Organizing Network (SON) paradigm aims to replace the classic manual configuration, post deployment optimization, and maintenance in cellular networks with self-configuration, self-optimization, and self-healing functionalities. A detailed review of the state-of-the-art SON functions for legacy cellular networks can be found in [1]. The main task within self-healing functional domain is autonomous cell outage detection and compensation. Current SON solutions generally assume that the spatio-temporal knowledge of a problem that requires SON-based compensation is fully or at least partially available; for example, location of coverage holes, handover ping-pong zones, or congestion spots are assumed to be known by the SON engine [2, 3]. Traditionally, to assess and monitor mobile network performance manual drive test have to be conducted. However, this approach cannot deliver the stringent resource efficiency and low latency, and cannot be used to construct dynamic models to predict system behavior in live-operation fashion. This ultimately results in pronounced reduction in capacity and quality of service, and coverage gap [4, 5]. Moreover, today the energy demand for mobile networks is in gigawatts per hour per year [6]. The traditional radio systems are optimized for maximum load, whereas excessive waste of energy occurs under conditions in which either the traffic is low or the system is not providing services to the users (i.e., under cell outages). With increase scale of networks, automatic detection and compensation of cell outage has become a necessity, and, it has been included in recent 3GPP releases [7]. The proposed COD framework aims to autonomously detect outage cells, i.e., cells that are not operating properly due to possible failures, e.g. external failure such as power supply or network connectivity, or even misconfiguration [4, 5, 7, 8]. The timely cell outage detection in a heterogeneous network not only ensure reliable services to the users, but it also significantly reduces the overall energy wastage, since the major source of power consumption in mobile networks stems from the radio base stations [9].

A few algorithms have already been proposed in literature, e.g. in [4, 8, 10–16] for COD. All these works have focused on the traditional homogeneous deployments, where only macro cells are deployed. However, future cellular deployments are expected to be heterogeneous and extremely dense. In this context, macro cells will provide the UEs with ubiquitous experience, while dense small cell deployments, operating in bandwidths with heterogeneous characteristics, will facilitate high data rate transmissions to a reduced number of UEs. At the same time, conventional heterogeneous deployments pose a number of challenges in terms of network management and energy consumption, as a result of the increased number of cells. In order to mitigate these challenges, a new HetNet architecture with split control and data planes has been recently proposed as a candidate architecture for 5G [17–21]. In such architecture, the control and data planes are separated and are not necessarily handled by the same node. Consequently, this gives the network operator more flexibility, since the small/data cells can be activated *on demand* to deliver UE-specific data only when and where needed, while the macro/control cells manage UEs connectivity and mobility [21]. Thus, the separated plane architecture allows for improved mobility management performance, since the RRC layers of the UEs and other control messaging, such as paging, will be handled by the control cells. In addition, the energy consumption is improved, since the proposed architecture also leads to longer data cell sleep periods, due to their on demand activation. Note that contrarily to the newly proposed HetNet architecture, the RRC layers of all UEs in the conventional HetNet are handled by their serving cell, which could be either a small or macro cell.

The control plane provides ubiquitous network access and is made up of macro base station (BS)s, which we refer to as control BSs. On the other hand, data plane supports high data rate transmission and is composed of the small BSs, which we call data BS [3–6]. Note that contrarily to the newly proposed HetNet architecture, the RRC layers of all UEs in the conventional HetNet are handled by their serving cell, which could be either a small or macro cell [19–21].

To the best of our knowledge, a complete COD solution particularly for HetNet with split control and data plane, is still missing. In this paper, we propose two distinct COD algorithms to cope with the peculiarities of data and control cells. Since control cells tend to have a large number of UEs, we exploit large scale collection of MDT reports, introduced by 3GPP in [22], and we apply machine learning based anomaly detection schemes for control COD. The reported studies in literature that addressed the problem of detecting outages in a macro cell environment are either based on quantitative models [10], which requires domain expert knowledge, or simply rely on performance deviation metrics for detection [11]. Until recently, researchers have applied methods from the machine learning domain such as clustering algorithms [23] as well as Bayesian Networks [14] to automate the detection of faulty cell behavior. Coluccia et al. [13] analyzed the variations in the traffic profiles for 3G cellular systems to detect real-world traffic anomalies, as well as network visibility graph approaches using *Neighbor Cell List (NCL)* reports [24] have also been considered for COD.

Compared to aforementioned approaches, our control COD adopts a model-driven approach that exploits MDT functionality [25] as specified by 3GPP. MDT mechanism allows control BS to request and configure UEs to report back the key performance indicators (KPIs) from the serving and neighboring cells along with their location information. To accurately capture the network dynamics, we first collect UE reported MDT measurements and further extract a minimalistic KPI representation by projecting them to a low-dimensional embedding space. We then employ these embedded measurements together with machine learning algorithm to autonomously learn the "normal"operational profile of the network. The learned profile leveraging the intrinsic characteristics of embedded space intelligently diagnose a outage cell situation. This is in contrast to state-of-the-art techniques that analyze one or two KPIs to learn the decision threshold levels and subsequently apply them for detecting network anomalies. In addition, the proposed solution further exploits the geo-location associated with each measurement to localize the position of the faulty cell, enabling the SON to autonomously trigger cell outage compensation actions.

Furthermore, to find the best detection model for control COD, we compare and evaluate the performance of density and domain based anomaly detection approaches: *Local Outlier Factor based Detector* (LOFD) and *One Class Support Vector Machine based Detector* (OCSVMD), respectively, while taking into consideration the acute dynamics of the wireless environment due to channel conditions as well as load fluctuations. To the best of our knowledge, no prior study examines the use of OCSVMD and LOFD in conjunction with embedded MDT measurements for autonomous cell outage detection.

However, the same COD scheme cannot be applied for data cells, as number of users will not be large enough to constitute reliable training models for underlying anomaly detection techniques. To overcome this problem, we take advantage of the following peculiarities, about data cells, to develop a heuristic, yet reliable data COD algorithm. The RRC layers of all UEs are handled by the control cells, as a result, the control BS is aware of: (1) every UE-data BS association within its coverage, (2) the state of each UE (idle or active), (3) every radio link failure between the UEs and data BSs, (4) every handovers to other data BSs in its coverage and (5) data link handover from the data BS to itself. Also, once the normal state of the control cells has been established, each UE associated to the data cells can periodically report the RSRP statistic between itself and its associated data cell to its serving control cell. Using these observations, we propose a heuristic data COD scheme, which works despite of small number of UEs in the data cell, by exploiting a GM for detecting data cell outage.

We design, evaluate and compare COD solutions with network simulations that are setup in accordance with 3GPP LTE standard. In addition to proposing a COD framework for the HetNet with split control and data plane, we believe the proposed solution provided paves a way towards developing a fully automated cell outage management solution via integrating self-healing functionality in the proposed architecture for the emerging (LTE) as well as future (5G) self organizing networks. The remainder of this paper is organized as follows. In Sect. 2, we present the system architecture, which includes description of the system model and an overview of the COD framework. In Sect. 3, we elaborate the control COD solution for detecting outages in control cells. In Sect. 4, we introduce a heuristic based data COD scheme. In Sect. 5, we present extensive simulations to substantiate the performance of our proposed COD framework for HetNets with separated control and data planes. Furthermore, we also discuss the impact of energy efficiency gain of the COD framework. Finally, Sect. 6 concludes this paper.

2 System Architecture

2.1 System Model

We consider the new paradigm of HetNets architecture where the control plane provides ubiquitous network access and is made up of macro\control BSs, and the data plane supports high data rate transmission and is composed of the small\data BSs. The control plane is responsible for handling UE connectivity, RRC connection management as well as different radio-specific functions. In contrast, the data plane handles UE specific data, and its functionlities are unicast and synchronization [20]. Consequently, UEs requiring high data rate transmission are connected to both the control and data BSs, while low rate UEs are connected to just the control BS.

We consider that the control and data BSs are operated on separate frequency carriers, so that there is no interference between the two planes. We assume that the HetNet is composed of a set of $\mathscr{M}$ control BSs and $\mathscr{F}$ data BSs, where $M = |\mathscr{M}|$ control BSs form a regular hexagonal network layout with inter-site distance D and provides coverage over the entire network. The $F = |\mathscr{F}|$ data BSs are randomly distributed. We also consider that U_m and U_f UEs are provided with service by the $\mathscr{M}$ control BSs and $\mathscr{F}$ data BSs, respectively. The multi user resource assignment to the R RBs in a plane is carried out by a FFR scheduler and each of the UEs in the plane is assigned a CQI value.

2.2 COD Framework

The COD framework aims to detect the network performance deterioration, whenever a problem occurs within a network either in control or data plane. This can be achieved by monitoring deviations from normal operational behaviour of the network. To do so, we first collect the KPI measurements report from the fault free network which are subsequently used to learn decision models by our control and data COD solutions. These decision models can then be employed to detect outage situations in the network during the monitoring period. An overview of our proposed COD framework, which is primarily consists of the control COD and data COD modules, is illustrated in Fig. 1.

As mentioned earlier, active high data rate UEs are served by both the data and control BS, while the low rate UEs are served by only the control BS. This implies that all UEs maintain connectivity with the control BS. Furthermore, as a result of the split of the control and data planes, the control and data cell outages are independent of each other, hence, the detection of a cell outage in each plane is executed independently of the other. As shown in Fig. 1, our framework has two distinct COD algorithms to cope with the idiosyncrasies of the control and data cell. The control cell tend to have large number of users, so a large scale data collection and machine learning is used for the control COD. Consequently, our control COD scheme is

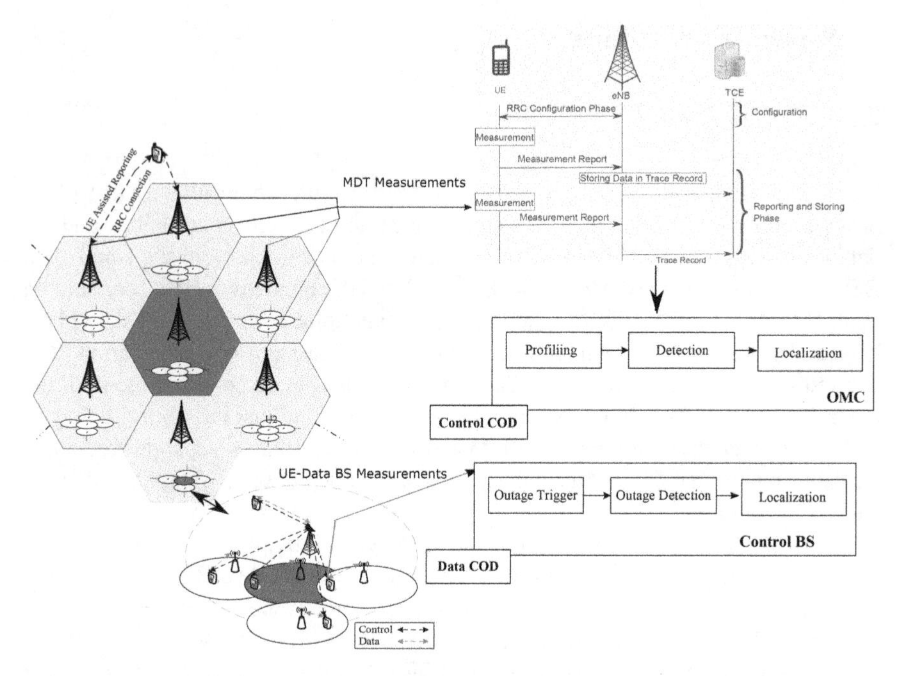

Fig. 1 System model for cell outage detection

based on the MDT functionality, where all UEs report their MDT measurements, which include RSRP of the serving and neighboring cells, to the OMC center vis respective BS. A normal network profile is built based on the measurements in a non-outage scenario. The control COD is then performed at the OMC, by using anomaly detection algorithms, such as OCSVMD and LOFD, on the actual network profile.

However, this approach is not applicable for data COD, where UE statistics are sparse, due to the small number of UEs connected to each data BS. The control BS knows the location of every data BS in its coverage and it can passively monitor the RSRP measurements of every UE-data BS association within its coverage. Consequently, a heuristic scheme, which can effectively leverage on the small number of reports effectively and the fact that the control BS can monitor the UE-data BS association, is used for the data COD. The data COD is executed at the control BS and is triggered when the control BS detects irregularities in UE-data BS associations, while the actual detection is performed by using a GM algorithm.

3 Control COD

The MDT reporting schemes have been defined in LTE Release 10 during 2011 [25]. The release proposes to construct a data base of MDT reports from the network using *Immediate* or *Logged* MDT reporting configuration. In this study, the UE's are configured to report the cell identification and radio-measurement data to the control BS based on immediate MDT configuration procedure as shown in Fig. 1. The signaling flow of MDT reporting procedure consists of configuration, measurement, reporting and storing phase. The UE is first configured to perform measurements periodically as well as whenever an A2 event (i.e., serving cell becomes worst than a *threshold*) occurs. Subsequently, it performs KPI measurements: serving and neighbors Reference Signal Received Power (RSRP), serving and neighbors Reference Signal Received Quality (RSRQ), and further reports it to the control BS. The control BS after retrieving these measurements further appends time and wide-band channel quality information (CQI) and forwards it to Trace Collection Entity (TCE). TCE collects and stores the trace reports which are subsequently processed to construct a MDT database. In this study, the trace records obtained from the reference scenario (i.e., fault-free) act as a benchmark data and is used by the anomaly detection models to learn the network profile. These models are then employed to autonomously detect and localize outage situations. The control COD solution as shown in Fig. 1 consists of profiling, detection and localization phases, as detailed in the following subsections.

3.1 Profiling Phase

The next step after collecting measurements from the network is to perform data transformation. Each trace record is processed to extract a KPI vector V that contains the RSRP and RSRQ KPIs of the serving as well as of the three strongest neighbouring cells along with the CQI augmented to form a measurement vector as shown in Eq. 1

$$\begin{aligned} V = \{&RSRP_S, RSRP_{n_1}, RSRP_{n_2}, RSRP_{n_3},\\ &RSRQ_S, RSRQ_{n_1}, RSRQ_{n_2}, RSRQ_{n_3}, CQI\} \end{aligned} \tag{1}$$

where S and n stands for serving and neighboring cells, respectively. The 9-dimensional feature vector V corresponds to one measurement sample which is further embedded to only three dimensions in the Euclidean space using Multi-Dimensional Scaling (MDS) method [26]. MDS provides a low-dimensional embedding of the target KPI vector V while preserving the pairwise distances amongst them. Given, a $t \times t$ dissimilarity matrix Δ^X of the MDT dataset, MDS attempts to find t data points $\psi_1 \ldots \psi_t$ in m dimensions, such that Δ^Ψ is similar to Δ^X. Classical MDS (CMDS) operates in Euclidean space and construct an m dimensional embedding of the data points, whereas the value of m is chosen to be 3 in our case. Further details on the mathematical formulation of MDS embedding can be found in [15]. The embedded KPI representation V^e has several advantages. First, it makes the framework generic allowing it to incorporate new KPI's and network-centric features such as call drop ratios, data traffic etc. without imposing higher computational requirements. Moreover, the interrelationships of high-dimensional databases can be explored in a lower-dimension space. Secondly, given the growing complexity of the networks, particularly in case of SON, it is challenging to identify few KPIs that accurately capture the behavior of the system. The network-level intelligence can be inferred through low-dimensional representation of large volume of network measurements. The embedded space reveals a hidden structure by mapping similar measurements close to each other and vice versa, that naturally isolates high and low data density regions. This makes it easier to detect the underlying patterns that are representative of network dynamics. The learning algorithm leverages embedded network measurements to learn a optimal decision rule and subsequently during the monitoring phase apply it to classify observed network measurements as anomalous or normal, as discussed in detail below.

To construct a reference database D_M, we apply an MDS based data transformation on the network measurements collected from a fault-free operating scenario. The D_M also includes samples of Radio Link Failure (RLF) events, in addition to periodical MDT measurements, as expected in a realistic environment. As shown in Fig. 2a, the D_M acts as a training database for the anomaly detection algorithm, enabling it to learn the "normal" network behaviour. This involves learning a decision function 'f' and a corresponding threshold 'θ', which is used to differentiate between normal and abnormal network measurements. Thus, it can be treated as a binary classification problem which can formally be expressed as follows:

$$f(x_i) = \begin{cases} Normal, & \text{if } f(x_i, D_M) \leq \theta \\ Anomalous, & \text{if } f(x_i, D_M) > \theta \end{cases} \tag{2}$$

where x_i is the test measurement. Two state-of-the-art algorithms from the machine learning domain: OCSVMD and LOFD are examined for modeling the dynamics of network operational behaviour. The brief working description of the two detection algorithms are summarized as follows.

3.1.1 Local Outlier Factor Based Detector (LOFD)

The LOFD method [27] adopts a density based approach to measure the degree of outlyingness of each instance. In comparison to nearest neighbor based approaches, it works by considering the difference in the local density ρ of the sample to that of its k neighbors; instead of relying on distance estimation alone. A higher score will be assigned to the sample, if ρ is highly different from the local densities of its neighbor. The algorithm starts by first computing the distance of the measurement x to its kth nearest neighbor denoted by d_k, such that

$$\begin{aligned} d(x, x_j) &\leq d(x, x_i) \quad \text{for at least } k \text{ samples} \\ d(x, x_j) &< d(x, x_i) \quad \text{for at most } k-1 \text{ samples} \end{aligned} \tag{3}$$

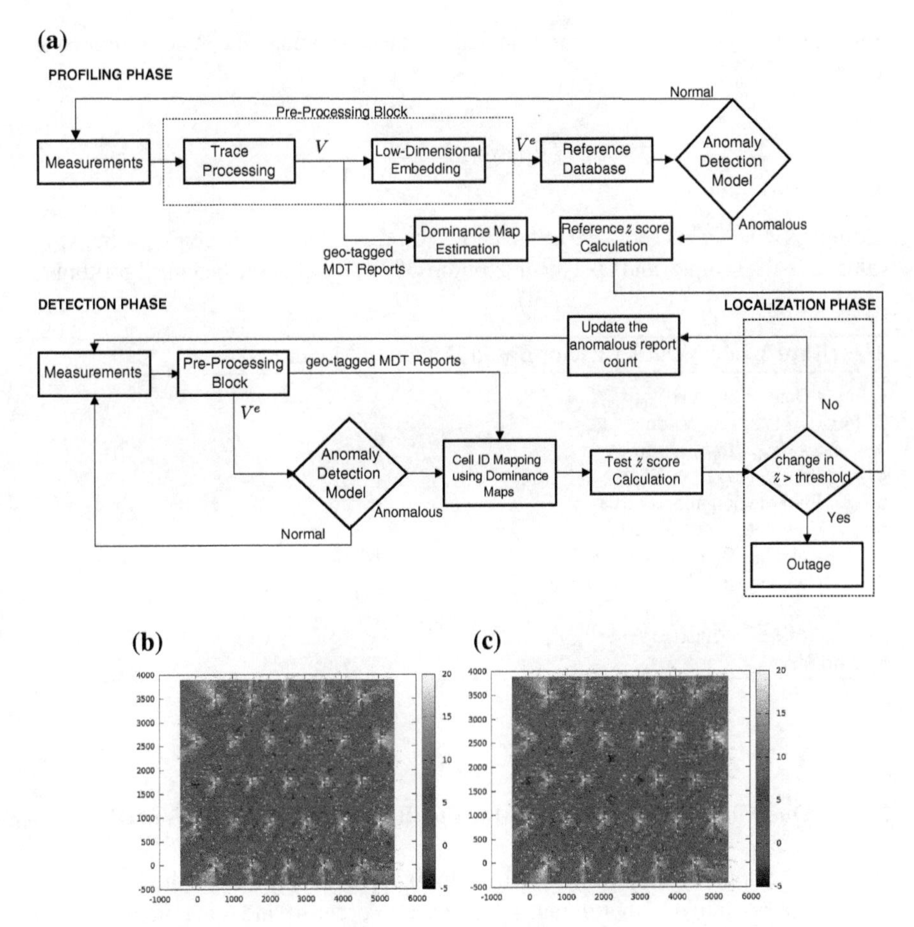

Fig. 2 An overview of COD framework. **a** Profiling, detection and localization phases of COD framework. **b** Normal/reference scenario. **c** Outage scenario

The subsequent step is to construct a neighborhood $\mathcal{N}_k(x)$ by including all those points that fall within the d_k value. The following step is to calculate the reachability distance of sample x with respect to rest of the samples

$$d_r(x, x_i) = \max\{d_k(x_i), d(x, x_i)\} \tag{4}$$

The local reachability density ρ is the inverse of average d_r and can be defined as

$$\rho(x) = \frac{\mid \mathcal{N}_k(x) \mid}{\sum_{x_i \in \mathcal{N}_k(x)} d_r(x, x_i)} \tag{5}$$

Finally, the $\mathscr{S}^{(LOFD)}$ represents a local density-estimation score whereas value close to 1 mean x_i has same density relative to its neighbours. On the other hand, a significantly high $\mathscr{S}^{(LOFD)}$ score is an indication of anomaly. It can be computed as follows:

$$\mathscr{S}^{(LOFD)}(x) = \frac{\sum_{x_i \in \mathscr{N}_k(x)} \frac{\rho(x_i)}{\rho(x)}}{\mid \mathscr{N}_k(x) \mid} \tag{6}$$

Since, $\mathscr{S}^{(LOFD)}$ is sensitive to the choice of k, we iterate between k_{min} and k_{max} value for each sample, and take the maximum $\mathscr{S}^{(LOFD)}$ as described in Algorithm 1.

Algorithm 1 Local Outlier Factor Based Detection Model

1: Input Data $\mathscr{X} = \{x_j\}_{j=1}^{N}, k_{min}, k_{max}$
2: **for** $j = 1, 2, \ldots, N$: **do**
3:    **for** $k = k_{min}$ to k_{max}: **do**
4:       Find $d_k(x_j)$ from Eq. 3
5:       Find the neighborhood $\mathscr{N}_k$ of x_j
6:       Calculate $d_r(x_j, x_i)$ from Eq. 4
7:       Calculate $\rho(x_i)$ from Eq. 5
8:       Calculate $\mathscr{S}^{(LOFD)}$ from Eq. 6
9:    **end for**
10:    $\mathscr{S}^{(LOFD)} = \max(\mathscr{S}^{(LOFD)}{}_{k_{min}}, \ldots, \mathscr{S}^{(LOFD)}_{k_{max}})$
11: **end for**

3.1.2 One-Class Support Vector Machine Based Detector (OCSVMD)

One-Class Support Vector Machine by Schölkopf et al. [28] maps the input data/feature vectors into a higher dimensional space in order to find a maximum margin hyperplane that best separates the vectors from the origin. The idea is to find a binary function or a decision boundary that corresponds to a classification rule

$$f(x) = <\mathbf{w}, \mathbf{x}> +b \tag{7}$$

The $\mathbf{w}$ is a normal vector perpendicular to the hyperplane and $\frac{b}{\|\mathbf{w}\|}$ is an offset from the origin. For linearly separable cases, the maximization of margin between two parallel hyperplanes can be achieved by optimally selecting the values of w and b. This margin, according to the definition is $\frac{2}{\|\mathbf{w}\|}$. Hence, the optimal hyperplane should satisfy the following conditions

$$\begin{aligned} minimize\ \frac{1}{2}\|\mathbf{w}\|^2 \\ subject\ to: y_i(\langle \mathbf{w}, x_i \rangle + b) \geq 1 \\ for\ i = 1, \ldots, N \end{aligned} \tag{8}$$

The solution of the optimization problem can be written in an unconstrained dual form which reveals that the final solution can be obtained in terms of training vectors that lie close to the hyperplanes, also referred to as support vectors. To avoid overfitting on the training data, the concept of *soft decision* boundaries was proposed, and slack variable ξ_i and regularization constant ν is introduced in the objective function. The slack variable is used to soften the decision boundaries, while ν controls the degree of penalization of ξ_i. Few training errors are permitted if ν is increased while degrading the generalization capability of the classifier. A *hard margin* SVM classifier is obtained by setting the value of $\nu = \infty$ and $\xi = 0$. The detail mathematical formulation for SVM models can be found in [28]. The original formulation of SVM is for linear classification problems; however non-linear cases can be solved by applying a kernel trick. This involves replacing every inner product of $x.y$ by a non-linear kernel function, allowing the formation of non-linear decision boundaries. The possible choices of kernel functions includes polynomial, Gaussian radial basis function (RBF), and sigmoid. In this study, we have used the RBF kernel: $\kappa(x, y) = exp(-\|x - y\|^2/2\sigma^2)$, and the corresponding parameter values of the model are selected using cross validation method, as discussed in Sect. 5.1.

As shown in Fig. 2a, using the benchmark data, we compute a reference z-score for each control BS in the network. The z-score is calculated as follows: $z_b = \frac{|n_b - \mu_n|}{\sigma_n}$ where n_b is the number of MDT reports labeled as anomalies for the eNB b, and variables μ_n and σ_n are the mean and standard deviation anomaly scores of the neighbouring cells. In the profiling phase, we also estimate the so called dominance area, i.e., for each cell, we define the area where its signal is the strongest. This is to establish the coverage range for each cell by exploiting the location information tagged with each UE measurement. The dominance estimation is required to determine a correct cell and MDT measurement association during an outage situation. This is because as soon as the SC situation triggers in the network, the malfunctioning control BS either becomes completely unavailable or experience severe performance issues. This triggers frequent UE handovers to the neighboring cells, and as a result the reported measurements from the affected area contains the neighbor cell E-UTRAN Cell Global Identity (CGI), instead of the target cell. Hence, CGI alone cannot be used to localize the correct position of faulty cell during an outage situation. The detection and localization phase of our control COD solution make use of estimated dominance map and reference z-score information established in the profiling phase to detect and localize faulty cell as discussed in the following subsection.

3.2 *Detection and Localization Phase*

In the detection phase, the trained detection model is employed to classify network measurements as normal or anomalous. The output of the detection models allow us to compute a test z-score for each control BS. To establish a correct cell measurement association, the geo-location of each report is correlated with the estimated dominance maps. In this way, we can achieve detection and localization by comparing the deviation of test z-score of each cell with that of reference z-score, as illustrated in Fig. 2a.

4 Data COD via Heuristic Approach

Contrarily to the control COD, which is performed at the OMC, the data COD is performed at each control BS. Hence, establishing the normal state of the control BS is a pre-requisite for data COD. The data COD process is organized into the trigger and detection phases, as illustrated in Fig. 3. The control BS receives a periodic update of the RSRP of each UE to its associated data BS and stores this value in a database, which is later used in the detection process.

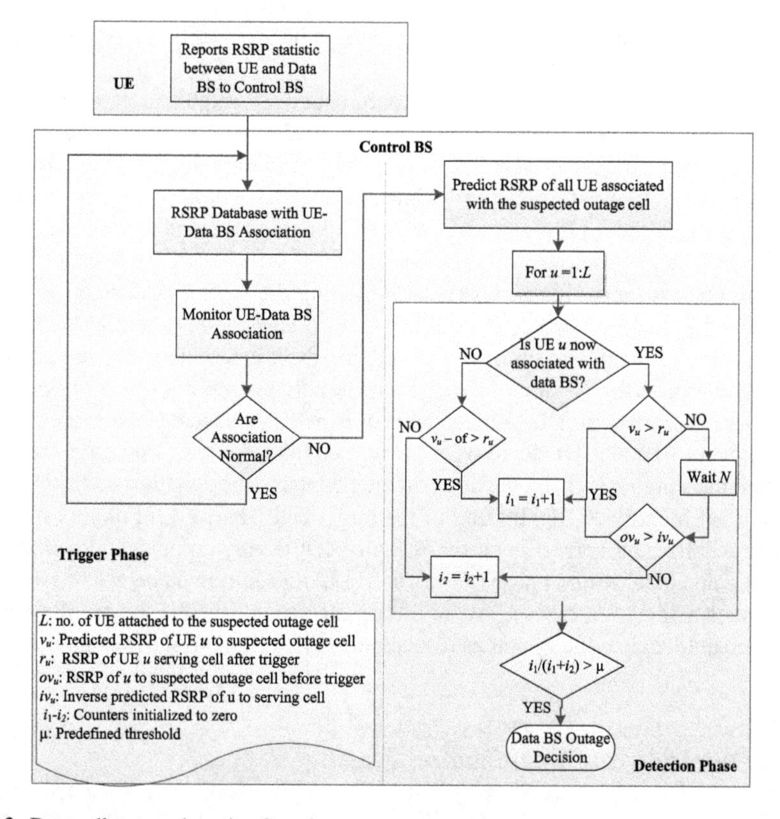

Fig. 3 Data cell outage detection flowchart

4.1 Outage Trigger Phase

As mentioned earlier in Sect. 2.2, the control BS is responsible for managing UE connectivity, as well as radio specific functions, such as: (1) RRC connection, (2) configuration and measurement reporting, (3) cell handover and network controlled mobility. Consequently, the control BS is aware of any change in UE-data BS association, as a result of handover or radio link failure. The control BS is also aware of any state change in the UE, such as a change from active to idle state, idle to detached state and vice versa. Furthermore, the conditions for data BS to enter the sleep mode is known to the control BS. For example, the data BS could be allowed to enter the sleep mode if the number of active UEs is lower than a certain predefined threshold during the last scheduling time interval.

In the outage trigger phase, the control BS monitors the UE-data BS association and triggers the outage detection when it discovers irregularities in UE-data BS association. Irregularities in UE-data BS association occur when all UEs attached to a particular data BS changes their association without any of the following: (1) prior handover initiation process, (2) change in state of all the UEs, (3) radio link failure notification from all the UEs, (4) the data BS going into sleeping mode.

4.2 Outage Detection Phase

Once the outage detection phase is triggered, the control BS can detect outage of the data BS by predicting the RSRP of all the UEs that were associated with it prior to the outage. We utilize the GM, which has been extensively used in handover, positioning and general forecasting algorithms [29–31], as prediction model.

4.2.1 GM Approach

In grey system theory, $GM(\bar{n}, \bar{m})$ denotes a grey model, where $\bar{n}$ is the order of the differential equation and $\bar{m}$ is the number of variables. Here we focus on $GM(1, 1)$, which is a widely used time series forecasting model. According to [29], the $GM(1, 1)$ model can only be used on positive data sequences. Note that the RSRP values are always positive, hence, the grey model can be used to predict the next RSRP value from data points obtained in the database.

The three basic operations in grey prediction are: (1) the AGO, (2) the IAGO, (3) Grey Modelling. By using AGO, an irregular raw data can be transformed into a regular data, which can be used to construct a model in grey differential equation. The non-negative RSRP data sequence of UE u prior to the outage is denoted as

$$r_u^{(0)} = \left(r_u^{(0)}(1), r_u^{(0)}(2), r_u^{(0)}(3), \ldots, r_u^{(0)}(n)\right), \forall n \geq 4. \tag{9}$$

The prediction value of the benchmark RSRP data at time $(c+1)$ can be calculated by an IAGO as

$$\hat{r}_u^{(0)}(c+1) = \left[r_u^{(0)}(1) - \frac{b}{a}\right] e^{-ac}\left(1 - e^{a}\right), \tag{10}$$

where a and b are coefficient defined in [29, 31],

4.2.2 GMF

According to [31] grey model prediction accuracy can be improved by the Fourier series of error residuals. Consider the uth UE RSRP sequence, $r_u^{(0)}$ in (9) and its predicted values obtained from (10), then the error of the sequence $r_u^{(0)}$ can be expressed as

$$\xi_u^{(0)} = \left(\xi_u^{(0)}(2), \xi_u^{(0)}(3), \ldots, \xi_u^{(0)}(n)\right), \tag{11}$$

where

$$\xi_u^{(0)}(c) = r_u^{(0)}(c) - \hat{r}_u^{(0)}(c), \forall c = 2, 3, \ldots, n. \tag{12}$$

The error residuals given in (12) can be re-expressed in Fourier series [31] such that the Fourier series correction can be expressed as

$$\hat{\hat{r}}_u^{(0)}(c) = \hat{r}_u^{(0)}(c) - \hat{\xi}_u^{(0)}(c), \forall\, c = 2, 3, \ldots, n+1. \tag{13}$$

4.3 *Outage Decision*

Firstly, the RSRP of all the UEs that were previously attached to the data BS, whose outage is being detected, i.e., data BS, d, are predicted according to (10) or (13). Then, for each UE the control BS compares its predicted RSRP value, $v_u = \hat{r}_u^{(0)}(c+1) \approx \hat{\hat{r}}_u^{(0)}(c+1)$, with the RSRP after the trigger, $r_u = r_u^{(0)}(c+1)$. If afterwards outage is triggered, the UE, u, is served by the control BS for data transmission and $v_u = \hat{r}_u^{(0)}(c+1) \approx \hat{\hat{r}}_u^{(0)}(c+1) > r_u - \Delta$, where Δ is the data cell range expansion offset, the counter, i_1, is incremented by 1, since the UE should be associated with data BS, d, based on the prediction. Otherwise, the counter, i_2 is incremented by 1. On the other hand, if another data BS is serving UE u, after the outage trigger and $v_u \approx \hat{r}_u^{(0)}(c+1) \approx \hat{\hat{r}}_u^{(0)}(c+1) > r_u$, the counter, i_1, is incremented by 1, otherwise an inverse prediction is performed on the RSRP to the serving data BS. The inverse prediction checks the RSRP to the data BS d and the RSRP to the serving data BS after the trigger, i.e. data BS $\bar{d}$, at the point just before the trigger. The control BS waits for the prediction window size, N, and performs an inverse prediction on the RSRP of each of the UEs associated with data BS $\bar{d}$ to obtain the predicted RSRP prior to the trigger decision, iv_u. Thus, if the RSRP of the uth UE to the serving data BS (d) before trigger, ov_u, is such that $ov_u > iv_u$ the counter i_1 is incremented by 1 otherwise, the counter i_2 is incremented by 1. The data cell outage is declared if the ratio $\frac{i_1}{i_1+i_2} > \mu$, where μ is a predefined threshold.

5 Simulation Results and Discussion

5.1 Simulation Setup

To simulate the LTE network based on 3GPP specifications, we employ a full dynamic system tool. We consider a HetNet architecture where the control and data BSs operate on separate frequency carriers. Each operation mode occupies 5 MHz of channel bandwidth.The scenario that we set up consists of 27 macro/control BSs with $U_m = 20$ UEs per control BS, and $FB = 5$ femtocell blocks per control BS, each one with $l = 40$ apartments, $t = 1$ floor, $d = 0.2$ small/data BS deployment ratio as per [32], and $c = 0.5$ data BS activation ratio, which results in 20 data BSs per control BS. Also, there are U_f UEs per data BS in the scenario. To model the variations in signal strength due to topographic features in an urban environment, the shadowing is configured to vary within a range for values. Normal periodical MDT measurements as well as RLF-triggered data due to intra-network mobility, reported by UE's to control BS, is used to construct a reference database for training outage detection models. To simulate a hardware failure in the network, at some point in the simulation the antenna gain of cell 11 is attenuated to −50 dBi that leads to a cell outage in a network. The measurements collected from the outage scenario are then used to evaluate the detection and localization performance of the proposed COD framework. The SINR plots of the reference and SC scenario has been already shown in Fig. 2b, c, respectively. The detailed simulation parameters are listed in Table 1. The detection performance of the outage detection models is also examined for different network configurations, obtained by varying the simulation parameter settings for ISD, load and shadowing.

Parameter Estimation and Evaluation

The parameter selection for LOFD and OCSVMD is performed using a combination of grid search and cross-validation (CV) method. Initially, a grid of parameter values are specified that defines the parameter search space. For example, the hyperparameters of OCSVMD ν and kernel parameter γ is varied from 0 to 1 with 0.05 interval to determine different combinations. Subsequently, for every unique parameter combination C_i, CV is performed as follows: The D_M is divided into training D_{train} and validation dataset D_{val}, and subsequently performance of the model is evaluated using K-folds approach. The value of K is chosen to be 10 in our framework.

Table 1 Simulation parameters

Parameter	Values
Cellular layout	27 Macrocell sites
Inter-site Distance (ISD)	1000 m
Sectors	3 Sectors per cell
User distribution	Uniform random distribution
Path loss	$L[dB] = 128.1 + 37.6 log_{10}(R)$
Antenna gain (Normal Scn)	18 dBi
Antenna gain (SC Scenario)	−50 dBi
Slow fading Std	8 dB
Simulation length	420s (1 time step = 1 ms/1TTI)
Control BS Tx Power	46 dBm
Data BS Tx Power	23 dBm
Network synchronization	Asynchronous
HARQ	Asynchronous, 8 SAW channels, Maximum Retransmission = 3
Cell selection criteria	Strongest RSRP defines the target cell
Load	20 users/cell
MDT reporting interval	240 ms
Traffic model	Infinite buffer
HO margin	3 dB
Detection threshold μ	0.5
Detection window size N	10
Grey weighting factor α	0.5
SINR threshold	−6 db

The performance estimate of the model over K folds is averaged and iteratively this process is repeated until all the parameter combinations are exhausted. The C_i yielding the highest performance estimate is selected as an optimal parameter combination for the target model. The value of k_{min} and k_{max} for LOFD is found out to be 5 and 14. In case of OCSVMD, RBF kernel is employed and the values of the hyperparamters ν and γ is found out to be 0.3 and 0.25, respectively. Finally, the test data D_{test} from the outage scenario, has been used to estimate the performance of the trained models.

In our study, the quality of the target models is evaluated using Receiver Operating Characteristic (ROC) curve analysis. The ROC curve plots the true positive rate or detection rate (DR) (i.e., a percentage of anomalous measurements correctly classified as anomalies) against the false positive rates (FPR) (i.e., a percentage of normal cell measurements classified as anomalies) at various threshold settings. An Area under ROC curve (AUC) metric is used for model comparison, whereas a AUC value of 1 or close to it, is an indicator of higher discriminatory power of the target algorithm.

5.2 Control COD Outage Detection Results

The training database D_M contains pre-processed embedded measurements from the reference scenario as discussed in Sect. 3.1. The database is subsequently used to model the normal operational behaviour of the network. The database measurement also includes RLF-triggered samples, since even in the reference scenario UE's experience connection failures due to intra-LTE mobility or shadowing. The test data collected from the outage scenario is used to evaluate the performance outage detection models.

The diagnosis process has been tested in twelve scenarios by changing the shadowing, user-density and inter-site distance (ISD) parameters of the baseline simulation setup as listed in Table 1. We have evaluated the detection performance of the OCSVMD and LOFD against every target network configuration. Figure 4a, illustrates the MDS projection of MDT measurements from the normal and the outage scenario using the baseline network operational settings. It can be observed that the abnormal measurements belonging to SC scenario lie far from the regular training observations. As discussed earlier in Sect. 3.1, MDS tries to maximize the variance between the data points and consequently dissimilar points are projected far from each other, allowing the models to compute a robust dissimilarity measure for outage detection. The goal of OCSVMD is to learn a close frontier delimiting the contour of training observations obtained from the non-outage scenario. In this way, any observation that lie outside of this frontier-delimited subspace (i.e. representative of the normal state of the network) is classified as an anomaly or an abnormal measurement. However, the inlier population (i.e. measurements that lie inside the OCSVMD frontier) is contaminated with RLF events, which ultimately elongates the shape of the learned frontier. As a result, during the detection phase, the observations from the outage scenario exhibiting similarity to RLF-like observations are positioned within the frontier-delimited space as shown in Fig. 4a, and hence wrongly classified as normal. The shape of the learned frontier determines the precision of the model for detecting anomalous network measurements.

To study the impact of different radio propagation environment on the detection performance, we varied the shadowing parameter from 8 to 4 dB and 12 dB cases. Under low-shadowing conditions (i.e., 4 dB), it can be observed from Fig. 4b that inlier population exhibits wider separation from anomalous observation in comparison to reference scenario. This is because higher shadowing conditions affects the spread of the KPI measurements, as indicated in Fig. 4c. It can be inferred from the ROC analysis of OCSVMD, that detection performance deteriorates as the shadowing effect is varied from low to high. As shown in Fig. 5a, at target false positive rate of 10 %, the model reports the highest detection rate (i.e. TPR) of 93 % under low-shadowing conditions. Likewise, the AUC score has also decreased from 0.98 to 0.94 for high-shadowing scenario (i.e., 12 dB). Moreover, we also analyzed the OCSVMD detection performance under varying traffic conditions. Figure 4d depicts the distribution of measurements in the MDS space for a user density of 20 per cell. The higher user density implies an increase in the number of training observations

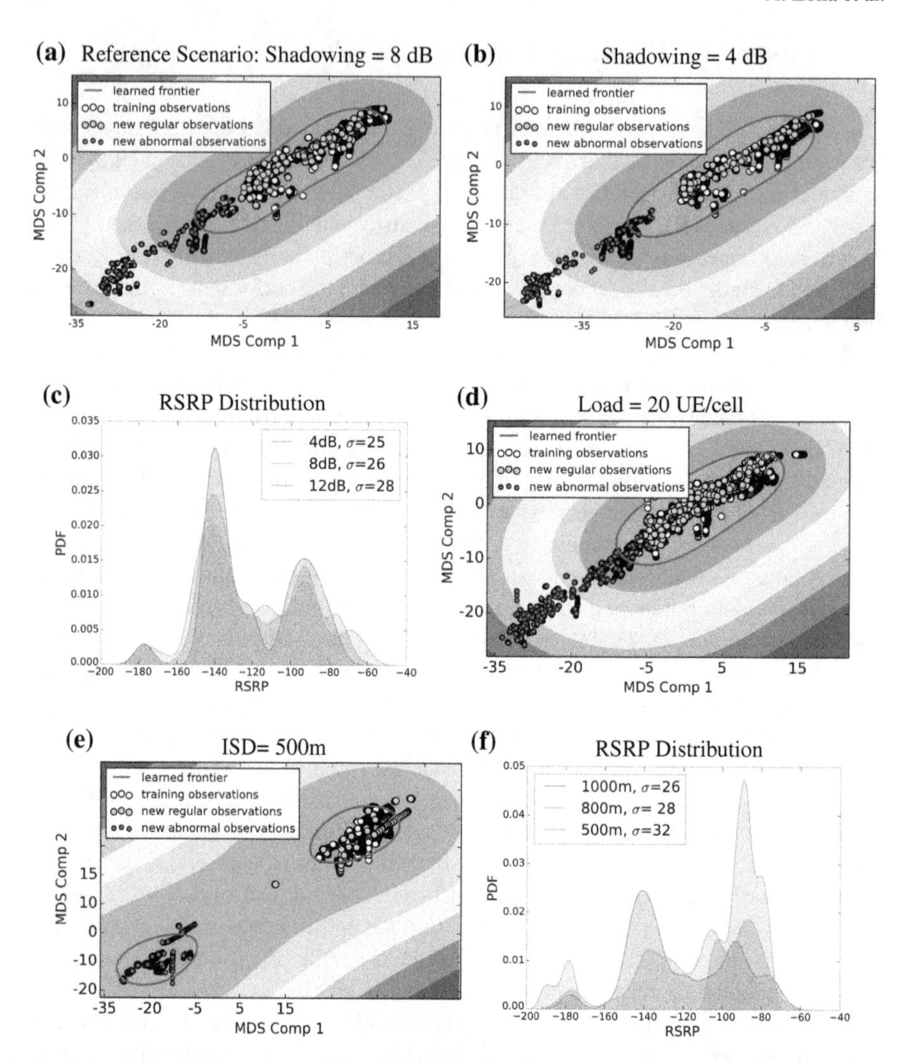

Fig. 4 **a** OCSVMD learned network profile for Reference Scenario. **b** Low-shadowing case. **c** Distribution of RSRP values for all shadowing cases. **d** Medium Traffic case. **e** Smaller ISD case **f** Distribution of RSRP values for all ISD cases

that leads to a more accurate estimate of the frontier shape. This explains the slight improvement in the AUC score for OCSVMD with the increase in the cell load as shown in Fig. 5b. A notable detection rate improvement of 10 % is observed for high traffic scenario (i.e., 30 users per cell) in comparison to the baseline OCSVMD.

As for different ISD configurations of a network, we see a significant change in the values of KPI measurements. This is expected since there is a strong correlation between UE reported KPI's and their distance from the eNB. Figure 4f shows the distribution of UE reported RSRP values for three different ISD cases. In case of

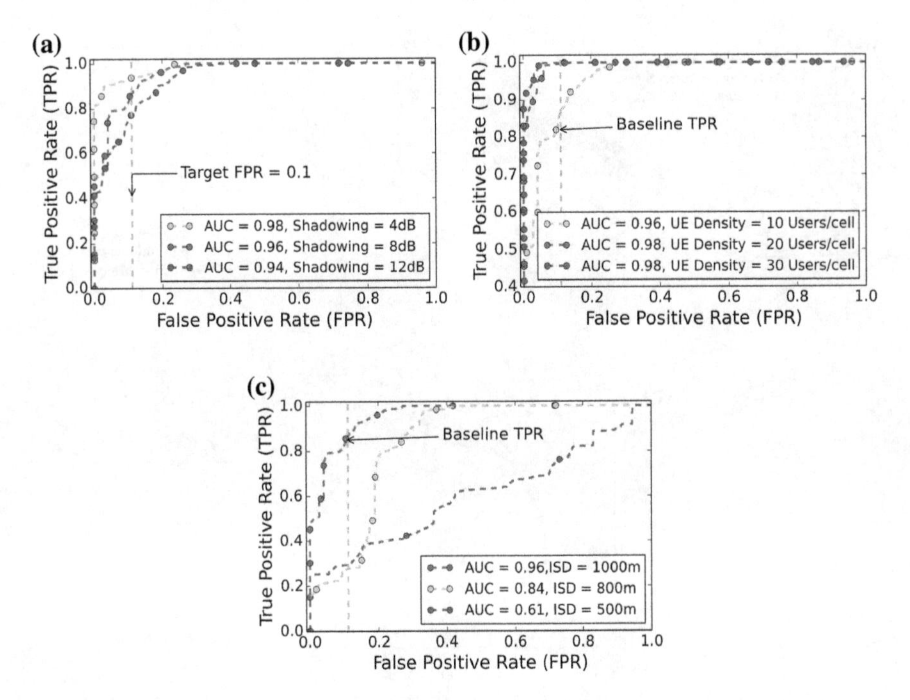

Fig. 5 OCSVMD ROC curves for shadowing, traffic and ISD cases. **a** Shadowing. **b** Cell load. **c** ISD

ISD = 500 m, we see a distinct peak of RSRP values around −90 dBm. Likewise, at the farther left end we see a small peak around −180 dBm that is mainly due to RLF-like observations. In contrast, when ISD = 1000 m, the highest peak value is observed at around −140 dBm, and the observed measurements have lower data spread as indicated in Fig. 4f. As already highlighted, the shape of the learned frontier by OCSVMD is directly affected by the distribution of observations in the embedded space. This becomes evident in Fig. 4e which shows that the OCSVMD learns two decision frontiers instead of one, since there exists two distinct modes in the data distribution, for the case of ISD = 500 m. As a result, OCSVMD interprets a region where RLF-like event are clustered, as inliers, which leads to an inaccurate network profile. The ROC analysis shown in Fig. 5c, clearly indicate the degradation of OCSVMD performance for lower ISD values.

Similar to OCSVMD, the performance of LOFD is also evaluated for all target network configurations. As explained in Sect. 3.1, LOFD derives a measure of outlyingness of an observation (i.e., $\mathscr{S}^{LOFD}$), based on the relative data density of its neighborhood. Figure 6a illustrates the labels assigned by LOFD to the observations obtained from the baseline scenario. It can be observed that LOFD classifies some of the test instances that even lie close to the vicinity of training observations as anomalous. Due to such instances LOFD receives a high outlying scores $\mathscr{S}^{LOFD}$, since the local density around them is highly different from the density of its neighborhood.

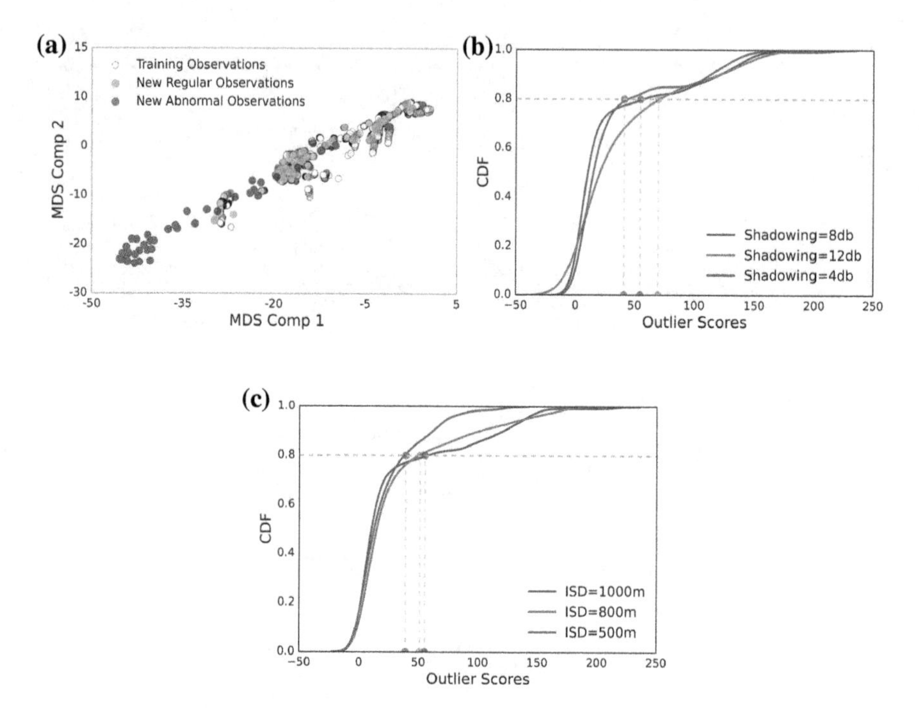

Fig. 6 Network profiling using LOFD. **a** Reference Scenario: Shadowing = 8 dB. **b** CDF of $\mathscr{S}^{LOFD}$ for shadowing cases. **c** CDF of $\mathscr{S}^{LOFD}$ for ISD cases

To further illustrate the impact of the variation and spread of the data on the values of $\mathscr{S}^{LOFD}$, we plot a cumulative distribution function (CDF) for different shadowing scenarios, as shown in Fig. 6b. It can be seen that for low-shadowing scenario, almost 80 % of the observations obtain $\mathscr{S}^{LOFD}$ value less than 50. However, as the shadowing increases we see a gradual increase in the value of $\mathscr{S}^{LOFD}$. Likewise, a similar behaviour is observed with the increase of ISD, as shown in Fig. 6c. The shadowing and ISD parameters influence the distribution and spread of the data as explained earlier, and consequently the value of $\mathscr{S}^{LOFD}$. This leads to a low detection performance of LOFD, since it generates an increased number of false alarms, as inferred from our ROC analysis.

As shown in Fig. 7a, the AUC score for LOFD decreases for high-shadowing scenario. On the other hand, the increase in the cell load also increase the spread of the data, which consequently affect the detection performance of LOFD. As shown in Fig. 7b, at false alarm rate of 10 %, the highest detection rate of 81 % is achieved for a network scenario in which load configuration is set to be 10 users per cell. Similarly, the change in the ISD has a severe effect on the model performance and low detection performance of 60 and 30 % is achieved for 800 and 500 m ISD configurations, as shown in Fig. 7c.

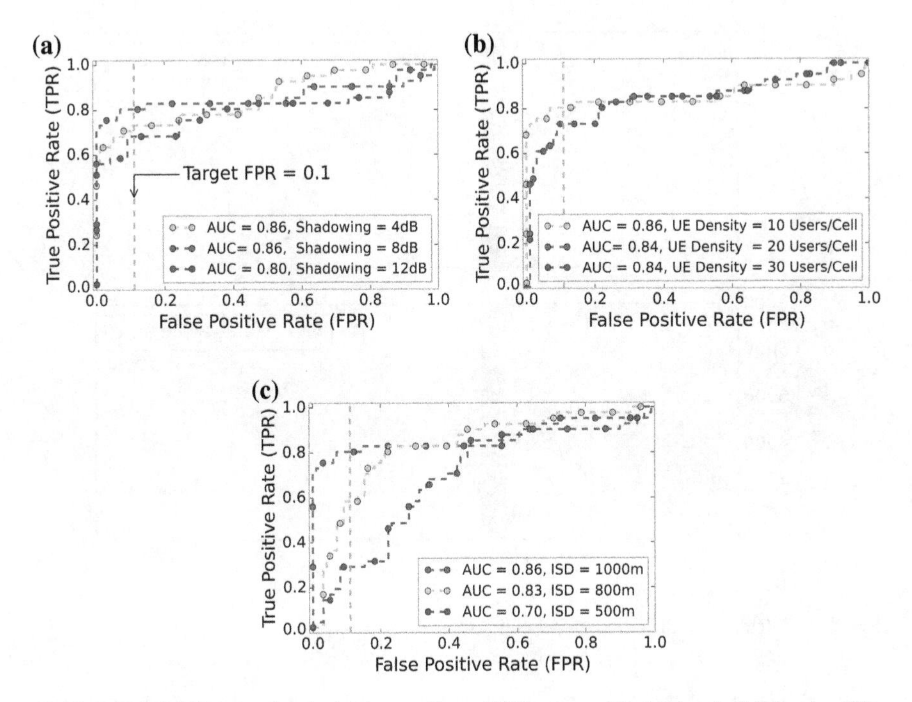

Fig. 7 LOFD ROC curves for shadowing, traffic and ISD cases. **a** Shadowing. **b** Cell Load. **c** ISD

In summary, we can conclude from the reported results that OCSVMD under most cases achieves a better detection performance in comparison to LOFD. The outage detection models yields worst performance scores particularly for network configuration with low ISD. The performance issue of the target outage detection models can be addressed as follows: For OCSVMD, in the pre-processing step the RLF-like events must be filtered before constructing a training database. This would also help decrease the spread of the data and the model would only learn frontier that corresponds to normal operational network behaviour. The performance issues of LOFD can be addressed by incorporating a concept dirft detection mechanism, that indicates when to re-tune model parameters to minimize the false alarm rate.

5.3 Localization

Since, OCSVMD model has outperformed LOFD for most test cases, it has been selected as a final model to compute per cell z-scores for the normal and SC scenario, as shown in Fig. 8. It can be observed from Fig. 8 that measurements are classified as anomalous even in the normal operational phase of the network due to occurrence of RLF events. This is particularly true for cell ID 1, 5, 11, 16, and 19 whose n_b values are found to be 700, 2000, 3000, 1500, respectively in the reference scenario.

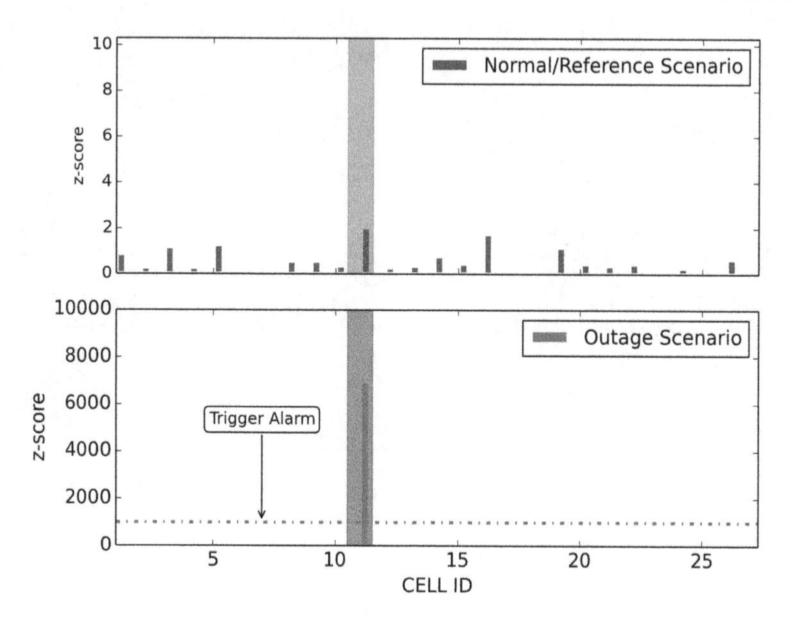

Fig. 8 Localization of SC based on per cell z-scores

However, during an outage scenario, since cell 11 is configured as a faulty cell, the corresponding z-scores are significantly higher than the rest of the network. A simple decision threshold can be applied on the computed z-scores to autonomously localize faulty cells, and consequently an alarm can be triggered. In addition to cell outage localization, the change in the z-score can be used to identify performance degradation issues or a weaker coverage problems. This information can act as an input to self-healing block of SON engine, which can then trigger automated recovery or optimization procedures.

5.4 Data COD Outage Detection Results

Figure 9 illustrates the performance of our data COD framework in terms of the DR. In Fig. 9a, we compare the performance of the GM and GMF, which are obtained from (10) and (13), respectively, by plottin their DR against the data cell UE density, U_f, and for shadowing fading standard deviation of, $\sigma = 2$ and 10 dB. We observe that the GMF scheme outperforms the GM as expected, since the former utilizes the prediction error in the later to improve its performance. Figure 9a clearly shows that increasing the UE density increases the DR. This is due to the fact that increasing UE density enables a better spatial correlation. Figure 9b depicts the DR for various data BS power levels and a data cell UE density of $U_f = 3(/100\,\text{m} \times 100\,\text{m})$. The result shows that low data BS transmission power results in degradation of the DR,

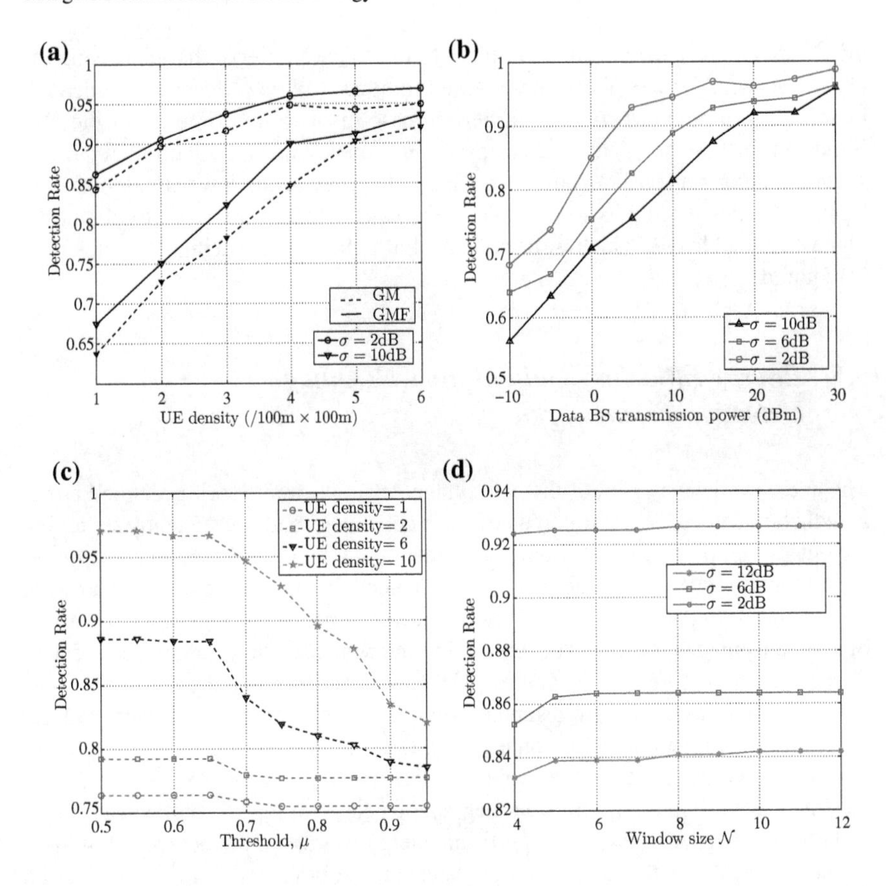

Fig. 9 performance of data COD framework. **a** GM versus GMF. **b** Effect of data BS transmission power on DR. **c** Effect of threshold setting on DR. **d** Effect of window size on DR

while increasing the transmission power leads to an improvement in the DR. This is because when the data BS transmission power increases, it becomes easier to distinguish between the predicted RSRP statistics of the outage case and normal case. We also observe in Fig. 9a, b that the DR becomes lower with larger shadowing fading standard deviation σ. This is because a high σ means a severe shadowing fading, which leads to a more random RSRP statistics.

Figure 9c, d investigate the impact of the predefined threshold μ and prediction window size $\mathcal{N}$, respectively, on the DR. We observe in Fig. 9c that the highest DR is obtained by setting $\mu = 0.5$. This setting implies that the RSRP prediction of more than half of the UEs that were associated with the data BS whose outage is being detected, i.e. d, must indicate the existence of an outage, before d can be declared to be in outage. The stepwise shaped plot is obtained since the number of UEs must be an integer value. We also observe that there is not much degradation in DR until $\mu > 0.67$, which implies more than two-third of UEs that were associated with d

must indicate the existence of an outage. In Fig. 9d, we observe that increasing the prediction window size above the required minimal ($\mathcal{N} = 4$) leads to an increase in DR up to a point where any further increase in $\mathcal{N}$ has no impact on the DR. Figure 9d further shows that increasing $\mathcal{N}$ has more impact on the DR for larger shadow fading standard deviation, σ. This is because of the lower randomness in RSRP statistics when σ is low; hence a low value of $\mathcal{N}$ is required to obtain the highest attainable DR, which is the contrary for higher σ where a higher value of $\mathcal{N}$ is required.

5.5 Energy Efficiency Gain of the Cell Outage Detection Framework

The energy efficiency gain of the cell outage detection framework is defined as the gain that is achieved as a result of detecting a cell outage and consequently switching the outage cell off as against leaving it in operational mode. As mentioned earlier in Sect. 1, cell outage can be due to network connectivity, misconfiguration, hardware failure and power supply failure. With the advancement in technology, only a few of the cell outages are of result power failure, which involves a complete cut in the power supply to the base stations [33]. Hence, we focus on gains from outages which are not based on power supply failures. In situation with hardware failure, misconfiguration and network connectivity, the base station of the outage cell still consumes significant power even when not transmitting data [9].

Consequently, in addition to the degradation in system performance caused by cell outage, it can also result in significant energy wastage if the outage cell remain undiscovered. Once the cell outage is detected, the base station in outage can be completely switched off until the cause of the outage is identified and correction measures are implemented. The basic power consumption model of the base station can be expressed as the sum of circuit power and the transmit power given as:

$$P^{TOT} = P_0 + \Delta \times P^t \tag{14}$$

where P^{TOT} is the total power consumed, P_0 is the circuit power drawn if the BS is active, P^t is the transmit power and Δ is the slope of load dependent power consumption. As assumed in our simulation setup in Sect. 5.1, if the cell goes into an outage due to antenna malfunction, from Eq. 14 we can see that it is still consuming P_0, even if it is not transmitting data. According to [34] a typical P_0 of a macro BS and femto BS is 130 and 4.8 W, respectively. In case of no automatic COD solutions in place, cell outages may remain undetected for days and weeks [35]. Assuming a 24 h outage duration, a typical macro BS would waste around 11.23×10^6 J of energy. This wastage of energy can be reduced by putting in place an effective COD solution as proposed in this study, that can reduce the outage detection time and consequently faulty BS can be repaired or shut down completely to minimize energy expenditure.

6 Conclusion

In this study, we have presented a Cell Outage Detection (COD) framework for HetNets with split control and data planes. Two distinct COD algorithms have been proposed taking into account expected large number of UEs in the control cells and small number of UEs in the data cells. For control COD, we have utilized the large scale data gathering of MDT reports, as recently standardized by 3GPP in release 10. The solution exploits multidimensional scaling techniques to reduce the complexity of data processing while retaining pertinent information to develop training models to reliably apply anomaly detection techniques. Furthermore, within the control COD, domain and density based outage detection models: OCSVMD and LOFD respectively, were examined for different network configurations. It was established that OCSVMD, a domain based model attained a higher detection accuracy compared to LOFD which adopts a density based approach to identify network abnormality. Finally, the UE reported coordinate information is employed to establish the dominance areas of target cells which are subsequently used to localize the position of the cell in outage. On the other hand, for data cell outage, we have utilized a heuristic Grey-Prediction approach, which can reliably work despite of small number of UEs in the data cells by exploiting the information stemming from the fact that the control BS manages the UE connectivity to the data BS within its coverage. The simulation results have shown that both control and data COD schemes can detect control and data cell outages, respectively, in a reliable manner. The proposed COD framework can act as a foundation for next generation network monitoring tools that aims to provide an autonomous self-healing functionality, as well as to detect other network problems including coverage holes, weak coverage and performance degradation problems.

Acknowledgments This work was made possible by NPRP grant No. 5-1047- 2437 from the Qatar National Research Fund (a member of The Qatar Foundation).The statements made herein are solely the responsibility of the authors.

References

1. Aliu, O.G., Imran, A., Imran, M.A., Evans, B.: A survey of self organisation in future cellular networks. IEEE Commun. Surveys Tutorials **15**(1), 336–361 (2013)
2. Imran, A., Zoha, A., Abu-Dayya, A.: Challenges in 5G: how to empower son with big data for enabling 5G. IEEE Network **28**(6), 27–33 (2014)
3. Akbari, I., Onireti, O., Imran, M.A., Imran, A., Tafazolli, R.: Effect of inaccurate position estimation on self-organising coverage estimation in cellular networks. In: Proceedings of 20th European Wireless Conference. VDE, pp. 1–5 (2014)
4. Mueller, C., Kaschub, M., Blankenhorn, C., Wanke, S.: A cell outage detection algorithm using neighbor cell list reports. In: International Workshop on Self-Organizing Systems, pp. 218–229 (2008)
5. Wang, W., Zhang, J., Zhang, Q.: Cooperative cell outage detection in self-organizing femtocell networks. In: IEEE INFOCOM, pp. 782–790, Apr 2013

6. Lister, D.: An operator's view on green radio. In: Proceedings of IEEE International Conference on Communicaiton (ICC Worshops'09), 1st International Worshop on Green Communication (2009)
7. 3GPP TS 32.541: 3rd Generation Partnership Project; Technical Specification Group Services and System Aspects; Telecommunications Management; Self-Organizing Networks (SON); Self-Healing Concepts and Requirements (Release 11), 2012-09, v11.0.0
8. Liao, Q., Wiczanowski, M., Stanczak, S.: Toward cell outage detection with composite hypothesis testing. In: IEEE ICC, June 2012, pp. 4883–4887
9. Auer, G., Giannini, V., Desset, C., Godor, I., Skillermark, P., Olsson, M., Imran, M.A., Sabella, D., Gonzalez, M.J., Blume, O., et al.: How much energy is needed to run a wireless network? IEEE Wireless Commun. **18**(5), 40–49 (2011)
10. Barco, R., Wille, V., Díez, L.: System for automated diagnosis in cellular networks based on performance indicators. Eur. Trans. Telecommun. **16**(5), 399–409 (2005)
11. Cheung, B., Fishkin, S.G., Kumar, G.N., Rao, S.A.: Method of monitoring wireless network performance. US Patent 10/946,255, 21 Sep 2004
12. Ma, Y., Peng, M., Xue, W., Ji, X.: A dynamic affinity propagation clustering algorithm for cell outage detection in self-healing networks. In: IEEE WCNC, Apr 2013, pp. 2266–2270
13. Coluccia, A., D'Alconzo, A., Ricciato, F.: Distribution-based anomaly detection via generalized likelihood ratio test: a general maximum entropy approach. Comput. Netw. **57**(17), 3446–3462 (2013)
14. Khanafer, R., Solana, B., Triola, J., Barco, R., Moltsen, L., Altman, Z., Lázaro, P.: Automated diagnosis for UMTS networks using Bayesian network approach. IEEE Trans. Veh. Technol. **57**(4), 2451–2461 (2008)
15. Zoha, A., Saeed, A., Imran, A., Imran, M.A., Abu-Dayya, A.: A son solution for sleeping cell detection using low-dimensional embedding of mdt measurements. In: IEEE 25th Annual International Symposium on Personal, Indoor, and Mobile Radio Communication (PIMRC), Sept 2014, pp. 1626–1630
16. Onireti, O., Imran, A., Imran, M.A., Tafazolli, R.: Cell outage detection in heterogeneous networks with separated control and data plane. In: Proceedings of 20th European Wireless Conference. VDE, pp. 1–6 (2014)
17. Liu, S., Wu, J., Koh, C.H., Lau, V.K.N.: A 25 Gb/s(/km^2) urban wireless network beyond IMT-Advanced. IEEE Commun. Mag. **49**(2), 122–129 (2011)
18. Capone, A., Filippini, I., Gloss, B., Barth, U.: Rethinking cellular system architecture for breaking current energy efficiency limits. In: Sustainable Internet and ICT for Sustainability (SustainIT), pp. 1–5 (2012)
19. Ishii, H., Kishiyama, Y., Takahashi, H.: A novel architecture for LTE-B :C-plane/U-plane split and phantom cell concept. In: IEEE Globecom Workshops (GC Wkshps), Dec 2012, pp. 624–630
20. Xu, X., He, G., Zhang, S., Chen, Y., Xu, S.: On functionality separation for green mobile networks: concept study over LTE. IEEE Commun. Mag. **51**(5), 82–90 (2013)
21. Mohamed, A., Onireti, O., Imran, M., Imran, A., Tafazolli, R.: Control-data separation architecture for cellular radio access networks: A survey and outlook. IEEE Commun. Surveys Tutorials **99**, 1–1 (2015)
22. 3GPP TS 37.320: Universal Mobile Telecommunications System (UMTS); LTE; Universal Terrestrial Radio Access (UTRA) and Evolved Universal Terrestrial Radio Access (E-UTRA); Radio measurement collection for Minimization of Drive Tests (MDT); Overall description; Stage 2, 2011-04, v10.1.0 Release 10
23. Ma, Y., Peng, M., Xue, W., Ji, X.: A dynamic affinity propagation clustering algorithm for cell outage detection in self-healing networks. In: Proceedings of IEEE Wireless Communications and Networking Conference (WCNC). IEEE, pp. 2266–2270 (2013)
24. Mueller, C.M., Kaschub, M., Blankenhorn, C., Wanke, S.: A cell outage detection algorithm using neighbor cell list reports. In: Self-Organizing Systems. Springer, pp. 218–229 (2008)
25. Hämäläinen, S., Sanneck, H., Sartori, C., et al.: LTE Self-Organising Networks (SON): Network Management Automation for Operational Efficiency. Wiley, New York (2012)

26. Cox, T., Cox, M.A.: Multidimesional Scaling. CRC Press, Boca Raton (2010)
27. Breunig, M.M., Kriegel, H.-P., Ng, R.T., Sander, J.: LOF: identifying density-based local outliers. In: ACM Sigmod Record, vol. 29, no. 2. ACM, pp. 93–104 (2000)
28. Schölkopf, B., Platt, J.C., Shawe-Taylor, J., Smola, A.J., Williamson, R.C.: Estimating the support of a high-dimensional distribution. Neural Comput. **13**(7), 1443–1471 (2001)
29. Deng, B.L.: Introduction to grey system. J. Grey Syst. **1**, 1–24 (1989)
30. Sheu, S.-T., Wu, C.-C.: Using grey prediction theory to reduce handoff overhead in cellular communication systems. In: IEEE PIMRC, vol. 2, pp. 782–786 (2000)
31. Kayacan, E., Ulutas, B., Kaynak, O.: Grey system theory-based models in time series prediction. Expert Syst. Appl. **37**(2), 1784–1789 (2010)
32. 3GPP R4-092042: Simulation Assumptions and Parameters for FDD HeNB RF Requirements (2009)
33. Understanding High Availability of IP and MPLS Networks. http://www.ciscopress.com/articles/article.asp?p=361409. Accessed 30 June 2015
34. Auer, G., et al.: D2.3: Energy efficiency analysis of the reference systems, areas of improvement and target breakdown. INFSO-ICT-247733 EARTH (Energy Aware Radio and NeTwork TecHnologies), Technical Report, 2010
35. Hämälïnen, S., Sanneck, H., Sartori, C.: LTE Self-Organising Networks (SON): Network Management Automation for Operational Efficiency. Wiley, New York (2012)

Towards Energy-Aware 5G Heterogeneous Networks

Hafiz Yasar Lateef, Mischa Dohler, Amr Mohammed, Mohsen Mokhtar Guizani and Carla Fabiana Chiasserini

Abstract Over the past decade, the telecommunication industry has witnessed excessive growth in the number of mobile users. Market forecasts envision that there will be nearly 8.6 billion mobile devices worldwide by 2017. This tremendous increase in the number of cellular users demands an expansion in the wireless Base Stations (BSs) for improved coverage and capacity. However, this hike in the deployment of base stations will lead to immense energy consumption, because in mobile networks 70–80 % of the power is consumed by BSs. This upsurge in the energy consumption of telecommunication networks implies an increase in CO_2 emissions in the environment. In addition, energy bills also represent a major chunk of wireless network operators' expenditures. These ecological and economical challenges have provoked the curiosity of telecommunication standardization bodies and researchers in an emerging research area termed 'energy-aware Heterogeneous Networks (HetNets)'. HetNets are a mix of various cell shapes and sizes, including high power macro cells and low power nodes such as micro cells, pico cells and relays. The large macro cells are responsible for the basic coverage of the cell users, and the small cells are effective in providing higher data rates to their nearby users in dense areas with reduced power consumption. The combination of various BSs with different cell sizes and a wide range of power levels can lead to substantial gains in network energy consumption by creating hotspots and enabling dense spatial reuse. It is envisioned that a dense deployment of low power BSs will take place in the near future. HetNets in particular are considered as a promising solution for Fifth Generation (5G) in order to meet the exponentially growing demand for multimedia traffic. The main focus of this chapter is to investigate optimal energy efficient deployment strategies

H.Y. Lateef (✉) · A. Mohammed · M.M. Guizani
Department of Computer Science, Qatar University, Doha, Qatar
e-mail: yasar.lateef@qu.edu.qa

M. Dohler
Wireless Communications, King's College London, London, UK
e-mail: mischa.dohler@kcl.ac.uk

C.F. Chiasserini
Telecommunication Networks Group, Politecnico di Torino, Turin, Italy
e-mail: chiasserini@polito.it

© Springer International Publishing Switzerland 2016

M.Z. Shakir et al. (eds.), *Energy Management in Wireless Cellular and Ad-hoc Networks*, Studies in Systems, Decision and Control 50,
DOI 10.1007/978-3-319-27568-0_2

for low power nodes such as relays and small cells in 5G HetNets. In this chapter, a comprehensive overview of remarkable small cell deployment schemes is presented in order to facilitate the debate on technical challenges in deploying HetNets. It goes on to discuss some useful techniques to mitigate the severe interference in 5G dense HetNets. Finally, a novel Long Term Evolution (LTE)-Advanced relay deployment scheme is introduced using graph theory, not only to address some of the identified deficiencies of existing solutions, but also to optimize the energy efficiency of 5G cellular networks.

1 Introduction

Over the past decade, the cellular industry has witnessed an unprecedented growth in the number of subscribers and traffic, particularly, video and multimedia. This tremendous increase in mobile subscribers calls for major investments in additional wireless infrastructure, namely base stations for enhanced coverage and capacity. However, such hike in the deployment of BSs will result in huge energy consumption. It is estimated that wireless networks currently consume approximately 60 billion kWh per year globally and these statistics are predicted to double by 2020 [1–3].

The economical challenges and high power consumption of conventional macro BSs have led standardization bodies and researchers to seek alternative, cost-effective and energy-efficient solutions. This has, in turn, shifted the focus on energy efficient Fifth Generation (5G) wireless networks in the research community. In this regard, some recent projects such as GreenTouch, Greenet, Towards Green 5G Mobile Networks (5GrEEn), Green Radio Excellence in Architecture and Technology (GREAT), Communicate Green (ComGreen), Energy Aware Radio and neTwork tecHnology (EARTH) and Towards Real and Energy Efficient Network Design (TREND) have started to realize the vision of, both eco-friendly and green 5G cellular networks [4–10].

The challenges associated with the deployment of traditional macro base stations can be overcome through the utilization of BSs with lower transmit power. Specifically, HetNets are the potential solution to achieve energy efficiency in future wireless networks. In a HetNet, macro BSs are deployed in a planned way to achieve required coverage (large area), while Low Power Nodes (LPNs) serve the purpose of coverage extension, throughput enhancement, and achieving overall lower energy consumption for the network [11].

The introduction of dense HetNets and Massive Multiple Input Multiple Output (MIMO) techniques are the key ideas for 5G technology in order to achieve both capacity gains and energy efficiency in future wireless networks [12, 13]. Typically, network operators place low power BSs at strategic areas to enhance the network performance while keeping the infrastructure deployment cost low [14]. However, the dense and random deployment of low power BSs raises fundamental challenges for the energy consumption of dense HetNets. Therefore, dense deployment of small BSs should be carefully designed in order to avoid undesired network behaviour

[15]. Key challenges for dense HetNets from the energy perspective include finding the optimal densities of small BSs and determining which infrastructure network nodes should be switched on/off depending on the user traffic patterns. Thus, optimal switch on/off policies for dense HetNets can play a key role in enhancing the energy efficiency and data rate towards 2020.

2 Energy Efficient Resource Allocation Schemes for 5G HetNets

This section presents a comprehensive survey of state-of-the-art work on energy efficient resource allocation and load balancing schemes for 5G HetNets. Several works have addressed energy-efficient sleep mode protocols and traffic offloading schemes for HetNets, e.g., [16–21]. In particular, the performance of a macro-pico network is studied in [16]. Specifically, the work highlights that the number of pico BSs, the user distribution and the fact that pico BSs can enter a sleep mode can lead to significantly high energy savings. The work in [17] has proposed an analytical framework for the performance evaluation of the energy saving that can be obtained by applying sleep mode to the network devices. Specifically, the authors formulated a theoretical model which allows to estimate that how much energy can be saved for different network topologies. The performance evaluation results reveal that highly connected networks, with high randomness tend to make the use of sleep modes more energy efficient.

The authors in [18] proposed an analytical model to determine the optimal set of BSs that can be switched off based on the daily traffic pattern. Specifically, the authors derived analytical expressions for the energy saving by first assuming that only a single BS can switch off per day and then considering that multiple BSs can switch off per day. The performance evaluation results indicate that substantial energy saving can be realized by switching off a single BS per day, while the advantage of switching off multiple BSs is minor. In [19], the authors have utilized stochastic geometry theory to analyze the optimal macro/micro BS density for energy efficient HetNets under Quality of Service (QoS) constraints. The authors have addressed the two important issues: capacity extension and energy saving, and they have proposed a rule to determine which type of BSs should be deployed or slept with higher priority. Yong et al. have investigated the impact of random sleeping and strategic sleeping on the power consumption and energy efficiency of HetNets [20]. On the other hand, energy minimization in macro-relay networks has been studied in [21] and [22], where minimum user data rate requirements are accounted for. In particular, the authors formulated an integer optimization problem and proposed a heuristic solution for energy minimization.

The effect of coverage area on energy efficiency of macro-pico HetNets has been studied in [23]. System-level simulation results reveal that the area energy efficiency of macro-pico networks can be substantially improved with interference reduction and adaptive power control. The authors in [24] have investigated the energy efficiency of pico nodes in HetNets by taking into account the effect of pico cell size on the overall energy efficiency of the network. The performance evaluation results reveal that energy efficiency of pico BS can improve substantially by applying efficient resource allocation and cross-tier interference mitigation scheme. The work in [25] has proposed a dynamic on/off switching algorithm for BSs based on the concept of network impact which is defined as how much can switching on/off a BS affect the whole network. Moreover, the authors proposed various heuristic algorithms for determining the on/off state of a BS with partial feedback or even no feedback. Shengrong et al. considered the role of smart grid in designing energy efficient cellular networks by taking into consideration, both real-time traffic conditions and the associated carbon emissions [26]. The authors proposed a scheme in which some of the base stations can be switched off to save energy while Coordinated Multi-Point (CoMP) scheme is used to increase the coverage of the active base stations.

3 System Model

In this section, we present a system model for the performance evaluation of an energy efficient macro-relay network consisting of macro eNBs, low power Relay Nodes (RNs) and User Equipment (UEs). An overview of a multi-cell macro-relay network is shown in Fig. 1. The eNBs, RNs and UEs are equipped with single antenna. We consider in-band Type 1 LTE-Advanced RNs, which use the same frequency resources for both backhaul (eNB to relay) and access (relay to UE) links. Moreover, the backhaul and access links are time division multiplexed in order to avoid interference between these links. The RNs must connect to a donor macro eNB either through a backhaul link, or, in a multi-hop fashion, to another RN as shown in Fig. 1. The users can connect to the network through macro eNB, either directly or through RNs using decode and forward technique. The UEs are uniformly distributed in the macro-relay network under consideration. The energy consumption of UEs varies depending on the distance and path loss from macro eNBs or RNs. We adopt the large-scale path loss propagation model that is endorsed by 3GPP [27–29].

The EARTH project [9] introduced a linear power model for different types of base stations, which details the relation between base station power consumption P_{in} and Radio Frequency (RF) output power P_{out}. According to the EARTH power model, we have;

$$P_{in} = P_o + \nabla_p \times P_{out} \qquad 0 \leq P_{out} \leq P_{max} \tag{1}$$

where P_{max} represents the maximum RF output power at full load, P_o is the minimum power consumption when the node is in idle mode and ∇_p denotes the power amplifier efficiency. The power consumption parameters for different types of base stations and relays, based on the EARTH project state-of-the art estimation, are presented in Table 1.

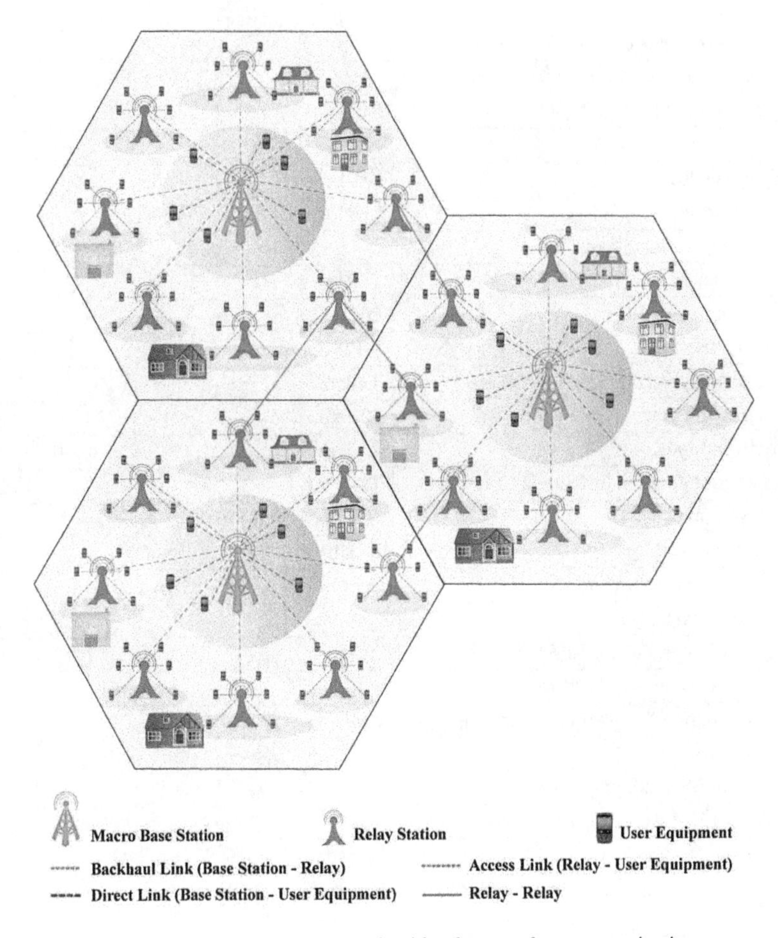

Fig. 1 An illustration of a macro-relay network with relay-to-relay communication

4 Optimization of Energy Efficient Relay Placement and Load Balancing

With the aim to minimize energy consumption, we now present an analytical model for optimizing relay placement and load balancing in a macro-relay heterogeneous network. We divide the network service area into a set τ of non-overlapping tiles. These tiles in general cannot be bigger than a cell and may differ in size and shape as shown in Fig. 2.

We define τ_t to be the amount of traffic or data (e.g., Megabits) requested by each user, for each tile t. We assume that an estimate of τ_t is already known. The set β represents macro eNBs and the set $\mathcal{L}$ represents candidate locations where RNs can be deployed. The eNBs, relay candidate locations and tiles constitute the vertices of a graph, representing our network. The edges of the graph represent the

Table 1 Performance evaluation parameters

Earth parameters for different base stations			
LTE base station type	$P_{max}(W)$	$P_o(W)$	∇_p
Macro	40	712	14.5
Relay urban 2014	1	19.91	5.6
Relay advanced	5 or 1	13.91	20.4
Performance evaluation parameters			
Carrier frequency		2 GHz	
Bandwidth		20 MHz	
Thermal noise PSD		−174 dbm/Hz	
eNB transmit power		43 dbm	
Relay transmit power		30 dbm	
User transmit power		23 dbm	
Antenna configuration (eNB, relay and User)		Tx-1, Rx-1	
User distribution		Hotspot + Uniform	
Distance and path loss		R (km) & PL (dB)	
3GPP Case 1 urban scenario		Inter Site Distance (ISD) = 500 m	
Direct link (Macro-UE)			
$PL_{LOS} = 103.4 + 24.2 log10(R)$			
$PL_{NLOS} = 131.1 + 42.8 log10(R)$			
$P(LOS) = \min\left(\frac{0.018}{R}, 1\right) * \left(1 - \exp\left(-\frac{R}{0.063}\right)\right) + \exp(-\frac{R}{0.063})$			
Access link (Relay-UE)			
$PL_{LOS} = 103.8 + 20.9 log10(R)$			
$PL_{NLOS} = 145.4 + 37.5 log10(R)$			
$P_{(LOS)} =$			
$P(LOS) = 0.5 - \min\left(0.5, 5\exp\left(\frac{-0.156}{R}\right)\right) + \min\left(0.5, 5\exp\left(\frac{-R}{0.03}\right)\right)$			
Backhaul link (Donor eNB-Relay and Relay-Relay)			
$PL_{LOS} = 100.7 + 23.5 log10(R)$			
$PL_{NLOS} = 125.2 + 36.3 log10(R) - b$			
$P_{(LOS)} = 1 - \left(1 - \min\left(\frac{0.18}{R}, 1\right) * \left(1 - \exp\left(-\frac{R}{0.072}\right)\right) + \exp\left(-\frac{R}{0.072}\right)\right)^{c} b = 5, c = 3$			

connectivity opportunities among the vertices. It is assumed that each macro base station provides single cell coverage. All the direct link connections from each macro eNB to its neighboring cells are considered redundant and therefore we neglect these connections.

For each edge connecting a pair of nodes (e_1, e_2), there is a corresponding weight $w(e_1, e_2)$, representing how much data we can transfer from e_1 (representing an eNB or a RN) to e_2 (representing a RN or a tile).

For all the edge points (e_1, e_2), representing any of the following pairs (relay candidate location, tile), (eNB, tile), (eNB, relay candidate location), (relay candidate location, relay candidate location) we know the associated transmit power $P(e_1, e_2)$,

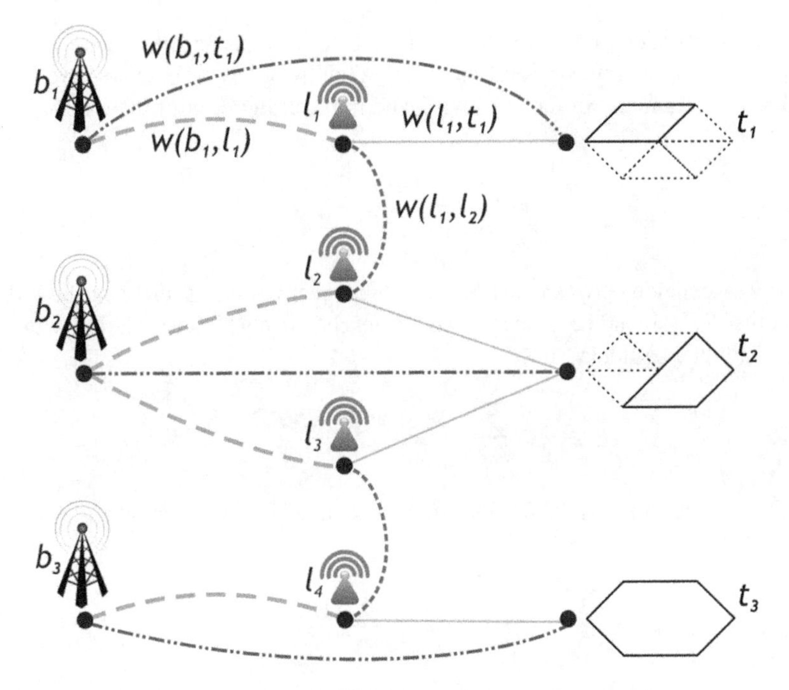

Fig. 2 A graph based representation of a LTE-Advanced macro-relay network

Furthermore, active eNBs and RNs consume $P_o(b)$ and $P_o(l)$ amount of power, respectively, which depend on the transceiver electronics, cooling, etc., i.e., they are independent of the node traffic load. Finally, due to operator's budget constraints, we have a maximum number R of RNs that can be deployed.

The energy efficient relay placement optimization algorithm is formulated as a MILP problem. First, we introduce a set of binary variables y_l, $y_b \in \{0, 1\}$, representing, respectively, whether we place a RN in a candidate location $l \in \mathcal{L}$, and whether eNB is $b \in \beta$ ON or OFF. Furthermore, we need to denote how much traffic we transmit between UEs, RNs and eNBs. We do so through real variables $x(e_1, e_2)$. At last, we introduce a set of binary variables $z(b, l)$, each expressing whether $b \in \beta$ is a donor eNB for RN in location $l \in \mathcal{L}$. To rationalize the notation, we also denote the donor eNB for RN in l as $D_l \in \beta$.

Constraints: The first constraint corresponds to the capacity. For each pair of nodes (UEs, RNs and eNBs) that can communicate with each other, the total amount of transmitted data x must not exceed the capacity (w) of the edge:

$$x(e_1, e_2) \leq w(e_1, e_2) \tag{2}$$

The exact value of the weight w can be calculated using the channel capacity formulas. Next, a flow conservation equation holds for RNs. These are purely relay nodes, so the amount of data receiving and transmitting each of them must be the same:

$$\sum_{e_1 \epsilon \beta U \mathcal{L}} x(e_1, l) = \sum_{e_2 \epsilon \mathcal{L} U \tau} x(l, e_2) \qquad \forall l \epsilon \mathcal{L} \tag{3}$$

The association between the RN and their respective donors (these can be eNB or any other RN) should be in such a way that each active RN is association with only one donor at a particular time.

$$\sum_{e \epsilon \beta U \mathcal{L}} z(e, l) = y_1 \quad \forall l \epsilon \mathcal{L} \tag{4}$$

The constraint in (5) defines that an inactive node (eNB or RN) can't be a donor to any RN.

$$\sum_{e \epsilon \beta U \mathcal{L}} + \sum_{l \in \mathcal{L}} z(e, l) \leq \sum_{e \epsilon \beta U \mathcal{L}} y_e \quad \forall e \in \beta U \mathcal{L} \tag{5}$$

Obviously no date can flow between inactive nodes. Therefore we modify the capacity constraint as follows:

$$x(e_1, e_2) \leq y_{e_1}.w(e_1, e_2). \quad \forall e_1 \in \beta U \mathcal{L} \tag{6}$$

When y_{e_1} is zero, i.e. the source node (eNB or RN) is not active, the right side of the equation becomes zero and no data can be transmitted.

The association variables $z(b, l)$ will also make sure that there is no data flow between an eNB and RN, if they are not associated with each other. This can be represented as;

$$x(e_1, e_2) \leq z(b, l).w(e_1, e_2) \quad \forall b \in \beta, l \in \mathcal{L} \tag{7}$$

As described earlier, each tile must receive the adequate traffic τ_t, in order to meet the minimum data rate requirements. This translates into the following constraint:

$$\sum_{b \in \beta} x(b, t) + \sum_{l \in \mathcal{L}} x(l, t) \geq \tau_t \tag{8}$$

Finally, the following constraint represents the limit on the maximum number of RNs that can be deployed in the network, due to budget constraints

$$\sum_{l \in \mathcal{L}} y_l \leq R \forall l \in \mathcal{L} \tag{9}$$

Objective: Our objective is to minimize the total power consumption of the network. This includes;

- The static or fixed power consumed by active source nodes (eNB and RN)
- The dynamic or traffic dependent power $P(e_1, e_2)$ based on communication end points.

$$\min \sum_{b \in \beta} \left(y_b P_0(b) + \sum_{l \in \mathcal{L}} P(b,l)x(b,l) + \sum_{t \in \tau} P(b,t)x(b,t) \right) + \sum_{l \in \mathcal{L}} \left(y_l P_0(l) + \sum_{t \in \tau} P(l,t)x(l,t) \right) \tag{10}$$

Clearly, the objective function and all the constraints are linear and the complexity stems from the binary variables y_b and y_l.

5 Performance Evaluation Parameters

In this section, we evaluate the performance of our proposed energy efficient algorithm. The parameter values we use in our analysis are reported in Table 1. The scenario we consider consists of 7 cells macro-relay network with hotspot and uniformly distributed users as shown in Fig. 3. In this scenario, we utilize the relay urban 2014 power model for performance evaluation.

Figure 4 depicts the effect of the density of RNs on the number of active macro base stations in operation. Specifically, from the plot, it can be seen that our proposed load balancing algorithm for a dense macro-relay network can offload traffic from macro BSs and switch off most of the lightly loaded macro BSs. Moreover, it is evident from Fig. 4 that more macro BSs are switched off by increasing the density of RNs in the network.

The effect of the density of RNs on the Area Energy Efficiency (AEE) of a macro-relay network is shown in Fig. 5. The bar labeled "Optimal macro-relay network" in Fig. 5 represents the scenario where we take into account transmission and circuit energy of both macro base stations and RNs in active mode which are obtained by solving the problem in (10) using CPLEX.

Similarly, the bar labeled "macro relay network without sleep mode" represents the scenario where we consider transmission and circuit energy of all macro base stations and RNs in the network. Finally, the bar "Macro only" refers to the scenario where we only deploy macro base stations and all macro base stations are in the active mode. From Fig. 5, we note that our proposed algorithm has the best AEE as compared to other cases. The rationale behind this fact is that most of the macro

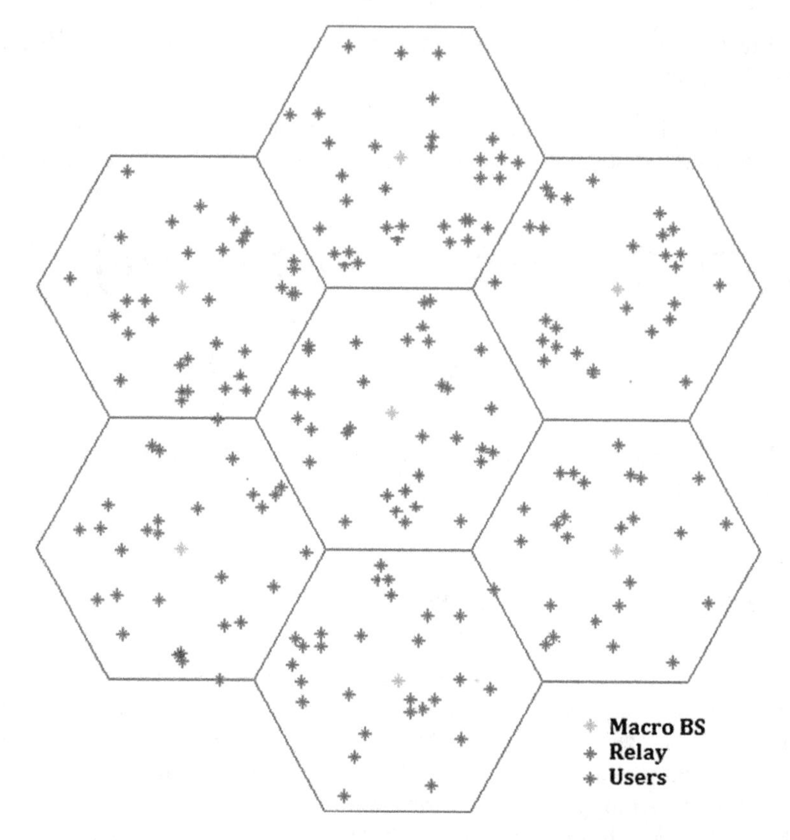

Fig. 3 An illustration of a 7 cell LTE-Advanced macro-relay network with uniformly distributed users

BSs are operating in "off" state in the optimal macro-relay configuration, as shown in Fig. 4. Moreover, it can be seen from Fig. 5 that the AEE of macro-relay network increases with an increase in the density of RNs.

Figure 6 depicts the effect of the density of RNs on the user association for a macro-relay network. It is evident from Fig. 6 that our proposed energy efficient load balancing scheme connects more users to RNs than macro base stations with an increase in the density of RNs in the network. As a result, it relaxes traffic load of some macro eNBs in order to allow them to switch into inactive mode and reduce the overall power consumption of the network.

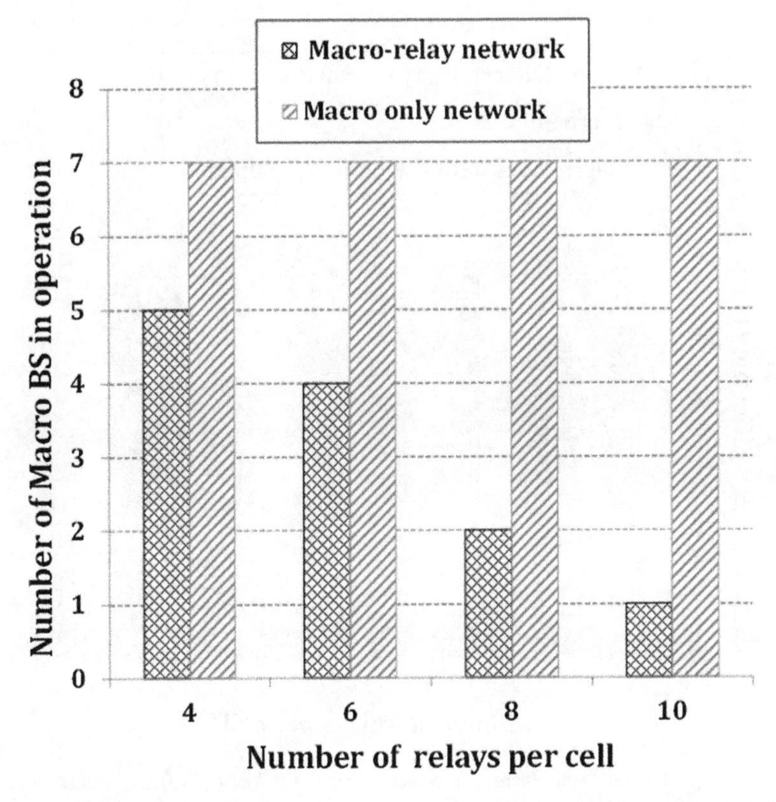

Fig. 4 An illustration of the number of active macro base stations versus relay density for a 7 cell LTE-Advanced macro-relay network

6 Conclusion

In this chapter, we present energy efficient and quality of service aware load balancing and sleep policy for Fifth Generation (5G) dense macro-relay networks. We formulate an energy minimization problem for dense macro-relay HetNets as a Mixed Integer Linear Programming (MILP) problem. Specifically, our proposed algorithm not only optimally connects users to macro BSs and RNs, but also enables lightly loaded macro BSs to switch into off state. Our extensive performance evaluation results reveal that our proposed algorithm for dense macro-relay network can switch off most of the macro BSs with an increase in the density of RNs. It is worth mentioning that our unique approach of relay-to-relay communication forms the basis for relays to act as donors for neighboring relays instead of macro BSs thus allowing the latter to enter the off state. As a result, our proposed energy efficient load balancing and sleep algorithm power consumption is significantly lower than a macro network

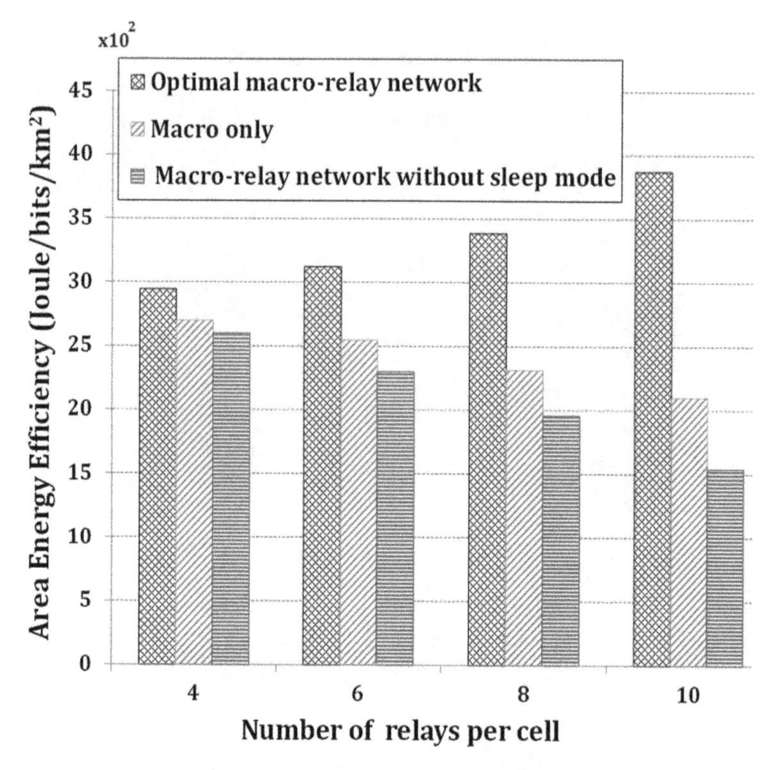

Fig. 5 A comparison of Area Energy Efficiency (AEE) for a 7 cell LTE-Advanced macro-relay network

without relays. Moreover, we have shown that the power consumption of the proposed optimal solution is also lower than a macro-relay network without inactive mode. Our performance evaluation results depict that most of the users connect to RNs with an increase in the density of RNs in the network. It is worth noting that optimally deploying relay nodes should yield communication over short ranges and, hence, lower-power transmissions, as well as enabling switching off some macro BSs. We demonstrated that our proposed algorithm has the best AEE as compared to other load balancing and sleep policies. In summary, we have shown that the proposed algorithm for 5G dense macro-relay networks can significantly reduce system energy consumption while guaranteeing the minimum required data rate in 5G wireless networks. In this chapter, we also present a comprehensive survey of state-of-the-art work on energy efficient resource allocation, load balancing and energy harvesting schemes. However, a number of open challenges suggest a variety of future research directions that can be pursued in order to design QoS aware energy efficient user association techniques for Fifth Generation (5G) cellular networks. One such direction would be to investigate optimal user association between Wi-Fi and LTE coexisted networks

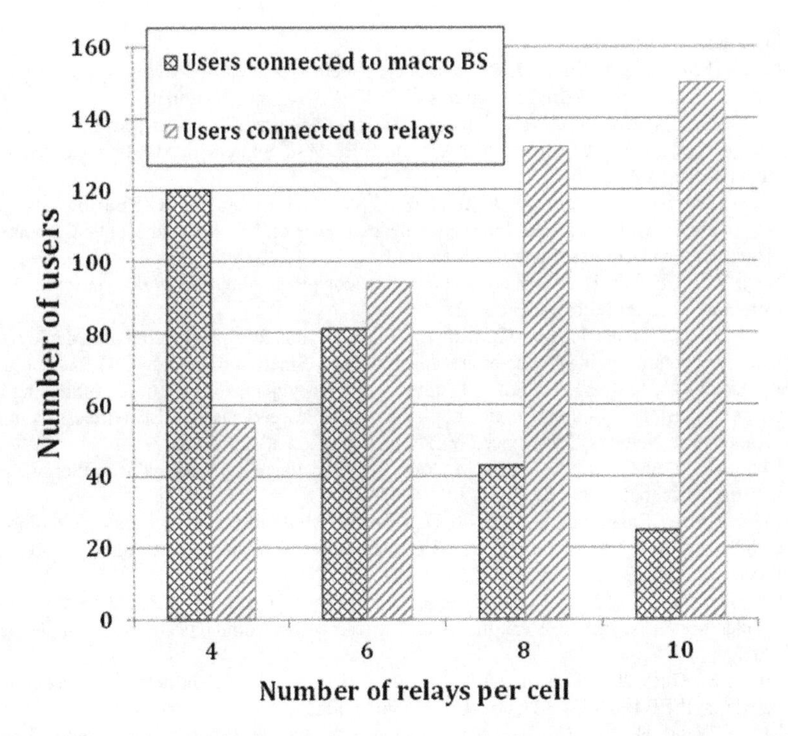

Fig. 6 An illustration of the user association for a LTE-Advanced Macro-pico-relay network

powered by renewable energy sources. Another dimension would be to determine that how much renewable energy should be utilized to power a base station during a specific period of the day based upon real-time weather forecasts.

References

1. Hu, R.Q., Qian, Y.: An energy efficient and spectrum efficient wireless heterogeneous network framework for 5G systems. IEEE Commun. Mag. **52**(5), 94–101 (2014)
2. Chin, W.H., Fan, Z., Haines, R.: Emerging technologies and research challenges for 5G wireless networks. IEEE Wireless Commun. **21**(2), 106–112 (2014)
3. Feng, D., et al.: A survey of energy-efficient wireless communications. IEEE Communic. Surveys Tutorials **15**(1), 167–178 (2013)
4. Andrews, J.G., Buzzi, S., Choi, W., Hanly, S.V., Lozano, A., Soong, A.C.K., Zhang, J.C.: What will 5G be? IEEE J. Sel. Areas Commun. **32**(6), 1065–1082 (2014)
5. http://www.greentouch.org
6. http://www.eitictlabs.eu/innovation-areas/future-networking-solutions/5green-towards-green-5g-mobile-networks/
7. http://www.fp7-greenet.eu
8. http://www.communicate-green.de/
9. Energy Aware Radio and NeTwork TecHnologies (EARTH). https://www.ict-earth.eu/

10. http://www.fp7-trend.eu
11. Leem, H., Baek, S.Y., Sung, D.K.: The effects of cell size on energy saving, system capacity, and per-energy capacity. In: Proceedings of IEEE WCNC, pp. 1–6 (2010)
12. Talwar, S., Choudhury, D., Dimou, K., Aryafar, E., Bangerter, B., Stewart, K.: Enabling technologies and architectures for 5G wireless. In: IEEE MTT-S International Microwave Symposium (IMS), pp. 1–4 (2014)
13. Olsson, M., Cavdar, C., Frenger, P., Tombaz, S., Sabella, D., Jantti, R.: 5GrEEn: towards green 5G mobile networks. In: IEEE International Conference on Wireless and Mobile Computing, Networking and Communications (WiMob), pp. 212–216 (2013)
14. Ling, B.J., Chizhik, D.: Capacity scaling of indoor pico-cellular networks via reuse. IEEE Commun. Lett. **16**(2), 231–233 (2012)
15. Fu, Y., Fei, Z., Wang, N., Xing, C., Wan, L.: An energy-efficient dense pico station deployment and power control strategy for heterogeneous networks. Smart Comput. Rev. **3**(1), 24–32 (2013)
16. Arshad, M.W., Vastberg, A., Edler, T.: Energy efficiency gains through traffic offloading and traffic expansion in joint macro pico deployment. In: Proceedings of IEEE Wireless Communications and Network Conference (WCNC 2012), pp. 2230–2235 (2012)
17. Chiaraviglio, L., Ciullo, D., Mellia, M., Meo, M.: Modeling sleep mode gains in energy-aware networks. Comput. Netw. **57**, 3051–3066 (2013)
18. Marsan, M.A., Chiaraviglio, L., Ciullo, D., Meo, M.: Multiple daily base station switch-offs in cellular networks. In: Proceedings of IEEE International Conference on Communications and Electronics (ICCE 2012), pp. 245–250 (2012)
19. Cao, D., Zhou, S., Niu, Z.: Optimal base station density for energy-efficient heterogeneous cellular networks. In: IEEE International Conference on Communications, pp. 4379–4383 (2012)
20. Soh, Y.S., Quek, T.Q.S., Kountouris, M., Shin, H.: Energy efficient heterogeneous cellular networks. IEEE J. Sel. Areas Commun. **31**, 840–850 (2013)
21. Li, X., Wang, H., Liu, N., You, X.: Dynamic user association for energy minimization in macro-relay network. In: Proceedings of IEEE Wireless Communications and Signal Processing (WCSP), pp. 1–5 (2012)
22. Li, X., Wang, H., Meng, C., Wang, X., Liu, N.: Total energy minimization through dynamic station-user connection in macro-relay network. In: Proceedings of IEEE Wireless Communications and Networking Conference (WCNC), pp. 697–702 (2013)
23. Wang, W., Shen, G.: Energy efficiency of heterogeneous cellular network. In: Proceedings of IEEE Vehicular Technology Conference (VTC Fall), pp. 1–5 (2010)
24. Jiang, J., Peng, M., Zhang, K., Li, L.: Energy-efficient resource allocation in heterogeneous network with cross-tier interference constraint. In: IEEE International Symposium on Personal, Indoor and Mobile Radio Communications (PIMRC Workshops), pp. 168–172 (2013)
25. Oh, E., Son, K., Krishnamachari, B.: Dynamic base station switching-on/off strategies for green cellular networks. IEEE Trans. Wirel. Commun. **12**(5), 2126–2136 (2013)
26. Bu, S., Yu, F.R., Cai, Y., Liu, X.P.: When the smart grid meets energy-efficient communications: green wireless cellular networks powered by the smart grid. IEEE Trans. Wirel. Commun. **11**(8), 3014–3024 (2012)
27. EARTH Project, Ed.: D2.3 V2—Energy efficiency analysis of the reference systems, areas of improvements and target breakdown. https://www.ictearth.eu/publications/publications.html
28. 3GPP TR 36.814 V9.0.0: Technical specification group radio access network; evolved universal terrestrial radio access (e-utra); further advancements for e-utra physical layer aspects (release 9), Mar 2009
29. Saleh, A.B., Bulakci, O., Redana, S., Raaf, B., Hamalainen, J.: Evaluating the energy efficiency of LTE-advanced relay and picocell deployments. In: Proceedings of IEEE Wireless Communications and Networking Conference (WCNC), pp. 2335–2340, France, 2012

An Overview of 4G System-Level Energy-Efficiency Performance

Kazi Mohammed Saidul Huq, Shahid Mumtaz and Jonathan Rodriguez

Abstract This chapter provides the state-of-the-art (SoA) analysis on the current energy efficient approaches for 4G/OFDMA systems. These concepts will act as benchmark for comparison against the energy-efficient approaches to be developed in subsequent chapters. In order to position our work, we review briefly the key components, and decision metrics that can influence the system-level performance from an energy-efficient perspective.

1 Introduction

The ever-growing energy consumption in information and communication technologies (ICT) stimulated by the expected growth in data traffic has provided the impetus for mobile operators to refocus network design, planning and deployment towards reducing the cost per bit, whilst at the same time providing a significant step towards reducing their operational expenditure. In fact, the ICT industry constitutes 3 % of the global energy consumption and contributes towards 2 % of the worldwide CO2 emissions [1]. This is comparable to the worldwide CO2 emissions by airplanes or one quarter of the worldwide CO2 emissions by cars [2]. According to [2, 3], the 57 % of the energy consumption of the ICT infrastructure is attributed to users and network devices in cellular and wireless networks, the scale of which is still growing rapidly [4].

Given the dramatic expansion of wireless networks worldwide, the development of energy-efficient solutions for wireless networks can significantly reduce the energy consumption in the ICT sector. From the viewpoint of telecommunication operators,

K.M. Saidul Huq (✉) · S. Mumtaz (✉) · J. Rodriguez (✉)
Instituto de Telecomunicações, 3810-193 Aveiro, Portugal
e-mail: kazi.saidul@av.it.pt

S. Mumtaz
e-mail: smumtaz@av.it.pt

J. Rodriguez
e-mail: jonathan@av.it.pt

© Springer International Publishing Switzerland 2016

M.Z. Shakir et al. (eds.), *Energy Management in Wireless Cellular and Ad-hoc Networks*, Studies in Systems, Decision and Control 50,
DOI 10.1007/978-3-319-27568-0_3

minimizing the energy consumption is not only a matter of being environmentally responsible, but can substantially reduce their operational expenditure. Furthermore, developing energy-efficient products will open up new business models, since end-users will enjoy enhanced mobile services with longer battery lifetime [5].

2 Energy-Efficient Radio Resource Management

Here, we provide state-of-the-art (SoA) on energy-efficient radio resource management (RRM) in wireless networks. Wireless communications are dynamic in nature. This dynamic nature arises from multiple dimensions: propagation conditions, cell load level, interference, etc. Thus, proper radio management of the available radio resources are needed.

Radio management is performed by an RRM entity with an associated number of parameters that need to be chosen, measured, analyzed and optimized. Efficient utilization of the radio resources leads to higher capacity, Quality of Service (QoS) guarantees, and better user experience. RRM functions should take into account the constraints imposed by the radio interface in order to make decisions regarding the configuration of the different elements and parameters (e.g., the cell size, antenna numbers, the number of users transmitting at the same time). It is pretty evident that the number of parameters to be considered as well as their nature identifies a set of RRM functions whose joint behavior should lead to an overall radio access network optimization [6]. In order to perform properly in a real network environment, RRM schemes should be low in complexity, and require low signaling overhead whilst delivering high performance. Furthermore, they must provide stability and overload protection to the network, in addition to allowing the network to autonomously adapt to dynamic traffic and environment changes. The following subsections cover the above mentioned aspects related to "energy-efficient" RRM , where we first describe the design requirement/trade-offs.

2.1 Fundamental Trade-Offs in RRM Protocol Design

RRM protocol design implies several trade-offs involving energy-efficiency (EE). In [7] the authors present four fundamental trade-offs for EE to drive the design of RRM in next generation cellular networks, which are briefly described herein.

- **Deployment efficiency versus energy-efficiency trade-off**
  Deployment efficiency (DE) is a performance indicator of a wireless networks which quantifies system throughput in terms of per unit of deployment cost. Deployment cost includes Capital expenditures (CAPEX) and operational expenditures (OPEX). Wireless engineers estimate the network CAPEX and OPEX during network planning and EE is mostly considered during network operation. As an example (see [8]) cell radius has a relevant impact on EE: the greater the radius means a reduction in the EE. As a consequence, to maximize EE we

need to deploy additional transmission points, which in turn could increase the deployment cost. This implies the need to identify the proper balance between the DE and EE requirements. LTE-Advanced adopted a heterogeneous networks paradigm, which could provide enhanced deployment functionalities (femto/small cells, coordinated multi-point (CoMP), etc.) to enable proper DE-EE trade-offs.

- **Spectral efficiency versus energy-efficiency trade-off**
  Traditional research on wireless networks mainly focuses on system capacity and spectral efficiency (SE), defined as the system throughput per unit of bandwidth. The spectral efficiency is a key performance indicator of wireless cellular networks and the peak value of SE is always among the key performance indicators of 3GPP evolution. On the other hand, EE accounts for energy consumption: i.e., using less energy to provide the same level of service or using same energy to accomplish improved services.
  For point-to-point transmission in an additive white Gaussian noise (AWGN) channel, the relationship between EE and SE is shown to be in general monotonically decreasing [7]. However, for next generation wireless networks (3GPP LTE, WiMAX), reference OFDM/OFDMA technology [9, 10] and non-Gaussian channel models make such relationship more complex. Rate adaptation (RA), which maximizes throughput and thus increases SE, and margin adaptation (MA), which minimizes total transmit power and thus increases EE [11], are the two main resource allocation schemes to control the SE-EE trade-off in such framework. Improving the SE-EE trade-off curves as a whole and tuning the operation point on the curve to balance the specific system requirements are expected to guide practical system designs toward energy compliant solutions. Moreover, an accurate closed-form approximation of the SE-EE trade-off has not been discussed for interference limited multi-cell scenario. Nevertheless, the concepts demonstrated in the existing literatures [12–14] can be used for evaluating the impact of MIMO, CoMP transmission/reception and relay on the SE-EE trade-off.
- **Bandwidth versus power trade-off**
  Bandwidth (BW) and power (PW) are the most important yet limited resources in wireless communications. From Shannon's capacity formula, the relationship between transmit power and signal bandwidth demonstrate a monotonic trend [7]. For future wireless system such as UMTS and LTE, the trend remains similar.
  Future wireless systems such as LTE-Advanced demonstrate more flexibility in spectrum usage compared to GSM and UMTS, since spectrum re-farming is built-in LTE-Advanced. The deployment of different heterogeneous networks in LTE, such as coordinated multiple point (CoMP) and distributed antenna system DAS, introduces additional infrastructure nodes into the network, which increases control on the BW-PW trade-off.
- **Delay versus power trade-off**
  Delay (DL) is defined as service latency, i.e. a measure of quality of service (QoS) and quality of experience (QoE) [7]. Design of wireless networks should cope with both channel and traffic uncertainties, which makes the characterization of DL-PW trade-off more complex.

Few published works deal with DL-PW trade-off in wireless cellular networks, even though wireless systems need to deal with service latency in order to support users' expectations. As a consequence, it is necessary to analyze when and how to trade service delay with power consumption.

2.2 *Cross-Layer Framework for Energy-Efficient Resource Allocation*

A key component in RRM is the scheduling approach that identifies how to map the available resources to the user queues according to a priority assignment. This assignment is based on a given scheduling policy that can exploits specific network parameters (or context information) to feed a predetermined cost function. This function is intellectual property and governed by the network operator, and we can immediately observe that the type of function will have a large bearing on the operating performance of the network, and therefore energy-efficient operation can be directly linked to the way we design this function.

An energy-efficient design can benefit from a cross-layer (CL) approach as several layers of system design have impact on power consumption ranging from silicon to applications. The authors in [15] particularly focused on a system-based approach towards optimal energy transmission and resource management across time, frequency, and spatial domains. A framework for EE is developed in [16] and depicted in Fig. 1. The mentioned paper focuses on improving device energy-efficiency. Cross-layer approaches exploit interactions between different layers and can significantly improve energy-efficiency as well as adaptability to service, traffic, and environment dynamics. Since wireless network is a shared medium, layering is not the best approach to create impact on device energy consumption comprising a point-to-point communication link, because it impacts the entire network due to the interaction between links. Therefore a system approach better suits energy-efficient wireless communications.

The medium access control (MAC) layer deals with wireless resources for physical (PHY) layer and directly affects overall network performance. Traditional wireless systems have no power adaptation. System-level energy-efficiency is determined by

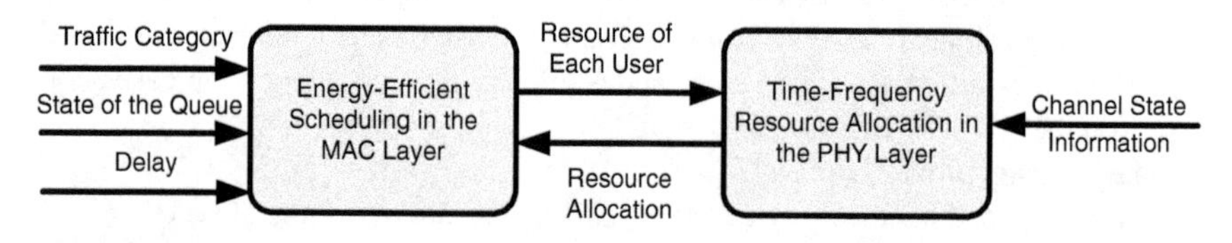

Fig. 1 Framework of EE based cross-layer resource allocation

a set of PHY parameters. The performance of the system ought to be adjusted to adapt the real user requirements (e.g. throughput and power consumption) and environments (such as propagation and multipath channel model) to trade off energy-efficiency and spectral efficiency. The MAC layer ensures that wireless resources are efficiently allocated to maximize network-wide performance metrics while maintaining user quality-of-service (QoS) requirements. According to [15] two types of access are discussed. In distributed access schemes, the MAC should be enhanced to reduce the number of wasted transmissions that are corrupted by other user interference or antenna elements, while in centralized access schemes, efficient scheduling algorithms should exploit the variations across users to maximize overall energy-efficiency of users in the network. From the Shannon capacity, energy-efficiency can only be obtained at the cost of infinite or huge bandwidth and results in zero or very low spectral efficiency.

The MAC layer can enhance energy efficiency using the following three methodologies [15].

- Energy can be saved in mobile devices by shutting down system components when inactive. The MAC can enable inactive periods by scheduling shutdown intervals according to buffer states, traffic requirements, and channel states.
- The MAC layer controls medium access to assure both individual QoS and network fairness. In distributed access schemes, the MAC should be improved to reduce the number of retransmissions; while in centralized access schemes, efficient scheduling algorithms should exploit the channel and traffic variations across users to maximize overall energy efficiency in the network.
- Power management at the MAC layer reduces the standby power by developing tight coordination between users such that they can wake up precisely when they need to transmit or receive data.

2.3 Load Adaptive Resource Management

To satisfy the users' QoS requirements most current network dimensioning is peak load oriented. As a matter of fact, the majority of the existing literatures [17–19] demonstrated that everyday traffic loads at base stations change widely over time and space. Therefore, a great deal of energy is wasted when the traffic load is low. Vendors and operators realized this problem and acted upon this. For example, Alcatel-Lucent proclaimed a new feature in their software upgrades called dynamic power save, which is quoted to save 27 % power consumption for BSs deployed by China Mobile [20]. Energy-saving solutions through cell-size breathing and sleep modes, based on traffic loads, were proposed by the OPERA-Net project [21].

In [21, 22], a measured traffic pattern was analyzed enabling optimal power-saving schemes using cell switch-off under a trapezoidal traffic pattern, where it is shown that a 25–30 % energy saving is feasible by merely switching off the active cells during the periods of low traffic activity. Nevertheless, the impact of switch-off on coverage was not investigated. In [18], the authors investigated the notion of blocking probability requirement enabling a traffic-aware BS mode (active or sleeping) switching algorithm. One minimum mode holding time was also recommended to avoid frequent BS mode switching. It was demonstrated that changing the holding time over a specified range will cause trivial performance change on either energy saving or blocking probability [18]. The effect of the traffic mean and variance as well as the BS density on the energy saving strategy with BS switching was extensively studied in [19], which proved that energy savings will increase with the BS density and the statistical ratio of the traffic load. In [23], they presented some possible approaches to establish energy consumption of the BS's scale with the traffic load across space, frequency and time domains. According to [24], joint reconfiguration of the bandwidth and the number of antennas and carriers according to the traffic load gained maximum energy saving. Similar energy-saving solutions based on user load variations on the terminal side were described in [25].

2.4 *Service Differentiation*

Service differentiation mainly deals with the trade-off between energy consumption and delay [26]. The trade-off between energy consumption and delay was extensively studied in the literature for wired non-cellular network. For cellular wireless networks, only few works had been done in the early systems (1G, 2G systems), because only limited service types (mainly voice communications) were available. However, the evolution of cellular systems provided the vehicle for more sophisticated services and devices (smart phones, iPhone, and the blackberry among others). To be precise, some applications, such as video conferencing, web-based seminars, and video games, require real-time service; and other applications, such as email, and downloading files for offline processing are delay tolerant services. Therefore, it is useful to separate the types of wireless traffic and build the energy consumption mechanism protocol with the traffic type.

Several researchers have targeted the efforts on the service latency of applications to reduce the energy consumption in cellular networks. In [27], energy-efficient power and rate control with delay QoS constraints using a game-theoretic approach was presented. The demonstration was based on CDMA system. They translated the delay constraint of a user into a lower bound on the user's output SIR (signal-to-interference ratio) requirement; afterwards the Pareto-dominant equilibrium solution is derived. The delay performance of users at the Nash equilibrium was also analyzed. Inspired by mobility-prediction-based transmission strategies, which are usually used in delay tolerant networks, a store-carry-and forward (SCF), relay-aided cellular architecture was proposed in [28, 29]. According to [24], in the SCF scheme, when the application data is not delay prone, a user can first transmit the data to a mobile relay (for instance, a vehicle) which conveys the message close to the BS, and then the mobile relay retransmits the data to the BS. Numerical results in [29] depicted that, for delay insensitive services, a factor of more than 30 in energy savings can be obtained by SCF compared to direct transmission.

3 Exploitation of Multi-user Diversity

The process of multiple users experiencing independent fading channels is known as multi-user diversity (MUD). In an energy-efficient context, it turns out that the sum capacity (sum of the simultaneous user capacities) is maximized if, for each time instant, the user with the best energy-efficient channel gain is scheduled. The gain achieved with such strategy can be defined as energy-efficient MUD gain. According to [30, 31], the most feasible solution takes into account a power control law which uses more transmit power for strong channels than weak channels. This solution is the opposite to conventional power control which uses transmit power to compensate weak channels.

A major problem for energy consumption multi-user diversity adaptation is how to design heuristic algorithms that achieve the multi user diversity gain while ensuring minimum energy consumption or increasing the QoS requirements using same amount of power under realistic conditions.

4 Relay Transmission

Relaying is widely acknowledged as a means to improve capacity and coverage in Wireless Broadband Networks [32]. The properties of the relay concept and the benefits that can be expected are as follows:

- Radio coverage can be improved in scenarios with high shadowing (e.g. bad urban or indoor scenarios). This allows to significantly increase the QoS of users in areas that are heavily shadowed. The extension of the radio range of BS by means of relay allows operating much larger cells with broadband radio coverage than with a conventional one-hop system.
- Using relaying can reduce the overall system-level energy consumption and pave the way to public acceptance, while in the case of mobile terminals it has the potential to increase battery lifetime.
- The fixed relay concept provides the possibility of installing temporary coverage in areas where permanent coverage is not needed (e.g. construction sites, conference-/meeting-rooms) or where a fast initial network roll-out has to be performed.
- The wireless connection of the relay to the fixed network substantially reduces infrastructure costs, which in most cases are the dominant part of the roll-out and operational costs; relay only requires a main supply. In cases where no main supply is available, relays could rely on solar power supply. A relay cellular network is illustrated in Fig. 2.

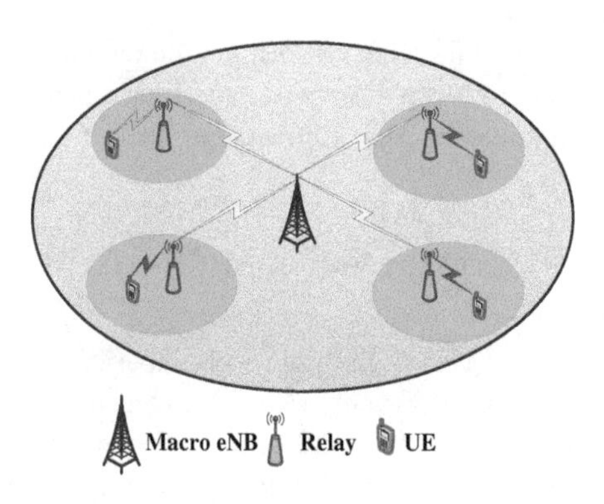

Fig. 2 Relay cellular network

A standard-conformant integration of the relays into any MAC frame based system would allow for a stepwise enhancement of the coverage region of an already installed system. Investments in new BS can be saved, and any hardware product complying with a wireless MAC frame based standard can be used without modifications.

An application of relay technology is the LTE-Advanced system. LTE-Advanced promises to provide improved performance with the aim of achieving high speed, high-capacity communication, and service capabilities beyond LTE. LTE Release-10 includes all the features of Rel-8/9 and several new ones, the most important of which are: carrier aggregation enhanced multi-antenna support, improved support for heterogeneous deployments, and relaying [6]. In LTE-Advanced, an important issue in addition to achieving high-speed and capacity, is to provide greater throughput for cell-edge users which could be accomplished by employing relay technology. Few of the main reasons for choosing relay technology for the LTE-Advanced system are given as follows.

- Lack of fairness: In a conventional cellular network (CCN), a base station (BS) controls a number of mobile stations (MS) within its own coverage area and all the terminals communicate directly with the BS. The current conventional deployment of cellular systems exhibit certain inherent problems such as low signal-to-noise-ratio (SNR) at the cell edge, lack of fairness, coverage holes that exist due to shadowing and non-line-of-sight (NLOS) connections.
- Energy consumption: The CCN has been primarily designed to meet the challenges of service quality. More recently, there is a growing focus on the importance of energy consumption, both from an operational expenditure (OPEX) point of view and from a climate change perspective. Over the past few years, the communication industry has pledged to reduce carbon emissions of wireless networks by up to 50 % by 2020 [33]. Minimizing energy consumption in LTE-Advanced network has been at the forefront of system design, and architectural approaches which recently have been proposed include femtocells, advanced spectrum management, efficient power amplifier, antenna technologies, etc.

However, using relay technology can amplify the energy gain further. It can be considered as an extension of the specific base station/eNB, and it uses the base-station air interface resources; therefore it does not require a separate backhaul connection while the femtocell and picocell act as separated base station using specific resources and hence require a separate backhaul connection. Picocell and Femtocell differ from each other in power range and capacity level [34].

Most existing work concentrates on single-point-to-single-point transmission; how to allocate resources in multi-point-to-single-point or multi-point-to-multi-point transmission, as in the multi-cell case needs further attention. Incremental time and power may be used for resource allocation during relay transmission. How to minimize the total energy consumption to ensure greater energy-efficiency taking the additional overhead into account is not known succinctly.

5 Energy Analysis: SISO Versus MIMO with Packet Scheduling

Han et al. analyzed and demonstrated energy-efficiency of SISO and MIMO [35] and use LTE standpoint as a case study. LTE already specified the Alamouti-based [36] space-frequency block coding (SFBC) technique for MIMO. They also considered spatial multiplexing (SM) as another MIMO approach. In SM, independent symbols are transmitted over different antennas as well as over different symbol times. They described for specific data rates, the energy-efficiency of SISO and MIMO schemes employing different levels of modulation order and coding rates. In [35], the authors described two types of energy analysis in the LTE system. These include:

1. Energy efficiency performance evaluation without considering overhead.
2. Energy efficiency performance evaluation with overhead.

The power level of the overhead shows a significant impact on the energy consumption ratio (ECR) of all schemes at low spectral efficiency range as the power required by transmitting user data is relatively low. As a result, the ECR of all schemes for low spectral efficiency transmission is significantly increased. These are the stepping stone of the energy efficiency analysis of SISO and MIMO.

Some of the open issues for multi-user and multi cell environments in MIMO still require attention, such as how to utilize the spatial resource to maximize EE while suppressing interference, since the existence of inter user and inter cell interference complicates the design of energy-efficient MIMO systems. Effective but simple algorithms need to be developed to obtain a trade-off between complexity and performance for MIMO-OFDMA system.

5.1 Energy Efficiency in SoA Packet Scheduling Techniques

The three major SoA resource scheduling algorithms which deal with downlink packet transmission are here discussed herein, these include maximum carrier-to-interference ratio (MCI) [37], proportional fairness (PF) [38] and round robin (RR) [39].

In the MCI method the users are scheduled to use radio resources based on maximum channel gain. This scheme is straight forward, in the sense that users are ranked according to their experienced channel gain. In other words, the user with the best channel quality indicator (CQI) has the highest ranking and is scheduled to utilize the physical resource blocks (PRB) for the specific time. The user with the next best CQI condition is then scheduled to utilize PRBs and so forth. The ranking 'U' can be found using the following equation:

$$U = \arg \max_{u} \left(\beta_{u,m}(t)\right) \quad for\, PRB\ m \tag{1}$$

where β is the vector of experienced channel gain of UE, u, for one PRB, m, in time t. The flowchart of the MCI scheduling is depicted in Fig. 3.

In order to perform scheduling, terminals send (in uplink) CQI to the BS. Basically in the downlink, the BS transmits the reference signal (downlink pilot) to the terminals. These reference signals are used by the UE for measuring the CQI; a high value for CQI means high quality channel condition. We should keep in mind that CQI is reported per PRB. MCI scheduling [41] can increase the cell capacity at the expense of fairness. For conventional cellular networks exploiting this scheduling strategy, terminals located far from the base station (i.e. cell-edge users) are unlikely to be scheduled.

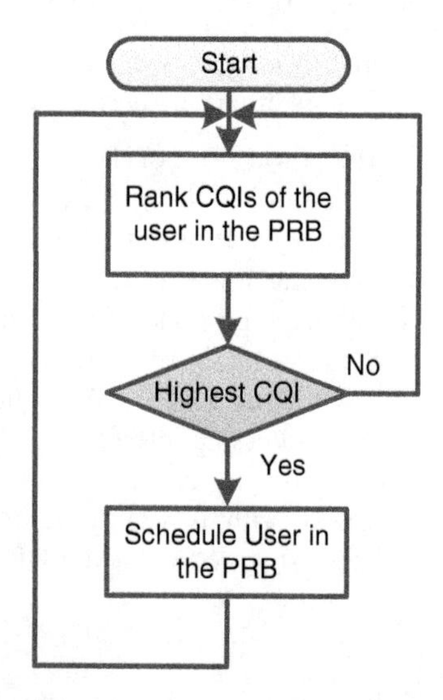

Fig. 3 MCI packet scheduling technique [40]

In the PF algorithm, for PRB m, the highest ranked user u' is scheduled to transmit according to the following:

$$u' = \arg\max_{u} \left(\frac{R_{u,m}(t)}{T_u(t)} \right) \tag{2}$$

where, $R_{u,m}(t)$ denotes the instantaneous achievable rate at PRB m and $T_u(t)$ is the user's average throughput. The average throughput, $T_u(t)$, is updated for each new time interval (after all PRBs are allocated).

This scheduler aims to combine throughput efficiency with long term resource-fairness. Practically, this scheduling policy provides the same fraction of resources for all the users in the long-term perspective. However, in each time-instance users are prioritized based on their normalized channel condition. The normalization factor is the past profile of each user, i.e. the exponential averaged data rate. As in Eq. (2), the numerator of this scheduling metric is in favor of the best-channel users, while the denominator tries to balance resource-fairness by penalizing the users with good past profile [39]. PF is throughput efficient and provides long term fairness through equalizing the resources allocated to different users in the system. Figure 4 depicts the flowchart of the PF algorithm. This policy does not provide any explicit bound on the QoS requirement of different users in the system [39].

In RR, the radio resources i.e., PRBs, are allocated to UEs in a round robin fashion irrespective of channel condition. The first opted UE is served for a specific time period and then these resources are revoked back and assigned to the next user for another time period. The previously served user is placed at the end of the waiting queue so that it can be served with radio resources in the next round. Newly arriving requests are also placed at the tail of the waiting queue. This scheduling continues in the same manner [39]. Thus every user is equally scheduled without taking the CQI into account as illustrated in Fig. 5. The principal advantage of Round Robin scheduling is the guarantee of fairness for all users, and it is easily implemented. Since Round Robin does not take the channel quality information into account, it results in low user throughput.

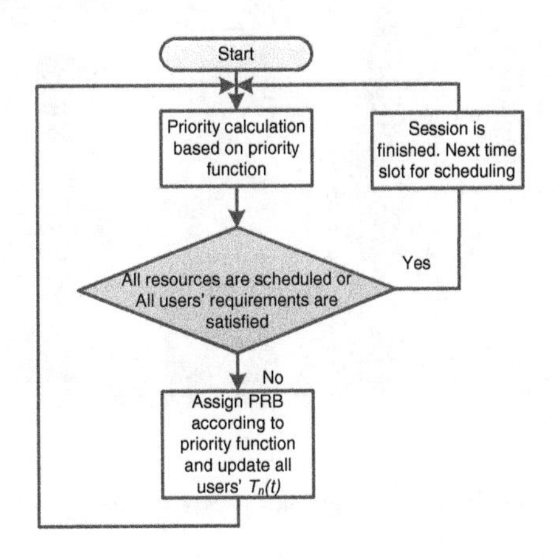

Fig. 4 PF packet scheduling technique [42]

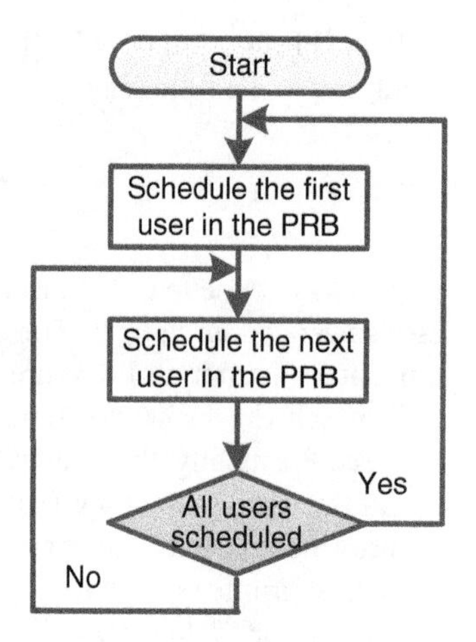

Fig. 5 RR packet scheduling technique [40]

In [43], the performance of an LTE system with various packet scheduling algorithms was studied from an energy-efficiency point of view. In this work, the performance of various classical scheduling algorithms such as RR, PF and MCI was used as a basis for the assessment of further innovative energy aware algorithms. They also analyze gains in terms of the energy consumption index (with respect to Round Robin scheduler). Figure 6 presents a benchmark of different packet scheduling in terms of EE in LTE system.

The paper [45] presents a link level analysis of the rate and energy-efficiency performance of the LTE downlink considering the unitary codebook based precoding scheme. The authors consider a multi-user environment to improve the performance gain by exploiting multi-user diversity in the time, frequency and space domains, and translating the gains to an energy reduction at the base station. Several existing and

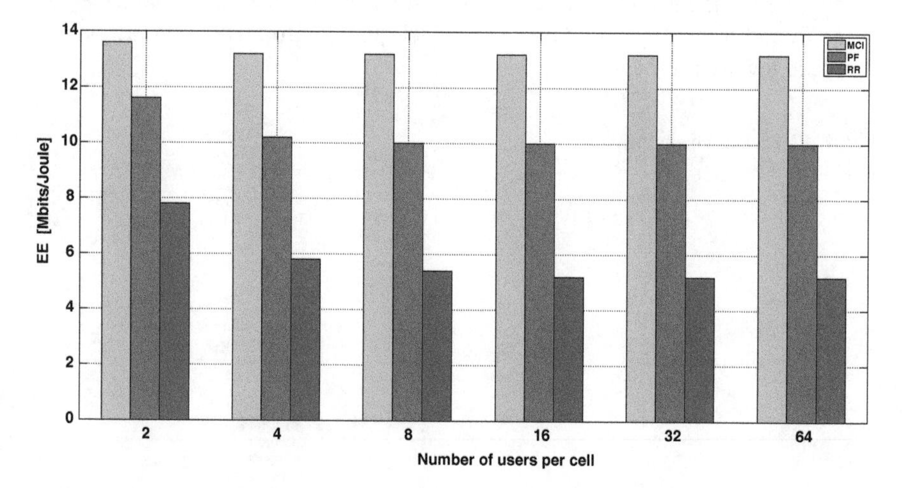

Fig. 6 Energy efficiency packet scheduling in LTE-Advanced [44]

novel dynamic resource allocation algorithms were studied, such as PF, FCA (Fair Cluster Algorithm), RSSA (Received Strength Scheduling Algorithm) and EG-DA (Equal Gain Dynamic Allocation) among others, for the LTE system. The authors mainly focus on the rate and power consumption performance of the 3GPP LTE-OFDMA downlink system employing SU-MIMO. Both of the above mentioned works employ standard transmission scheme using no coordination or cooperation between cells.

On the other hand, some works have already analyzed CoMP concepts. According to [46], a gain in the downlink cell-edge throughput as well as cell average throughput can be achieved in LTE-Advanced network with the CoMP transmission architecture. It refers to the possibility to coordinate the downlink transmission towards the same user adopting multiple base stations.

Similarly, our works in [44, 47] show important benchmark EE analysis of different CoMP techniques. As presented in Fig. 7, the EE increases with the number of users, which is due to multi user diversity, and is more improved in techniques that exploit a larger number of users with more antenna diversity.

To clarify this, Joint Transmission (JT) and Dynamic Point Selection (DPS) are more energy-efficient when compared to Coordinated Beamforming (CB) due to greater antenna diversity of coherent transmission of multi antenna and base stations, that is muted in the case of DPS. Both these techniques outperform CB thanks to their capability to transmit more reliable bits per unit of energy consumed. From our

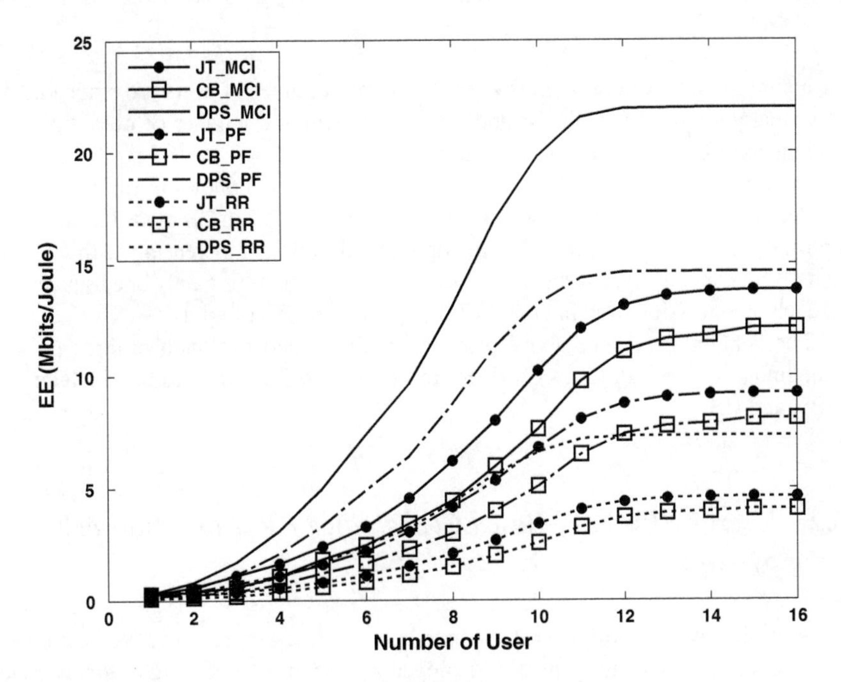

Fig. 7 Different packet scheduling in various CoMP transmission

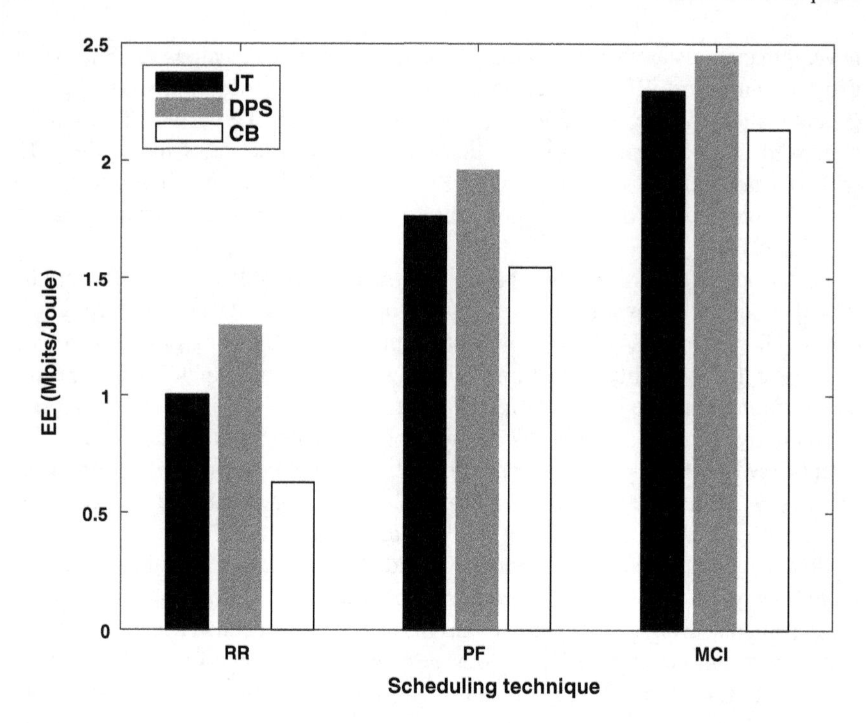

Fig. 8 Optimal energy-efficiency for different packet scheduling in various CoMP transmission

own simulations, we observed that until a certain number of users are reached, EE is virtually similar for both CB and DPS. Increasing the number of users beyond, widens the EE gap between CB and DPS.

Figure 8 shows the optimal EE for different scheduling policies in various CoMP modes, where the number of users is 20. It is shown that RR in CB mode has the least optimal EE. We also observe that the optimal EE value is increasing with CB, JT and DPS, respectively, irrespective of scheduling. Transmitting only one transmitter in each transmission time interval (TTI) improves the EE of the DPS.

The benchmark results above provides us the impetus to analyze three packet scheduling methods in terms of EE for different CoMP techniques that include JT, DPS, and CB.

5.2 Energy Efficiency Based Coordinated RRM for Multi-cell Systems

Here we deal with coordinated RRM based on multi-cell scheduling. Wireless communication networks are generally deployed and adapted according to the average expected traffic load, by carefully designing the cell radius and the reuse factor.

However, this static approach underperforms and fails to reach the accepted limits in the context of spectral efficiency, in particular at low reuse rates; since there is no mechanism in place for coordinating the sector bandwidth allocation to match the instantaneous spatial distribution of the users, their QoS requirements, and link quality.

In general, mobile user distribution within a cell and their channel condition are responsible for the reuse factor selection. Thus, using a single reuse factor within a cell is not particularly efficient. For instance, users that are close to their serving BS can reuse resources since interference is low, whereas cell-edge users, which are close to other sectors, should rely on an exclusive allocation of spectrum policy. The scheduling policy, which determines the users to be served, impacts on the suitability of the reuse factor and, therefore, the reuse factor determination should be included in the scheduling process. The study in [48] demonstrated a large potential for improvement in terms of spectral efficiency, but at the cost of increased information overhead for coordinating sectors of adjacent antennas. However, it remains low in comparison with the CoMP approach where full cooperation is utilized and, thus, it can be seen as a practical approach. In EARTH [48], one coordinated RRM was proposed for uplink scenario from the point of view of energy-efficiency. However, nothing was done for downlink multi-cell scenario which is still an open issue.

6 Interference Management for Heterogeneous Networks

Heterogeneous networks phenomenon was proposed in LTE-Advanced framework as a means to increase the spectral efficiency [49], and provide seamless coverage. According to [50], a multi-tier network topology appears to increase system performance due to the achievable radio link performance, providing a system-level gain close to the theoretical limit of 3G. In this strategy, macro base stations are used to provide blanket coverage, on the other hand, small low power base stations are introduced to eliminate the coverage holes and at the same time increase the system capacity in hotspots [49]. Recently these heterogeneous networks are investigated to increase the energy-efficiency of the network, however this new scenarios requires more stringent interference management.

The Interference problem can arise in difference forms. One type of interference is defined as intra-cell interference, which is defined as the interference from users within the same cell, whereas interference emanating from other cells refers to inter-cell interference; both of which lead to reduced system coverage and degrade the delivered QoS. Sometimes we misinterpret fading and interference. Fading is a phenomenon that is created by the natural random process from different copies of the signal after traveling through a time-variant multipath environment. In comparison, interference is mainly caused by artificially created signals that coexist with the desired signal along the same physical dimensions: code, frequency, space and time.

The term interference cancellation is commonly used in the literature for signal processing applications that exploit algorithms in which the "interference signals" can be estimated and emulated in a reliable way, and canceled from the desired signal [51]. Various kinds of interference are present depending on the type of communication systems, the source of interference, and who is being subject to such interference. For instance, in WLAN [52], the interference occurs when neighboring systems work in adjacent, or in the same frequency bands. This type of interference is called inter-system interference; and, as expected, its reduction or complete mitigation implies using complex and expensive devices.

One of the best known types of interference in wireless channels arises from transmitting a finite alphabet symbol through a multipath or band-limited channel, which is called inter-symbol interference [51]. Intersymbol interference had been the subject of substantial research efforts during the last decades or so. As another example, OFDM systems, experience inter-carrier interference that is caused by carrier frequency offset and phase noise due to the imperfect nature of the transmitter and receiver ends, thus causing the signal at a particular subcarrier to being affected by the superposition of several other subcarriers. A cyclic prefix is designed to combat this [53], different users might be assigned different subcarriers when OFDM technology is used as the multiple access technique, (i.e. OFDMA), hence, intercarrier interference is also known as inter-user or multiple access interference.

One of the main objectives of energy-efficient BS cooperation is to ensure that all the energy that is spent by the base stations is fully used to transport data [48]. In EARTH they demonstrated that interference can be seen as a waste of energy if it is uncontrolled. Several methods were proposed for BS cooperation using the backhaul links [48]. Depending on the capacity of these links, the cooperation can be implemented in the data plane by using joint or distributed processing algorithms, or in the control plane by coordinating the allocated resources for the users that are impacted by the inter-BS interference. They denoted a mechanism based on fractional frequency reuse (FFR) that exploits the reuse planning strategy. It consists in adjusting dynamically the parameters of the FFR strategy depending on the density of the served users. Still this system needs more research from an energy-efficiency perspective.

Reference [54] describes the femto cellular networks from both the technical and business aspects. They also emphasized the challenges of implementing these types of networks and focused on some potential research opportunities. It is indicated in [54] that femto cellular networks must deal with additional timing and synchronization, as well as interference management issues, which result in additional signaling overhead and potentially greater energy consumption. Thus, how to design and manage energy-efficient femto cellular networks is still an open research issue. Reference [55] emphasized one major problem for future research regarding the interference of heterogeneous networks. How to manage interference and design algorithms with respect to EE for heterogeneous networks. Since there will be more transmitter sources and access points with heterogeneous deployment, there is the potential greater interference and more frequent handoffs.

7 Conclusions

In this chapter, we provided the state-of-the-art analysis on the current energy-efficient approaches for 4G/OFDMA systems. We provided a brief overview of the key engineering trade-offs in energy-efficient RRM design and then summarized existing fundamental works and advanced techniques to promote energy-efficiency for OFDMA based cellular networks. The discussions and the results in this chapter will act as benchmark for comparison against the energy-efficient approaches to be developed in subsequent chapters.

Acknowledgments This work was carried out under the E-COOP project (PEst-OE/EEI/LA0008/2013—UID/EEA/50008/2013), funded by national funds through FCT/MEC (PIDDAC).

References

1. De Sanctis, M., Cianca, E., Joshi, V.: Energy efficient wireless networks towards green communications. Wirel. Pers. Commun. **59**(3), 537–552. http://www.springerlink.com.miman.bib.bth.se/content/e764hq2h2h83081t/ (2011)
2. Fettweis, G., Zimmermann, E.: ICT energy consumption—trends and challenges. In: Communications. WPMC, 2008, no. Wpmc 2008, pp. 2006–2009 (2008)
3. McLaughlin, S.: Green radio: the key issues. In: Mobile VCE (2008)
4. Hrault, L.: Green wireless communications, eMobility, Technical Report (2008)
5. CISCO.: CisCo visual netowrking index: global mobile, Cisco Systems, Inc., Technical Report (2010)
6. Sallent, O.: A perspective on radio resource management in B3G. In: 3rd International Symposium on Wireless Communication Systems, 2006. ISWCS '06, pp. 30–34 (2006)
7. Chen, Y., Zhang, S., Xu, S., Li, G.Y.: Fundamental trade-offs on green wireless networks. IEEE Commun. Mag. **49**(6), 30–37 (2011)
8. Badic, B., O'Farrell, T., Loskot, P., He, J.: Energy efficient radio access architectures for green radio: large versus small cell size deployment. In: Vehicular Technology Conference Fall (VTC 2009-Fall): IEEE 70th. Sep. 2009, pp. 1–5 (2009)
9. Akyildiz, I.F., Gutierrez-Estevez, D.M., Reyes, E.C.: The evolution to 4G cellular systems: LTE-advanced. Phys. Commun. **3**(4), 217–244 (2010)
10. Parkvall, S., Furuskar, A., Dahlman, E.: Evolution of LTE toward IMT-advanced. IEEE Commun. Mag. **49**(2), 84–91 (2011)
11. Bohge, M., Gross, J., Wolisz, A., Meyer, M.: Dynamic resource allocation in OFDM systems: an overview of cross-layer optimization principles and techniques. IEEE Network **21**(1), 53–59 (2007)
12. Onireti, O., Heliot, F., Imran, M.: On the energy efficiency-spectral efficiency trade-off in the uplink of comp system. IEEE Trans. Wirel. Commun. **11**(2), 556–561 (2012)
13. Heliot, F., Imran, M., Tafazolli, R.: On the energy efficiency-spectral efficiency trade-off over the mimo rayleigh fading channel. IEEE Trans. Commun. **60**(5), 1345–1356 (2012)
14. Onireti, O., Heliot, F., Imran, M.: On the energy efficiency-spectral efficiency trade-off of distributed mimo systems. IEEE Trans. Commun. **61**(9), 3741–3753 (2013)
15. Miao, G., Himayat, N., Li, Y., Swami, A.: Cross-layer optimization for energy-efficient wireless communications: a survey. Wirel. Commun. Mob. Comput. **9**(4), 529–542, (2009). (Geoffrey)
16. Xiao-hui, L., Ming, H., Ke-chu, Y., Chang-xing, P., Nai-an, L.: Radio resource management algorithm based on cross-layer design for OFDM systems. In: Seventh International Conference

on Parallel and Distributed Computing, Applications and Technologies: PDCAT '06, vol. 2006, pp. 311–314 (2006)

17. Willkomm, D., Machiraju, S., Bolot, J., Wolisz, A.: Primary users in cellular networks: a large-scale measurement study. In: 3rd IEEE Symposium on New Frontiers in Dynamic Spectrum Access Networks, 2008. DySPAN, vol. 2008, pp. 1–11 (2008)
18. Gong, J., Zhou, S., Niu, Z., Yang, P., Traffic-aware base station sleeping in dense cellular networks. In: 18th International Workshop on Quality of Service (IWQoS). June 2010, pp. 1–2 (2010)
19. Oh, E., Krishnamachari, B.: Energy savings through dynamic base station switching in cellular wireless access networks. In: 2010 IEEE Global Telecommunications Conference (GLOBECOM 2010), pp. 1–5 (2010)
20. Alcatel-Lucent: Alcatel-lucent demonstrates up to 27 percent power consumption reduction on base stations deployed by china mobile: Software upgrades can offer exceptional power and cost savings for mobile operators worldwide. In: Technical Report, Feb 2009
21. Opera-Net: Optimising power efficiency in mobile radio networks. In: OPERA-NET Project, vol. 42. http://opera-net.org/Documents/5026v1Opera-Nete-NEM%20Event%20Barcelona%202010Demos%20Presentation290710.pdf (2010)
22. Marsan, M.A., Chiaraviglio, L., Ciullo, D., Meo, M. Optimal energy savings in cellular access networks. In: IEEE International Conference on Communications Workshops, 2009. ICC Workshops 2009. IEEE, June 2009
23. Grant, P.: MCVE Core 5 Programme, Green radio -the case for more efficient cellular basestations. In: MCVE (2010)
24. Chen, T., Zhang, H., Zhao, Z., Chen, X.: Towards green wireless access networks. In: 2010 5th International ICST Conference on Communications and Networking in China (CHINACOM), pp. 1–6 (2010)
25. Ge, X., Cao, C., Jo, M., Chen, M., Hu, J., Humar, I.: Energy efficiency modelling and analyzing based on multi-cell and multi-antenna cellular networks. KSII Trans. Internet Inf. Sys. vol. 4(4). http://www.pubzone.org/dblp/journals/itiis/GeCJCHH10 Aug 2010
26. Feng, D., Jiang, C., Lim, G., Cimini, J., Feng, L.J.G., Li, G.: A survey of energy-efficient wireless communications. IEEE Commun. Surv. Tutor. 15(1), 167–178 (2013)
27. Meshkati, F., Poor, H.V., Schwartz, S.C., Balan, R.V.: Energy-efficient resource allocation in wireless networks with quality-of-service constraints. IEEE Trans. Commun. **57**(11), 3406–3414 (2009)
28. Kolios, P., Friderikos, V., Papadaki, K.: Ultra low energy store-carry and forward relaying within the cell. In: 2009 IEEE 70th Vehicular Technology Conference Fall (VTC 2009-Fall), pp. 1–5 (2009)
29. Kolios, P., Friderikos, V., Papadaki, K.: Store carry and forward relay aided cellular networks. In: Proceedings of the seventh ACM international workshop on VehiculAr InterNETworking, ser. VANET '10, pp. 71–72. ACM, New York. http://doi.acm.org/10.1145/1860058.1860071 (2010)
30. Prez-Neira, A.I., Campalans, M.R.: Cross-Layer Resource Allocation In Wireless Communications. Academic Press, New York (2009)
31. Perre, L.V.D., Craninckx, J., Dejonghe, A: Energy-aware cross-layer radio management. In: Green Software Defined Radios, ser. Integrated Circuits and Systems, pp. 115–133. Springer Netherlands. http://dx.doi.org/10.1007/978-1-4020-8212-2_6 (2009)
32. ETSI: Universal mobile telecommunications system (UMTS); selection procedures for the choice of the radio transmission technologies of the UMTS (UMTS 30.03 version 3.2.0), TR 101 112 v3.2.0, Apr 1998
33. Guo, W., O'Farrell, T.: Green cellular network: Deployment solutions, sensitivity and tradeoffs. Wirel. Adv. (WiAd) **2011**, 42–47 (2011)
34. Elkourdi, T., Simeone, O.: Femtocell as a relay: an outage analysis. IEEE Trans. Wirel. Commun. **10**(12), 4204–4213 (2011)
35. Han, C., Armour, S.: Energy efficient radio resource management strategies for green radio. IET Commun. **5**(18), 2629–2639 (2011)

36. Alamouti, S.: A simple transmit diversity technique for wireless communications. IEEE J. Sel. Areas Commun. **16**(8), 1451–1458 (1998)
37. Pokhariyal, A., Kolding, T., Mogensen, P., Performance of downlink frequency domain packet scheduling for the UTRAN long term evolution. In: IEEE 17th International Symposium on Personal, Indoor and Mobile Radio Communications, pp. 1–5, Sep 2006
38. Norlund, K., Ottosson, T., Brunstrom, A.: Fairness measures for best effort traffic in wireless networks. In: 15th IEEE International Symposium on Personal, Indoor and Mobile Radio Communications, 2004. PIMRC 2004. vol. 4, pp. 2953–2957, Sep 2004
39. Shariat, M., Quddus, A., Ghorashi, S., Tafazolli, R.: Scheduling as an important cross-layer operation for emerging broadband wireless systems. Commun. Surv. Tutor. IEEE **11**(2), 74–86 (2009)
40. Dikamba, T., LO, A.: Downlink Scheduling in 3gpp Long Term Evolution (LTE), Master Thesis, Wireless and Mobile Communication (WMC) Group, Faculty of Electrical Engineering, Mathematics and Computer Science, Delft University of Technology. http://resolver.tudelft.nl/uuid:0ba3daac-8697-4045-a52a-c67c4b9a47d9 Mar 2011
41. Schwarz, S., Mehlfuhrer, C., Rupp, M.: Low complexity approximate maximum throughput scheduling for LTE. In: 2010 Conference Record of the Forty Fourth Asilomar Conference on Signals, Systems and Computers (ASILOMAR), pp. 1563–1569, Nov 2010
42. Tang, Z.: Traffic Scheduling for LTE Advanced, Master's thesis, Linkping University, Department of Electrical Engineering, Communication Systems. http://www.diva-portal.org/smash/record.jsf?dswid=7952&pid=diva2%3A393400&c=5&searchType=SIMPLE&language=en&query=Traffic+Scheduling+for+LTE+Advanced&af=%5B%5D&aq=%5B%5B%5D%5D&aq2=%5B%5B%5D%5D&aqe=%5B%5D&noOfRows=50&sortOrder=author_sort_asc&onlyFullText=false&sf=all&jfwid=7952 (2010)
43. Sabella, D., Caretti, M., Fantini, R.: Energy efficiency evaluation of state of the art packet scheduling algorithms for LTE. In: Wireless Conference 2011–Sustainable Wireless Technologies (European Wireless), 11th European, pp. 1–4, Apr 2011
44. Huq, K.M.S., Mumtaz, S., Saghezchi, F.B., Rodriguez, J., Aguiar, V.: Energy efficiency of downlink packet scheduling in CoMP. Trans. Emerg. Telecommun. Technol. **26**(2), 131–146 (2015). http://onlinelibrary.wiley.com/doi/10.1002/ett.2686/abstract
45. Han, C., Beh, K.C., Nicolaou, M., Armour, S., Doufexi, A.: Power efficient dynamic resource scheduling algorithms for LTE. In: Vehicular Technology Conference Fall (VTC 2010-Fall): IEEE 72nd, pp. 1–5. IEEE, Sep 2010
46. Wang, Q., Jiang, D., Liu, G., Yan, Z., Coordinated multiple points transmission for LTE-Advanced systems. In: 5th International Conference on Wireless Communications, Networking and Mobile Computing, 2009. WiCom '09, pp. 1–4. IEEE, Sep 2009
47. Huq, K., Mumtaz, S., Rodriguez, J., Aguiar, R.: Comparison of energy-efficiency in bits per joule on different downlink CoMP techniques. In: 2012 IEEE International Conference on Communications (ICC), pp. 5716–5720, June 2012
48. EARTH (Energy Aware Radio and neTwork Technologies) FP7 Project: Deliverable 3.1: Most promising tracks of green network technologies, EU-FP7 Project, Deliverable D3.1 INFSO-ICT-247733 EARTH. https://www.ict-earth.eu/publications/deliverables/deliverables.html Dec 2010
49. Tombaz, S., Usman, M., Zander, J.: Energy efficiency improvements through heterogeneous networks in diverse traffic distribution scenarios. In: 2011 6th International ICST Conference on Communications and Networking in China (CHINACOM), pp. 708–713, Aug 2011
50. Khandekar, A., Bhushan, N., Tingfang, J., Vanghi, V.: LTE-Advanced: heterogeneous networks. In: Wireless Conference (EW). European, vol. 2010, pp. 978–982 (2010)
51. Andrews, J.: Interference cancellation for cellular systems: a contemporary overview. IEEE Wirel. Commun. **12**(2), 19–29 (2005)
52. Chiasserini, C., Rao, R.: Coexistence mechanisms for interference mitigation in the 2.4-GHz ISM band. IEEE Trans. Wirel. Commun. **2**(5), 964–975 (2003)
53. Mounir Ghogho and Ananthram Swami: Carrier frequency synchronization for OFDM systems, Interim Report, Sep 2003

54. Chandrasekhar, V., Andrews, J., Gatherer, A.: Femtocell networks: a survey. IEEE Commun. Mag. **46**(9), 59–67 (2008)
55. Li, G., Xu, Z., Xiong, C., Yang, C., Zhang, S., Chen, Y., Xu, S.: Energy-efficient wireless communications: tutorial, survey, and open issues. Wirel. Commun. IEEE **18**(6), 28–35 (2011)

Part II
Energy Management and Energy Efficiency in Cellular Systems

Energy Harvesting Oriented Transceiver Design for 5G Networks

Marco Maso

Abstract One of the biggest challenges for future wireless and cellular network deployment is to achieve the expected overall spectral efficiency enhancement, all the while minimizing both the additional capital expenditure for wireless operators and the carbon footprint of the information and communication technology infrastructure. In this context, it appears more and more clear that *energy efficiency* will be one of the key metrics to assess the performance of the future network. Thus, it is of utmost importance that both wired and wireless devices are designed for optimal energy efficiency, while satisfying the target performance for future networks in terms of quality of service. In this chapter, we start by providing an overview of this problem and discuss its implications for both the network and the end user. Along similar lines, we discuss the potential of a very promising approach to increase the energy efficiency of the network that has recently gained momentum, i.e., *energy harvesting*. In particular, we focus our attention on two of the most intriguing new technologies to provide energy to mobile devices, such as the so–called *wireless power transfer* and *simultaneous information and power transfer*. We introduce two wireless–empowered transceiver designs that can harvest energy from the received signals to increase their energy efficiency. More specifically, we first describe an orthogonal frequency division multiplexing transceiver capable of harvesting in–band interference, discussing its potential as a means of realizing self–sustainable transmissions and studying its performance. Subsequently, we propose a self–interference harvesting full–duplex radio architecture and shows that it can deliver both spectral and energy efficiency gains over its state–of–the–art counterparts. Our results confirm both the lack of optimization of the current technology in terms of energy efficiency and the potential of the proposed approaches to increase it. Naturally, our findings are far from being conclusive, and lot is yet to be done. However, they offer a set of interesting arguments to substantiate the idea of energy harvesting oriented transceiver design as a means to realize more energy efficient future wireless and cellular networks.

M. Maso (✉)
Mathematical and Algorithmic Sciences Lab, Huawei France Research Center, Boulogne-Billancourt, France
e-mail: marco.maso@huawei.com

© Springer International Publishing Switzerland 2016

M.Z. Shakir et al. (eds.), *Energy Management in Wireless Cellular and Ad-hoc Networks*, Studies in Systems, Decision and Control 50,
DOI 10.1007/978-3-319-27568-0_4

1 Introduction

The deployment of every new generation of cellular networks is typically preceded by both an analysis of the issues affecting the current generation and the identification of new technological challenges. In this context, a question about the paradigm to adopt to shape the transition from the current to the next generation typically arises: should the next step be an *evolution* or a *revolution*? The importance of this question is extremely evident at the dawn of the development of the coming Fifth generation (5G) network [43]. In this context, one of the biggest concerns for both the telecommunication industry and the operators is related to the *energy efficiency* of future networks. The reason is very simple. Let us take a step back for a moment and focus on the world's information and communication technology (ICT) ecosystem as a whole.

A mid–range estimate of its annual electricity consumption is around 1500 TWh. In order to contextualize this impressive figure, it is sufficient to consider that this quantity is actually equal to all the electric generation of Japan and Germany combined, or alternatively to the consumption the global illumination system in 1985 [40]. Aa a matter of fact, ICT approaches 10 % of the current world electricity generation, or in other terms 150 % of what is generated for global aviation.

The penetration of new technologies such as cloud computing, internet of things, ubiquitous mobile internet and new devices such as tablets, smart phones or smart watches, in all areas of human activity, will keep increasing at very high pace in the coming years. The trend that will be experienced by ICT's energy consumption is not expected to be different. A constant annual increase will be ineluctable if no adequate countermeasures are identified and implemented in the process. In this context, it is evident that *energy efficiency* will be one of the key metrics to assess the performance of future networks. Designing more energy efficient information acquisition, processing and distribution strategies/algorithms for wired and wireless networks has recently become a major challenge for researchers.

2 Why Energy Efficiency?

Several tests have been performed in the last years in order to quantify the contribution of wireless networks to the energy efficiency (or rather inefficiency) of ICT [45]. The results have been striking. Energy consumption in mobile networks is dominated by the radio access network (RAN), which accounts for more than 50 % of the total consumed energy by ICT [20, 33]. The objective for 5G networks is to significantly reduce this figure. A reduction of 90 % in network energy usage is expected and up to ten year battery life for low power machine–type devices is envisioned [51].

In this scenario, all the generations of the next network technology should be characterized by a special attention to energy consumption reduction in order to improve their sustainability. However, at present it is not clear how a net reduction of the power consumption could be achieved. Intuitively, this requirement is opposite to 5G major requirements on data rate [51]. In fact, the next network technology will possibly make use of larger bandwidths and antenna arrays, higher carrier frequencies, and greater complexity in terms of air interface and waveform design (e.g., filter–bank based [14]). This could result in the deployment of more power hungry components on top of pre–existing network equipment in an overlay fashion.

Accurate network planning will clearly play an important role in the process of network energy consumption reduction. The current trend in this context is the so–called network densification. According to this paradigm, the overall energy consumption of the cellular networks could be greatly reduced, all the while increasing the network spectral efficiency, if the average distance between the mobile user and the base station (BS) is reduced [19]. The rationale behind this argument is that if a massive deployment of BS were to be performed, then their size and power consumption could be made substantially smaller as compared to a legacy macro–BS. Unsurprisingly, such small form factor BS, commonly called small–cells, are already deployed by operators worldwide [19].

Now, consider the power consumption of a current generation small-cell, i.e., around 6–10 W. As previously said, the population of small-cells is forecast to grow significantly in the coming years, i.e., their number should increase up to 100 millions by 2020 [13]. Assume now that the no new technologies were to be developed to substantially reduce the power consumption of these devices. Thus, the 100 million small-cells could consume up to 4.4 TWh in 2020, de facto increasing the overall energy consumption of the existing BS infrastructure by 5 %, if no suitable countermeasure were to be adopted [25]. Thus, we cannot expect to achieve the envisioned network energy consumption reduction only by performing an accurate network planning.

The aforementioned considerations sketch a rather clear picture. The identification and development of novel paradigm–shifting technologies is by now paramount to achieve the 5G target performance, and cannot be postponed. The importance of this aspect is evident to researchers of both academic and industrial background. Efforts in this direction are already being pursed within standards organizations. Cell size adjustment schemes, power saving protocols, massive deployment of relays, use of renewable energy sources are just some of the examples of solutions whose impact on both spectral and energy efficiency of the network as a whole is currently under evaluation.

2.1 The Impact of Energy Efficiency on the User Experience

It is a common belief that a more efficient network energy management would not bring benefits only in terms of energy consumption of the RAN. Its potential and effects can actually be seen at many levels. For instance, the main actors in a cellular network, i.e., the mobile users, could certainly enjoy several benefits if their devices were to operate with a higher energy efficiency. This could offer the possibility for both new business opportunities for the operators, e.g., thanks to a larger battery lifetime that could keep each device active for a longer time, and a better end user experience.

In an era in which the interactivity offered by the most popular wireless applications is constantly coupled with a higher power consumption, the limited lifetime of current generation batteries is a crucial limitation for the mobile users [30]. Unfortunately, the capacity of batteries for mobile devices increases at a much slower pace as compared to the increasing trend of the energy requirements with respect to the previous network generations. This energy bottleneck constrains the performance of mobile user equipment, inducing a detriment in the overall user experience [23]. In this context, the finite, and possibly short, lifetime of the mobile devices is a hindering factor for future large-scale deployments of highly performing 5G networks.

The development of strategies to cope with the aforementioned problem is a prominent challenge for researchers in the wireless communication community. Very active research lines are pushing towards the development of the new techniques to extend the battery life of wireless devices. One of the main difficulties faced by researchers in this domain is to achieve a device lifetime enhancement, without increasing the size of the battery. In fact, a larger battery would necessarily reduce both the portability and the commercial value of the mobile device. This is not always desirable and very often cannot be afforded by manufacturers. The ideal goal in this sense, would be to design every aspect of the network, from the architecture to the end user device, in order to achieve a self–sustainable system that can maintain itself by independent effort.

3 Energy Harvesting

A very promising approach to move away from the energy bottleneck impasse advocates the adoption of harvesting techniques at the mobile devices. The rationale behind this approach is based on the observation that our environment provides a plethora of virtually cost–free sources of energy. In general, energy can be transferred and harvested in many ways. Notable examples of sources of energy to harvest are:

- **Vibrational**: energy can be harvested by exploiting the oscillation of a mass resonantly tuned to the environment's dominant mechanical frequency [44]. Interestingly, mechanical stimuli of different frequency and amplitude can be experienced in many situations.
- **Photovoltaic**: solar energy can be harvested both indoor and outdoor, even though with different efficiencies, and offers a virtually inexhaustible sources of power with little or no adverse environmental effects.
- **Thermoelectric**: energy can be scavenged by exploiting the temperature difference between two objects or environments. In fact, a thermal gradient formed between two dissimilar conductors produces a voltage. The fundamental limit to the energy obtained from a temperature difference is given by the so–called Carnot cycle [7].

- **Wind**: a green and renewable source of energy is provided by wind. The process to harvest energy from the wind is typically based on electromagnetic, piezoelectric and triboelectric mechanisms [41, 50, 61]

These sources are already exploited to scavenge power[1] for human activities, hence it seems legit to envision their exploitation in the context of wireless communications as well. Recent advancements in energy harvesting techniques offer a means to achieve this goal. This is the result of decades of research efforts, which made energy harvesting techniques grow from long–established concepts into devices for powering ubiquitously deployed sensor networks and mobile electronics [44].

3.1 Wireless Power Transfer

A crescent attention has been lately devoted in the wireless communications community to another potential source of energy to harvest, i.e., radio frequency (RF) signals. The intuition behind this interest is rather simple. Consider the current wireless network, populated by massively deployed radio transmitters which typically broadcast a significant amount of RF energy to remote devices. If we forget for a moment their primary role in the network and consider their physical behavior, we clearly see that they actually represent an abundance of possible sources of energy, available for harvesting. Thus, although commonly seen only as information carriers, it is legit to wonder if the role of RF signals could be enlarged to encompass a new dimension and act as energy bearers for the mobile devices. This intuition dates back to Nikola Tesla, who first dreamed of electrical equipments ubiquitously supplied by wireless power or ambient energy, by means of an out–and–out wireless power transfer (WPT) [35].

Technological limitations postponed the implementation of WPT for several decades, yet recent breakthroughs have shown that its actual realization is more than just a dream. This allowed to open new research fronts for both the electronics and wireless communications research community. Nowadays, WPT can be performed by means of several technologically different techniques. The main difference between the possible WPT techniques lies in the nature of the propagation of the electromagnetic wave whose energy is supposed to be harvested, i.e., *non–radiative* (i.e., in the near–field region) and *radiative* (in the far–field region). As a matter of fact, the three most prominent solutions to realize a WPT are the following:

- **Magnetic resonance coupling**: this technique exploits the non–radiative propagation of the wave in the near–field region and allows an effective reach ranging from few centimeters to few meters. It is suitable for plug–in hybrid electric vehicles (PHEV) and cell phone charging. Its efficiency ranges from around 30 % to around

[1]The words power and energy are used interchangeably in this chapter, despite their conceptual difference, for the sake of simplicity.

90 % for distance between the transmitter and the receiver varying between 0.75 to 2.25 m [29].

- **Resonant inductive coupling**: this technique exploits the non–radiative propagation of the wave in the near–field region and allows an effective reach ranging from few millimeters to few centimeters. It is suitable for cell phone charging, contact–less smart cards and passive RF identification (RFID) cards. Its efficiency ranges from around 6 % to around 90 % for wave frequency between 10 KHz and 30 MHz, respectively [48].
- **RF energy transfer**: this technique exploits the radiative propagation of the wave in the far–field region and allows an effective reach ranging from several meters to several kilometers. It is suitable for wireless body and wireless sensor networks. Its efficiency ranges from around 0.4 % to over 50 % for input power varying between −40 and −5 dBm, respectively [39].

The growing maturity of this technology opens the road for the definition of new scenarios and paradigms for energy management in cellular and, in general, wireless networks. This could lead to the deployment of wireless powered communication networks (WPCN), in which the wireless nodes are supplied with wireless energy harvesting and transfer capability [24]. In such a context, we could envision a scenario where a wireless sensor can harvest energy from RF signals coming from ambient sources (e.g., cellular BS and TV towers), or end users could replenish the battery of their smart phones automatically, when in the proximity of a wireless charging facility (e.g., in a coffee shop), and without physical connections. In this context, harvesting ambient RF energy seems has the potential of yielding not only a larger energy efficiency but also longer network lifetime. Several studies have been performed to assess its merit as compared to alternative forms of harvesting. Remarkably, recent measurements campaigns have demonstrated for the first time not only the practical feasibility of exploiting existing freely available sources of RF energy but also that RF energy harvesting can indeed represent a competitive solution within urban and semi–urban environments [47].

3.2 Simultaneous Information and Wireless Power Transfer

Many ambitious ideas and visions for the application of WPT to wireless networks have been proposed in the research community. In this context, the idea of using the same electromagnetic field to deliver both information and energy to the end users/devices, realizing the so–called simultaneous wireless information and power transfer (SWIPT), is certainly one of the most intriguing applications [16, 55]. From a practical perspective, the energy transfer within the SWIPT falls into the category of the RF energy transfer. In this context, the energy is captured by the receiver and converted into functional direct current (DC) voltage by means of a specialized circuit directly connected to a receiving antenna, called a *rectenna* [6]. This process is generally referred to as RF to direct current (RF–to–DC) conversion. The most advanced models of rectenna can be adopted for both indoor and outdoor applications and are able to deliver a remarkable RF–to–DC conversion efficiency, e.g., up to almost 80 % for certain configurations, [2, 49, 49, 56, 58].

By construction, SWIPT could enable both a simultaneous and efficient *on-demand* delivery of information and energy to the wireless devices in the network. This would result in a longer battery lifetime for the devices, and in a lower overall net power consumption for the network. For these reasons, SWIPT is considered an extremely appealing approach to tackle the problem of energy-constrained wireless networks. Information and energy flows can be transported within a SWIPT according to two paradigms:

1. **Orthogonal**: the two flows are explicitly separated in one or more domains by the transmitter, e.g., time and/or frequency.
2. **Parallel**: the two flows need to be separated by the terminal by means of specific approaches [38, 63]:
   - *Power splitting (PS)*: the terminal splits the received signal into two streams of different power, by means of an adjustable power divider [59, 64] for decoding information and harvesting energy separately [65];
   - *Time switching (TS)*: the terminal switches between information decoding and RF energy harvesting phase according to static or dynamic patterns [65].

In practice, each device in a SWIPT-based setting would have at least two options to mitigate the depletion speed of its battery:

- Receiving both energy and information from another device;
- Recycling resources used in the transmission with other devices.

Despite its promises, some important technological issues still need to be addressed in order for the SWIPT to become a viable alternative for future wireless network deployment. Consider for instance the nature of the two transfers involved in the SWIPT. The amount of information conveyed by an RF signal depends on the ability of the receiver to correctly detect its variations. The extent of this ability hinges on the ratio between the power of the signal and the power of all the disturbances that affect its decoding, i.e., the so–called signal to interference plus noise ratio (SINR). In practice, the information rate of a signal could be very high even if the power of the latter was very low. The same is not true for the amount of energy that can be extracted from an RF signal, which uniquely depends on its magnitude. Thus, a trade–off exists between the amount of transferred energy and the information rate in a SWIPT, and the "rate" of the two transfers cannot be maximized at the same time.

A direct consequence of the previous observations is that an effective implementation of SWIPT will not be possible in the future, unless current wireless networks were to be redesigned. Accordingly, many research issues related to the design, analysis, and optimization of architectures and protocols for SWIPT–based networks arise in this context. In this regards, examples of relevant topics include:

- Modeling and analysis of large scale and relay–based energy harvesting networks (e.g., multi–tier cellular networks);
- Optimization of the harvesting time for the devices;
- Definition of optimal scheduling policies for users data transmission and energy harvesting;
- Design of energy cooperation strategies among harvesting devices;
- Design and optimization of self–sustainable transmissions and networks;
- Novel wireless–empowered transceiver architectures.

Remarkably, they are all already matter of intense both academic and industrial research [60].

4 Interference Harvesting for Energy Efficiency

The full potential of WPT cannot be unveiled if we restrain our attention only to SWIPT. As previously discussed, an inherent feature of dense and heterogeneous networks is the likely presence of a plethora of RF signals, occupying large portions of the spectrum. Such undesired signals are usually seen by mobile devices as interference to mitigate. However, these signals may provide an abundance of free resources, in the form of *in–band* or *out–of–band* RF energy, i.e., present inside or outside the band occupied by the useful signal, respectively, that is wasted if unharvested. For instance, a harvesting device capable of sensing the spectrum and reconfiguring both its RF chain and antenna for the optimum voltage standing wave ratio (VSWR) could effectively harvest out–of–band interference and prolong its lifetime [58]. In this case, the harvesting would be simplified by the opportunistic approach to spectrum access and usage that provides an explicit separation between the useful and interference signal.

Unfortunately, the approaches developed for out–of–band interference harvesting cannot be directly extended to the in–band case, where the useful and interference signal coexist in the same band. Interestingly, strategies to realize this harvesting are already subjects of current research [31, 36–38]. In particular, recent efforts in the direction of *self–sustainability*, i.e., a condition in which each wireless device receives from the BS both the information and the energy to retrieve it, have studied this aspect for different network configurations and proposed the design of novel wireless–empowered transceivers, obtaining encouraging results [36–38]. In this context, in–band interference harvesting may actually be a key factor for realizing self–sustainable transmissions in 5G networks.

4.1 Wireless–empowered Transceiver for Block Transmissions

As previously said, in–band interference harvesting is a challenging process due to the coexistence of both interference and useful signal in the same band. Latest communication standards, often based upon physical–layer strategies known as block–transmission schemes, may offer a means to circumvent this problem. In these strategies, the transmitted signal is constructed as a block with two sections: (1) a useful signal component, and (2) a redundant signal, e.g., prepended in form of a cyclic prefix (CP). The main purpose of this redundancy is to combat the self–induced in–band inter–block interference (IBI) and inter–symbol interference (ISI) that affect the received signal. In particular, the CP is the only portion of such signal affected by IBI at the receiver. This entails a neat separation between the self–induced in–band interference and the useful signal. Accordingly, the structure of the received block can be exploited to harvest the interference, as follows.

Let us first focus on the receiver and take a classic orthogonal frequency division multiplexing (OFDM) reception as a reference model. After the time and frequency synchronization of the received signal, the CP of each block is discarded and its content and nature typically neglected. Now, consider a receiver architecture such as the one in Fig. 1. Therein, the CP removal element used in the legacy OFDM receiver is substituted by a CP retrieval element and an energy harvester (EH). Furthermore, we note that the dashed gray lines are just symbolical illustration of the interaction between the energy harvester and the digital signal processing (DSP) blocks of the OFDM receiver. In this sense, they are meant to represent the additional energy that can be used by the receiver to operate the DSP blocks, obtained as a result of the harvesting operation. An important prerequisite for the OFDM interference harvesting receiver to correctly work is the absence of sampling during the down–conversion of the received signal in the RF chain. This ensures that the input signal

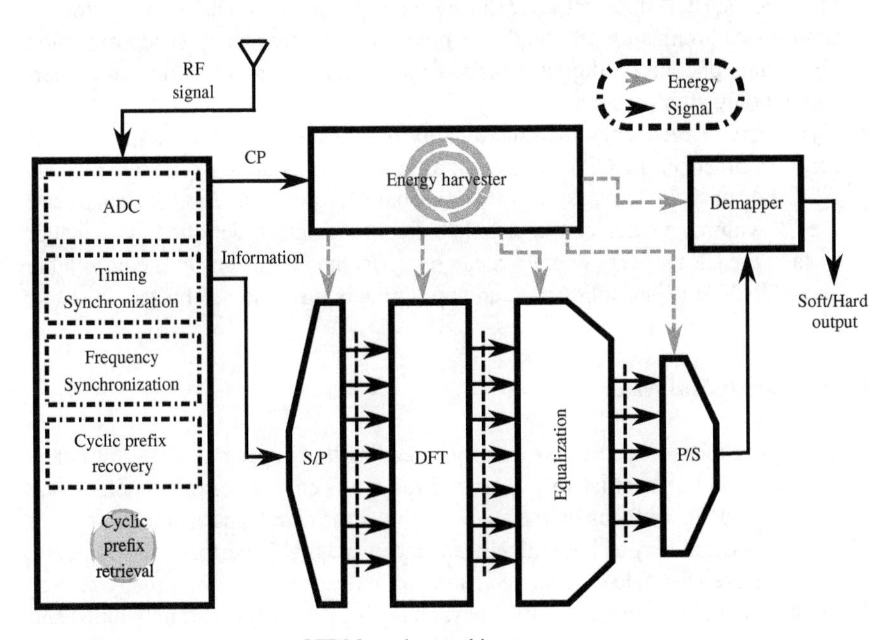

Fig. 1 Interference harvesting OFDM receiver architecture

to the receiver (RX) chain is still in analogue form. This way the EH could operate on a signal that still carries its energy.

4.1.1 CP Retrieval Element

We first focus on the CP retrieval element, which has to provide the following functionalities:

1. **Analog–to–digital conversion (ADC)**: this functionality serves a two–fold purpose. In practice, it is useful both to identify the position of the CP within a received frame and to sample the useful part of the received signal once the CP has been recovered in analogue form. Accordingly, since the state–of–the–art timing synchronization algorithms operate on digital signals, first the ADC samples the received signal until the first useful frame is identified. Subsequently, it samples only the portion of the signal corresponding to its useful part, in order to obtain a digital signal on which the OFDM DSP blocks can operate.
2. **Timing synchronization**: this functionality is necessary to identify both the beginning of the frame and the portion of the latter corresponding to the CP of each OFDM symbol. Without this operations it would be impossible to provide an analogue version of the CP to the EH.
3. **Frequency synchronization**: this functionality is fundamental to preserve the orthogonality of the sub–carriers in the OFDM system. Its main role is to estimate the carrier frequency offset (CFO) and compensate it, such that inter–carrier interference (ICI) does not arise during the subsequent OFDM DSP. As for the timing synchronization, it should be noted that the frequency synchronization algorithms operate on digital signals. Consequently, this operation can be performed only after the ADC.
4. **Cyclic prefix recovery**: this functionality requires the signal to be in analogue form. In practice, the CP is recovered after the first sampling performed by the ADC, i.e., once the timing synchronization has been performed and the position of the CP within the received frame can be deterministically identified. As a matter of fact, the CP recovery operation can be performed by means of a controllable and adjustable signal splitter and an appropriately dimensioned buffer.

4.1.2 Energy Harvester

Switching our focus to the EH, we start by observing that in general the performance of a any EH can be quantitatively characterized by its energy conversion efficiency $\eta \in [0, 1]$. Now, if we focus on the state–of–the–art research and manufacturing of RF EH, we see that very technologically advanced and highly performing composite components are available. In practice, state–of–the–art devices are already able to deliver a remarkable conversion efficiency, i.e., $\eta > 50\,\%$, if appropriately tuned and if the input power of the signal is sufficiently high. Now, let us consider the structure and operations performed by a state–of–the–art EH [56–58]:

- First, the input signal is rectified by means of a device commonly referred to as *rectifier*. In practice, this operation is performed by p–n junction diodes when RF radiation is in the kHz-MHz low frequency range. Conversely, semiconductors and devices with shorter transit times and lower intrinsic capacitances like GaAs Schottky diodes are used to perform this operation when the RF radiation has frequencies in the GHz-THz range [15];
- The rectified signal is then filtered by means of a low–pass filter to obtain a DC voltage;
- Subsequently, a DC–to–DC converter, e.g., an *unregulated buck–boost converter* operating in the discontinuous conduction mode, usually adapts the voltage levels of the filter output to the level required by the application load, e.g., a storage device.
- The obtained voltage by means of this procedure may charge a battery within a range of few Volt.

At this stage, it is evident how an appropriate tuning of the EH corresponds to the adoption of a diode operating at a suitable frequency. In practice, every EH can be appropriately tuned as long as its components are suitably chosen.

Unlike traditional approaches for WPT, in which the frequency of the signal to be converted is in the order of few GHz [34, 42, 58, 63, 66], in the proposed receiver the energy is harvested from a low frequency signal, e.g., intermediate frequency (IF) or baseband, after the amplification provided by the RF front–end. This positively impacts both the amount of energy harvested by the EH and its energy conversion efficiency. In particular:

1. The distance-dependent path loss has a minor impact on the power of the signal at the input of the EH as compared to state–of–the–art solutions. Thus, the potential amount of energy harvested by a receiver such as the one depicted in Fig. 1 is larger.
2. As previously discussed, in this case, which correspond to the so-called low frequency and high power (LFHP) case, the rectifier is a p–n junction diode. In particular, this device yields a half-wave rectification with an efficiency given by

$$\eta = \frac{1}{1 + \frac{v_{\mathrm{d}}}{2v_{\mathrm{dc}}}}; \tag{1}$$

with v_{d} and v_{dc} defined as the voltage drop across the diode and the output rectified DC voltage, respectively [15]. It is worth noting that the value of η typically grows with the power of the input signal, when the power of the latter is within the typical range of RF/IF signals [49, 57]. As a consequence, harvesting after the amplification provided by the RF front–end ensures high energy conversion efficiency.

4.1.3 Self–Sustainability of the Transmission

As a matter of fact, the receiver in Fig. 1 transforms the role of the CP from an unavoidable redundancy into a useful source of energy. In this context, the CP can be seen both as a means to implement the SWIPT and as a tool to perform a resource recycling. In the first interpretation, the CP is the portion of the signal used by BS to perform the WPT towards the OFDM receiver. In the second interpretation, the CP carries a portion of the energy invested by the power amplifier (PA) in the RF chain to amplify the received signal. In fact, the PA amplifies the entirety of the signal in the RF chain, CP included. Quantitatively, the cost of this inefficiency for an OFDM receiver is non–negligible, provided that the duration of the CP can be up to 20 % of the total symbol duration. As a consequence, by both retrieving and harvesting the CP, the proposed receiver recycles resources otherwise wasted in legacy OFDM receivers with the CP removal operation. In practice, the impact of the DSP on the power consumption of the receiver is reduced. As a result, both the energy efficiency of the device and its battery's lifetime could be significantly increased. The ideal goal in this sense would be to diminish the impact of the DSP on the receiver's battery as much as possible. This observation leads directly to the following definition.

Definition 1 Let P_{H} and P_{C} be the harvested and consumed power by the OFDM interference harvesting receiver for its DSP, respectively. A *fully self–sustainable* transmission is achieved whenever $P_{\mathrm{H}} = P_{\mathrm{C}}$. Similarly, a *partially self–sustainable* transmission is achieved whenever $P_{\mathrm{H}} = \psi P_{\mathrm{C}}$, with $0 < \psi < 1$.

Naturally, realizing a fully self–sustainable transmission in terms of power consumption of the OFDM DSP would strongly depend on both the adopted TX/RX parameters and the quality of the DSP components at the OFDM RX. In this sense, achieving a full self–sustainability may not be feasible. However, this does not prevent to envision an adaptive approach to system parameters dimensioning in order to maximize the achievable level of partial self–sustainability, as discussed in the following.

4.1.4 Performance Evaluation

In this section the performance of the proposed interference harvesting OFDM receiver is assessed by means of numerical simulations. For the sake of completeness, we note that no assumption related to the link budget is made in our simulations. Inappropriate choices in this sense could strongly undermine the relevance of the obtained results. Thus, we will assume that the considered link may meet the link budget requirements and perform correctly without being over–designed at extra cost. This is a safe assumption considering the careful link budget analysis typically performed in real systems to guarantee their operability. Naturally, the considered setting could be suitably extended in order to account for a detailed link budget model. However, this seems an unnecessary complication at this stage of our study. In this regards, it is worth noting that this approach is by no means different from what is typically done in the context of research works related to DSP in wireless communication literature, for the same reasons.

From the previous discussion, it is straightforward to infer that the amount of energy that can be harvested from the CP clearly depends on the size of the latter. However, an unnecessary increase of the CP size may result in a non–negligible loss in terms of achievable rate of the information transfer. For this reason, in practical settings, the system designer may want to set a CP size not greater than that root mean square delay spread of the channel between the transmitter and the receiver. In practice, this could minimize the spectral efficiency loss due to the presence of the CP, all the while guaranteeing an IBI–free transmission.

Now, consider the link between a small–cell and its associated receiver. Assume that the BS performs an OFDM transmission over $M = \{64, 128\}$ sub–carriers, with variable CP size K chosen from the set $\left[\left\lceil\frac{9M}{128}\right\rceil, \frac{M}{4}\right]$, according to the parameters defined for LTE/LTE–A [1], and expressed in terms of baseband samples for simplicity. In particular, we note that herein $\lceil x \rceil$ denotes the smallest integer number not smaller than x. Let the multi–path channel between the transmitter and the receiver, whose sampled baseband channel impulse response has duration $l + 1$ taps, with $l = 5$, be perfectly known at the transmitter and characterized by Rayleigh fading. Accordingly, a CP size of $K = l = \left\lceil\frac{9M}{128}\right\rceil$ samples would be sufficient to accommodate the channel impulse response, if perfect timing synchronization was achieved at the receiver and the goal of the BS were the maximization of the transmission rate.

As previously said, increasing the CP size in this scenario would impact the spectral efficiency of the considered link. In order to characterize the resulting loss we define β as the ratio between the achievable spectral efficiency when $K = l$ and $l < K \leq \frac{M}{4}$, respectively, i.e., the measure of the portion of the maximum spectral efficiency that can be achieved when the CP size increases. The variation of β as the CP size varies is depicted in Fig. 2, for $M = 64$ and $M = 128$ and signal to noise ratio (SNR) at the receiver of 10 dB.

As evident from Fig. 2, the spectral efficiency loss for both the considered values of M is slightly lower than 25 %, when the CP is at its maximum size, i.e., $K = \frac{M}{4}$. This is a significant loss. However, as a matter of fact the CP is typically over–dimensioned in real OFDM implementations w.r.t. the number of channel paths [54], to compensate for practical imperfections and impairments. In other words, a portion of the aforementioned spectral efficiency loss is already accounted for during the system design, even when no energy–related consideration is made. Therefore, it seems reasonable to assume a moderate level of flexibility in the choice of K in real implementations, that is $K > l$. Now, if we admit such an over–dimensioning and we focus on the lower part of the considered range of values of K, we see that a CP size twice as long as l reduces a spectral efficiency loss to only 10 %, i.e., $\beta \approx 90\,\%$. Interestingly, in such context, the benefit resulting from harvesting the CP at the receiver could provide an additional and relevant motivation to relax the constraint $K = l$, to favor a more flexible choice of the CP size and achieve more efficient WPT.

Provided the rationality of the flexibility assumption, we switch our focus to the amount of power that can be scavenged from the CP by the proposed OFDM receiver.

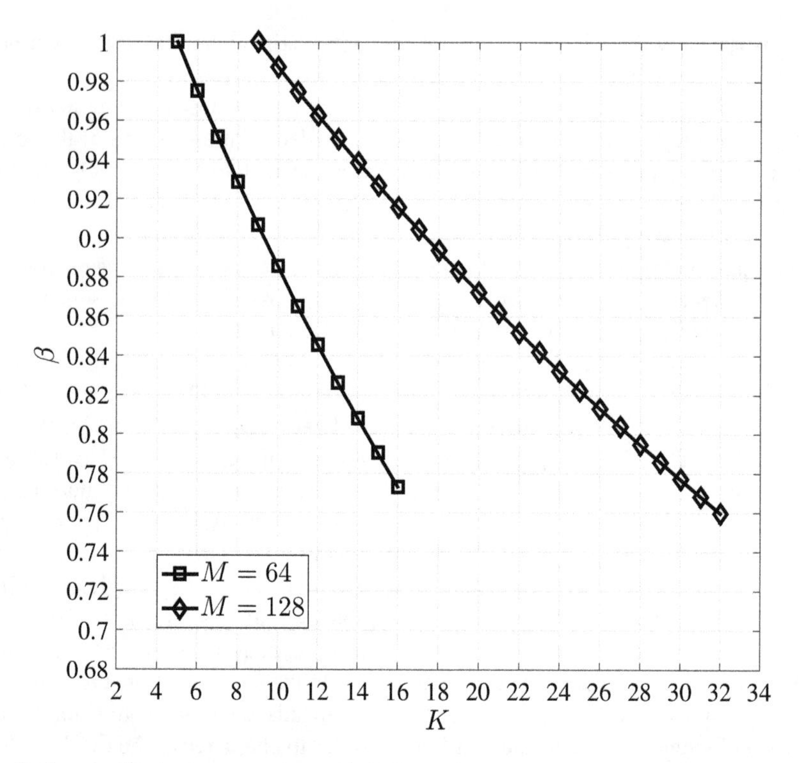

Fig. 2 β as the CP size increases, $M \in \{64, 128\}$ and SNR at the receiver of 10 dB

For ease of representation, we measure this quantity by means of $\gamma(K) \in \mathbb{R}$, defined as the level of self–sustainability of the transmission, i.e., the ratio between the power obtained from the CP and the power consumed by the receiver to perform the DSP of an OFDM block. This allows us to perform a set of Monte–Carlo simulations of the target system, for all its considered parameter configurations, and study the average achievable level of self–sustainability of the transmission, denoted by $\overline{\gamma}(K)$.

Practically relevant values are assumed for both the total transmit power budget at the BS, i.e., P_{T} and the power consumption of the OFDM DSP at the receiver, i.e., P_{C} as per Definition 1. Thus, we let $P_{\mathrm{T}} \in [0.9, 2.25]$ W and $P_{\mathrm{C}} = 500\,\mathrm{mW}$, in compliance with realistic implementations i.e., [8, 10, 62]. Analogously, we let $\eta = 0.5$ in order to mimic the performance of state–of–the–art commercial products [49] and align our tests with what is typically done and assumed in the literature [34, 63]. In Fig. 3, $\overline{\gamma}(K)$ is computed as a function of $\varphi = \frac{P_{\mathrm{T}}}{P_{\mathrm{C}}} \in [1.8, 4.5]$, for ease of representation, for several values of the CP size.

The results in Fig. 3 show that, given the considered parameters, remarkable levels of average self–sustainability are achieved for $M = 64$. We observe that a full self–sustainability of the transmission (i.e., $\overline{\gamma}(K) \geq 1$) is achieved for $\varphi \geq 5$, and for a CP size lower than $\frac{M}{4}$, i.e., the maximum allowed in modern standards. In particular, we note that $\overline{\gamma}(K) \geq 1$ also implies that the average power that is harvested by the

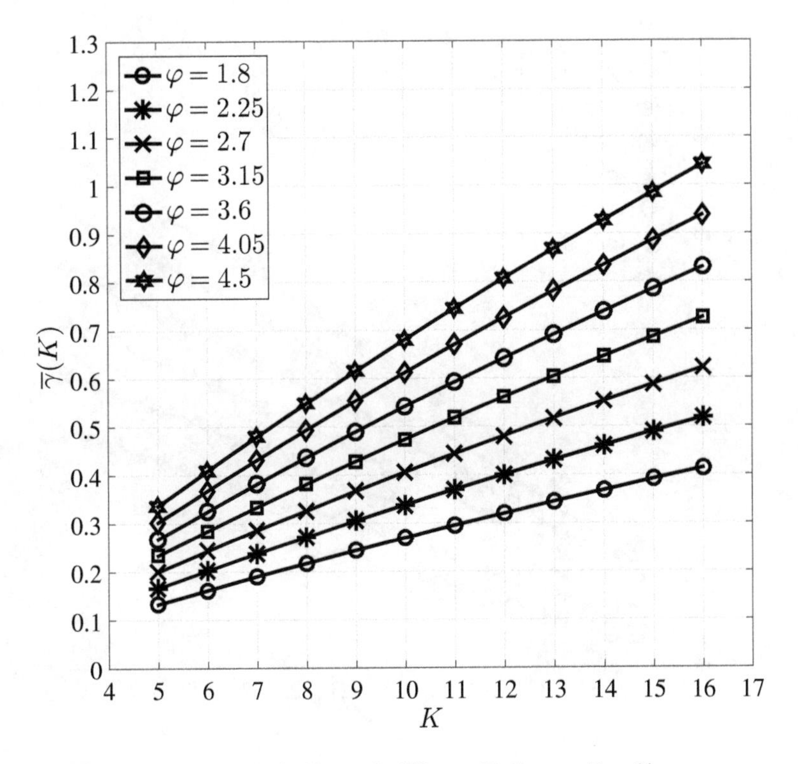

Fig. 3 Average achievable level of self–sustainability, as K changes, $N = 64$

proposed OFDM receiver not only is sufficient to provide sufficient energy for the DSP blocks to operate, but also offers the possibility to enjoy an energy surplus that can be stored for future usage. This aspect could be a very relevant added value in case of variable power consumption of the receiver, or if the latter moved rapidly enough to experience shadowing events, i.e., random signal attenuation caused by occasionally interposed obstacles between the antenna of the receiver and the BS. Naturally, as expected, the full self–sustainability is a not achievable on average for all the considered values of φ. Nevertheless, rather remarkable values of $\overline{\gamma}(K)$ are achievable even for the lowest value of φ, i.e., around 40 % for $\varphi = 1.8$ when $K = \frac{M}{4}$. At this stage, if we focus on the impact of a transition from higher to lower values of φ, we observe that the CP size for which a level of self–sustainability is achieved decreases as φ grows. In practice, a higher spectral efficiency would be achieved while keeping the self–sustainability of the transmission constant.

Now, we increase the number of sub-carriers to $M = 128$ and perform the same test as done for $M = 64$. A modification to the range of φ is in order for this test, before computing $\overline{\gamma}(K)$. In fact, the power consumption of the OFDM DSP increases with the number of sub–carriers M [21]. In particular, as detailed in [21], if the number of sub–carriers goes from $M = 64$ to $M = 128$, then the power consumption

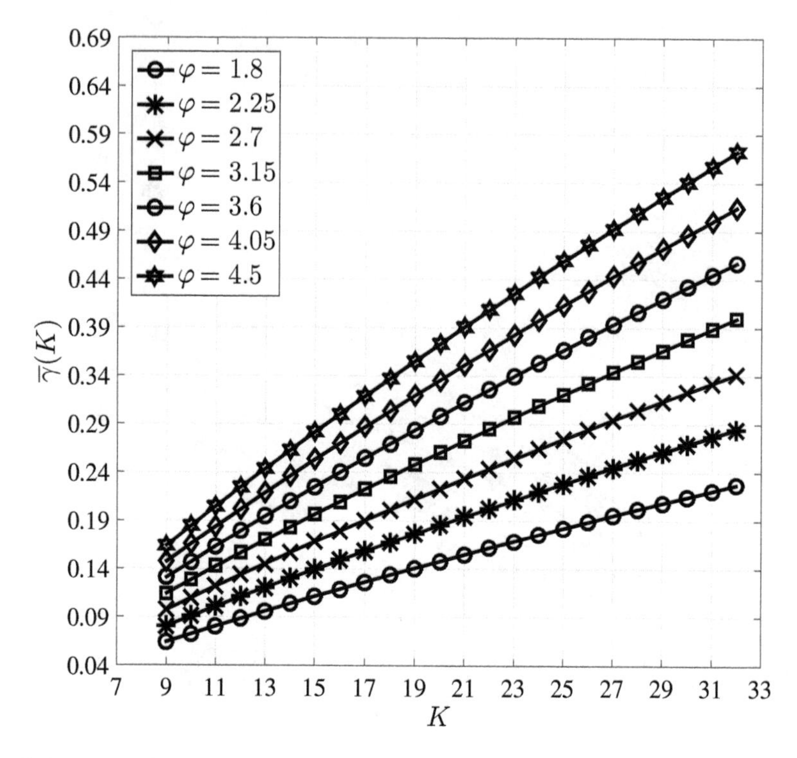

Fig. 4 Average achievable level of self–sustainability, as K changes, $N = 128$

approximately grows by a factor 1.8. Thus, we rescale φ accordingly and depict $\overline{\gamma}(K)$ in Fig. 4 for several values of the CP size.

The difference between the two cases is qualitatively negligible. Conversely, as could have been expected, the same is not true from a quantitative point of view. As a matter of fact, non–negligible values of $\overline{\gamma}(K)$ are achievable in this case as well, i.e., around 23 and 57 % when $K = \frac{M}{4}$ for $\varphi = 1$ and $\varphi = 2.5$, respectively. However, the OFDM transmission never achieves the full self–sustainability when $M = 128$, regardless of the considered φ. In other words, a fully self–sustainable transmission could be achieved in this case only if the transmit power at the BS, i.e., P_{T} were higher. In particular, $\overline{\gamma}(K)$ would be equivalent in the two considered cases only if the power increase at the BS was sufficient to compensate for the greater power consumption at the receiver, due to the larger number of sub–carriers over which the OFDM transmission is performed. Interestingly, this condition does not seem extremely unlikely in a real scenario. Consider a macro–BS and a small–cell in a real system. Therein, the macro–BS must serve a greater number of users, cover a larger cell and typically enjoys higher computational capabilities. Accordingly, the bandwidth of its transmission, i.e., the number of adopted sub–carriers, can be safely assumed to be larger than the bandwidth of the transmission performed by the small–cell. Furthermore, we observe that by both construction and standard definitions, the transmit power of a macro–BS is typically larger than the transmit power of a small–cell, i.e., up to 2 orders of magnitude [1]. Thus, it does not seem unreasonable to assume that P_T can moderately grow with the number of sub–carriers in a practical system implementation. Nevertheless, this should not downplay the fact that a significant dependence of the feasibility of the full self–sustainability on the number of sub–carriers is clear from the obtained results. In a scenario in which a larger bandwidth was adopted, and a constant P_T were to be used, lower levels of self–sustainability would be achieved by the OFDM transmission, with $\overline{\gamma}(K)$ diminishing as the number of sub–carriers increases. This highlights a potential limitation of the proposed scheme, and offers interesting elements for future research on the subject.

4.2 Wireless–Empowered Transceiver for Full–Duplex Communications

In the previous section we have discussed the design of an interference harvesting OFDM receiver, and assessed its performance. Therein the harvested interference was in the form of in–band IBI, i.e., an external source of interference. However, the nature of the interference that can be harvested by a wireless–empowered transceiver is certainly multifarious. In this sense, restraining our focus only to external sources of interference could bring us to narrow the potential of this approach quite significantly. In order to substantiate this statement, in this section we provide an example of another type of interference that can be harvested by a wireless–empowered transceiver, and propose the corresponding transceiver architecture to perform this operation. Accordingly, we depart from the previously discussed block transmission setting and switch our focus to another very promising technology for the next generation of cellular networks, i.e., to so–called full–duplex (FD) radio.

4.2.1 Prior-Art and Problems of Current Full–Duplex Technologies

As a matter of fact, FD radios have been drawing a growing level of attention lately. Their popularity is due to the promising potential that an adoption of FD radios instead of their legacy half–duplex (HD) counterparts seems to have in terms of network spectral efficiency increase. In practice, the capability of performing simultaneous bidirectional in–band communications makes this approach one of the most promising innovative solution to further enhance the performance of current networks. The most attractive and challenging solution inside this family of radio devices is the so–called in–band single antenna FD implementation. In this case, the device not only transmits and receives simultaneously over the same time and frequency resource, but also does it by means of the same antenna. The advantage of such a solution is mainly economical. In fact, an in–band single antenna FD radio needs neither a separate RF circuitry for its TX and RX parts nor an additional antenna to realize the FD communication. This can positively impact the cost of the device. However, a very severe problem affects such FD radio mainly due to practical hardware limitations, i.e., the so–called self–interference (SI).

Main responsible for the presence of the SI is the non–ideality of the circulator by means of which TX chain, antenna and RX chain are connected [12]. This non–ideality results in the leakage through the circulator of a portion of the transmit signal which, as a result, transits from the TX chain to the RX chain of the FD radio. In this regards, we define α_{L} as the ratio between the power of the leaked signal and P_{T}, TX power of the FD radio. The rest of the SI is given by the reflections that signal transmitted by the antenna of FD radio experiences during its propagation in the multi-path environment, before coming back to the antenna [26]. Nevertheless, it should be noted that the difference in terms of magnitude between these two components can be extremely relevant. In practice, the SI is very often dominated by the signal leakage at the circulator.

If unmanaged, the SI can irreversibly compromise the performance of the radio device. A two-fold consequence occurs as a result of its unavoidable presence:

- The signal received by the FD radio, referred to as *desired signal* henceforth, typically experiences a significant signal to interference plus noise ratio (SINR) reduction. In fact, the power of the SI could be several orders of magnitude larger than the power of the desired signal, received by the FD radio severely attenuated by the wireless propagation. This can significantly degrades the overall throughput. A correct decoding of the desired signal may not be possible, unless the SI can be significantly reduced, de facto nullifying the potential of the FD radio over a HD counterpart.
- A non–negligible portion of the power invested by the FD radio to transmit the signal, i.e., up to 10–15 % [5, 26] in real implementations, is wasted with the leakage. This implies that the radiated power by the antenna of the FD radio is lower than what it should theoretically be, net of both the inefficiency of the PA and thermal dissipation in the TX chain. As a consequence, the FD radio experiences a significant energy efficiency reduction, given that its power consumption could be lower if the SI were not present.

In practice, the aforementioned spectral efficiency increase depends on how much SI can be subtracted from the received signal [52]. The design of self–interference cancellation (SIC) algorithms for in–band FD radios has been the subject of many research efforts in the last years. Typically, SIC is partially/fully accomplished by either digital [3, 27, 28, 32, 53] or analogue [5, 9, 22, 26, 46] signal processing. The best performers in this sense are certainly the hybrid SIC algorithms that process the signal both in the analogue and the digital domain [5, 22]. In particular the authors in [5] propose a joint analogue–digital cancellation techniques for OFDM single antenna FD radios. Therein, the SI is cancelled to the receiver noise floor, provided that the FD radio is using a limited transmit power, herein referred to as P_{Th}, above which a residual SI remains and affects the performance of the device. As compared to other state–of–the–art solutions, the quantitative result achieved by [5] in terms of SIC is extremely remarkable. In fact, this algorithm, based on an accurate model of the impact of both linear and non–linear components on the SI affecting the reception of the desired signal, delivers 110 dB of SIC over a bandwidth of 80 MHz. Motivated by these achievements, practical implementations have been proposed to assess the feasibility of FD transmissions in real–life scenarios [4, 12], confirming that real–time FD radios can effectively operate in different environmental conditions.

Despite its evident potential and merit, the FD architecture proposed in [5] is far from being perfect. In this sense, it has two main practical disadvantages. First, it is worth recalling that the presence of the leaked signal contributing to the SI represents an evident inefficiency in terms of energy management of the FD device. Second, it should be noted that an implicit TX power limitation is imposed on the FD device to preserve the excellent SIC capability of the proposed algorithms. In practice, for the latter to be able to deliver their highest effectiveness, the difference between the TX power of the FD radio and its noise floor, expressed in dB, should be lower or equal than the maximum SIC that the architecture can deliver, e.g., 110 dB in [5]. It is straightforward to see that if realistic values for the noise floor of modern equipment are considered, i.e., typically ranging between -90 and -115 dBm [17], the resulting P_{Th} could be rather low, especially for outdoor applications/implementations. This would entail rather stringent upper bounds on both the achievable rate and the range of coverage of the outgoing transmission. In this context, disposing of an FD transceiver that could cope with the aforementioned issues would be highly desirable in view of the possible adoption of the FD technology in 5G. Starting from these observations, a novel FD architecture that can provide this features is proposed in the next section.

4.2.2 Self–Interference Harvesting in Full-Duplex Radios

Consider a FD radio as described in [5]. As previously discussed, such a device would inevitably experience a performance detriment whenever $P_{\mathrm{T}} > P_{\mathrm{Th}}$. Interestingly, an effective solution to this problem can be found by modifying the FD architecture proposed in [5] as follows.

We first assume that the FD radio may arbitrarily attenuate the signal coming from the circulator such that the power of the SI component of the resulting signal is lower than $\alpha_{\mathrm{L}} P_{\mathrm{th}}$. If this were possible, the full capability of the subsequent SIC algorithms would be restored, and non–negligible SINR gains could be experienced by the desired signal as compared to state–of–the–art solutions, thanks to the combined effect of the aforementioned attenuation and SIC algorithms. Now, consider the FD radio architecture illustrated in Fig. 5 where, differently from the state–of–the–art architecture, a RF EH operating according to the PS paradigm [63] is added between the circulator and the RX chain. This completely passive component does not require additional energy expenditure for the FD radio to operate and consists of a cascade of an adjustable power splitter followed by a rectifier performing the RF–to–DC conversion. In the proposed FD architecture, the signal coming from the circulator is split in two parts by the EH, referred to as information component (IC) and energy component (EC) henceforth, for simplicity. In practice, IC and EC can be modeled as attenuated versions of the received signal, with attenuating factor $\sqrt{\sigma}$ and $\sqrt{1-\sigma}$

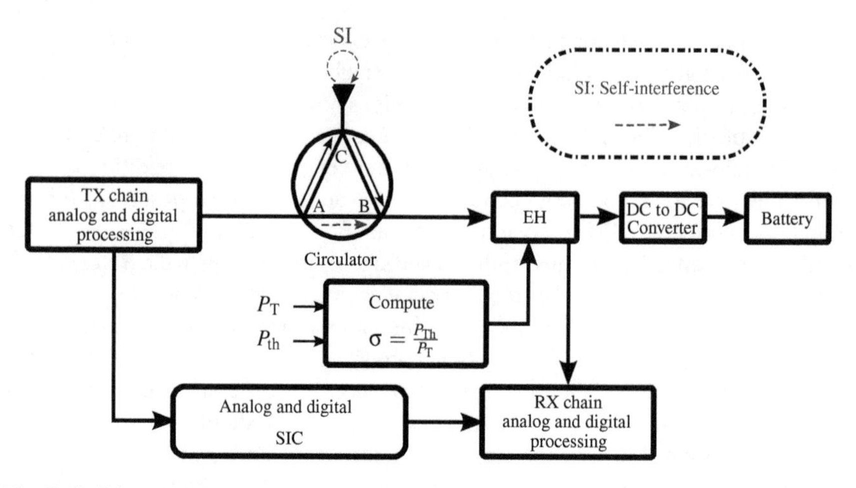

Fig. 5 Self–interference harvesting FD architecture

(with $\sigma \leq 1$), respectively. We note that σ is usually known as the power–splitting factor in the literature on SWIPT. After the split, the EC is suitably rectified as previously said, to convert the received microwave power into DC power, i.e., by means of a RF–to–DC conversion. Subsequently, it is fed as input to the DC–to–DC converter to proceed according to the legacy harvesting process discussed in Sect. 4.1. The interested reader may refer to [56], and references therein, for more detailed information about this aspect.

We switch now our focus to the IC which, after the split, is fed as input to the RX chain, where it is processed by state–of–the–art SIC algorithms and decoded [5]. If we look at the operation performed by the EH on its input signal, we observe that the IC is actually seen by the RX chain as an attenuated version of the signal coming from the circulator. In this context, from the point of view of the RX chain, the EH can be modeled as a component providing the aforementioned variable attenuation, despite not technically being an attenuator. As a consequence, the full effectiveness of the state–of–the–art SIC algorithms would be restored whenever $\sigma \leq \frac{P_{\mathrm{Th}}}{P_{\mathrm{T}}}$. Conversely, residual SI with power $\frac{\rho P_{P_{\mathrm{T}}} N_0}{P_{\mathrm{Th}}}$, where N_0 denotes the noise floor of the device, would occur in the RX chain whenever $\sigma > \frac{P_{\mathrm{Th}}}{P_{\mathrm{T}}}$, in turn inducing a SINR reduction for the IC.

At this stage, it is worth recalling that, by construction, the IC is composed of both the SI and the desired signal. Hence, it is evident that the splitting operation performed by the EH affects the power of the latter as well, whenever $\sigma < 1$, in turn reducing its SNR as compared to what could have been measured between the circulator and the EH. Consequently, a trade–off between the SINR increase and SNR decrease exits, and σ should be optimized accordingly. We will discuss this aspect in the following section. At this stage, it is worth observing that a further constraint needs to be satisfied in order to guarantee the effectiveness of the operations performed in the RX chain. This is due to the aforementioned upper bound in terms of input power for the circuitry of the RX chain, which can easily saturate if the input power exceeds a certain threshold, e.g., due to an over–abundance of SI. In practice, if we let P_S be this value, this implies that the full theoretical operativeness of the RX chain of the FD radio can be guaranteed whenever $\sigma \leq \min\{1, \frac{P_S}{P_T}\}$, regardless of the power of the residual SI.

Summarizing, the goal of the EH in the proposed architecture is two–fold. One the one hand, it targets a suitable equivalent power reduction for the received signal such that the subsequent SIC algorithms can operate on the IC at their full effectiveness. On the other hand, it aims at recycling at least a portion of the wasted resources at the circulator, i.e., the energy carried by the leaked signal, by harvesting the EC and converting it into DC for charging the battery. Finally, it is worth observing that the proposed architecture can be flexibly adopted, and provide interesting performance enhancement, not only in purely FD scenarios, but also in hybrid half/full-duplex scenarios. For instance, in the context of future 5G networks, this flexibility could offer an effective solution to perform FD device–to–device (D2D) communications, FD machine–to–machine (M2M) transmissions, FD–based in–band wireless backhauling solutions, just to name a few. This is an extremely appealing feature, especially considering the likely heterogeneity that will characterize the next network technology.

4.2.3 Spectral and Energy Efficiency Increase

The performance of the proposed SI harvesting FD architecture is assessed following the same method adopted to study the performance of the interference–harvesting OFDM receiver in Sect. 4.1.4. Consider an outdoor scenario in which a FD BS, implemented according to the proposed architecture, communicates with two HD user terminal (UT), one served in the downlink and one in the uplink. We assume that an OFDM transmission is performed in the downlink as in [5] and a single–carrier frequency division multiplexing (SC-FDMA) is performed in the uplink, as in LTE/LTE-A [1]. The number of sub–carriers over which the OFDM is performed is $M = 128$, with CP size $K = 16$. Furthermore, the channels between the devices (and the multi–path channel seen by the SI reflected by the environment towards the antenna of the FD BS) are modeled as Rayleigh fading channels with $l + 1$ taps, and $l = 16$. Finally, for simplicity, and without loss of generality we assume that the distance between the BS and the two UTs is the same, i.e., d. The rest of the parameters of the simulations, whose values are reported in Table 1, are set in order to frame a practically relevant and realistic scenario.

Before proceeding with our tests, we define two quantities that will be functional for a compact representation of the results. Accordingly, we first let ξ_R be the ratio between the achievable uplink spectral efficiency for the proposed FD architecture and the achievable uplink spectral efficiency for the state–of–the–art solution. Similarly, we define ξ_P as the ratio between the harvested power by the proposed FD radio and the TX power of the device. In this context, we note that the proposed architecture outperforms the state–of–the–art when $\xi_R > 1$ or $\xi_P > 0$. We start our

Table 1 Parameters of the considered system

	Parameter value/configuration
UT TX power	23 dBm [18]
FD radio TX power	$P_T \in [23, 40]$ dBm [18]
Distance between FD radio and UT	$d \in [0.1, 0.55]$ Km
Path-loss model	ITU COST-231 Hata model (urban and suburban) [11]
Noise floor	$N_0 = -90$ dBm [5]
SIC capability	100 dB [5]
Leakage at the circulator	−10 dB [26]

analysis by computing ξ_R as σ varies, as function of the TX power of the FD device. In particular, we assume distance between the BS and the two UTs equal to $d = 250$ m. Moreover, no markers will be used in the plot, for the sake of clarity. The results of this test are illustrated in Fig. 6. We first observe that for $\sigma = \frac{P_{Th}}{P_T}$, i.e., when the optimal σ is adopted, $\xi_R > 1$ for all the considered values of P_T. This result clearly shows a first remarkable advantage brought by the proposed FD architecture over the state of the art. In particular, it confirms that the proposed radio is able to cope with a larger set of TX powers, all the while consistently outperforming the state–of–the–art solution. Quantitatively, the spectral efficiency increase, at the considered distance between the devices, ranges from 2 % (for $P_T = 35$ dBm) and 24 % (for $P_T = 23$ dBm). In practice, the smaller P_T the higher the resulting ξ_R. This result is trivially due to the decreasing trend of the optimal σ as P_T increases, by construction. In other words, the attenuation induced on the desired signal becomes stronger as the TX power increases. As such, the SNR loss experienced by the desired signal is more severe when the TX power is high and the penalty paid in terms of spectral efficiency increases. Naturally, this behavior strongly depends on the value of the noise floor. A lower value of N_0 would allow to compensate for the effect of a smaller optimal σ, inducing higher values for ξ_R. In this sense, a FD radio with high quality components would experience a more consistent performance enhancement as compared to a lower quality counterpart, even if both were implemented according to the proposed architecture.

At this stage, it is worth observing that the proposed approach does not yield any downlink spectral efficiency increase over the state–of–the–art, when the same TX power is adopted. In this regards, the only added value brought by the proposed FD architecture is that it enables a more flexible choice of the TX power of the device. Consequently, the FD radio can increase its power beyond P_{Th} and increase the downlink spectral efficiency, with a lower impact on the uplink spectral efficiency as compared to the state–of–the–art solution. We now switch our focus to the behavior of the system as the distance between the FD BS and the two UTs varies. Accordingly, we compute ξ_R as a function of the distance between the two devices in Fig. 7. As before, only a subset of values of P_T is considered, for ease of representation.

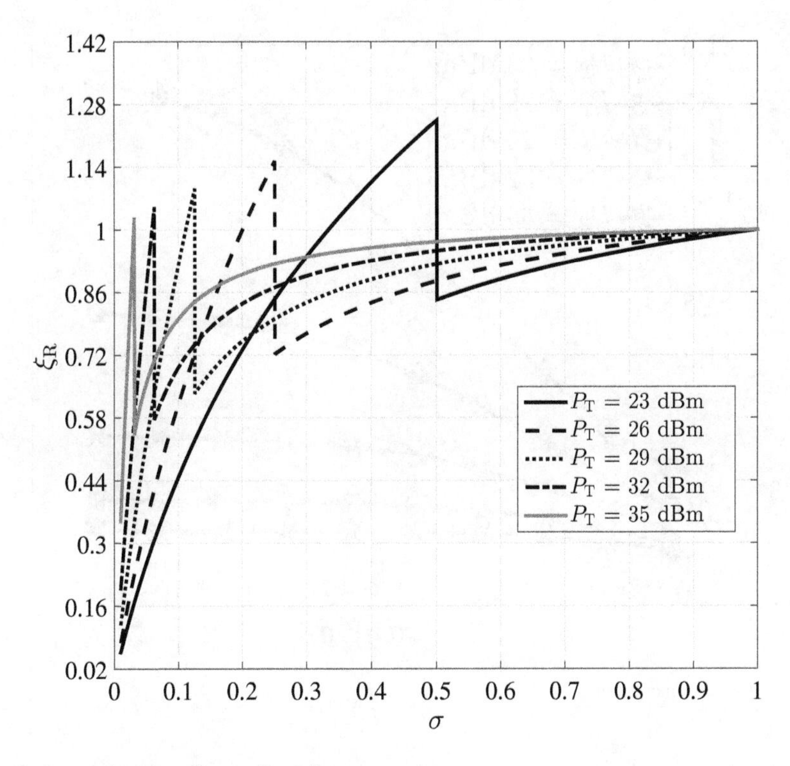

Fig. 6 ξ_R as a function of σ, as P_T varies

It is evident from Fig. 7 that a higher spectral efficiency of the uplink is achievable thanks to the proposed FD architecture for all the considered values of d, i.e., $\xi_R > 1$, $\forall d$. In particular, we observe two trends. First, ξ_R is a non–decreasing function of d, for all the considered values of P_T. This behavior is straightforward to explain. In fact, the larger the distance between the FD and the transmitting UT, the greater the attenuation experienced by the signal in its propagation. In this context, the impact of a more effective SIC increases as the SINR decreases, and the effectiveness of the proposed architecture is naturally higher. The second trend observable from Fig. 7 is related to how ξ_R reacts to a variation of P_T. In practice, the smaller P_T the higher the resulting ξ_R. This finding is compliant with what has been observed in the previous test, and thus can be explained analogously. Quantitatively, the obtained results show a remarkable spectral efficiency increase as compared to the state of the art, ranging from 2.5 % (for $d = 100$ m and $P_T = 35$ dBm) to 47 % (for $d = 550$ m and $P_T = 23$ dBm).

Our final study is related to the portion of the TX power that is actually recycled by the proposed FD architecture as P_T increases. In this regard, we note that the optimal σ, i.e., $\sigma = \frac{P_{Th}}{P_T}$, is adopted for each of the considered values of P_T, and that such value does not depend on d. Accordingly, and differently from the previous

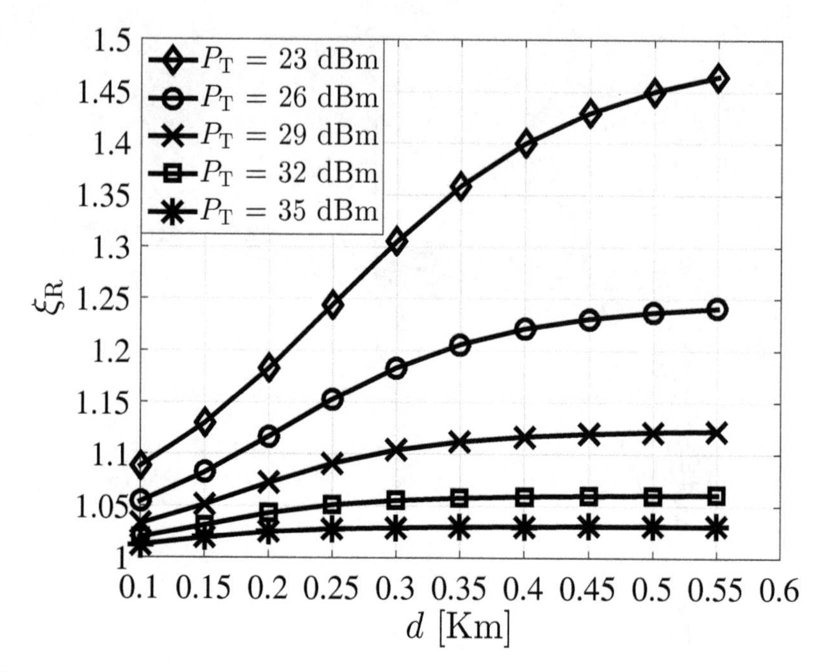

Fig. 7 ξ_R as a function of d, as P_T varies

tests, only one curve in present in Fig. 8, where ξ_P is illustrated as P_T varies in the considered range.

As could have been expected after the previous results, ξ_P increases with P_T. Once again, this behavior is rather intuitive to explain, since a TX power increase can only result in a smaller optimal σ, i.e., the majority of the signal reaching the EH is harvested. Quantitatively, the FD radio can recycle around 2.5 % of the TX power (i.e., around 25 % of the power of the leaked signal) for $P_T = 23$ dBm, and around 5 % of the TX power (i.e., around 50 % of the power of the leaked signal) $P_T = 35$ dBm. This performance may not look impressive at first glance, however it should be noted that the energy recycling comes at no cost for the spectral efficiency of the uplink (and thus of the downlink as well). As a matter of fact, this non–negligible portion of the otherwise wasted energy can be collected and re–used thanks to the proposed architecture, which realizes an energy saving. The extent of the saving increases with the transmit power. This highlights the potential of the proposed FD architecture as a means to increase both the energy efficiency of the device and the overall throughput of the ongoing transmissions between the FD BS and the UTs.

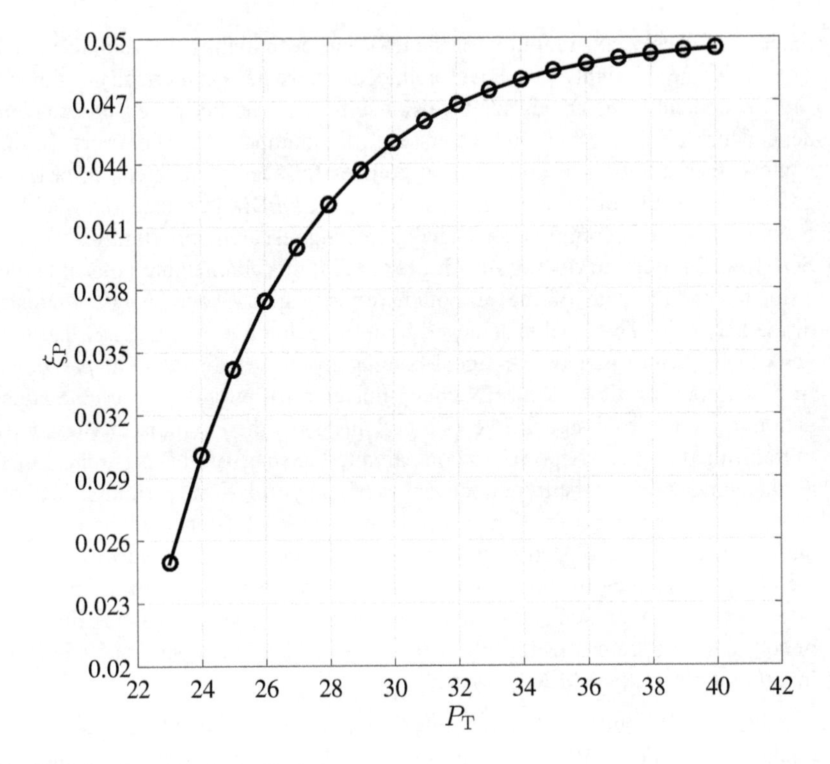

Fig. 8 ξ_P for the optimal σ as P_T varies

5 Conclusion

At the dawn of the deployment of the 5G network, energy efficiency is rightfully considered as one of the most important metrics to measure the performance of future wireless and cellular networks. In this chapter, we started from this consideration to discuss the positive impact that energy harvesting oriented technologies, such as WPT and SWIPT, may have on this metric. Subsequently, we introduced the concept of energy–harvesting oriented transceiver design and discussed the new perspectives that this approach can unveil, both in terms of potential reduction of network energy consumption and increase of the quality of the user experience. Two novel architectures have been proposed:

- **Interference–harvesting OFDM receiver**: this architecture aims at reducing the impact of the OFDM DSP on the lifetime of the receiver's battery. This goal is achieved by replacing the CP removal element with a combination of a CP retriever and an EH. This way, the energy carried by the CP, prepended to each OFDM block by construction, can be harvested and either used to reduce the overall net energy consumption of the OFDM DSP or stored. The potential of the proposed architecture as a means to realize self–sustainable transmissions, in terms of energy consumption for the DSP, has been highlighted and discussed. The advantages brought by the novel architecture have been numerically evaluated. More specifically, the achievable levels of self–sustainability have shown a clear dependence on the system parameters, e.g., the number of sub–carriers. In this context, the definition of optimization policies for such parameters, in order to guarantee target levels of self–sustainability of the OFDM transmission, is still an open problem and certainly opens new interesting research opportunities.

- **Self–interference harvesting full-duplex radio**: this architecture aims at taking a step beyond the state–of–the–art both in terms of spectral and energy efficiency of the FD radio. This goal is achieved by introducing a parametric EH that harvests energy from part of the signal coming from the circulator. In particular, from the point of view of the RX chain, this element provides a variable attenuation to the received signal. Interestingly, this gives the FD radio the flexibility to transmit at higher TX power as compared to the state–of–the–art architecture, all the while enjoying the full effectiveness of its hybrid SIC algorithms. Furthermore, this allows to recycle a portion of the energy of the leaked signal at the circulator, otherwise wasted in state–of–the–art architectures, in turn increasing the energy efficiency of the device. Our numerical results confirm the potential of the proposed approach and motivate further research activities in the direction of self–interference harvesting FD radios, especially in the context of multiple–input/multiple–output (MIMO) systems.

The solutions and results presented in this chapter serve a two-fold purpose. On the one hand, they confirm the lack of optimization of the current network technology in terms of energy efficiency. On the other hand, they highlight the potential of the proposed approaches for increasing this metric, and open new interesting research directions. Naturally, our findings are far from being conclusive, and lot is yet to be done. However, they offer a set of interesting arguments to substantiate the idea of energy harvesting oriented transceiver design as a means to realize more energy efficient future wireless and cellular networks.

References

1. 3GPP: TR 36.814, Further advancements for E-UTRA physical layer aspects, v.9.0.0. Technical report, 3GPP (2010)
2. Agrawal, S., Pandey, S.K., Singh, J., Parihar, M.S.: Realization of efficient RF energy harvesting circuits employing different matching technique. In: 15th International Symposium Quality Electron Design (ISQED), pp. 754–761 (2014). doi:10.1109/ISQED.2014.6783403
3. Ahmed, E., Eltawil, A.M.: All-digital self-interference cancellation technique for full-duplex systems. CoRR abs/1406.5555 (2014). arXiv:1406.5555
4. Bharadia, D., Joshi, K., Katti, S.: Robust full duplex radio link. In: Proceedings of ACM SIGCOMM Conference, pp. 147–148 (2014)
5. Bharadia, D., McMilin, E., Katti, S.: Full duplex radios. In: Proceedings of ACM SIGCOMM Conference, pp. 375–386 (2013)
6. Brown, W.C.: The history of power transmission by radio waves. IEEE Trans. Microw. Theory Tech. **32**(9), 1230–1242 (1984). doi:10.1109/TMTT.1984.1132833

7. Çengel, Y., Boles, M.: Thermodynamics: An Engineering Approach. McGraw-Hill Series in Mechanical Engineering. McGraw-Hill Higher Education, Salt Lake (2006)
8. Chang, C.C., Su, C.H., Wu, J.M.: A low power baseband OFDM receiver IC for fixed WiMAX communication. In: IEEE Asian Solid-State Circuits Conference, pp. 292–295 (2007). doi:10.1109/ASSCC.2007.4425688
9. Chen, S., Beach, M.A., McGeehan, J.P.: Division-free duplex for wireless applications. IEEE Electron. Lett. **34**(2), 147–148 (1998)
10. Chiueh, T.D., Tsai, P.Y.: OFDM Baseband Receiver Design for Wireless Communications. Wiley, New York (2008)
11. COST Action 231: Digital mobile radio. towards future generation systems final report. Technical report, European Communities, Technical report. EUR 18957, Ch. 4 (1999)
12. Duarte, M., Sabharwal, A.: Full-duplex wireless communications using off-the-shelf radios: feasibility and first results. In: Conference Recreation 44th Asilomar Conference on Signals, Systems, and Computers, pp. 1558–1562 (2010). doi:10.1109/ACSSC.2010.5757799
13. Energy Aware Radio and Network Technologies (EARTH): http://www.ict-earth.eu (2012). Accessed 25 Apr 2014
14. Farhang, A., Marchetti, N., Figueiredo, F., Miranda, J.P.: Massive MIMO and waveform design for 5th generation wireless communication systems. In: 1st International Conference 5G Ubiquitous Connectivity (5GU), pp. 70–75 (2014). doi:10.4108/icst.5gu.2014.258195
15. Gabriel Abadal Javier Alda, J.A.: ICT—energy—concepts towards zero—power information and communication technology, chapter electromagnetic radiation energy harvesting the rectenna based approach. In: Tech (2014)
16. Grover, P., Sahai, A.: Shannon meets Tesla: Wireless information and power transfer. In: IEEE International Symposium on Information Theory Processing (ISIT), pp. 2363–2367 (2010). doi:10.1109/ISIT.2010.5513714
17. Holma, H., Toskala, A.: LTE for UMTS—OFDMA and SC-FDMA Based Radio Access. Wiley (2009). http://books.google.co.uk/books?id=AHr43Lh-roQC
18. Holma, H., Toskala, A.: LTE for UMTS—OFDMA and SC-FDMA Based Radio Access. Wiley (2009)
19. Hoydis, J., Kobayashi, M., Debbah, M.: Green small-cell networks. IEEE Veh. Tech. Mag. **6**(1), 37–43 (2011). doi:10.1109/MVT.2010.939904
20. Huawei Technologies: Improving energy efficiency, lower CO2 emission and TCO. Huawei energy efficiency solution, White Paper (2010). www.mobilontelecom.com/Huawei-Energy-Efficiency-White-Paper.pdf
21. Isheden, C., Fettweis, G.: Energy-efficient multi-carrier link adaptation with sum rate-dependent circuit power. In: IEEE Global Telecommun. Conference (GLOBECOM), pp. 1–6 (2010). doi:10.1109/GLOCOM.2010.5683700
22. Jain, M., Choi, J.I., Kim, T., Bharadia, D., Seth, S., Srinivasan, K., Levis, P., Katti, S., Sinha, P.: Practical, real-time, full duplex wireless. In: Proceedings of ACM 17th Annual International Conference on Mobile Computing and Networking, pp. 301–312 (2011)
23. J.D. Power and Associates: 2010 wireless smartphone customer satisfaction study. http://www.jdpower.com/Electronics/ratings/Wireless-Smartphone-Ratings-(Volume-1)/ (2010)
24. Ju, H., Zhang, R.: Throughput maximization in wireless powered communication networks. In: IEEE Global Communications Conference (GLOBECOM), pp. 4086–4091 (2013). doi:10.1109/GLOCOM.2013.6831713
25. Khan, S., Mauri, J.: Green Networking and Communications: ICT for Sustainability. CRC Press (2013). https://books.google.fr/books?id=0obNBQAAQBAJ
26. Knox, M.: Single antenna full duplex communications using a common carrier. In: Proceedings of IEEE 13th Annual Wireless and Microwave Technology Conference, pp. 1–6 (2012)
27. Korpi, D., Anttila, L., Syrjälä, V., Valkama, M.: Widely-linear digital self-interference cancellation in direct-conversion full-duplex transceiver. CoRR abs/1402.6083 (2014). arXiv:1402.6083
28. Korpi, D., Anttila, L., Valkama, M.: Reference receiver aided digital self-interference cancellation in MIMO full-duplex transceivers. CoRR abs/1405.2202 (2014). arXiv:1405.2202

29. Kurs, A., Karalis, A.S., Moffatt, R., Joannopoulos, J.D., Fisher, P., Soljai, M.: Wireless power transfer via strongly coupled magnetic resonances. Science **317**(5834), 83–86 (2007)
30. Lahiri, K., Raghunathan, A., Dey, S., Panigrahi, D.: Battery-driven system design: a new frontier in low power design. In: Proceedings of the 15th International Conference VLSI Design Automation Conference, pp. 261–267 (2002). doi:10.1109/ASPDAC.2002.994932
31. Lee, S., Zhang, R., Huang, K.: Opportunistic wireless energy harvesting in cognitive radio networks. IEEE Trans. Wireless Commun. **12**(9), 4788–4799 (2013). doi:10.1109/TWC.2013.072613.130323
32. Li, N., Zhu, W., Han, H.: Digital interference cancellation in single channel, full duplex wireless communication. In: Proceedings 8th International Conference Wireless Communications, Network Mobile Computing, pp. 1–4 (2012). doi:10.1109/WiCOM.2012.6478497
33. Lister, D.: An operators view on green radio. Vodafone Group Research & Development, Presented at the Proceedings of the IEEE International Workshop on Green Communications (2009)
34. Liu, L., Zhang, R., Chua, K.C.: Wireless information and power transfer: a dynamic power splitting approach. IEEE Trans. Commun. **61**(9), 3990–4001 (2013). doi:10.1109/TCOMM.2013.071813.130105
35. Lumpkins, W.: Nikola Tesla's dream realized: wireless power energy harvesting. IEEE Consum. Electron. Mag. **3**(1), 39–42 (2014). doi:10.1109/MCE.2013.2284940
36. Maso, M., Lakshminarayana, S., Quek, T.Q.S., Poor, H.V.: A composite approach to self-sustainable transmissions: rethinking OFDM. IEEE Trans. Commun. **62**(11), 3904–3917 (2014). doi:10.1109/TCOMM.2014.2361124
37. Maso, M., Lakshminarayana, S., Quek, T.Q.S., Poor, H.V.: Energy harvesting for self-sustainable OFDMA communications. In: IEEE Global Communications Conference (GLOBECOM), pp. 3168–3173 (2014). doi:10.1109/GLOCOM.2014.7037293
38. Maso, M., Lakshminarayana, S., Quek, T.S.Q., Poor, H.V.: The price of self-sustainability for block transmission systems. IEEE J. Sel. Areas Commun. **PP**(99), 1–1 (2015). doi:10.1109/JSAC.2015.2391752
39. Mikeka, C., Arai, H.: Sustainable Energy Harvesting Technologies—Past, Present and Future, Yen Kheng Tan Dr. (Ed.), chap. Design Issues in Radio Frequency Energy Harvesting System. InTech (2011). doi:10.5772/25348
40. Mills, M.P.: The cloud begins with coal - Big data, big networks, big infrastructure, and big power—An overview of the electricity used by the global digital ecosystem. Digital Power Group. http://www.tech-pundit.com/wp-content/uploads/2013/07/Cloud_Begins_With_Coal.pdf?c761ac&c761ac (2013)
41. Myers, R., Vickers, M., Kim, H., Priya, S.: Small scale windmill. Appl. Phys. Letters **90**(5), 054–106 (2007)
42. Ng, D.W.K., Lo, E.S., Schober, R.: Energy-efficient resource allocation in multiuser OFDM systems with wireless information and power transfer. In: IEEE Wireless Commun. Networking Conf. (WCNC), pp. 3823–3828 (2013). doi:10.1109/WCNC.2013.6555184
43. Osseiran, A., Boccardi, F., Braun, V., Kusume, K., Marsch, P., Maternia, M., Queseth, O., Schellmann, M., Schotten, H., Taoka, H., Tullberg, H., Uusitalo, M., Timus, B., Fallgren, M.: Scenarios for 5G mobile and wireless communications: the vision of the METIS project. IEEE Commun. Mag. **52**(5), 26–35 (2014). doi:10.1109/MCOM.2014.6815890
44. Paradiso, J.A., Starner, T.: Energy scavenging for mobile and wireless electronics. IEEE Pervasive Comput. **4**(1), 18–27 (2005). doi:10.1109/MPRV.2005.9
45. Pentikousis, K.: In search of energy-efficient mobile networking. IEEE Commun. Mag. **48**(1), 95–103 (2010). doi:10.1109/MCOM.2010.5394036
46. Phungamngern, N., Uthansakul, P., Uthansakul, M.: Digital and RF interference cancellation for single-channel full-duplex transceiver using a single antenna. In: Proceedings of the IEEE 10th International Conference Electrical Engineering/Electronics, Computer, Telecommunications and Information Technology, pp. 1–5 (2013)
47. Pinuela, M., Mitcheson, P., Lucyszyn, S.: Ambient rf energy harvesting in urban and semi-urban environments. IEEE Trans. Microwave Theory Tech. **61**(7), 2715–2726 (2013). doi:10.1109/TMTT.2013.2262687

48. Pinuela, M., Yates, D.C., Mitcheson, P.D., Lucyszyn, S.: Maximising the link efficiency of resonant inductive coupling for wireless power transfer. In: 1st International Workshop Wireless Energy Transport Harvesting, pp. 1–14 (2011)
49. Powercast Corp.: P2110–915MHz RF powerharvester receiver. Product Datasheet pp. 1–12 (2010)
50. Priya, S.: Modeling of electric energy harvesting using piezoelectric windmill. Applied Physics Letters **87**(18), 184,101 (2005)
51. Public Private Partnership in Horizon 2020: Creating a Smart Ubiquitous Network for the Future Internet (2013)
52. Sabharwal, A., Schniter, P., Guo, D., Bliss, D., Rangarajan, S., Wichman, R.: In-band full-duplex wireless: challenges and opportunities. IEEE J. Sel. Areas Commun. **32**(9), 1637–1652 (2014). doi:10.1109/JSAC.2014.2330193
53. Shao, S., Quan, X., Shen, Y., Tang, Y.: Effect of phase noise on digital self-interference cancellation in wireless full duplex. In: Proceedings of the International Conference on Acoustics, Speech and Signal Processing, pp. 2759–2763 (2014). doi:10.1109/ICASSP.2014.6854102
54. Tarighat, A., Sayed, A.H.: An optimum OFDM receiver exploiting cyclic prefix for improved data estimation. In: International Conference on Acoustics, Speech and Signal Processing (ICASSP), vol. 4, pp. IV–217–20 (2003). doi:10.1109/ICASSP.2003.1202598
55. Varshney, L.R.: Transporting Information and Energy Simultaneously. In: International Symposium on Information Theory (ISIT), pp. 1612–1616 (2008). doi:10.1109/ISIT.2008.4595260
56. Visser, H.J.: Indoor wireless RF energy transfer for powering wireless sensors. Radioengineering **21**(4), 963–973 (2012)
57. Visser, H.J., Pop, P., Op het Veld, J.H.G., Vullers, R.J.M.: Remote RF battery charging. In: IET 10th International Workshop on Micro and Nanotechnology for Power Generation and Energy Conversion Applications (PowerMEMS) (2011)
58. Visser, H.J., Vullers, R.J.M.: RF energy harvesting and transport for wireless sensor network applications: principles and requirements. Proc. IEEE **101**(6), 1410–1423 (2013). doi:10.1109/JPROC.2013.2250891
59. Wu, Y., Liu, Y., Xue, Q., Li, S., Yu, C.: Analytical design method of multiway dual-band planar power dividers with arbitrary power division. IEEE Trans. Microw. Theory Tech. **58**(12), 3832–3841 (2010). doi:10.1109/TMTT.2010.2086712
60. Xiao, L., Wang, P., Niyato, D., Kim, D., Han, Z.: Wireless networks with RF energy harvesting: a contemporary survey. IEEE Commun. Surveys Tuts. **PP**(99), 1–1 (2015). doi:10.1109/COMST.2014.2368999
61. Yang, Y., Zhu, G., Zhang, H., Chen, J., Zhong, X., Lin, Z.H., Su, Y., Bai, P., Wen, X., Wang, Z.L.: Triboelectric nanogenerator for harvesting wind energy and as self-powered wind vector sensor system. ACS Nano **7**(10), 9461–9468 (2013)
62. Zhang, J., Lopez Perez, D., Song, H., de la Roche, G., Liu, E., Chu, X.: Small Cells—Technologies and Deployment, Second and Expanded Edition. Wiley (2014)
63. Zhang, R., Ho, C.K.: MIMO broadcasting for simultaneous wireless information and power transfer. IEEE Trans. Wireless Commun. **12**(5), 1989–2001 (2013). doi:10.1109/TWC.2013.031813.120224
64. Zhang, W., Liu, Y., Wu, Y., Shen, J., Li, S., Yu, C., Gao, J.: A novel planar structure for implementing power divider or balun with variable power division. Prog. Electromagn. Res. C **48**, 111–123 (2014)
65. Zhou, X., Zhang, R., Ho, C.K.: Wireless information and power transfer: architecture design and rate-energy tradeoff. IEEE Trans. Commun. **61**(11), 4754–4767 (2013). doi:10.1109/TCOMM.2013.13.120855
66. Zhou, X., Zhang, R., Ho, C.K.: Wireless information and power transfer in multiuser OFDM systems. CoRR (2013). arXiv:1308.2462

Multi-operator Collaboration for Green Cellular Networks

Hakim Ghazzai, Elias Yaacoub, Abdullah Kadri and Mohamed-Slim Alouini

Abstract This chapter investigates the collaboration between multiple mobile operators for optimizing energy efficiency in cellular networks. Most of the works in the literature optimize the performance of a given cellular network, without considering the existence of mobile networks belonging to other operators. This leads to suboptimal results, compared to the case where the optimization of the joint performance of mobile networks of multiple operators is considered. However, incentives need to be created to allow multiple (and generally competing) operators to collaborate for the purpose of energy efficiency. Indeed, random collaboration can cause certain unfairness among cooperative operators. Therefore, additional parameters should be considered to perform a fair green networking between mobile operators. This chapter aims to provide answers for similar situations. We start by investigating the case of uniform cooperative mobile operators having the same objectives and we establish cooperation decision criteria based on derived roaming prices and operators' profit gains. Afterwards, we consider the case of non-uniform operators where

H. Ghazzai (✉) · A. Kadri
Qatar Mobility Innovations Center (QMIC), Qatar Science & Technology
Park Tech 2 Building, Suite No: 201, Qatar University QSTP-B QSTP Free Zone,
Education City, Doha, Qatar
e-mail: hakimg@qmic.com

A. Kadri
e-mail: abdullahk@qmic.com

E. Yaacoub
Faculty of Computer Studies, Arab Open University (AOU), Beirut, Lebanon
e-mail: eliasy@ieee.org

E. Yaacoub
Strategic Decision Group (SDG), Jallad Bldg, 4th Floor, Banks Street,
Beirut Central District, Beirut, Lebanon

M.-S. Alouini
King Abdullah University of Science and Technology (KAUST),
Al-Khawarizmi Applied Math. Building (Building 1), Level 3, Office 3123,
Thuwal 23955-6900, Makkah, Kingdom of Saudi Arabia
e-mail: slim.alouini@kaust.edu.sa

© Springer International Publishing Switzerland 2016

M.Z. Shakir et al. (eds.), *Energy Management in Wireless Cellular and Ad-hoc Networks*, Studies in Systems, Decision and Control 50,
DOI 10.1007/978-3-319-27568-0_5

a green operator focuses on exploiting the infrastructure of non-green operators to achieve CO_2 emissions saving. A two-level Stackelberg game is formulated to optimize the utilities of both types of operators.

1 Introduction

Over the years, mobile user demand is witnessing an unprecedented rise that is leading to an enormous growth of energy consumption of wireless networks as well as the greenhouse gas (GHG) emissions which are estimated currently to be around 70 million tons over a year [1]. The huge increase of number of connected terminals, in addition to the deployed infrastructure necessary to serve them, impels network companies to pay enormous bills which represent about 50 % of their operating expenditures (OPEX). Therefore, the reduction of the network energy consumption and the limitation of their CO_2 emissions and their energy expenses becomes more and more attracting the researchers' attentions [2]. Many studies were proposed to develop green communication systems which can save energy consumption and reduce CO_2 emissions [3, 4]. Most of them tackle the radio access network part as base stations (BSs) consume more than 70–80 % of the total power consumption [5]. Half of this energy is redundant especially during off-peak periods when these BSs are underused [6]. To overcome this issue, new system level features were designed to help decide which parts of the redundant BS should be turned off. Therefore, subscribers, who were covered by an underutilized BS, will be served by another BS [7] or even by other operator's infrastructure [8] in the context of green mobile operator networking.

Several schemes have focused first on the optimization of the energy consumption of the radio access network part by turning off redundant BSs during low traffic period while considering the network quality of service (QoS). In [9, 10], the BS ON/OFF strategy is applied in order to eliminate underutilized BSs while optimizing QoS utility functions. The authors of [11] proposed an efficient green planning method based on the spatial and temporal traffic variations where the BSs needed to be activated during a certain period of the day are identified since the network planning phase. To turn off a BS, the authors of [12] calculated the joint Signal to Interference plus Noise Ratio (SINR) corresponding to the sum of SINRs of all subscribers in the network and compared it to a fixed SINR threshold: If it is higher than the threshold, the selected BS to be switched off is underutilized and is maintained off. Another BS sleeping strategy is presented in [13] where an approximate solution to the problem was proposed. The main idea is to establish a relationship between the traffic load (user arrival) and energy savings. An optimization problem aiming to minimize the number of active BSs subject to two constraints: Maintaining the user connection in the cell and covering the same initial area is solved.

Another approach to ensure energy savings for wireless cellular networks is to investigate the energy-efficient communications while considering the dynamics of the smart grid that depend on the traffic, real-time price and the pollutant level

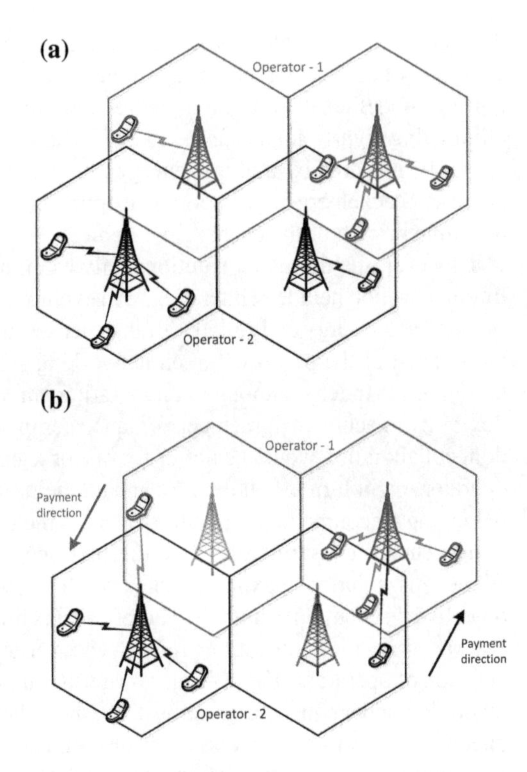

Fig. 1 Example of mobile operator collaboration: **a** without collaboration; **b** with collaboration

associated with the generation of the electricity. The authors in [14] introduced the use of coordinated multipoint communication (CoMP) to ensure acceptable QoS in cells whose BSs have been shut down to save energy. Meanwhile, the active BSs decide from which energy sources and how much energy they need to procure in order to ensure the safe operation of the network. This is performed while taking into account the pollutant level of each retailer and the proposed price. The scheme proposed in [14] could reduce operational expenditure and CO_2 emissions in green wireless cellular networks.

Optimizing the joint utilities of different mobile networks serving the same area provides more degrees of freedom for mobile operators to achieve green communication [7]. The fundamental idea was to completely switch off the equipment of a service provider while serving the corresponding subscribers by another infrastructure belonging to another operator under some fairness constraints. In Fig. 1, we illustrate an example of mobile operator collaboration where an operator (Operator 1 in red) exploits the other mobile operator's infrastructure (Operator 2 in black) to serve its subscribers while turning off its own BS and vice versa. However, this operation would eventually require the introduction of certain incentives in order to ensure fairness among competitive mobile operators. In literature, few research work had focused on the mobile operator cooperation for green purposes. One of the first studies in this field was proposed in [8, 15]. In [15], the authors identified four different sleeping strategy schemes such as balanced roaming costs and balanced energy savings. The authors in [8, 16] have improved the operator cooperation problem by modeling it in a game-theoretical strategy that ensures energy saving by eliminating lightly loaded BSs. In [17], the green collaboration between mobile operators powered by multiple energy retailers existing in the smart grid is investigated. A Stackelberg game theoretical approach is employed to model the smart grid real-time pricing of the energy procurement.

However, the discussed solutions only examine one aspect of the problem each time and do not include neither long term evolution advanced (LTE-A) nor the aspects of renewable energy and collaboration expenses in the problem formulation. Furthermore, most of the proposed green networking schemes do not consider the collaboration cost. Indeed, although collaboration among mobile operators provide more flexibility in achieving green performance compared to the traditional scenario, random collaboration would be unfair for one or a group of operators. For instance, one operator might turn off all the BSs and all the users are roamed to the infrastructure belonging to competitive operators. Hence, the serving operator might suffer from a high energy consumption while the first operator is enjoying its profit increase. While this solution maximizes the overall objective of operators, the individual objective distribution is not fair. Therefore, it is important to introduce some fairness criteria during the cooperation process. These criteria will influence the collaboration decision of operators. These fairness criteria can take different forms such as the collaboration under equal charge allocation where the total cost is equally shared among operators [8]. Another fairness criterion could be the equal share of the collaboration cost. In this case, only the cost due to collaboration is considered and shared among operators. Note that these fairness criteria force operators to be a member of the collaboration group and share the cost with other competitive operators. In our study, we propose a fairness criterion based on roaming prices defined by each operator who is willing to serve users of competitive operators. The proposed method will also allow operators to decide whether to enter in collaboration or not. This decision is made after checking whether the total profit of this operator is affected due to collaboration or not.

In this chapter, we propose to investigate the collaboration among multiple mobile operators deploying LTE-A networks in the same area. The objective is to study the interactions among competitive mobile operators collaborating together in order to achieve green goals without compromising their profits and QoS. Two scenarios are investigated in this chapter: The first one considers the case of uniform mobile operators having the same green objectives, while the second one investigates the case of non-uniform operators having different objectives. In the first scenario, a practical and low complexity iterative algorithm is applied to determine the efficient active BS combination that ensures energy saving while respecting the network QoS. The BSs are assumed to be powered by either traditional retailer and/or renewable energy equipment (e.g., solar panel or wind turbine) owned by mobile operators and placed on BS sites. During this cooperation, extra charge can be added to operators that keep their BSs active as they are serving other mobile operators' subscribers. Therefore, a fairness criterion for mobile operator cooperation based on their profits before and after cooperation is introduced. Finally, the roaming prices for all operators and conditions for cooperation decision are also proposed.

Afterwards, we investigate the case of non-uniform mobile operators having different objectives: One green operator (GO) that aims to achieve a tradeoff between its profit and its network CO_2 emissions and other non-green operators (NGOs) available to serve green operator's users in order to enhance their profits. The GO's BSs are powered by either a traditional electricity retailer or renewable energy equipment. We employ a two-level Stackelberg game that helps the GO (playing the role of the follower) reduce its energy consumption and CO_2 emissions by roaming some of its users to one or many NGOs (playing the role of leaders). The GO can either offload all the users of a BS and switch it off or just offload some of them. However, during cooperation, extra charge will be imposed on the GO when exploiting another NGO's infrastructure to serve its subscribers. The NGO's objective is to maximize its own profits by attracting the maximum number of GO roamed users. In this competition, the leaders focus on offering the best roaming prices while taking into account multiple system parameters (e.g., energy cost and pollution level, service fee, GO renewable energy availability, etc.). We solve this problem by achieving the Stackelberg equilibrium and investigate the player behaviors for various system parameters.

In our simulation result section, we investigate the impact of several parameters on the system performance such as the traffic volume, availability of locally generated green energy, and energy cost. We also show that mobile operator collaboration can significantly contribute in reducing the carbon footprint of cellular networks.

This chapter is organized as follows. Section 2 investigates the case of uniform mobile operators collaboration. Section 3 discusses the case of the collaboration of non-uniform mobile operators. In Sect. 4, we present our simulation results for both studied settings and provide some insights about the future challenges of green networking. Finally, Sect. 5 summarizes the chapter.

2 Collaboration of Uniform Mobile Operators

In this section, we focus on the collaboration among uniform mobile operators having similar goals, i.e., reduce their fossil fuel consumption. Each mobile operator tries to turn off the maximum number of BSs in order to achieve energy saving but without affecting the QoS. The QoS is maintained thanks to intra and inter operator collaboration. Indeed, the users previously connected to a turned off BS are offloaded to neighbor BSs which belong either to the same operator or to a competitive one. However, in order to avoid random collaboration, which might lead to negative impacts on one of the operator performance, a fairness criterion based on the roaming price is proposed. This criterion will determine whether collaboration is beneficial for all operators or not. A thorough version of this work with additional mathematical details is presented [18].

2.1 System Model

We assume N_{op} mobile operators are deploying N_{op} LTE networks that satisfies the traffic demands of its customers and covers a geographical area of interest. We denote by $N_{BS}^{(n)}$ the number of BSs that are deployed uniformly by the mobile operator n in that area, $n = 1, \ldots, N_{\text{op}}$. We consider that the area is divided into cells of equal size where a BS is placed in the center of each cell. The access scheme for the LTE downlink (DL) is the orthogonal frequency division multiple access (OFDMA) while in the uplink (UL), the single carrier frequency division multiple access (SC-FDMA) is used. In fact, the DL and UL available spectrums are divided into N_{RB} resource blocks (RBs) that contain a fixed number of consecutive subcarriers ($N_{\text{RB}} = N_{\text{RB}}^{(\text{UL})} = N_{\text{RB}}^{(\text{DL})}$). RBs are assigned to users according to the resource allocation procedure followed by each operator. We assume that the mobile operators are using different frequency bands such that there is no inter-operator inference. However, intra-operator interference is taken into account (i.e., frequency reuse of 1 is assumed within each operator's network). In this section, the considered channel gain for both directions (UL and DL) captures the pathloss, shadowing, and fading effects. More details about the channel model and the data rate expressions in DL and UL for LTE can be found in [4]. We assume that the subcarriers constituting a single RB are subjected to the same fading and hence the channel gain on the subcarriers of a single RB is considered to be the same. In addition, the fading is assumed to be independent and identically distributed (iid) across RBs. In this section, we allocate one UL RB and one DL RB for each user. First, we start by allocating DL subcarriers in order to save BS power usage, since usually the DL traffic is much heavier than UL traffic. Then, the DL and UL rates are computed using the typical Shannon rate expression.

2.1.1 Energy Consumption Model for Base Stations

We consider that each BS is equipped with a single omni-directional antenna. The consumed power of an active BS j belonging to mobile operator n, $P_j^{(n)}$, can be computed as follows [4]:

$$P_j^{(n)} = aP_{n,j}^{(\text{tx})} + b, \tag{1}$$

where the coefficient a corresponds to the power consumption that scales with the radiated power due to amplifier and feeder losses, and the term b models an offset of site power that is consumed independently of the average transmit power and is due to signal processing, battery backup, and cooling. In (1), $P_{n,j}^{(tx)}$ denotes the radiated power of the jth BS belonging to operator n and can be expressed as follows:

$$P_{n,j}^{(\text{tx})} = \sum_{r=1}^{N_{\text{RB}}^{(\text{DL})}} P_{n,r}, \tag{2}$$

where $P_{n,r}$ is the power consumed per one RB and depends on the RB state. If the RB r of BS j is allocated to a certain user, then $P_{n,r} = \frac{P_{tot}}{N_{RB}^{(DL)}}$, else $P_{n,r} = P_{idle} \approx 0.19$ dBm, [19]. If a BS j is completely switched off, we assume that its power consumption $P_j^{(n)} = 0$. To power its BSs, the mobile operator either procures energy from a traditional electricity provider or uses renewable energy generators installed on BS sites, e.g., solar panels or wind turbine. The amount of energy procured from the fossil fuel retailer and the auto-generated amount of energy consumed by BS j of mobile operator n are denoted by $q_j^{(n,f)}$ and $q_j^{(n,g)}$, respectively, where f and g stands for fossil fuel and green energy, respectively. The amount of green energy generated locally varies from one BS to another depending on technical and environmental reasons. For instance, the solar rating depends essentially on the size of photovoltaic (PV) panels and whether they experience any shading during the day.

Note that the locally generated energy is free of charge whereas the electricity procured from the external retailer is evaluated by $\pi^{(f)}$ where $\pi^{(f)}$ is the cost of one unit of energy. That is, the fossil fuel, $q_j^{(n,f)}$, procured by BS j belonging to mobile operator n is equal to the total power $P_j^{(n)}$ consumed by this BS multiplied by its operation time Δt minus the amount of renewable energy generated locally $q_j^{(n,g)}$. The objective of each mobile operator is to minimize the consumption of its fossil fuel in order to reduce its energy cost.

2.1.2 Operator Services

In our study, we consider M different services are offered by mobile operators to their subscribers. Each service is identified by the data rate thresholds $R_{m,th}^{(UL)}$ and $R_{m,th}^{(DL)}$ for UL and DL, respectively, and a unitary price $p^{(m)}$ with $m = 1, \ldots, M$. We suppose that each subscriber associated to the network n is using one of the M offered services. For simplicity, we assume that all mobile operators offer similar services to their corresponding subscribers.

The main objective of this study is to formulate an optimization problem that minimizes the total fossil fuel consumption of cellular networks operating in the same area of interest. The BS ON/OFF strategy in a cooperative fashion will be applied in order to achieve green goals. In this setting, we also aim to at least not degrade the network QoS and the profit of each mobile operator but rather enhance them. However, in some cases, although it helps in reducing the CO_2 emissions, cooperation might lead to a negative impact on the profit of one of the mobile operators. Therefore, we establish a fairness condition that indicates whether green networking is favorable to operators or not. Finally, we compare the performance of our proposed scheme with the traditional scenario where cellular companies operate individually without cooperations.

2.2 Green Uncooperative Operators

We start by evaluating the gain of applying the BS sleeping strategy separately for each operator in terms of energy saving and profit. Let $\varepsilon^{(n)}$ be a binary vector that indicates the states the nth mobile operator BSs during the period Δt. Its elements $\varepsilon_j^{(n)}$ indicates whether a BS j of cellular company n is turned off or not as follows:

$$\varepsilon_j^{(n)} = \begin{cases} 1, & \text{if BS } j \text{ is turned on,} \\ 0, & \text{if BS } j \text{ is turned off.} \end{cases} \tag{3}$$

The number of ones and zeros in this vector indicates the number of active and inactive BSs, respectively. Thus, the fossil fuel consumption and the corresponding total cost of the nth mobile operator, denoted by $\mathcal{E}^{(n)}$ and $\mathcal{C}^{(n)}$ respectively, are given as follows:

$$\mathcal{E}^{(n)} = \sum_{j=1}^{N_{\text{BS}}^{(n)}} \varepsilon_j q_j^{(n,f)}, \text{ and } \mathcal{C}^{(n)} = \pi^{(f)} \mathcal{E}^{(n)}. \tag{4}$$

On the other hand, we compute the mobile operator profit provided by the operating BSs in the area. It is exclusively computed from the number of served customers and the corresponding service. In fact, during Δt, each mobile operator n is serving $N_U^{(n)}$ mobile stations connected to the network and enjoying one of the M proposed services. We denote by $N_{\text{out}}^{(n)}$ the number of users in outage during a period of Δt where $N_{\text{out}}^{(n)} \ll N_U^{(n)}$. A user i using the mth service communicates successfully with a BS if its UL and DL data rates, denoted by $R_i^{(\text{UL})}$ and $R_i^{(\text{DL})}$, are higher than the service data rate thresholds, $R_{m,th}^{(\text{UL})}$ and $R_{m,th}^{(\text{DL})}$ respectively. By denoting a binary parameter $\gamma_i^{(n)}$, $i = 1 \dots N_U^{(n)}$, we can express this assumption as follows:

$$\gamma_i^{(n)} = \begin{cases} 1, & \text{if } R_i^{(\text{UL})} \geq R_{m,th}^{(\text{UL})} \text{ and } R_i^{(\text{DL})} \geq R_{m,th}^{(\text{DL})}, \\ 0, & \text{if } R_i^{(\text{UL})} < R_{m,th}^{(\text{UL})} \text{ or } R_i^{(\text{DL})} < R_{m,th}^{(\text{DL})}. \end{cases} \tag{5}$$

In other words, if $\gamma_i^{(n)} = 0$, then the ith user fails to achieve its QoS during Δt. Let the vector $\boldsymbol{\gamma}^{(n)} = [\gamma_1^{(n)} \cdots \gamma_{N_U}^{(n)}]$, then the number of ones and zeros in $\boldsymbol{\gamma}^{(n)}$ corresponds to the number of served users and the number of users in outage, respectively. Consequently, only the served users pay the equivalent of the proposed service. Hence, the profit $\mathcal{P}_u^{(n)}$ of the nth mobile operator corresponding to its individual operation in this area is expressed as follows:

$$\mathcal{P}_u^{(n)} = \sum_{i=1}^{N_U^{(n)}} \gamma_i^{(n)} p_i^{(n,m)} + R_{\text{op}}\left(N_U^{(n)}\right) - \mathcal{C}^{(n)}, \tag{6}$$

where $p_i^{(n,m)}$ is the unitary cost of the service m used by the ith user of the nth cellular company and R_{op} is a constant extra revenue due to fixed subscription fees paid by the mobile operator subscribers. Hence, the optimization problem for a single mobile operator n is expressed as follows:

$$\underset{\boldsymbol{\gamma}^{(n)},\boldsymbol{\varepsilon}^{(n)}}{\text{Minimize}} \quad \mathscr{E}^{(n)} = \sum_{j=1}^{N_{\text{BS}}^{(n)}} \varepsilon_j q_j^{(n,f)}, \tag{7}$$

$$\text{Subject to:} \quad \frac{N_{\text{out}}^{(n)}(\boldsymbol{\gamma}^{(n)})}{N_U^{(n)}} \leq P_{\text{out}}. \tag{8}$$

The unique constraint of the problem is (8) which forces the percentage of users in outage to be lower than an outage probability threshold P_{out}. This constraint is an output of the resource allocation algorithm applied for a given $\boldsymbol{\varepsilon}^{(n)}$. It is very complex to find the optimal solution of this problem since the decision variables correspond to large binary vectors that depend on $N_U^{(n)}$ and $N_{\text{BS}}^{(n)}$. In [4], the authors proposed and compared deterministic and heuristic algorithms used to solve similar optimization problems for a single mobile operator scenario. They showed that the low complexity iterative algorithm achieves close performance to the evolutionary algorithms (e.g., genetic algorithm and particle swarm optimization approach) with a certain gain in terms of computational time. Hence, we employ the iterative algorithm to solve the formulated optimization problems in this section. Once optimization problem (7) is solved for each mobile operator n, we can deduce the corresponding profit $\mathscr{P}_u^{(n)}$ by computing (6) which has to be at least maintained in case of cooperative operation.

2.3 Green Cooperative Operators and Cooperation Decisions

In the cooperative mode, mobile operators can exploit the existence of other competitive providers in order to ensure energy saving and additional profit as well. In fact, instead of keeping lightly loaded BSs on, the mobile operator can turn them off and the subscribers may maintain their communication active using the radio access network of another operator serving the same area and vice versa. We propose to perform this by solving the following optimization problem where BS sleeping strategy is applied in order to achieve energy saving:

$$\underset{\boldsymbol{\gamma}^{(n)},\boldsymbol{\varepsilon}^{(n)},\, n=1,\ldots,N_{\text{op}}}{\text{Minimize}} \quad \sum_{n=1}^{N_{\text{op}}} \mathscr{E}^{(n)}, \tag{9}$$

$$\text{Subject to:} \quad \frac{N_{\text{out}}^{(n)}(\boldsymbol{\gamma}^{(n,1)},\ldots,\boldsymbol{\gamma}^{(n,N_{\text{op}})})}{N_U^{(n)}} \leq P_{\text{out}}. \tag{10}$$

Once this optimization problem is solved using the iterative algorithm presented in [4], we obtain the optimal energy consumption under cooperative operation for each network and thus the optimal vectors $\boldsymbol{\gamma}^{(n,t)}$ and $\boldsymbol{\varepsilon}^{(n)}$, $n, t = 1, \ldots, N_{\text{op}}$. Comparing to (8), we notice that, in the cooperative case, $N_{\text{out}}^{(n)}$ depends also on the allocation over other mobile operators which are determined using $\boldsymbol{\gamma}^{(n,t)}$, $t \neq n$. Indeed, thanks to the cooperation between mobile operators, some of users of mobile operator n can be served by another mobile operator t and vice versa. For this reason, we have introduced new binary vectors $\boldsymbol{\gamma}^{(n,t)}$ of size $1 \times N_U^{(n)}$ that indicates whether a user of mobile operator n is served successfully by mobile operator t or not. This way, during the resource allocation algorithm, more degrees of freedom are provided for all users because of the increase of the number of RBs in the DL and UL directions. Thus, higher channel gains can be allocated and energy saving can be achieved. However, random cooperation may lead to the increase of a certain mobile operator profit at the expense of other competitive operators. This can cause a high energy consumption and a very low profit for the active network. For instance, a mobile operator A may switch off all its BSs while all its users are served by BSs owned by mobile operator B which pays all energy bills. For this reason, we enforce fairness by introducing the notion of roaming price that will allow any mobile operator to decide whether to cooperate or not.

In our study, we assume that the roaming price, denoted by p_{nt}, corresponding to the cost of serving users belonging to another operator is equal for every pairs of cooperative operators (n, t), i.e., $p_{nt} = p_{tn}$. In our framework, the profit of the cooperative mobile operator n denoted by $\mathscr{P}_c^{(n)}$ is expressed as follows:

$$\mathscr{P}_c^{(n)} = \sum_{i=1}^{N_U^{(n)}} \gamma_i^{(n,n)} p_i^{(n,m)} + \sum_{\substack{t=1\\t\neq n}}^{N_{\text{op}}} \sum_{i=1}^{N_U^{(n)}} \gamma_i^{(t,n)} \left(p_i^{(n,m)} - p_{nt}\right)$$
$$+ \sum_{\substack{t=1\\t\neq n}}^{N_{\text{op}}} \sum_{i=1}^{N_U^{(t)}} p_{nt} \gamma_i^{(t,n)} + R_{\text{op}}\left(N_U^{(n)}\right) - \mathscr{C}_c^{(n)}, \tag{11}$$

where the first term in (11) corresponds to the operator revenue coming from serving its own users while the second term is the revenue coming from users served by other mobile operators after paying the roaming cost. The third term in (11) is the gain obtained from serving users belonging to other networks which depends on p_{nt}. Finally, R_{op} is the constant revenue and $\mathscr{C}_c^{(n)}$ is the network energy consumption cost obtained after solving (9)–(10). A mobile operator n cooperates only if its cooperative profit $\mathscr{P}_c^{(n)}$ is greater than or equal to the uncooperative profit $\mathscr{P}_u^{(n)}$ expressed in (6). Thus, the operators have to solve the following non-homogenous system of N_{op} linear inequalities with $N_p = \frac{N_{\text{op}}(N_{\text{op}}-1)}{2}$ unknown variables:

$$\mathscr{P}_c^{(n)} \geq \mathscr{P}_u^{(n)}, \forall n = 1, \ldots, N_{\text{op}}. \tag{12}$$

We distinguish here two cases depending on the number of operators N_{op}:

–$N_{\text{op}} = 2$: In this particular case, we have two inequalities with one unknown variable p_{12}. For simplicity, let us denote $A_1 = \sum_{i=1}^{N_U^{(1)}} \gamma_i^{(1,1)} p_i^{(1,m)} + R_{\text{op}}\left(N_U^{(n)}\right) - \mathscr{C}_c^{(1)} + \sum_{i=1}^{N_U^{(1)}} \gamma_i^{(1,2)} p_i^{(1,m)}$, $A_2 = \sum_{i=1}^{N_U^{(2)}} \gamma_i^{(2,2)} p_i^{(2,m)} - \mathscr{C}_c^{(2)} + \sum_{i=1}^{N_U^{(2)}} \gamma_i^{(2,1)} p_i^{(2,m)}$, $B = \sum_{i=1}^{N_U^{(2)}} \gamma_i^{(2,1)}$ and $D = \sum_{i=1}^{N_U^{(1)}} \gamma_i^{(1,2)}$. Then, the system of inequalities can be written as follows

$$
\begin{aligned}
(B - D)p_{12} &\geq \mathscr{P}_u^{(1)} - A_1, \\
(D - B)p_{12} &\geq \mathscr{P}_u^{(2)} - A_2.
\end{aligned}
\tag{13}
$$

Note that B corresponds to the number of users belonging to operator 2 served by operator 1 while D corresponds to the opposite situation. Thus, the problem solution depends on these variables. Indeed, if $B = D$, mobile operators do not need to impose a roaming price to each other and their profits are equal to A_1 and A_2, respectively. A simple comparison between $\mathscr{P}_u^{(n)}$ and $\mathscr{P}_c^{(n)}$ let them decide either they cooperate or no. Else (i.e., $B \neq D$), from (13), we distinguish two sets of possible solutions of p_{12}. If they are disjoint, cooperation is impossible. If there is an intersection interval, the operator collaboration is favorable for energy saving and profit enhancement. A fair choice of p_{12} is to maintain a close percentage change as follows:

$$
\frac{\mathscr{P}_u^{(1)} - \mathscr{P}_u^{(2)}}{\mathscr{P}_u^{(1)}} \approx \frac{\mathscr{P}_c^{(1)}(p_{12}) - \mathscr{P}_c^{(2)}(p_{12})}{\mathscr{P}_c^{(1)}(p_{12})}. \tag{14}
$$

–$N_{\text{op}} \geq 3$: In this case, the system can be written in the following matrix form:

$$
\boldsymbol{A}_{N_{\text{op}} \times N_p} \boldsymbol{p}_{N_p \times 1} \leq \boldsymbol{b}_{N_{\text{op}} \times 1}, \tag{15}
$$

where $\boldsymbol{A}$ is a matrix that contains the coefficients of the system of linear inequalities while $\boldsymbol{b}$ is a vector that contains constant terms. $\boldsymbol{p}$ is the decision vector that is constituted by the roaming price $p_{nt}, n, t = 1, \ldots, N_{\text{op}}$. Each of the inequalities determines a certain half-space while all the inequalities together determine a certain region in the N_p-dimensional space which is the intersection of a finite number of half-spaces [20]. If this system admits a feasible solution (i.e., the system is said compatible), the mobile operator can cooperate safely without degrading neither their QoS nor their individual profits. If the system is incompatible, then the multi-operator collaboration is impossible. A system is said compatible if and only if, its concomitant system is compatible. Indeed, from the system (15), we can construct a concomitant system involving $N_p - 1$ unknowns after discarding the last unknown, and for this new system, we can construct another concomitant system involving $N_p - 2$ unknowns and so on. This way, after a number of steps, we construct a system consisting of inequalities of one unknown. Thus, the compatibility of the original system is determined from the compatibility of the last constructed concomitant system. Using the same steps detailed above, we can find a solution of the problem in case of compatibility. The set of solutions of this non-homogenous system can be also determined via different methods. (For more details, see [20]).

3 Collaboration of Non-uniform Mobile Operators

After investigating the cooperation between uniform mobile operators having the same green objectives, we propose to investigate the collaboration between a green mobile operator and non-green operators with different utility functions. The green mobile company aims to achieve a tradeoff between its carbon dioxide emissions saving and its profit by exploiting the infrastructure of the non-green mobile companies existing in the same area. For the uniform scenario, the roaming price was based on mutual interests while in the non-uniform scenario, the non-green operators try to increase the roaming prices as much as possible in order to maximize their own profits. In [21], we present a thorough version of this work including more mathematical details.

3.1 System Model

We consider a geographical area served by $N_{\mathrm{op}} + 1$ mobile operators. Each mobile operator is deploying an LTE network with N_{BS} BSs that satisfies the traffic demand of its customers and covers the total area (i.e., $N_{\mathrm{BS}}^{(1)} = \cdots = N_{\mathrm{BS}}^{(N_{\mathrm{op}})} = N_{\mathrm{op}}^{(0)}$; we set the index of the GO to 0). We assume that each cell is controlled by $N_{\mathrm{op}} + 1$ BSs, each of which is owned by one operator. Thus, the BSs of the different operators are identically distributed and each operator controls N_{BS} of them. Although this is not generally the case, this assumption is used to simplify the problem.

3.1.1 Energy Consumption Model for Base Stations

We adopt the same power model given in (1) but we define $P_{n,j}^{(\mathrm{tx})}$, the radiated power of the jth BS of mobile operator n, as a function of the number of users served by this BS, denoted by N_j, multiplied by a constant power and can be expressed as follows:

$$P_{n,j}^{(\mathrm{tx})} = P_{\mathrm{T}} N_j, \tag{16}$$

where P_{T} is a constant power and is defined such that

$$P_{\mathrm{T}} = \frac{P_{\min}}{K} R^{\upsilon}, \tag{17}$$

where $P_{\min}$ denotes the minimum received power required by each mobile station (i.e., it represents the user QoS), K is a parameter accounting for several effects including BS antenna settings, carrier frequency and propagation environment, υ is the path loss exponent, and R denotes the inter-cell distance. If a BS j is completely switched off, we assume that its power consumption $P_j = 0$.

3.1.2 CO_2 Emission Penalty Function

Recall that BSs are powered either from a traditional electricity provider or from renewable energy generators installed on BS sites. The consumption of fossil fuels causes a harmful impact on the environment due to the emission of GHGs. The amount of this damage depends on the nature of the energy source. The CO_2 emission penalty function of a network can be modeled as a quadratic function of the consumed fossil fuel by a BS as it is given in [22]:

$$\mathscr{I} = \sum_{j=1}^{N_{\mathrm{BS}}} \alpha_f \left(q_j^{(n,f)}\right)^2 + \beta_f q_j^{(n,f)}, \tag{18}$$

where α_f and β_f are the emission coefficients related to the energy source of the electricity provider.

3.2 *Utility Functions and Problem Formulation*

In our framework, we investigate the cooperation between the non-uniform mobile operators. We assume that one of them is considered as a green mobile operator. Its objective is to minimize its network CO_2 emissions, maximize its profit or achieve a tradeoff between both objectives. The other mobile operators, denoted by (NGO_n) $n = 1, \ldots, N_{\mathrm{op}}$, are considered as traditional mobile operators having the goal of the maximization of their own profit regardless of their impact on the environment. The NGOs cooperate with the GO by offloading its users when needed. For instance, GO might switch off some of its BSs during low traffic period and the corresponding subscribers can connect to the NGOs infrastructure. In return, NGOs may impose on the GO to pay extra charge per number of roamed users. Thus, the GO aims to determine the number of users per BS to be offloaded to the NGO networks in order to maximize its objective while the NGOs seek the optimal roaming prices to impose in order to attract GO users and maximize their profits. In the sequel, in order to differentiate between the GO and NGO parameters, the notation $x^{(\mathrm{GO})}$ and $x^{(\mathrm{NGO}_n)}$ will be used, respectively.

3.2.1 Green Operator

The first objective of the GO is to maximize its profit, $\mathscr{P}^{(\mathrm{GO})}$, expressed as

$$\mathscr{P}^{(\mathrm{GO})} = \sum_{j=1}^{N_{\mathrm{BS}}} p^{(\mathrm{GO})} N_{T,j}^{(\mathrm{GO})} - \sum_{l=1}^{N_{\mathrm{op}}} \pi_n^{(\mathrm{r})} N_{j,n}^{(\mathrm{r})} - \pi^{(\mathrm{GO})} q_j^{(\mathrm{GO},f)}, \tag{19}$$

where $p^{(\mathrm{GO})}$ denotes the service fee of the GO per user while $\pi_n^{(\mathrm{r})}$ corresponds to the roaming price per user imposed by the nth NGO. $N_{T,j}^{(\mathrm{GO})}$ denotes the total number of GO users covered by BS j and $N_{j,n}^{(\mathrm{r})}$ is the number of users belonging to GO covered by BS j and served by NGO l. By definition, $q_j^{(\mathrm{GO},f)} = \max(P_j^{(\mathrm{GO})} \Delta t - q_j^{(\mathrm{GO},g)}, 0)$ where $P_j^{(\mathrm{GO})} = a P_{\mathrm{T}} N_j^{(\mathrm{GO})} + b$. Also, $N_{T,j}^{(\mathrm{GO})} = N_j^{(\mathrm{GO})} + \sum_{n=1}^{N_{\mathrm{op}}} N_{j,n}^{(\mathrm{r})}, \forall j = 1, \ldots, N_{\mathrm{BS}}$. However, if all users of BS j are roamed to neighbor BSs of other operators $N_{T,j}^{(\mathrm{GO})} = \sum_{n=1}^{N_{\mathrm{op}}} N_{j,n}^{(\mathrm{r})}$ (i.e., $N_j^{(\mathrm{GO})} = 0$), then the BS j is turned off and $P_j^{(\mathrm{GO})} = 0$, Finally, $\pi^{(\mathrm{GO})}$ is the unitary cost of fossil fuels per kWh paid by the GO. The GO's second objective is to reduce the CO_2 emissions, $\mathscr{I}^{(\mathrm{GO})}$, defined in (18). GO might target to achieve a tradeoff between both objectives. For this reason, we introduce a Pareto parameter, denoted by ω, in its utility function $\mathscr{U}^{(\mathrm{GO})}$ which will be maximized using the following optimization problem:

$$\max_{N_{j,n}^{(\mathrm{r})}} \quad \mathscr{U}^{(\mathrm{GO})} = \omega\, \mathscr{P}^{(\mathrm{GO})} - (1-\omega)\, \mathscr{I}^{(\mathrm{GO})} \tag{20}$$

$$\text{subject to:} \quad N_j^{(\mathrm{GO})} + \sum_{n=1}^{N_{\mathrm{op}}} N_{j,n}^{(\mathrm{r})} = N_{T,j}^{(\mathrm{GO})}, \quad \forall j = 1, \ldots, N_{\mathrm{BS}}, \tag{21}$$

$$0 \le N_{j,n}^{(\mathrm{r})} \le N_{T,j}^{(\mathrm{GO})}, \quad \forall j = 1, \ldots, N_{\mathrm{BS}}, \forall n = 1, \ldots, N_{\mathrm{op}}. \tag{22}$$

When $\omega \to 1$, we are dealing with the utility function given in (19). This corresponds to a selfish network operator that aims to maximize its own profit $\mathscr{P}^{(\mathrm{GO})}$ regardless of its impact on the environment. When $\omega \to 0$, we deal with the utility function given in (18), which corresponds to an environmentally friendly network operator that aims to reduce CO_2 emissions regardless of its own profit. Other values of ω constitute a tradeoff between these two extremes.

3.2.2 Non-green Operators

On the other hand, each NGO n tries to maximize its profit by serving as many roamed users as possible. Its utility function $\mathscr{U}^{(\mathrm{NGO}_n)}$ can be optimized using an optimization problem formulated as follows

$$\max_{\pi_n^{(r)}} \quad \mathcal{U}^{(\mathrm{NGO}_n)} = \sum_{j=1}^{N_{\mathrm{BS}}} p^{(\mathrm{NGO}_n)} N_j^{(\mathrm{NGO}_n)} + \pi_n^{(r)} N_{j,n}^{(r)} - \pi^{(\mathrm{NGO}_n)} q_j^{(\mathrm{NGO}_l, f)} \tag{23}$$

$$\text{subject to} \quad \pi_n^{(r)} \geq a\pi^{(\mathrm{NGO}_n)} P_{\mathrm{T}} \Delta t, \tag{24}$$

where $q_j^{(\mathrm{NGO}_n, f)} = \left(a\, P_{\mathrm{T}} \left(N_j^{(\mathrm{NGO}_n)} + N_{j,n}^{(r)}\right) + b\right) \Delta t$. Note that constraint (24) was added to ensure that the NGOs will always choose a profitable roaming price. In other words, if serving GO users is not beneficial, then NGOs will prefer to not cooperate.

3.3 *Analysis of the Stackelberg Equilibrium*

In order to solve the problem formulated in Sect. 3.2, we propose to model it as a Stackelberg game where the GO plays the role of the follower and NGOs play the role of the leaders. We apply a backward induction approach to derive the solution of the Stackelberg Equilibrium.

3.3.1 Green Operator Level Game: The Follower

The objective of the follower is to determine how many users per BS are needed to be roamed for NGO l in order to maximize its utility function. As the leaders aim to maximize their utility functions anticipating the predicted response of the follower, we should start first by deriving the best response of the follower with respect to the numbers of roamed users per BS $N_{j,n}^{(r)}$, $\forall j = 1, \ldots, N_{\mathrm{BS}}$, $\forall n = 1, \ldots, L$. It is known that the problem solution is an integer solution; however, we propose to relax the problem by transforming the integers to real non-negative variables. Then, we round the obtained solution to find the exact number of roamed users. Thus, the number of roamed users to the lth NGO can be obtained by computing the first derivative of the Lagrangian function with respect to $N_{j,n}^{(r)}$ and equating to zero. We showed in [21] that the second derivative of the utility function with respect to the number of roamed users is negative. Thus, $U^{(\mathrm{GO})}$ is concave with respect to $N_{j,n}^{(r)}$. Finally, the optimal number of roamed users per BS is expressed as follows

$$N_{j,n}^{(r)\,(*)} = \min\Bigg\{\Bigg[N_{T,j}^{(\mathrm{GO})} - \sum_{\substack{k=1\\k\neq l}}^{N_{\mathrm{op}}} N_{j,k}^{(r)} + \frac{b\Delta t - q_j}{a P_{\mathrm{T}} \Delta t} + \frac{\beta_f}{2\alpha_f a P_{\mathrm{T}} \Delta t} + \left(\frac{\omega}{1-\omega}\right)\frac{\pi^{(\mathrm{GO})}}{2\alpha_f a P_{\mathrm{T}} \Delta t} - \left(\frac{\omega}{1-\omega}\right)\frac{\pi_n^{(r)}}{2\alpha_f (a P_{\mathrm{T}} \Delta t)^2}\Bigg]^+, N_{T,j}^{(\mathrm{GO})}\Bigg\}, \tag{25}$$

where $\min(., N_{T,j}^{(\mathrm{GO})})$ and $[.]^+ = \max(., 0)$ are added to fulfill constraints (21) and (22), respectively. From the expression above, we can notice that the number of roamed users per BS decreases with the increase of the NGO roaming price. Moreover, we can see that this decrease depends on the GO's Pareto weight. For instance, when $\omega \rightarrow 1$, the GO is more and more concerned by its profit and thus the decrease of the number of roamed users is more important.

3.3.2 Non Green Operator Level Game: The Leader

The objective of the leader l in this Stackelberg game is to maximize its profit by attracting the maximum number of GO users. Therefore, the NGO l has to find the best roaming price depending on the system parameters in order to optimize its Stackelberg Equilibrium by injecting the relationship given in (25) in its utility function and deriving its first derivative with respect to the roamed price $\pi_n^{(\mathrm{r})}$ and equating it to zero. Hence, the optimal roaming price of NGO l is given as follows

$$\pi_n^{(\mathrm{r})(*)} = \max\Bigg\{ \left(\frac{1-\omega}{\omega}\right)\Bigg[\frac{\alpha_f (aP_{\mathrm{T}}\Delta t)^2}{N_{\mathrm{BS}}} \sum_{j=1}^{N_{\mathrm{BS}}} \left(N_{\mathrm{T},j} - \sum_{\substack{k=1\\k\neq l}}^{N_{\mathrm{op}}} N_{j,k}^{(\mathrm{r})} \right) + \alpha_f a P_{\mathrm{T}}\Delta t \left(b\Delta t - \sum_{j=1}^{N_{\mathrm{BS}}} \frac{q_j}{N_{\mathrm{BS}}} \right) + \frac{aP_{\mathrm{T}}\Delta t}{2}\beta_f \Bigg] + \frac{aP_{\mathrm{T}}\Delta t}{2}\left(\pi^{(\mathrm{GO})} + \pi^{(\mathrm{NGO}_n)}\right), a\pi^{(\mathrm{NGO}_n)} P_{\mathrm{T}}\Delta t \Bigg\}. \quad (26)$$

Note that the $\max\{., a\pi^{(\mathrm{NGO}_n)} P_{\mathrm{T}}\Delta t\}$ is added to ensure that the profit of a leader will not decrease below its profit obtained without cooperation as it is given in constraint (24). $U^{(\mathrm{NGO}_n)}$ is also concave with respect to $\pi_n^{(\mathrm{r})}$ as its second derivative with respect to $\pi_n^{(\mathrm{r})}$ is also negative.

From expressions (25) and (26), it can be noticed that the determination of the SE of the nth leader depends on the number of roamed users to the other $(N_{\mathrm{op}} - 1)$ NGOs as well as their respective roaming prices. Therefore, we propose to employ a fixed point algorithm to determine the optimal number of roamed users and the corresponding roaming prices.

4 Results and Discussion

In this section, we investigate the performance of the proposed approach for uniform mobile operator collaboration detailed in Sect. 2. Then, we discuss the results obtained for the non-uniform mobile operator collaboration setting.

4.1 Performance of the Collaboration Between Uniform Mobile Operators

We consider $N_{\text{op}} = 2$ mobile operators, denoted by A and B, serving a $5 \times 5\,(\text{Km}^2)$ LTE coverage area. A and B are placing uniformly $N_{\text{BS}}^{(1)} = 16$ and $N_{\text{BS}}^{(2)} = 9$ BSs, respectively. We assume the nonexistence of inter-operator interference and both networks are operating in disjoint 10 (MHz) bandwidths that are subdivided into $N_{\text{RB}} = 50$ RBs. The LTE and channel parameters are obtained from [19]. All BSs and all mobile stations have the same power model with the same maximal transmit power 46 dBm, $a = 21.45$ and $b = 354.44$ W. We set $\upsilon = 3.76$, $\kappa = -122.1$ dB, $\sigma_\xi = 8$ dB and the tolerance $P_{out} = 2\,\%$. The mobile station transmit power is set to 23 dBm. In addition, we suppose that the network operators offer similar $M = 3$ services. Each one is characterized by its cost (unitary price) $p^{(m)}$, expressed in monetary units (MU), DL and UL data rate thresholds ($R_{m,th}^{(\text{DL})}$ and $R_{m,th}^{(\text{UL})}$ respectively), and the occurrence probability of the service as it is shown in Table 1. The occurrence probability of a given service corresponds to the percentage of users in the network using that service.

Mobile operators are procuring energy either from electricity retailer, which provides enough energy to cover the network operation, or from renewable energy generated locally. We assume that amount of energy available at each BS varies between 0 and 100 Watt which corresponds to the maximum amount of energy that can be stored locally during the operation time $\Delta t = 1$ s. We set the unitary price of the fossil fuel to $\pi^{(f)} = 0.1$ (MU). Finally, we assume that $N_U^{(1)} = \alpha N_U^{(2)}$, $0 \leq \alpha \leq 1$ and that mobile operators are engaged to serve 98 % of the connected users simultaneously (i.e., $P_{\text{out}} = 0.02$). In our results, we compare our approach, denoted by "coop", with the traditional case, denoted by "uncoop", when both cellular companies operate individually in addition to the case when all BSs are assumed to belong to a single virtual network operator, denoted by "virtual".

In Table 2, we study the performance of mobile operator collaboration versus the number of subscribers connected to the networks for $\alpha = \frac{2}{3}$. In all scenarios, the amount of renewable energy generated by the BSs is the same. We notice that the proposed cooperative scheme achieves almost the same performance as the virtual scenario by activating almost the same number of BSs and consuming a slightly higher amount of fossil fuels. This small difference is due to the QoS constraints separately imposed on each operator as it is given in (10) while, in the

Table 1 Service parameters

Services	Service 1	Service 2	Service 3
$p^{(m)}$(MU)	10	5	1
$(R_{m,th}^{(\text{DL})}, R_{m,th}^{(\text{UL})})$ (kbps)	(1000, 384)	(384, 384)	(64, 64)
Occurrence probability (%)	15	25	60

Table 2 Approach performance versus total number of users ($\alpha = 2/3$)

Number of A users $N_U^{(1)}$	20	80	140
Uncoop. fossil fuels (kW) [Active BSs]	1.8 [4.3]	5.1 [10.7]	8.5 [16]
Coop. fossil fuels (kW) [Active BSs]	1 [2.7]	4 [7]	6.9 [11]
Virtual fossil fuels (kW) [Active BSs]	1 [2.7]	3.7 [6]	6.3 [9.2]
Uncoop. profit (kMU): A, B	0.10, 0.06	0.98, 0.62	1.86, 1.21
Coop. profit (kMU): A, B	0.12, 0.08	1.04, 0.68	1.96, 1.30
Roaming price (MU)	5.37	2.84	1.19

virtual scenario, there is only one constraint as it is a single big network. Compared to the traditional case, an important energy saving is obtained thanks to cooperation. For instance, for $N_U^{(1)} = 130$, the fossil fuel consumption is reduced by more than 23 %. Concerning the profit, we notice that the results satisfy the condition imposed in (12) that forces the cooperation profits to be higher than the individual ones by choosing an appropriate roaming price. For instance, the gained profit when collaborating is greater by 5 % than the uncooperative case for both operators when $N_U^{(1)} = 70$. This is because with collaboration, mobile operators are offered more degrees of freedom in order to reduce the energy cost and maximize their profit. To achieve this gain, an appropriate choice of roaming price has to be determined by solving (12). We can see that the higher traffic densities are, the lower the roaming price is. Finally, our simulation experiments indicate that the percentage of successful cooperation is 97 %. In other words, the probability that mobile operators decide to not cooperate is $\approx$3 %.

Figure 2 investigates the impact of generating renewables by mobile operators on the cooperation performance for $N_U^{(1)} = 50$ and $\alpha = \frac{2}{3}$. To do this, we introduce a parameter β_{RE} that represents the percentage of green energy generated by A while $100 - \beta_{RE}$ corresponds to the percentage of green energy generated by B. In other words, if $\beta_{RE} = 0$ %, then only B possesses renewables and vice versa. We assume here that all BSs owned by an operator are storing the same amount of renewables. In Fig. 2a, we plot the consumed fossil fuels. We notice that the operator that is controlling renewable energy is able to reduce its CO_2 emissions more when there is no cooperation with a gain in terms of profit (Fig. 2b). However, when cooperating, most of its BSs are kept active to serve most of the users of the competing provider (as it is shown in Fig. 3a, 90 % and 95 % for $\alpha = \frac{2}{3}$ and $\alpha = \frac{1}{3}$, respectively). However, the optimal value is when $\beta_{RE} = 50$ %. At this equilibrium, all BSs of both operators have the same characteristics and thus the BS selection set is larger. The curves are unbalanced because of the difference in the number of connected users and the number of available BSs per each operator. Finally, we can notice that the roaming price is higher when A is controlling the renewable energy. Indeed, as the number of subscribers of B is lower, A is forced to increase the roaming price in order to maximize its profit when cooperating, while the inverse can be deduced for B.

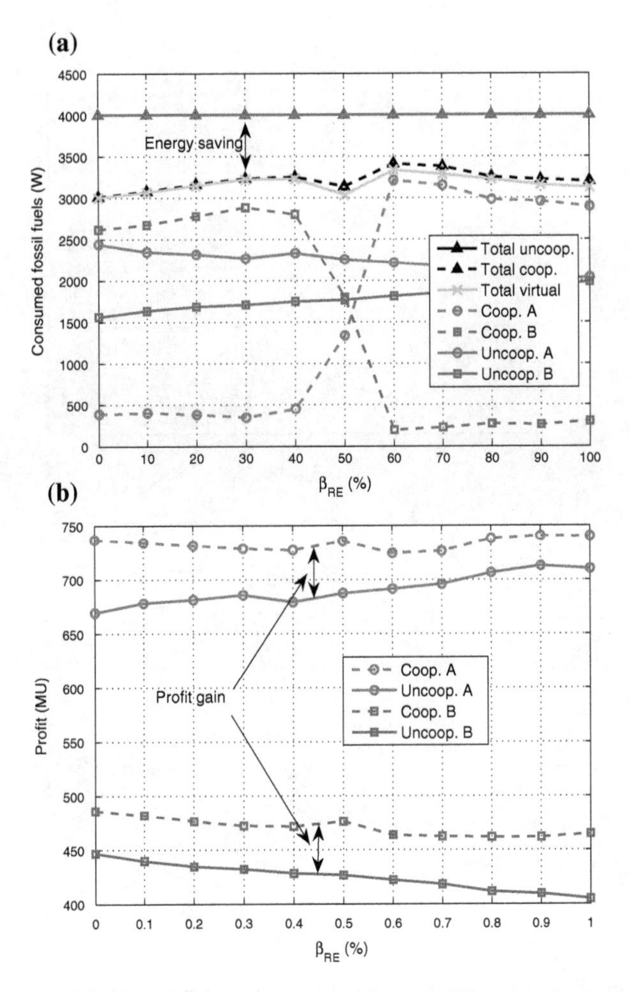

Fig. 2 **a** Consumed fossil fuels **b** Profit of mobile operators versus the distribution of green energy over cellular networks β_{RE}

Note that in all our simulations, the network QoS is satisfied for all operators, i.e., $P_{\text{out}} = 2\,\%$.

It is inpractical to assume that the roaming price varies instantaneously and dynamically with each channel in the network. It should have a pre-defined fixed average value for a given traffic density, or range of traffic densities in the network (e.g., there can be a price during the day corresponding to high density and another during the night corresponding to relatively lower density). This value can be set through collaboration agreements between mobile operators. The results derived in these simulations are averaged over 1000 channel realizations using Monte Carlo simulations. Hence, these results provide insights about the average roaming price

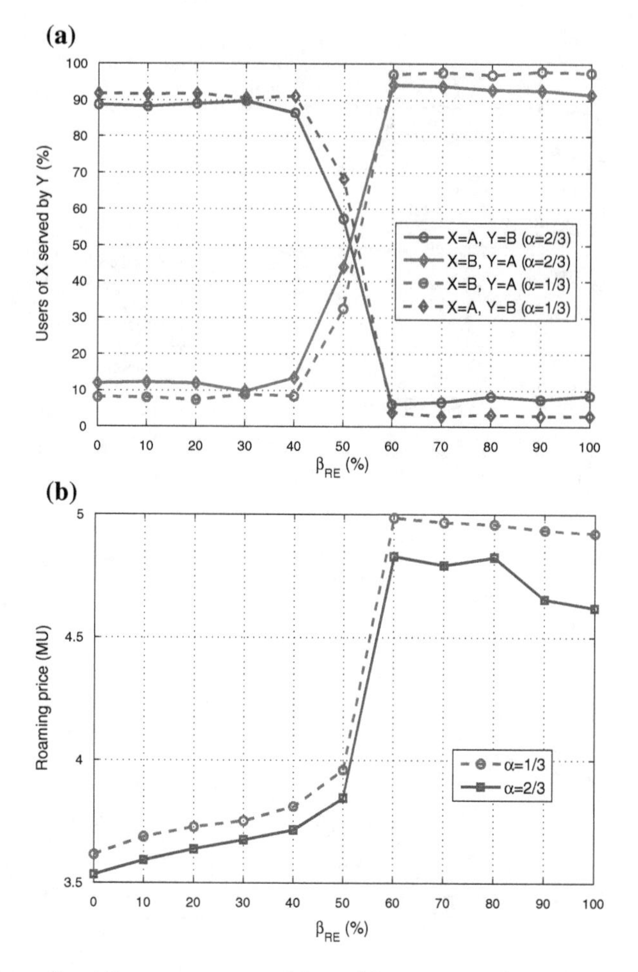

Fig. 3 **a** Users of mobile operator x served by mobile operator y **b** roaming price versus the distribution of green energy over cellular networks β_{RE}

that should be imposed between mobile operators for different traffic densities in order to ensure mutual benefit.

4.2 *Performance of the Collaboration Between Non-uniform Mobile Operators*

In this section, we present some numerical results for one-follower one-leader setting and one-follower two-leaders setting as an example of the scheme proposed in Sect. 3. We consider an area of interest where the $N_{\mathrm{op}} + 1$ mobile operators are deploying

Table 3 System parameters

Parameter	Value	Parameter	Value
P_{min}	−120 dBm	υ	3.76
K	0.0001	R	1000 m
(a, b)	(7.84, 71.5)	(α_f, β_f)	(0.02, 0.1)

$N_{BS} = 10$ identical BSs. All the BSs are powered by traditional electricity providers except the BSs of the GO network which are also supplied via green energy equipment deployed in BS sites. The amount of the auto-generated green energy differs form a BS to another. This can be explained essentially by the fact that PV panels in BS sites have different sizes and whether they experience any shading during the day. Finally, we consider that $p^{(GO)} = p^{(NGO_n)} = 5$ MU, $\forall n = 1, \ldots, N_{op}$ (MU stands for monetary unit). In our simulations, we set the channel and power parameters as it is detailed in Table 3.

First, we investigate the one-leader one-follower scenario where only one NGO is available to serve the users of the GO. In Table 4, we study the cooperation between the operators for three cases when $\pi^{(GO)} = 0.2 < \pi^{(NGO)} = 0.4$ and $\pi^{(GO)} > \pi^{(NGO)} = 0.4$. For each case, we vary the Pareto weight ω representing the behavior of the GO towards the environment and we provide some information about the number of roamed users and the NGO roaming price. The amount of energy $q_j^{(GO)}$ available in each GO's BS and the number of users served by each BS $N_{T,j}^{(GO)}$ are also given in Table 4. We assume that $N_{T,j}^{(GO)} = N_{T,j}^{(NGO)}$. We can first deduce that there are three categories of BSs in this roaming setting. BSs that offload all their users to the NGO (e.g., $j = 2, 6, 8$): These BSs, having very low amounts of renewable energy, prefer to be turned off instead of serving users using fossil fuel independently of the value of $\pi^{(GO)}$. The second category is the BSs that do not offload any users as they have sufficient amount of green energy to serve all of them (e.g., $j = 4, 9, 10$). The final category encloses the BSs that offload some of their users depending on the available amount of green energy (e.g., $j = 1, 3, 5, 7$). Another remark is that as ω increases the roaming price $\pi^{(r)}$ decreases. Indeed, as it is more concerned by its profit, the GO tries to avoid the maximum to pay extra roaming fees. Thus, NGO are obliged to reduce their roaming price to attract GO users. We can see that the higher $\pi^{(GO)}$ is, the higher $\pi^{(r)}$ is.

Let us now study the cooperation behavior case by case. When $\pi^{(GO)} < \pi^{(NGO)}$, we notice that as ω increases the number of roamed users decreases. In the case, $\omega = 0.1$, GO offloads the maximum number of users such that it minimizes its CO_2 emissions. This means that most of GO users are either served by green energy or by NGO infrastructure. When ω is close to 1, we can see that GO does no more offload users since it prefers to serve them using its BSs even with fossil fuels as its $\pi^{(GO)} < \pi^{(NGO)}$. Now, if $\pi^{(GO)} > \pi^{(NGO)}$, we can see as ω increases, the number of roamed users increases too. In this case, the GO prefers to offload its users as its

Table 4 Performance of the proposed scheme for one NGO one GO case

			$\pi^{(GO)} = 0.2 < \pi^{(NGO)} = 0.4$					$\pi^{(GO)} = 0.6 > \pi^{(NGO)} = 0.4$				
ω			0.1	0.4	0.6	0.9	0.98	0.1	0.4	0.6	0.9	0.98
$\pi^{(r)}$(MU)			10.53	2.03	1.08	0.45	0.44	10.75	2.25	1.30	0.67	0.57
j	$q_j^{(GO,g)}$	$N_{T,j}^{(GO)}$	$N_j^{(r)}$					$N_j^{(r)}$				
1	90	47	9	8	6	0	0	9	11	13	30	47
2	9	21	21	21	21	21	0	21	21	21	21	21
3	74	65	41	40	38	21	0	42	43	45	62	65
4	145	49	0	0	0	0	0	0	0	0	0	49
5	71	51	30	29	27	10	0	31	32	34	51	51
6	4	39	39	39	39	39	0	39	39	39	39	39
7	57	80	72	71	69	52	0	72	74	76	80	80
8	13	59	59	59	59	59	0	59	59	59	59	59
9	105	17	0	0	0	0	0	0	0	0	0	17
10	208	71	0	0	0	0	0	0	0	0	0	37
Total	776	499	271	267	259	202	0	273	279	287	342	465
$\mathscr{P}^{(GO)}$			−379.5	1932	2192	2368	2378	−497	1815	2073	2262	2227
$\mathscr{I}^{(GO)}$			53.9	58.6	68.6	164.6	1147	51.7	45.1	37.1	0.12	0
$\mathscr{U}^{(NGO)}$			4725	2413	2156	1992	1988	4805	2494	2237	2069	2050

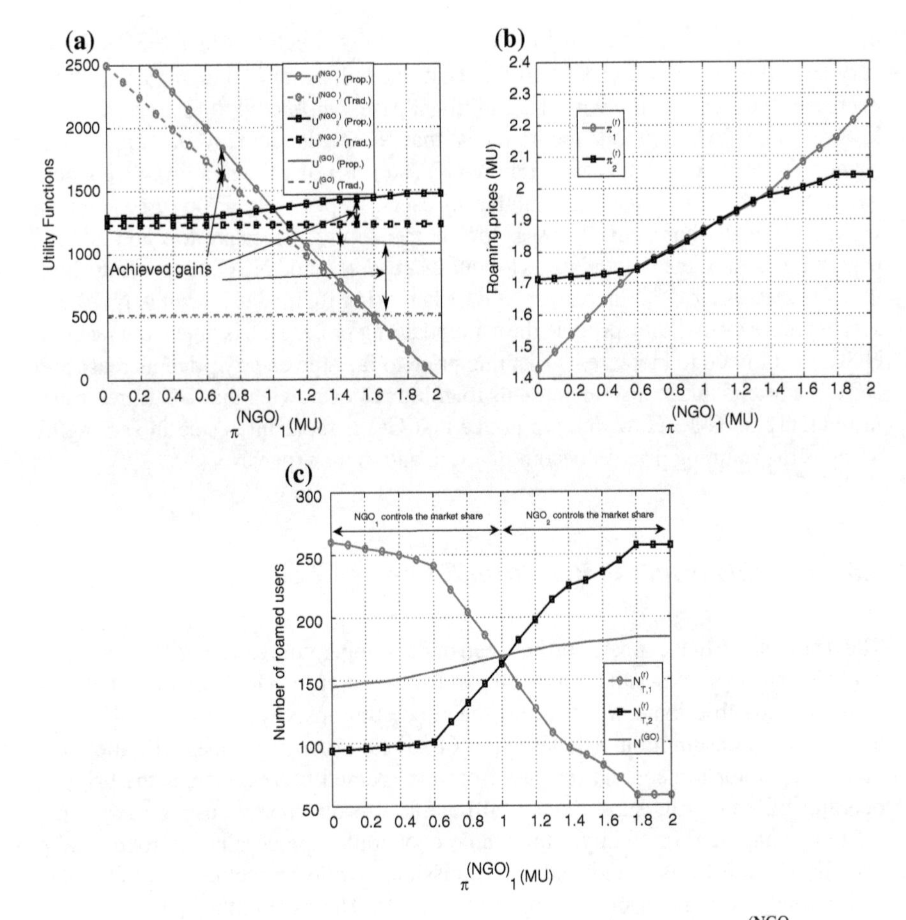

Fig. 4 Performance of the proposed scheme for one GO two NGOs versus $\pi_1^{(\mathrm{NGO})}$, **a** utility functions, **b** NGO roaming prices, and **c** number of users (roamed $N_{T,n}^{(r)}$, not roamed $N^{(GO)}$)

fossil fuel cost is more expensive than the NGO one and is more and more concerned by its profit.

In Fig. 4, we investigate the performance of the proposed scheme for one follower and two leaders scenario. In this case, we set $\omega = 0.6$ which belongs to the Pareto efficiency region as it is given in Table 4 and we set $\pi^{(\mathrm{GO})} = 1.5$ (MU) and $\pi_2^{(\mathrm{NGO})} = 1$ (MU). Finally, we vary $\pi_1^{(\mathrm{NGO})}$ between 0 and 2 (MU). In Fig. 4a, we investigate the performance of all operators under the proposed cooperation mode (denoted by Prop.) when varying the fossil fuel cost of NGO 1 by plotting their utilities functions. Also, we compare them with the performance of the non-cooperation mode (denoted by Trad.) where all operators serve their own users without roaming. The figure shows that, thanks to their collaboration, all operators are able to enhance their performance comparing to the traditional scenario. Indeed, independently of the value of $\pi_1^{(NGO)}$, GO is able to double its utility function while NGO utilities vary according to the NGO 1 fossil fuel cost. Indeed, as $\pi_1^{(NGO)}$ increases, $U^{(NGO_1)}$ decreases until coinciding with the traditional case, while $U^{(NGO_2)}$ increases with a lower scale. NGO 2 exploits the high cost that NGO 1 is facing to provide a lower roaming user price and thus attract more GO users as it is shown in Fig. 4b, c where we plot the roaming prices and number of users, respectively. From these figures, we can see that when NGO 1 cost is low, both roaming prices are low and NGO 1 is gaining most of the roamed users (about 260 users) while NGO 2 is serving about 95 GO users. As $\pi_1^{(NGO)}$ increases, NGO 1 is loosing roamed users while NGO 2 is serving more even if they provide the same roaming price. This is due to the fact that NGO 1 is obliged to increase its roaming price to face the energy price increase and NGO 2 exploits this to also increase its roaming price knowing that GO is interested in reducing its CO_2. However, we notice that GO is more interested in serving its users as the roaming price is becoming more and more expensive.

4.3 Insights from the Results and Future Challenges

The results show that cooperation between multiple operators can be beneficial, and it could lead to a win-win situation for all parties. Operators can increase their revenues from roamed subscribers, while cutting their operating costs by reducing significantly their energy consumption. The selection of suitable roaming prices will allow the serving operator to generate revenue from the roamed users, whereas the original operator offloading its subscribers will be able to save energy costs by switching off redundant BSs. In addition, the whole cooperation process is environmentally friendly since it leads to reduced CO_2 emissions, while respecting users QoS and maintaining, even increasing, the operators profits. Throughout the operation of the network, the roles of the operators will be reversed, depending on the network dynamics. Hence, each operator will at certain times (or at different locations, at the same time) save energy by switching off some of its BSs, while at other times (and/or locations), it will be serving the roamed users of other operators.

In practice, this multi-operator collaboration is in line with the active research area of network function virtualization (NFV) and service orchestration. In fact, mobile virtual network operators (MVNOs) are a market reality, where a physical operator lends its infrastructure (e.g., the mobile access network) to be used by several other virtual operators. Having two existing operators unite the operation of their networks for the benefit of all can thus be implemented in practice. However, the main challenge is in pricing and billing issues. Operators need to find the best roaming price that can lead to benefits for all involved parties. This requires an assessment of the value of the savings obtained by switching a certain BS off and of the costs incurred by the new serving operator to serve subscribers of other operators. Once studies are made to estimate these values, suitable billing agreements can be signed between concerned operators. The whole process remains transparent to subscribers, who will pay their bills to their initial operator, and will receive their expected QoS seamlessly across the networks of the collaborating operators.

5 Summary

In this chapter, we investigated the performance of the green networking approach for multi-operator collaboration for two different settings: The uniform mobile operator collaboration and the non-uniform mobile operator collaboration. In the first scenario, we have formulated an optimization problem that aims to reduce the total CO_2 emissions by eliminating redundant base stations while respecting the network QoS. We have also derived a system of linear inequalities to decide whether to cooperate or not by determining the roaming price. Our approach leads to an important saving in terms of fossil fuel consumption while it enhances the cooperative mobile operator profit. In addition, it shows that the roaming price is inversely proportional to the number of subscribers of the network as well as the number of BSs generating renewables.

In this second scenario, we investigated the performance of a green networking system where one green operator interested in minimizing its CO_2 emissions cooperates with several non green operators interested in maximizing their own profits by serving the green operator subscribers. The problem was formulated as a two-level Stackelberg game that leads to the maximization of both player utility functions. A Stackelberg equilibrium was derived and the optimal roaming prices and number of offloaded users are determined. Our simulation results showed the behavior of each mobile operator in this competition game and showed that the green operator is able to ensure a significant reduction in terms of CO_2 emissions compared to the traditional case.

At the moment, collaboration between mobile companies are still not applied in reality. Therefore, there is a pressing need to propose additional and new approaches in order to encourage telecommunication leaders and regulators to discuss and focus more on such approaches for possible implementation in next cellular network generation. The work can be extended and enriched by formulating differently the problem. Game theoretical approaches could be used to model the coalition and/or the competition among mobile operators while including the dynamic traffic variation.

References

1. Mobile's Green Manifesto 2012, Groupe Speciale Mobile Association (GSMA), Technical Report, (Jun. 2012)
2. Cai, L., Poor, H., Liu, Y., Luan, T., Shen, X., Mark, J.W.: Dimensioning network deployment and resource management in green mesh networks. IEEE Wirel. Commun. **18**(5), 58–65 (2011)
3. Bianzino, A., Chaudet, C., Rossi, D., Rougier, J.: A survey of green networking research. IEEE Commun. Surv. Tutor. **14**(1), 3–20 (2012). First Quarter

4. Ghazzai, H., Yaacoub, E., Alouini, M.-S., Abu-Dayya, A.: Optimized smart grid energy procurement for LTE networks using evolutionary algorithms. IEEE Trans. Veh. Technol. **63**(9), 4508–4519 (2014)
5. Oh, E., Son, K., Krishnamachari, B.: Dynamic base station switching-on/off strategies for green cellular networks. IEEE Trans. Wirel. Commun. **12**(5), 2126–2136 (2013)
6. Louhi, J.: Energy efficiency of modern cellular base stations. In: Proceedings of the 29th International Telecommunications Energy Conference, INTELEC, pp. 475–476, October (2007)
7. Khan, A., Kellerer, W., Kozu, K., Yabusaki, M.: Network sharing in the next mobile network: TCO reduction, management flexibility, and operational independence. IEEE Commun. Mag. **49**, 134–142 (2011)
8. Militano, L., Molinaro, A., Iera, A., Petkovics, A.: Introducing fairness in cooperation among green mobile network operators. In: 20th International Conference on Software, Telecommunications and Computer Networks (SoftCOM 2012), September (2012)
9. Yaacoub, E.: Performance study of the implementation of green communications in LTE networks. In: Proceedigns of the 19th International Conference on Telecommunications (ICT 2012), Jounieh, Lebanon, April (2012)
10. Bousia, A., Kartsakli, E., Alonso, L., Verikoukis, C.: Dynamic energy efficient distance-aware base station switch On/Off scheme for LTE-Advanced. In: Proceedings of the IEEE Global Telecommunications Conference (GLOBECOM 2012) Anaheim. CA, USA, December (2012)
11. Ghazzai, H., Yaacoub, E., Alouini, M., Dawy, Z., Abu Dayya, A.: Optimized LTE cell planning with varying spatial and temporal user densities. To appear in IEEE Transactions on Vehicular Technology (2015)
12. El-Beaino, W., El-Hajj, A.M., Dawy, Z.: A proactive approach for LTE radio network planning with Green considerations. In: Proceedings of the 19th IEEE International Conference on Telecommunications (ICT 2012), Jounieh, Lebanon, April (2012)
13. Xiang, L., Pantisano, F., Verdone, R., Ge, X., Chen, M.: Adaptive traffic load-balancing for green cellular networks. In: Proceedings of the 22nd IEEE International Symposium on Personal Indoor and Mobile Radio Communications (PIMRC 2011), Toronto. Canada, September (2011)
14. Bu, S., Yu, F.R., Cai, Y., Liu, P.: When the smart grid meets energy-efficient communications: Green wireless cellular networks powered by the smart grid. IEEE Trans. Wirel. Commun. **3**(4), 2252–2261 (2012)
15. Marsan, M.A., Meo, M.: Energy efficient wireless internet access with cooperative cellular networks. Elsevier J. Comput. Netw. **55**(2), 386–398 (2011)
16. Bousia, A., Kartsakli, E., Antonopoulos, A., Alonso, L., Verikoukis, C.: Game theoretic approach for switching off base stations in multi-operator environments. In: IEEE International Conference on Communications (ICC 2013), pp. 4420–4424, June (2013)
17. Ghazzai, H., Yaacoub, E., Alouini, M.-S.: A game theoretical approach for cooperative environmentally friendly cellular networks powered by the smart grid. In: Proceedings of the IEEE Online Conference on Green Communications (GreenCom'2014), November (2014)
18. Ghazzai, H., Yaacoub, E., Alouini, M.-S.: Multi-operator collaboration for green cellular networks under roaming price consideration. In Proceedings of the 80th IEEE Vehicular Technology Conference (VTC Fall 2014) Vancouver. Canada, September (2014)
19. 3rd Generation Partnership Project (3GPP), 3GPP TS 36.211 3GPP TSG RAN Evolved Universal Terrestrial Radio Access (E-UTRA) Physical Channels and Modulation, version 11.4.0, Release 11, April (2013)
20. Solodovnikov, A.S.: Systems of Linear Inequalities. Mir Publishers, Moscow (1979)
21. Ghazzai, H., Jardak, S., Yaacoub, E., Yong, H.-C., Alouini, M.-S.: A game theoretical approach for cooperative green mobile operators under roaming price consideration. In: IEEE International Conference on Communications (ICC 2015), London, UK, June (2015)
22. Senthil, K.: Combined economic emission dispatch using evolutionary programming technique. IJCA Special Issue on Evolutionary Computation for Optimization Techniques **2**, 62–66 (2010)

An Integrated Approach for Functional Decomposition of Future RAN

Zainab R. Zaidi, Vasilis Friderikos, Oluwakayode Onireti, Jinwei Gang and Muhammad A. Imran

Abstract Software-defined radio access networks (SD-RAN), dense deployment of small cells with possible macro-overlay for users with high mobility, decoupled signaling and data transmissions, or beyond cellular green generation (BCG2) architecture for enhanced energy efficiency, etc. are some of the very active research themes and most promising technologies for future RAN architecture. In this chapter, we present the idea of an integrated deployment solution for energy efficient cellular networks combining the strengths of the above mentioned themes. While SD-RAN envisions a decoupled centralized control plane and data forwarding plane for flexible control, the BCG2 architecture calls for decoupling coverage from capacity and coverage is provided through always-on low-power signaling node for a larger geographical area; capacity is catered by various on-demand data nodes or small cells for maximum energy efficiency. We identify that a combined approach bringing in both decompositions together can, not only achieve greater benefits, but also facilitates the faster realization of both technologies. We propose the idea and design of a *signaling controller* which acts as a signaling node to provide always-on coverage, consuming low power, and at the same time also hosts the control plane functions for the SD-RAN through a general purpose processing platform. Phantom cell concept is also a similar idea where a normal macro cell provides interference control to densely

---

Z.R. Zaidi (✉) · V. Friderikos · J. Gang
Kings College London, London, UK
e-mail: Zainab.Zaidi@gmail.com

V. Friderikos
e-mail: vasilis.friderikos@kcl.ac.uk

J. Gang
e-mail: jinwei.gang@kcl.ac.uk

O. Onireti · M.A. Imran
Institute of Communication Systems (ICS), University of Surrey,
Guildford GU2 7XH, UK
e-mail: o.s.onireti@surrey.ac.uk

M.A. Imran
e-mail: m.imran@surrey.ac.uk

© Springer International Publishing Switzerland 2016

M.Z. Shakir et al. (eds.), *Energy Management in Wireless Cellular and Ad-hoc Networks*, Studies in Systems, Decision and Control 50,
DOI 10.1007/978-3-319-27568-0_6

deployed small cells, although, our preliminary results show that the proposed integrated architecture has much greater potential of energy savings in comparison to phantom cells as a signaling controller is supposed to consume minimal power in comparison with the normal macro cell BS.

1 Introduction

Continuous miniaturization of computing units will yield smarter and more capable mobile devices with even greater demand for capacity. This growth is also a major factor in predicted doubling of global ICT emissions from 2 to 4 % by 2020 [1]. Addition of many small cells, with different architectural variations, emerged as a major solution for enhancing capacity but only if interference and mobility are properly managed. Mobile communication systems witnessed growth at a much slower pace than the user-end devices mainly due to inflexible and expensive equipment, complex control plane protocols, and vendor specific configuration interfaces. Software defined networking (SDN) suggests hardware agnostic programmable platform for development of protocols, applications, etc., hiding all complexity of execution through separation of control and data plane [3]. This decoupling will introduce unparalleled flexibility for innovation and future growth and will also reduce CAPEX and OPEX through ideas like network virtualization.

Moreover, with the growth in user data, more and more base stations (BSs), currently consuming over 80 % of the total network energy [1], are added into the system substantially increasing the energy consumption and carbon footprint of cellular networks. The state-of-the-art energy management schemes exploit the redundant capacity during the low traffic scenarios and put a fraction of the BSs in sleep mode. However, they might cause coverage holes and in order to achieve the real benefits of energy management, it is needed to separate capacity and coverage via logical decoupling of the data and control or signaling transmissions in the future systems, also known as BCG2 or cell on-demand architecture [1]. The signaling nodes provide coverage and always-on connectivity and will be designed for low rate services, for system access and paging, consuming very small fraction of power; whereas the data nodes can only be used on-demand depending on the traffic. The decoupling is expected to provide 85–90 % energy saving potential compared to the current systems [2].

Although, both of the approaches have different technical objectives, i.e., SDN focuses on inducing flexibility through programmable hardware and BCG2 architecture tries to get linear relationship between energy consumption and user traffic. The end goal, however, has a lot in common in terms of physical realization. The centralized controller in the state-of-the-art proposals of SD-RAN, either resides in the core network or in a centralized data centre [3], an idea migrated from Cloud RAN (Radio Access Network). We argue that the signaling node providing coverage and system access can also be a suitable host for the centralized controller, or virtual big BS of [3], in SD-RAN containing major functionalities of control plane, such as, coordination and resource allocation, for a number of BSs in a geographical area. In this chapter, we present the idea of a *signaling controller* which provides always-on system access, contains control plane functionalities of interference management, resource allocation, etc. Since, the control plane can be implemented using general purpose processors; the signaling node does not need additional power consuming elements and can still conform to the low-power consumption attribute as required by the BCG2 architecture. The signaling controller can use dedicated microwave links to connect to the BSs in its coverage area or fibre if cost-effective.

Moreover, SDN can also be viewed as the enabling technology for BCG2 architecture. Such basic architectural change in contemporary cellular systems is very expensive to implement as it requires redesigning of several components and hardware; a major reason to delay the decoupling for later standards after a full feasibility investigation. A cellular SDN experimentation platform can, however, facilitates the performance evaluation and quick implementation of BCG2 architecture. It is high time to re-think the basic architecture of the cellular systems and not only make it energy efficient but also make them flexible and amenable for future growth. These two areas are getting huge interest of the research community and in our view their intersection also creates a unique space with immense potential of performance improvement.

Our architecture is inline to phantom cell concept [4] which was introduced for realizing true potential of dense deployment of small cells as suggested for LTE Release 12. In this idea, many small cells, called the phantom cells as they contain only LTE user plane, are overlaid with a normal macro cell which provides interference coordination. However, a macro cell consumes over 100 times more power than a pico/femto cell as described in the EU FP7 EARTH project [2] and keeping it on all the time would lead to severe power inefficiency. In the proposed integrated architecture however, the signaling controller is used instead of the macro cell which is a low-rate signaling-only node consuming a small fraction of network energy. A signaling controller can also host a logically separated data node but it can be treated as any other data node in the area. Our simulation results show the exceptional potential of energy savings using our proposed architecture in comparison to phantom cell concept.

Moreover, with less than 1 % emissions, mobile telecom infrastructure is not among the biggest polluters [1] and higher capacity and faster data rates are still the high priority and valued design objectives for future mobile networks than enhanced energy efficiency. However, our integrated approach provides improved energy efficiency while facilitating dense deployments of small cells and functional decomposition for software defined networking.

The rest of the chapter is organized as follows: in Sect. 2, we review SDN and BCG2 themes and their salient features, the new integrated design is presented in Sect. 3 along with the important outstanding issues. Prominent advantages of the integrated architecture are presented in Sect. 4. Finally, Sect. 5 concludes the chapter.

2 Background

2.1 BCG2 Architecture

The logical decoupling of data transmissions and control signaling paradigm is one of the key directions being explored by GreenTouch[1] under the project Beyond Cellular Green Generation (BCG2) [1]. GreenTouch is a consortium of leading ICT industry and academic experts working towards 1000 times enhanced network energy efficiency compared to 2010 by delivering specific designs and recommendations by 2015. In BCG2 architecture, the signaling nodes are responsible for the coverage and are usually assumed to deliver low rate services, such as, random access and paging, over long ranges; whereas the data nodes can be activated and deactivated depending on the traffic demand and it is designed for high rate and small ranges. The decoupling is logical in nature and a single location can host both types of nodes. A preliminary study of BCG2 architecture is presented in [1], where it is shown, through statistical modeling, that energy efficiency of current systems can be improved by more than 50 times depending on the daily load profile. A set of studies regarding the BCG2 architecture is performed under the EU FP7 IP project EARTH [2]. The study shows that up to 85–90 % saving potential is possible with this revolutionary changed architecture compared to the current systems. The results from these studies show the promise and potential of the BCG2 architecture. Although, considering the transition cost involved in moving to a new cellular architecture, a comprehensive feasibility study is necessary to clearly identify the cost-benefit trade-offs.

Coverage is also separated from data processing in a Cloud RAN architecture [5], where a centralized BBU (Base Band Unit) pool serves several RRH (Radio Remote Heads) in the area and not-in-service BBUs can be put in sleep mode to save energy. The prohibitive aspects of Cloud RAN are the expensive fibre needed to connect RRHs to the BBU pool and the high bandwidth requirements for this fronthaul.

Currently, the base stations consume over 80 % of total network energy and are designed for high-data rate services. During low-traffic scenarios, such as, night time, they can only be put in sleep mode if there are redundant BSs covering the area, which is the usual case in dense urban environment. The decoupling will provide tremendous opportunity for improving energy consumption. Data nodes or the BSs responsible for data transmissions can employ more efficient energy management schemes with sleep modes and green radio technologies, such as optimized beam forming, in the BCG2 architecture compared to the contemporary systems. They are potentially the major power consumer.

An incoming session should be allocated to an active data node which can provide the service, although it might not be the best BS for the job, as the incremental cost for serving an additional session is much smaller then activating a BS [1]. However, sending the transmission through a low SNR (Signal to Noise Ratio) path will affect the spectral efficiency of the system. Moreover, the channel between data BS and

[1]http://www.greentouch.org/.

the mobile terminal cannot be estimated before selection as done in the classical approach. More sophisticated mechanisms are required to predict and estimate the channel condition between the mobile terminal and any potential data BS. Signaling nodes are responsible for providing coverage and always-on connectivity, paging the mobile terminals, and providing access to them when required [2]. They are supposed to be designed for low rate and long range transmissions and consuming low power.

2.2 *Software Defined Radio Access Networks*

While the definitions of SDN are still evolving, it mainly focuses towards decoupling of the software-based control plane from the hardware-based data plane (e.g., packets forwarding) of networking and switching pieces of equipment [3]. The logically centralized controllers contain the control logic to translate the application requirements down to the data plane and are responsible in providing an abstract network view to the application plane. The major issue is to create appropriate mapping of the existing network functionalities to the decoupled control and forwarding planes. While a lot of the work is done for wired or optical networks, some proposals are also presented for cellular SDN architecture [3, 6].

For the radio access part of the mobile communication system, the major questions are: 1—How to decouple the control plane from the base stations and 2—where the control plane will be located? The centralized controller in the state-of-the-art proposals of SD-RAN, either resides in the core network [6] or in a centralized data centre [3]. The design in [6] tries to push all the control plane functionality into a centralized controller in the core network and proposes the use of local switch agents for scalability. The proposition in [3] is focused towards re-factoring the control plane into a virtualized big base station controller for a geographical area and local controllers within each base station for latency sensitive decision making.

The LTE architecture also distinguishes user plane, dealing with the data packet forwarding, and control plane, focusing on signaling and management messages and operations, using the same physical infrastructure. Both planes reside in the firmware of the system. This demarcation is much clearly designed in mobile core network EPC (Evolved Packet Core), but the radio access part of LTE consists of only base station node, eNodeB, performing data forwarding and control functions. The SDN based core network requires transporting the control plane into software along with the control logic required for the data forwarding plane, e.g., routing rules, mobility anchoring, etc. We also remark that user plane as defined by LTE is not exactly the same as data forwarding plane of SDN and similarly both control planes also differ slightly. In SDN, control plane relates to all control logic required to manage the network, connections, and forwarding the data packets. From SDN's perspective, base stations, serving gateway (S-GW), and packet gateway (P-GW) of LTE architecture are also performing some control plane functions along with data forwarding in addition to the designated control plane nodes MME (Mobility

Management Entity), HSS (Home Subscriber Server), and PCRF (Policy Control and Charging Rules Function) [6].

Although not within the scope of SDN, but it is worthwhile to discuss an extreme approach of re-designing radio access architecture, i.e., the Cloud RAN [5]. Cloud RAN centralizes all functionalities, control as well as data plane, into a centralized BBU pool, or a data center, for easier management and coordination while leaving only antennas and some active RF components, i.e., RRH, on the cell sites. Cloud RAN is proposed as a mechanism to realize small cell deployment in LTE through proper coordination for interference management. LTE small cell, however, assumes distributed control with self-organizing (SON) capabilities. The backhaul from the cell site to the serving gateway (S-GW) could be through wired or wireless links (http://scf.io/). On the other hand, Cloud RAN assumes a high bandwidth fibre link between RRH and the centralized data centre, which is also the most prohibitively expensive aspect of this proposal.

SoftRAN [3] observes that it is cost-effective to leave data plane functionality to the base stations along with some part of control plane for delay-sensitive decisions, but stressed on the coordination of closely-deployed BS in a dense network though a centralized virtual big base station. The idea of big BS or controller is very close to the signaling node in the BCG2 architecture and motivates us to explore the integration of both approaches. The authors of SoftRAN also presented a decomposition of protocols for data plane realization in cellular network in an earlier publication [7].

2.3 The Phantom Cell Concept

The phantom cell concept is presented by DOCOMO [4], where the small cells, called phantom cells as they are data-only BSs just containing LTE user plane, are overlaid with a normal macro cell which provides interference management and hosts both LTE user and control planes. Unlike Cloud RAN, phantom cell may perform baseband processing and does not need high capacity link between phantom cell and macro cell. The phantom cell uses new carrier type without cell-specific signal and a special BS discovery procedure managed by macro cell. Since the channel between UE (User Equipment) and potential phantom BS is unknown, the macro BS stores mean SNR (Signal to Noise Ratio) map for each Phantom BS to select the best one for a UE at a specific location [11]. In addition to training phases and excessive memory requirements, this method assumes constant transmission power and over-averaging of SNR values from UE which in fact can only measure SINR (Signal to Interference and Noise Ratio) instead. Moreover, a macro cell consumes over 100 times more power than a pico/femto cell [2] and keeping it on all the time would lead to severe power inefficiency.

A summary of all the approaches is provided in the Table 1 along with the respective design objectives, key themes and important related issues. The last row introduces our integrated approach which combines the benefits of SD-RAN and BCG2.

Table 1 Summary of architectural approaches for future mobile networks

Design concept	Objective	Key ideas	Comments
BCG2	Energy efficiency	Low-rate, low-power always on signaling nodes for coverage and on-demand data nodes	Major transition from current architecture
SD-RAN	Flexible innovation	Programmable control plane and hard-ware based data plane	Major paradigm shift and concepts are still evolving
Phantom cell	Interference management in dense deployments	Phantom/small/data cells and on-demand with an always-on macro cell for coverage	Macro cell consumes 100 times more power than small cells, a major cause of power inefficiency
Cloud RAN	Easier management and coordination	Centralized baseband processing with remote radio heads as nodes	High capacity fronthaul is required
Our approach	Energy efficiency, interference management and flexible innovation	Low-power, low rate signaling controller for coverage in a macro cell area and hosting control plane functionalities with on-demand data cells	Combined approach of SD-RAN and BCG2 which is more energy efficient than phantom cell concept

3 The Integrated Architecture

The evolution of contemporary LTE into the new integrated architecture is shown in Fig. 1. As depicted in [6], we also believe that the control plane of SDN based core network should contain some functional capabilities currently residing in S-GW (Serving Gateway) and P-GW (Packet Gateway), e.g., modification of routing rules, etc., along with the dedicated control nodes of EPC (Evolved Packet Core) shown with green dashed lines in Fig. 1. The functional decomposition of S-GW and P-GW is shown in Fig. 1 by means of control elements containing necessary APIs to EPC control nodes and also control logic required for data forwarding. These programmable control elements for routing modification etc., can be housed in control nodes in future architectures with specific APIs only in S-GW and P-GW. The interfaces between nodes in the core network, as shown in Fig. 1, are the same as EPC; but in a programmable SDN domain, they most probably be realized as software APIs. The data forwarding pipe is shown with a solid blue line passing through both gateways into the internet.

The new radio access system includes signaling controller, shown with big towers in Fig. 1, for a larger geographical area containing several data BSs shown with smaller towers in the figure. The eNodeB functionalities will be split between signaling controller and data BSs. The data BS can only be used on-demand and if there

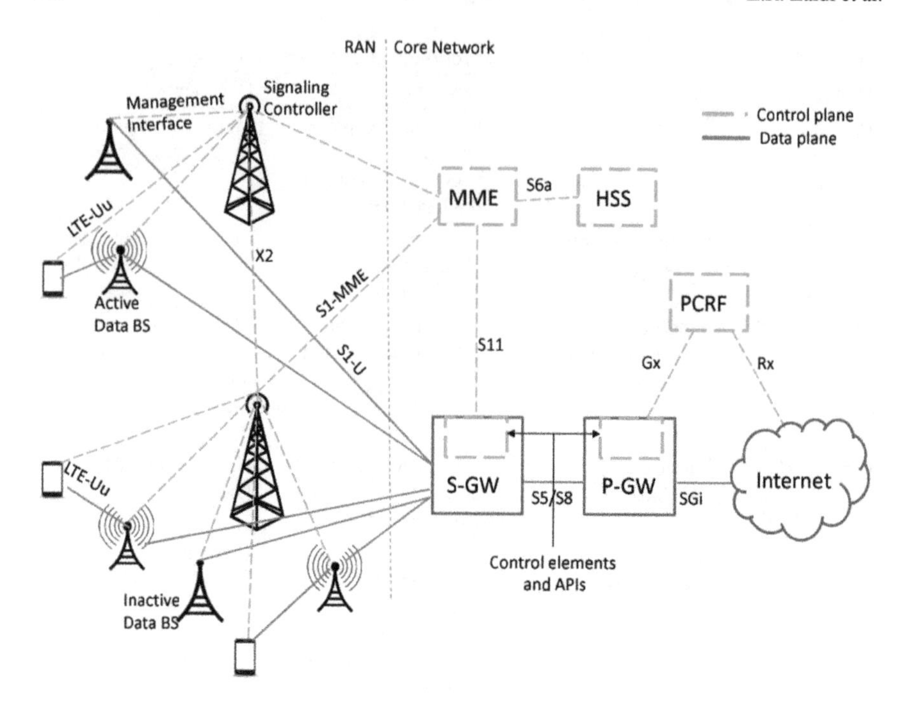

Fig. 1 The new integrated architecture

is no active session, the data BS will be put in sleep mode as also shown in Fig. 1. The control plane interfaces are shown with green dashed lines and the data paths are shown with blue solid lines in Fig. 1. We propose that only signaling controller is required to have an interface with MME (Mobility Management Entity), performing similar functions as S1-MME of EPC. The data BS requires connections to S-GW for data forwarding with the interface similar to S1-U of EPC.

Moreover, when mobile user is crossing the signaling controller coverage boundaries, a coordinated handover can be realized through communication between two signaling controllers. The interface between two signaling controller is labeled as X2 in Fig. 1, as it requires almost similar specifications as the interface X2 between eNodeBs in LTE but excluding functions related to data transmission and tunneling. Within the coverage area of a signaling controller, the handoff between data BS will be managed by the single associated signaling controller, most probably as part of the resource management function. The UE (User Equipment), also requires interfaces to both signaling controller for system access and data BS for data transmissions. The interfaces are similar to LTE-Uu interface which is defined to both control and data traffic in LTE. The only new interface in our architecture is the connection between signaling controller and the data BS, denoted as management interface in Fig. 1. This interface is similar to X3 interface in phantom cell concept [4] though implemented through software APIs. The management interface will be used for resource allocation by signaling controller and periodic updates from data BS for interference management. This connection can also be realized through dedicated wireless links or by any other wired technology if available.

Table 2 Combined requirements for signaling controller

	SDN
1	Hosts control logic for RAN, i.e., interference management, resource management, coordinated handover, etc.
2	Programmable on general purpose processors
3	Provides APIs to BSs and also to core network
	BCG2
1	Provides signaling for system access and paging
2	Has interfaces to UE, data BSs, and core network
3	Performs resource allocation
4	Consumes low power

3.1 Signaling Controller

The major component of our new integrated architecture is the signaling controller which performs the functions of a signaling node in BCG2 architecture and also of the centralized controller for SDN based radio access network. The combined specifications for the signaling controller conforming to the SDN principles and BCG2 architecture are given in Table 2. Since control plane implementation will be done using general purpose processors, the power consumption of a signaling node is not expected to substantially increase because of additional controller functionalities.

3.1.1 Resource Management and Assignment

In the BCG2 architecture, the possibility of sub-optimal channel allocation is one of the major concerns [2], for two reasons: (1) the channel between data BS and the mobile terminal cannot be estimated before selection as done in the classical approach and more sophisticated mechanisms are required to predict and estimate the channel condition between the mobile terminal and any potential data BS, such as, location of the user, etc. (2) It is less costly to allocate an in-coming request to an already active BS rather than waking up an inactive one; although it might not have the best possible channel to the user. However, the signaling controller can, not only null the ill effects described above but can actually optimize system capacity and energy efficiency trade-off with efficient resource management and interference mitigation using periodic updates from data BSs under its coverage and re-allocation of resources if it improves the capacity/energy trade-off.

In order to predict channel between user and inactive BS, phantom cell concept proposes to save SNR (Signal-to-Noise Ratio) map with each phantom cell [11]. In addition to the requirement of an initial training phase and excessive memory requirements, this method assumes constant transmission power and over-averaging of SNR values from UE which in fact can only measure SINR (Signal to Interference

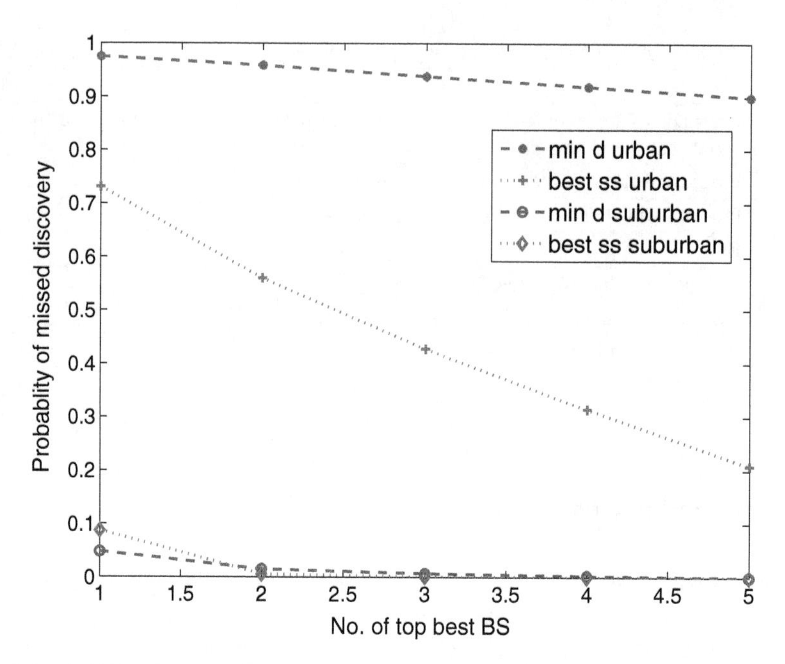

Fig. 2 Probability of missed discovery of best BS with minimum distance (d) and best signal strength (ss) methods for urban and suburban scenarios

and Noise Ratio) instead. We evaluated the performance of two methods of BS selection, i.e., 1—with best signal strength and 2—with minimum distance to the user, using real drive-test data from cellular networks in urban and suburban scenarios. We collected around 1000 observations with GPS location and signal strengths of nearby BSs. Although, the BSs were not small BSs but the data still provides insight into the performance of both methods. We remark that in dense deployments, it suffices if we select a BS among the top n best BSs by any possible method. We calculated mean signal strength over a moving window of 100 observations to find the most consistent or best BS. The results in terms of probability of missed detection are given in Fig. 2. We plotted the probability against the acceptable number of best BSs. i.e., $n = 1, \ldots, 5$. The results in Fig. 2 shows that both method are good and comparable for suburban settings, although, for urban scenario, both has large errors specially minimum distance selection has unacceptable performance. With smaller distances between users and BSs in dense deployments and lesser chances of obstacles in between, minimum distance discovery is expected to perform better than the performance shown in Fig. 2. BS discovery while in sleep mode is still an open problem and an important part of our on-going work.

Resource management is a multi-objective optimization problem which maximizes capacity $r_i(\eta) = \log(1 + \text{SINR}(\eta))$, where $\text{SINR}(\eta)$ is signal to interference and noise ratio, and minimizes power consumption $p_i(\eta)$ for all i users. If transmission power is kept constant, the $\text{SINR}(\eta)$ would depend on the specific resource block allocation to closely spaced users. The optimization function is

$$\min_{\eta} \mathrm{ECI}(\eta) = \min_{\eta} \left\{ \frac{\sum_i p_i(\eta)}{\sum_i r_i(\eta)} \right\}$$

where $\mathrm{ECI}(\eta)$ is the commonly used energy consumption index with units of W/bps or J/bit [2] and η is the parameter vector. The parameter space is a 3D resource grid containing the resource blocks, i.e., frequency carrier and time slot and data BSs. Under the new structure, the resource block mapping to the data BS does not need to be static and may be calculated by the resource management entity. Exploiting the ideas developed under self-organizing networks, the resource management parameter space can be extended to include BS variables, such as, transmission power, antenna tilt, etc.

A very simplified case study is presented in Fig. 3 to explain our idea of optimal resource management or re-allocation. Here one data BS is assigned 2 voice calls (64 Kbps each) and 1 video streaming session (384 Kbps). The streaming video session requires approximately 6 times more resource blocks than the voice calls. The relative distances of each user are also given with the serving data BS for both scenarios in Fig. 3. If we ignore all other effects and assume that transmit power for each resource block should compensate the respective path loss $L \propto d^{\gamma}$ (urban path loss exponent γ is 8), the required total transmit power needed for all resource blocks in scenario 1 will be approximately 4.5×10^3 times more than the power consumption for scenario 2 for the same required data rate and it saves a lot of energy if the systems moves from scenario 1 to 2. Undoubtedly, the above described calculations are based on a generic model; in a more realistic scenario issues regarding the actual deployment scenarios of the macro BS and more importantly allocation of the resources for the sporadic, in its nature, control based traffic need to be considered in a more detailed manner.

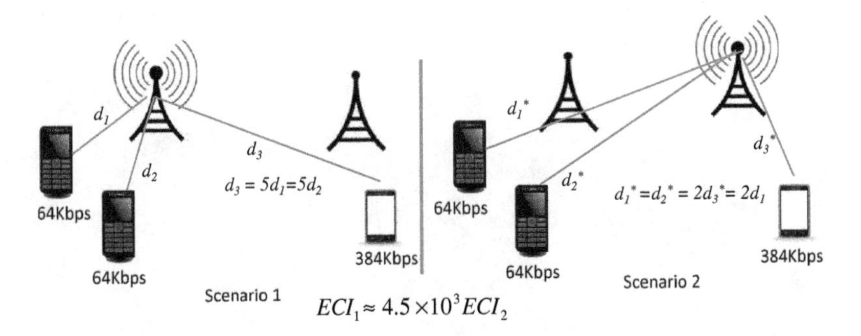

Fig. 3 An example of optimal resource management

3.1.2 Mobility Management

There are also some important implications in terms of managing mobility which implicitly also relate to energy consumption. The two salient features of IP based mobility management protocols, such as, fast handovers for MIPv6 [8], hierarchical MIPv6, and proxy MIPv6, are: 1—they are centralized and based on a mobility anchor located within the network domain and 2—these mobility protocols do not separate control and data planes. The former issue propelled discussions towards the Distributed Mobility Management (DMM) [9] solutions where the anchoring is done at the Access Point (AP) or BS; hence these functionalities are distributed at the points of attachment of the mobile nodes. Decentralization would alleviate network bottlenecks enabling better routing decisions at the same time. Nevertheless, clear benefits of such decentralization would emerge especially for the cases where the session duration time is significantly less than the cell residence time. The performance for high mobility users and/or delay sensitive flows under a DMM framework is very much topology dependant since the flows have to be tunneled form the old AP to the new AP under DMM. Regarding the second salient feature, in order to be compliant with an SDN-like separation of data and control planes and in the case of DMM, the APs will act as forwarding plane anchors and mobility related control functionalities will be logically centralized. In the proposed architecture these planes are *physically* decoupled, as shown in Fig. 4, and therefore localized mobility control will be required to take place via the signaling controller node. The signaling controller node will acquire all the control functionalities of a Mobile Access Gateway (MAG) and with control exchanges with the Local Mobility Anchor (LMA) will encapsulate the IPv6 address of allocated data BS for the mobile node. As shown in Fig. 4, the signaling node will act as an MAG for the mobile node which in essence means it has to provide a proxy Care of Address (pCoA) from the pool of available resources of the data BS in which the UE will be connected to (both in the case of new calls and handover calls).

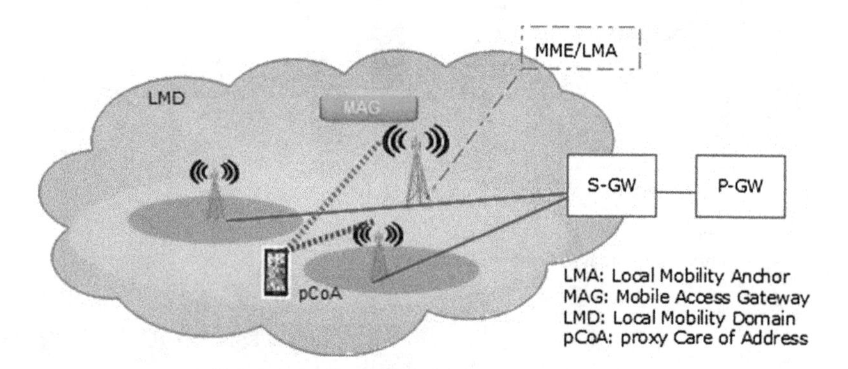

Fig. 4 Mobility management in the new architecture

3.1.3 Inter Macro Base Station Handover and Load Management

In the proposed, as well as, the nominal C/U wireless split architectures user handover management between small cells that are associated (i.e., controlled) by different macro controllers entail a significant overhead in terms of handover signaling. Signaling plane functionalities of the macro BS include, inter alia, the control plane interface with the Evolved Packet Core (EPC) entities such as the Mobility Management Entity (MME) via the 3GPP standardized S1-MME interface. The C-plane also includes all LTE signaling and control related functionalities. Such as for example, the radio resource control (RRC—establishment, modification, and release of mobile users Radio Resource Control (RRC) layers) network controlled mobility and functionalities related to measurement, configuration and reporting.

For a handover between small cells that are managed by different controllers both macro base stations will have to be involved in the handover process. This is shown in Fig. 5 below depicting the case of a handover between two pico-cells that are controlled by different macro-BSs. In addition to the overhead that this entail, an inter macro base station handover might result in a potential poor handover performance since, as mentioned above, macro controllers have limited information about the channel conditions between the mobile user and the candidate small cell.

In order to ease the above two issues dynamic algorithms for small cell to macro cell association would need to be implemented when deploying C/U based split wireless networks with a significant high density of small cells. Also, and in addition to the above mentioned system level aspects that need to be carefully taken into account it has to be reiterated that in terms of handover management C/U split architectures pose different set of problems compared to traditional cellular networks

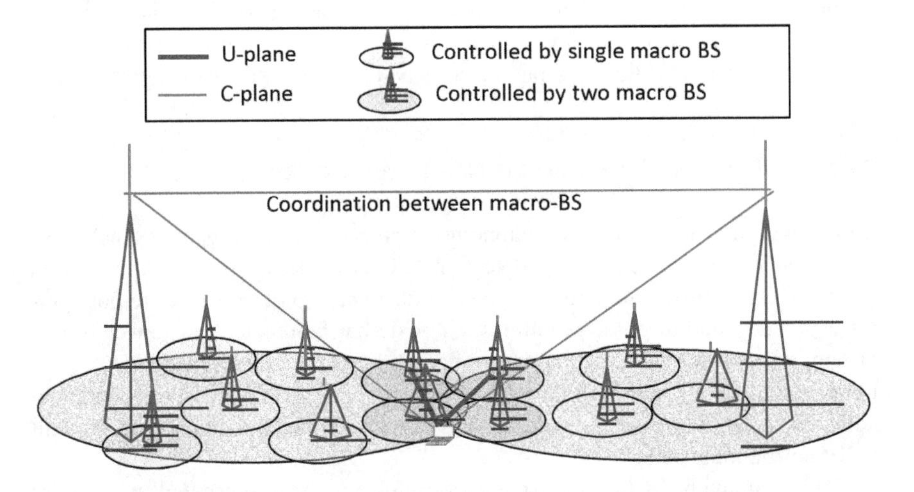

Fig. 5 Inter macro-BS handovers and load balancing in C/U split wireless architectures

since coverage of small cells can be non-continuous and scattered within the coverage area of a macro-controller.

The issue of handover management in such architectures influence some other mechanisms in the network such as for example load balancing. In current cellular networks macro cell load balancing is considered through various versions of the 'cell breathing' principle in which a base station adjusts the coverage area by changing its transmission power based on the load requirements. Another, in some sense complementary, technique for load balancing is the so-called inter-cell steering. The aim of steering is to decrease the coverage area of a cell which operates at high utilization levels by increasing at the same time the coverage of adjacent cell with relative low load profile. By doing this, load distribution could be adjusted between neighboring cells. However, such load balancing actions should not be considered in isolation since they might cause increased levels of handovers among the macro cells. For example in the case shown in Fig. 5 above depending on the number of small cell handovers that can be controlled from different macro-BSs a trade-off should be considered between balancing the scarce available resources of the macro-BSs and the reduction of quality for handovers between small cells managed by different macro-BS. Hence, inter macro cell handovers and load balancing are a pair of issues; they strongly influence each other and should be considered jointly.

3.2 *Data Base Stations*

Data BS belongs to the data forwarding plane. Under the new structure, they require to perform baseband processing and other transmission tasks as defined by the control plane. They will only be used on-demand as they are the most energy consuming component of the system. If not in use, they should be put in sleep mode to save energy. The cost of activation and de-activation should also be considered along with the activation delay while assessing the costs and benefits of the new architecture.

3.2.1 Cell Outage Management in New Integrated Architecture

Cell outage management is an autonomous process in self-organizing-networks (SON) which entails cell outage detection (COD) and cell outage compensation (COC). COD aims to autonomously detect outage cells, i.e. cells that are not operating properly due to possible failures, e.g. external failure such as power supply or network connectivity, or even misconfiguration. On the other hand, COC refers to the automatic mitigation of the degradation effect of the outage by appropriately adjusting suitable radio parameters, such as the pilot power, antenna tilt and azimuth of the surrounding cells.

Though individual COC and COD algorithms have been presented in literature for conventional deployments, a complete COM framework is still missing for the new integrated architecture [12–16]. The main difference in the COM framework of conventional heterogeneous network (HetNet) and the new integrated approach is in their architecture. In the new integrated architecture, active high data rate UEs are served by both the data and control BS, while the low rate UEs are served by only the control BS. This implies that all UEs maintain connectivity with the control BS. Furthermore, as a result of the split of the control and data planes, the control and data cell outages are independent of each other; hence, the detection of a cell outage in each plane is executed independently of the other. In the conventional HetNet, control and data functionalities are provided to the UE by the same node, whereas these functionalities are provided by separate nodes in the new integrated architecture. Hence, in the conventional architecture outage to a node can be compensated by any other node, whereas in the new integrated architecture, an outage to a node can only be compensated by another node that provides the same functionalities.

Hence two distinct COD algorithms have to be developed to cope with the peculiarities of data and control cells [17]. Since control cells tend to have a large number of UEs, machine learning based anomaly detection schemes can be applied for control COD. Control COD thus can be implemented at the operation and maintenance center (OMC) level. However, the same COD scheme cannot be applied for data cells, as the number of users will not be large enough to constitute reliable training models for underlying anomaly detection techniques. To overcome this problem, we take advantage of the following peculiarities, about data cells, to develop a heuristic, yet reliable data COD algorithm. The radio resource control (RRC) layers of all UEs are handled by the control cells, as a result, the control base station (BS) is aware of: (1) every UE-data BS association within its coverage, (2) the state of each UE (idle or active), (3) every radio link failure between the UEs and data BSs, (4) every handovers to other data BSs in its coverage and (5) data link handover from the data BS to itself. Also, once the normal state of the control cells has been established, each UE associated to the data cells can periodically report the reference signal received power (RSRP) statistic between itself and its associated data cell to its serving control cell. Hence, the control BS monitors the UE-data BS association and triggers the outage detection when it discovers irregularities in UE-data BS association. Irregularities in UE-data BS association occur when all UEs attached to a particular data BS changes their association without any of the following: (1) prior handover initiation process, (2) change in state of all the UEs, (3) radio link failure notification from all the UEs, (4) the data BS going into sleeping mode. Once the outage detection is triggered, the control BS can detect outage of the data BS by predicting the RSRP of all the UEs that were associated with it prior to the outage using a prediction model such as Grey Prediction Model [18].

Once the outage is properly detected, an online automatic COC scheme has to be implemented to continue serving the UEs in the outage area. A reinforcement learning (RL) algorithm which works by optimizing the coverage and capacity of the identified outage zone, by adjusting the gains of the antennas through electrical tilt and downlink transmission power of the surrounding BSs in that plane, is developed. The proposed COC algorithm can be applied independently in each plane, as different spectrum resources are allocated to each plane.

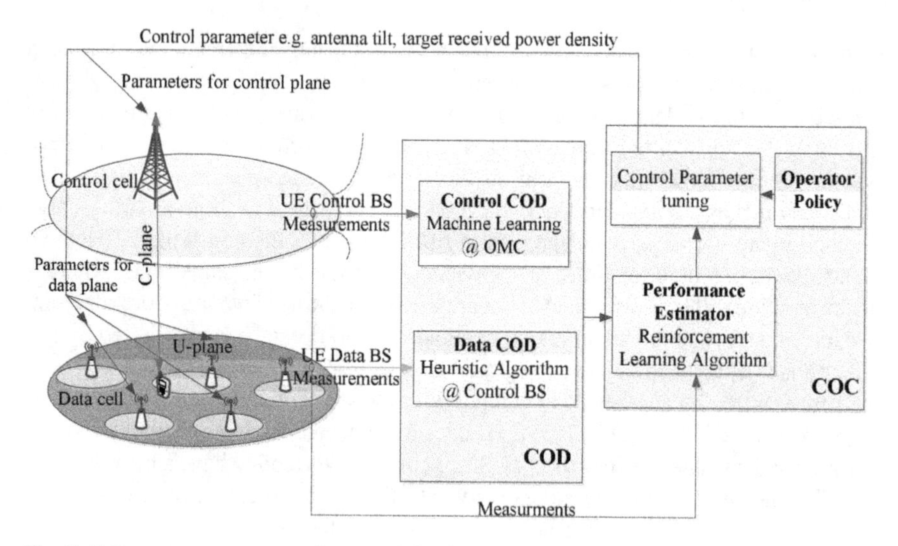

Fig. 6 Cell outage management framework for the new integrated architecture

Figure 6 shows our COM framework which has two distinct COD algorithms to cope with the idiosyncrasies of the control and data cell. After the control or data cell outage has been detected by the OMC or by the control BS, respectively, a RL based COC scheme is implemented independently in both planes. The outputs from the COC are the control parameters of the plane of interest, which are feedback to the plane and rightly applied to the neighboring cells to the outage cell.

4 Advantages

4.1 Energy Efficiency

The major objective of phantom cell concept is to provide interference coordination to dense deployments of small cells through a macro cell. Although, as the macro cell also takes care of the coverage, the small or phantom cell can be put in sleep mode if not in use. A macro cell consumes over 100 times more power than a pico/small cell as estimated in EU FP7 EARTH project [2] reducing the energy efficiency of phantom cell architecture. In Fig. 7, we show the mean daily consumption of phantom cell and our proposed architecture and compare it with the baseline scenario of all active nodes. It is clear that the reduction in phantom cell consumption only becomes prominent in very dense settings, whereas, our proposed architecture based on the decoupling of signaling and data improves energy efficiency greatly.

The simple simulation scenario consists of 1 macro cell and a number of homogeneous small cells. In our architecture, the macro BS is a signaling controller which consumes negligible energy [1]. We assumed uniformly distributed small BS and UEs in the macro cell radial range of 2 km. The total number of UEs in peak hour are assumed to be 10 times the number of BSs and proportional number of UEs are calculated for each 2 hour slot of the day using daily traffic profile introduced by EARTH [1, 2]. Each UE connects to the BS with lowest path-loss, i.e., minimum distance. The BSs are assumed to have sufficient capacity to serve the allocated UEs. The nodes without any associated user are assumed to be in sleep mode with zero energy consumption. A better resource allocation algorithm can be designed for both architectures to optimize the capacity/energy trade-off which should benefit both in the similar manner. Moreover, sleep mode also consumes small amount of energy [2] but it is ignored in this evaluation. The small cell BS consumption is normalized to 1 watt.

The average daily energy consumption of phantom cell and our architecture is shown in Fig. 7, where it is clear that the consumption of always-on macro cell in phantom cell concept offsets the energy savings of small cells in sleep mode. We also tried to estimate the average daily consumption when macro cell consumes 70 % more power than small cells. The results are labeled as All on-70 and Phantom cell-70, whereas the results when macro cell's consumption was 100 times more, as described in [2], are labeled as All on-100 and Phantom cell-100. In both scenarios, phantom cells become really beneficial in very dense deployments of small cells. In our on-going work, we are estimating the energy requirements of a signaling-controller for

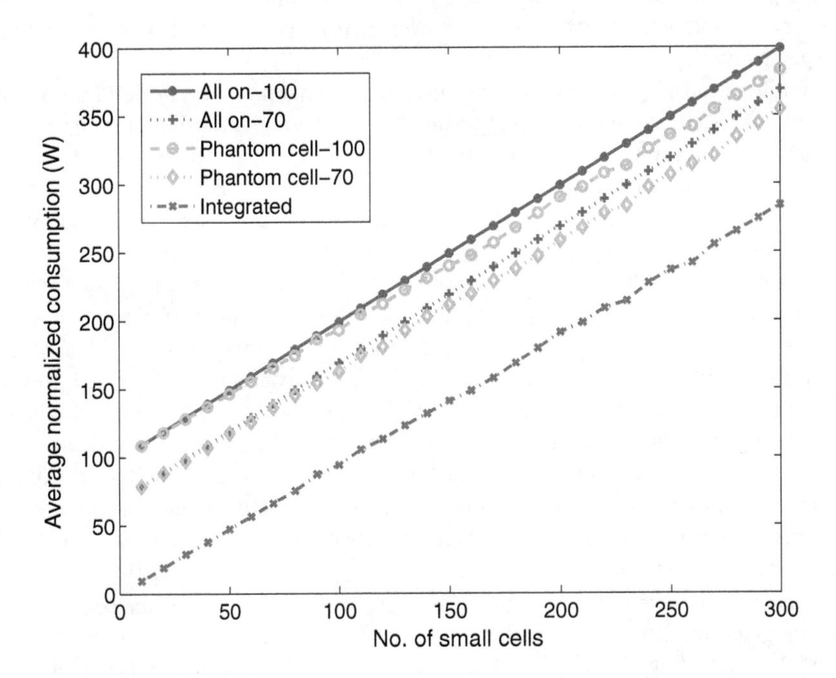

Fig. 7 Energy efficiency of the proposed architecture in comparison with all-nodes-on scenario and phantom cell concept

clear identification of the margin of benefit. In Fig. 7, our proposed architecture is slightly favored as the consumption of signaling controller is ignored, however, it will be much smaller than the consumption of a normal macro cell, still making our architecture the most energy efficient architecture for 5G which also comes with unprecedented ability of interference management to realize true benefits of dense deployments.

4.1.1 Power Consumption Comparison

A major figure of merit for BCG2 type architecture, while comparing with phantom cell concept, is the reduction in energy consumption of the system when a signaling controller is used instead of a normal macro base station to provide connectivity to a macro-cell size area. Assume P_T is the total power consumption of a base station. According to the EARTH power model [19],

$$P_T = N_{TRX} \frac{\frac{P_{out}}{\eta_{PA}(1-\sigma_{feed})} + P_{BB} + P_{RF}}{(1-\sigma_{DC})(1-\sigma_{MS})(1-\sigma_{cool})},$$

where N_{TRX} is the number of transceiver chains, P_{out} is the output power, P_{BB} and P_{RF} are the power consumption values for baseband and RF units respectively. η_{PA} is the power amplifier efficiency and, σ_{DC}, σ_{MS}, and σ_{cool} are the loss factors (loss factor = 1 − efficiency) for the DC-DC power supply, main supply, and active cooling respectively. σ_{feed} is the feeder loss.

While comparing the power consumption of signaling controller and a normal macro base station, we note that P_{BB} and P_{RF} are the major dissimilar components of the model. Any technological improvement in power amplifier, DC-DC, main supply, and cooling components can enhance both types of base stations in the same way. Moreover, we do not need MIMO antennas at a signaling controller due to very low capacity requirements, i.e., $N_{TRX} = 1$ for a signaling controller.

The modern baseband unit of cellular base stations comprises of multiple smaller sub-units according to the required capacity of the base station for better energy efficiency. Signaling traffic rate for a user will be in the neighborhood of 1–2 Mbps with bandwidth 1–5 MHz as opposed to 1.5–3 Gbps peak user rate of LTE-Advanced. The signaling controller will need order of magnitude less number of baseband processing units than a normal macro base station. If $P_{BB,m}$ and $P_{BB,s}$ denote the baseband power consumption values of a normal LTE-Advanced macro base station with $n_{BB,m}$ baseband processing units and that of a signaling controller with $n_{BB,s}$ baseband units respectively, then the excess baseband power consumption of a normal macro base station can be estimated as $\frac{P_{BB,m}}{P_{BB,s}} = \frac{n_{BB,m}}{n_{BB,s}} \leq (750 \cdots 3 \times 10^3)$.

Another estimate can be obtained through working out the power scaling formulae given in [20] for a signaling controller with 5 MHz, single antenna, QPSK modulation, coding rate 4/5, 100 % time-domain duty-cycling and 100 % frequency occupation. The reference scenario is of a macro base station with 20 MHz, single antenna, 64 QAM modulation, coding rate 1 and 100 % time-domain duty cycling and 100 % frequency occupancy. Estimating $\frac{P_{BB,m}}{P_{BB,s}}$ using the GOPS (Giga Operations Per Second) figures given in Table 2 of [2]

$$\frac{P_{BB,m}}{P_{BB,s}} = \frac{(850 + 0.1 \times 1060)\left(\frac{20}{20}\right)(2) + 10\left(\frac{20}{20}\right)2^2 + 200 \times 2}{(480 + 0.1 \times 1060)\left(\frac{5}{20}\right) + 380\left(\frac{5}{20}\right)\left(\frac{2}{6}\right)\left(\frac{4}{5}\right)} = \frac{2352}{171.83} = 13.69,$$

where the macro base station in comparison is supposed to be using peak user rate of 64-QAM modulation, coding rate 1, 2 × 2 MIMO antennas, bandwidth of 20 MHz, 100 % time-domain duty-cycling and 100 % frequency occupation [20]. Assuming 40 GOPS/W (for 65 nm CMOS technology [20]), the absolute baseband power consumption values are $P_{BB,m} = 58.8$ W and $P_{BB,s} = 4.3$ W.

According to [20], bandwidth, modulation, and coding rate has no effect on the RF power scaling, however, RF power consumption linearly increases with the number of antennas considering 100 % time-domain duty-cycling and 100 % frequency occupation. $\frac{P_{RF,m}}{P_{RF,s}} = 2$, if 2 × 2 MIMO antennas are used with normal macro base station and signaling controller is supposed to be using a single antenna. From [20], the estimated consumptions will be $P_{RF,m} = 11.4$ W and $P_{RF,s} = 5.7$ W.

Estimating $\frac{P_{T,m}}{P_{T,s}}$ as given in the above expression with the assumption of $P_{out} = 38.9$ W (46 dBm) and $\eta_{PA} = 31.1\,\%$ [19] for both normal macro-cell base station and signaling controller and ignoring feeder losses, we have $\frac{P_{T,m}}{P_{T,s}} = \frac{2(\frac{38.9}{0.311}+58.8+11.4)}{\frac{38.9}{0.311}+4.3+5.7} = 2.9$. The analysis concludes that a signaling controller can reduce the power consumption by approximately 200 % comparing with a normal macro-cell base station providing coverage for the similar size area. Also note, that the above analysis used power instead of energy consumption and if we consider the energy savings when signaling controller provides very short-term signaling to a larger macro-cell area and long-term data transmissions are handled by nearby small cells, the difference between BCG2 and current architecture would become even more distinctive.

4.2 *Virtualization and RAN Sharing*

Network function virtualization is one of the prime benefits of SDN where various virtual networks can use the same physical infrastructure. For example, the signaling controller can be shared by various operators. Each can control its radio resources with or without collaboration with other operators. RAN sharing has also been proposed and practiced in 3GPP systems. The most comprehensive framework is developed by NEC [10], which is based on minimum guaranteed reservation of resources for each operator.

In our on-going work, we are developing the idea of *opportunistic RAN sharing* with our proposed architecture. Opportunistic RAN sharing is a novel idea where infrastructure and radio resources are opportunistically shared among operators only if it improves the capacity/energy efficiency trade-off which can then be mapped into proportional gains for sharing operators. Since all access requests will be coming to the signaling controller, it is much easier to design radio resource sharing algorithm with optimal performance over a finer time scale. As an example, we consider night scenario where around 15 % of the traffic is expected compared to the peak load [1]. Two operators in an area, each individually serving 15 % of its subscribed users, can provide service to all the clients with better utilization of the BSs and resource blocks providing better capacity/energy trade-off. The gains can be shared among the operators in proportion to their costs.

4.3 Enabling Technologies

BCG2 architecture proposes a basic structural change in contemporary mobile communication system which requires re-designing of several components, such as, mobility management, resource allocation, etc., affecting the whole system which is the main reason to delay its implementation till the future standards. An SDN programmable control plane could be the enabling technology to realize cell on-demand architecture by facilitating development and implementation of appropriate modifications in protocols. On the other hand, signaling node as defined by the BCG2 architecture could be the best location to host the control plane for a particular geographical area of a radio access network.

The integration of SD-RAN and BCG2 architecture enables an energy efficient mobile communication system with optimal capacity. The proposed architecture provides clear demarcation of control and data plane specially in the radio access part enabling the realization of various novel aspects of SDN, such as, network virtualization and network function virtualization.

5 Conclusion

We present an integrated approach combining SDN based RAN and BCG2 architecture, by means of a *signaling controller* which provides always-on system access and contains control plane functionalities. The major takeaway message of this chapter is that both visions carry more potential than the benefits they are aiming for; and a union of both sets of requirements and specifications will result in an even richer and more commercially viable design of mobile communication systems. Our architecture is also very close to the phantom cell concept, although, our architecture has the potential to reduce energy consumption over multiple order of magnitude in comparison to phantom cells. Future avenues of research include a detailed functional view of the architecture where various components such as mobility and topology control can be envisioned within a Network Function Virtualization (NVF) paradigm in a cloud empowered RAN.

Acknowledgments We like to thank Australian Endeavour Research Fellowship for funding Zainab Zaidi's visit to King's College London, and for partial support by the FP7 CROSSFIRE project (FP7-PEOPLE-317126).

References

1. Capone, A. et al.: Rethinking cellular system architecture for breaking current energy efficiency limits. In: SustainIT 2012, pp. 1–5, 4–5 October 2012
2. Godor, I. (ed.): D3.3: Final report on green network technologies, INFSO-ICT-247733 EARTH, Technical Report, June 2012
3. Gudipati, A., Perry, A., Li, L.E., and Katti, S.: SoftRAN: software defined radio access network. In: Proceedigns of ACM SIGCOMM workshop HotSDN'13, 2013, pp. 25–30. ACM, New York, NY, USA (2013)
4. Ishii, H., Kishiyama, Y., Takahashi, H.: A novel architecture for LTE-B :C-plane/U-plane split and Phantom Cell concept. In: Globecom Workshop, 2012, pp. 624–630. IEEE, 3–7 Decmber 2012
5. Bhaumik, S., Chandrabose, S.P., Jataprolu, M.K., Kumar, G., Muralidhar, A., Polakos, P., Srinivasan, V., Woo, T.: Cloudiq: A framework for processing base stations in a data center. In: Proceedings of the 18th Annual International Conference on Mobile Computing and Networking, Mobicom'12, pp. 125–136. ACM, New York, NY, USA (2012)
6. Li, L., Mao, Z., Rexford, J.: Toward software-defined cellular networks. In: 2012 European Workshop on Software Defined Networking (EWSDN), pp. 7–12. October 2012
7. Bansal, M., Mehlman, J., Katti, S., Levis, P.: Openradio:a programmable wireless dataplane. In: HotSDN'12, pp. 109–114, ACM, New York, NY, USA (2012)
8. Koodli, R.: Fast handovers for mobile IPv6. In: RFC IETF RFC 4068, IETF Secretariat, July 2005
9. Chan, H., Liu, D., Seite, P., Yokota, H., Korhonen, J.: Requirements for distributed mobility management. In: Internet-Draft draft-ietf-dmmrequirements-04 (work in progress), IETF Secretariat, May 2013
10. Mahindra, R., Khojastepour, M.A., Zhang, H., Rangarajan, S.: Radio Access Network sharing in cellular networks. In: IEEE Network Protocols (ICNP) 2013, pp. 1–10. 7–10 October 2013
11. Ternon, E., Agyapong, P., Hu, L., Dekorsy, A.: Database-aided energy savings in next generation dual connectivity heterogeneous networks. In: IEEE WCNC 2014, Istanbul, Turkey. April 6–9
12. Rumney, M.: Taking 5G from vision to reality. In: IEEE ComSoc Tutorials. http://www.cambridgewireless.co.uk/Presentation/FWIC2014_MorayRumney.pdf (2014)
13. Mueller, C., Kaschub, M., Blankenhorn, C., Wanke, S.: A cell outage detection algorithm using neighbor cell list reports. In: International Workshop on Self-Organizing Systems, pp. 218–229 (2008)
14. Wang, W., Zhang, J., Zhang, Q.: Cooperative cell outage detection in self-organizing femtocell networks. In: IEEE INFOCOM 2013, pp. 782–790. April 2013
15. Amirijoo, M., Jorguseski, L., Litjens, R., Schmelz, L.-C.: Cell outage compensation in LTE networks: algorithms and performance assessment. In: IEEE VTC Spring, pp. 1–5. (2011)
16. Xue, W., Zhang, H., Li, Y., Liang, D., Peng, M.: Cell outage detection and compensation in two-tier heterogeneous networks, Int. J. Antennas Propag. (2014)
17. Onireti, O., Zoha, A., Moysen, J., Imran, A., Giupponi, L., Imran, M.A., Abu-Dayya, A.: A cell outage management framework for dense heterogeneous networks. In: IEEE Transactions on Vehicular Technology (2015), no. 99, May 2015
18. Onireti, O., Imran, A., Imran, M.A., Tafazolli, R.: Cell outage detection in heterogeneous networks with separated control and data plane. european wireless 2014, May 2014

19. Imran, M.A. (ed.): D2.3: Energy efficiency analysis of the reference systems, areas of improvements and target breakdown, INFSO-ICT-247733 EARTH (Energy Aware Radio and NeTwork TecHnologies), Technical Report, November 2010. https://www.ictearth.eu/publications/deliverables/deliverables.html
20. Desset, C., Debaillie, B., Giannini, V., Fehske, A., Auer, G., Holtkamp, H., Wajda, W., Sabella, D., Richter, F., Gonzalez, M.J., Klessig, H., Godor, I., Olsson, M., Imran, M.A., Ambrosy, A., Blume, O.: Flexible power modeling of LTE base stations. In: Wireless Communications and Networking Conference (WCNC) 2012, pp. 2858–2862. IEEE, 1–4 April 2012

Part III
Energy Management in Ad-hoc Networks

Cross-Layer Designs for Energy-Efficient Wireless Ad-hoc Networks

Auon Muhammad Akhtar and Xianbin Wang

Abstract Energy consumption is an important design criteria for wireless networks, not least because it directly impacts the cost of network operation and maintenance. Already, the information and communication technology (ICT) industry is being labeled as a substantial contributor to the total CO_2 emissions on the planet. Moreover, due to the slow improvement in battery technology, battery-operated wireless networks face a fundamental challenge, since there is an exponential increase in the gap between the demand for energy and the offered battery capacity. For these reasons, green ICT has become a critical issue world wide. This chapter adopts a cross-layer approach for enhancing the energy efficiency of wireless ad-hoc networks. Initially, the chapter discusses the importance of cross-layered designs for energy-efficient wireless networks. Following this, commonly used techniques for modeling energy consumption in wireless networks are outlined. Lastly, cross-layer designs based on cooperative physical layer network coding and hybrid automatic repeat request (HARQ) are presented. In this cross-layered design approach, energy-efficient transmission strategies are initially proposed for the physical (PHY) and medium access control (MAC) layers. Then, these optimized strategies are utilized as basic building blocks for energy-efficient routing at the network layer. All of the theoretical results are verified through computer simulations.

1 Introduction

Over the last decade, mobile data traffic has witnessed an explosive growth [1]. One of the important challenges resulting from this unprecedented growth is the increased energy consumption of the wireless networks. As a figure of merit, it was reported

A.M. Akhtar · X. Wang (✉)
Department of Electrical and Computer Engineering, Thompson Engineering Building,
The University of Western Ontario, London, ON N6A 5B9, Canada
e-mail: xianbin.wang@uwo.ca

A.M. Akhtar
e-mail: aakhtar8@uwo.ca

© Springer International Publishing Switzerland 2016

M.Z. Shakir et al. (eds.), *Energy Management in Wireless Cellular and Ad-hoc Networks*, Studies in Systems, Decision and Control 50,
DOI 10.1007/978-3-319-27568-0_7

in [2] that the data volume increases by a factor of 10 every five years, which corresponds roughly to an annual increase of around 16–20 % in energy consumption. This trend is set to continue, since recent forecasts suggest a further 10 fold increase in data traffic between 2014 and 2019 [1]. In such scenarios, it comes as no surprise that the ICT industry, in particular, the wireless industry, is being labeled as one of the major contributors to the total CO_2 emissions on the planet. To compound this problem further, battery operated wireless terminals face the challenge of diminishing battery life due to the exponential increase in the gap between the demand for energy consumption, due to media-rich smart devices, and the slow growth in battery technology. For these reasons, green communications and energy-efficient wireless networks have become a critical research issue across the globe [3].

Traditionally, energy consumption in wireless networks has been investigated in a layer wise manner, with particular emphasis on the physical (PHY) and medium access control (MAC) layers. At the PHY layer, research has focused on various design aspects, ranging from the modulation schemes to exploitation of the physical channel characteristics. For example, authors of [4] find the optimal modulation schemes which minimize the energy consumed in transferring a given amount of data between the source and the destination nodes. Similarly, the authors of [5] derive the optimal design parameters for M-ary Quadrature Amplitude Modulation (MQAM), M-ary Phase Shift Keying (MPSK) and non-coherent Multiple Frequency Shift Keying (MFSK), with the aim of minimizing energy consumption per information bit. Both the aforementioned works take the transmitted signal power as well as the circuit power consumption into account while designing the optimized solutions. On the other hand, the works in [6, 7] utilize distributed beamforming to achieve highly directional transmissions, resulting in significant energy saving gains compared to independent signal transmissions. More recently, the authors in [8] applied distributed beamforming in multi-source multi-destination clustered systems with the aim of minimizing transmit power while maintaining a minimum signal-to-interference-plus-noise-ratio (SINR) at the destinations. Other methods for energy-efficient PHY layers look into the multi-antenna transmissions, coding, power control, adaptive resource allocation, etc., [9].

The MAC layer plays a significant role in minimizing the overall energy consumption. One of the major energy consuming components in a wireless system is the radio itself, which is controlled by the MAC layer. Consequently, energy efficient MAC layer algorithms mostly focus on appropriately turning a node's radio on or off, depending on the given situation [10]. For example, the authors of [11] propose an energy-efficient MAC protocol for the battery operated wireless sensor networks, whereby, the sensor nodes save battery power by periodically going to sleep modes. On the other hand, the algorithm proposed in [12] minimizes the energy consumption of sensor nodes while meeting the user-specified delay constraint in low-duty-cycle sensor networks. In [13], the authors propose a distributed approach which extends the network lifetime by dynamically adapting various controllable parameters, including MAC duty cycling, while maintaining a minimum quality of service (QoS) within the network. In addition to the the aforementioned functionality, the MAC layer also decides the schedule for packet transmissions. To this end, the study in [14] proposes a scheduling scheme to optimize the energy consumption of a multiuser multi-access system such that the QoS constraints, in terms of packet loss rates, are fulfilled while the maximizing the advantages emerging from multiuser diversity.

While the aforementioned algorithms and designs showed a lot of potential for minimizing energy consumption, nevertheless, the fact they focused only on specific layers of the protocol stack made them sub-optimal when the overall system design was taken into consideration. Different layers can have conflicting requirements for minimizing energy consumption. For example, while minimizing transmission energy consumption at the PHY layer improves its performance in terms energy efficiency, however, this improvement comes at the cost of increased transmission attempts at the MAC layer, especially when the state of the channel between the communicating devices is unknown or only partially known. For such reasons, optimal energy consumption requires joint optimization across all layers. Consequently, recent works on energy minimization have focused mainly on cross-layer designs to improve energy efficiency [15, 16]. For example, the authors of [17] derive closed form expressions for optimal transmission energy and frame length such that the performance of the PHY and MAC layers is jointly optimized. On the author hand, the work in [18] jointly optimized the performance of PHY and network layers where distributed beamforming was utilized to minimize the energy consumption of the PHY layer while a routing algorithm for the network layer was proposed which worked in conjunction with the transmission strategy at PHY layer.

In this chapter, we present two cross-layered design algorithms which jointly optimize the performance of multiple layers. The first algorithm jointly optimizes the performance of the PHY and network layers. Cooperative physical layer network coding (CPLNC) is initially utilized as an energy-efficient transmission strategy at the PHY layer. Then, CPLNC is incorporated into energy-efficient routing at the network layer and it is shown how the network and PHY layers can work in tandem to improve the overall energy consumption of the system. The second algorithm focuses on PHY, MAC and network layers and utilizes cooperative automatic repeat request (ARQ) to improve energy efficiency. The energy consumption at the PHY and MAC layers is optimized by finding a balance between the number of transmission attempts at the MAC layer and the energy consumption per transmission attempt at the PHY layer. To achieve this goal, relay nodes are employed and cooperative link costs are derived. Then, routing algorithms are presented which utilize the derived cooperative link costs as basic building blocks. The chapter presents in depth analysis of the outlined methods and computer simulations are used to verify the performance gains.

The rest of the chapter is organized as follows: Sect. 2 discusses in detail the various design considerations that are taken into account when modeling energy consumption in wireless ad-hoc networks. Section 3 introduces a cross-layer design method based on physical layer network coding, which jointly optimizes the performance of the PHY and network layers. Theoretical and simulation results for the proposed approach are presented in the same section. Section 4 presents a cooperative ARQ based algorithm which jointly optimizes the performance of PHY, MAC and network layers. Finally, Sect. 5 summarizes the chapter.

2 Modeling Energy Consumption in Wireless Ad-hoc Networks

Energy consumption models, upon which the design of various energy efficient algorithms is based, form a key component of the research on energy-efficient communications. Generally, the energy consumption models consist of a variable component and a fixed component. The variable component of the model is based on the wireless channel conditions, e.g., distance-dependent pathloss, fading, shadowing, etc. It reflects the energy consumed by the power amplifier at the transmitter side since the radiated transmission power originates at the output of this power amplifier. Accordingly, the variable component is also referred to as the transmission energy consumption. On the other hand, the fixed component is used to model the energy consumed by the electronic circuitry of the transmitter and the receiver devices. This energy consumption results from the signal processing, coding/decoding, modulation/demodulation, etc., at the transceiver devices. With these design considerations, the general model for the energy consumed in transmitting a symbol between two devices a_u and z_v, separated by a distance x_{uv}, can be written as

$$E_{uv} = \beta x_{uv}^{\alpha} + 2E, \tag{1}$$

where α reflects the pathloss exponent and E represents the aforementioned radio electronics/hardware energy consumption at a transmitter or a receiver with the factor of 2 accounting for both of them. Finally, the scaling factor β is based on the physical parameters for a given scenario.

It is quite obvious from (1) that the scaling factor β is an important element of the overall energy consumption model. It can be used to capture the effects of a variety of physical parameters. For example, it can be used to model the impact of channel fading conditions. Specifically, for slow varying fading channels, where the channel coherence time is much larger than block coherence time and channel inversion based power control is used at the transmitter [19], β can be used for modeling the fading coefficients of the channel. On the other hand, β also incorporates the effect of the specific modulation in use by the transceivers, as done in [4, 17, 24]. To this end, the peak to average ratio of an M-ary modulation signal is an important design criterion which is accounted for within β. Other than the aforementioned physical parameters, some other important transmission parameters that are incorporated within β include the efficiency of the power amplifier, the targeted signal-to-noise-ratio (SNR) at the receiver, the effect of receiver noise power and the wavelength of transmissions, channel coding rate, etc. A simplified example for β can be obtained from the formulations presented in [18], where it is given as

$$\beta = \gamma_v P_\eta \frac{1}{\mu} \left[\left(\frac{4\pi}{\lambda} \right)^2 x_o^{2-\alpha} \right]. \tag{2}$$

In the above equation, γ_v and P_η account for the target SNR and receiver noise power, respectively. The term μ represents the power amplifier efficiency, and $1/\mu$ implies that the overall energy consumption increases when μ is small while it decreases for larger values for μ. Lastly, λ represents the wavelength while x_o represents the reference distance for antenna far field. The term inside the large brackets represents the constant terms from the free-space pathloss model utilized in [18]. Note that the free space pathloss formulation can be easily replaced by more accurate models, depending on the specific scenario under consideration, but the general structure of (1) still remains the same. Next, as far as the hardware energy consumption is concerned, it is modeled by a constant value, which is obtained through practical measurements. The measured values differ based on the chip design and manufacturer-specific components installed on the communications devices.

At this stage, it is imperative to mention that Eq. (1) can be manipulated based on the unit of the size of the signal being considered in the specific design. For example, it can be used in its current form if the unit being considered is a symbol. On the other hand, by multiplying the same equation with an appropriate scaling factor, it can be used to represent the energy consumption per information bit. To this end, two use cases of the above equation under different design units are presented in Sects. 3 and 4 of this chapter. More specifically, Sect. 3 utilizes (1) in its current form to represent the energy consumption per transmitted symbol. The whole design presented in Sect. 3 is based on this assumption. On the other hand, Sect. 4 manipulates the same equation to model average energy consumption per information bit by multiplying the equation by a factor of $(N + N_c)/N$, where N and N_c represent the number of information and overhead bits, respectively.

As a concluding remark for this section, it must be pointed out that the discussions made thus far in this section pertain to single hope point to point transmissions. However, wireless ad-hoc networks utilize multiple modes for transmissions and communications, for example, broadcast, multicast, multipoint to point transmissions, multihop communications, etc. For broadcast and multicast transmissions, the formulation in (1) remains the same as far as transmission energy is concerned. However, for the hardware part of E_{uv} above, the constant E is multiplied by $N_r + 1$, where N_r is the number of receivers while the additional 1 accounts for the transmitter. On the other hand, for multipoint to point transmissions, the transmission energies of the involved transmitters are simply added together while the constant term E is multiplied by $N_t + 1$, where N_t is the number of transmitting devices while the additional 1 accounts for the single receiver. Lastly, for multihop communications, the energy consumption of all the hops are added together to get the overall energy consumption of the multiphop path.

3 Cross-Layer Design Using Physical Layer Network Coding

As mentioned previously, cross-layered designs for energy-efficient communications are critical to optimizing the overall energy-efficiency performance of the wireless ad-hoc networks. To this end, the current section introduces a cross-layered design for energy efficient communications in wireless ad-hoc networks. The presented approach jointly optimizes the performance of the PHY and network layers. To achieve this goal, CPLNC is presented as an energy-efficient transmission strategy at the PHY layer while CPLNC-based energy-efficient routing algorithms are proposed to optimize the energy-efficiency performance of the network layer. CPLNC works for slow varying channels, where the channel coherence time is much larger than the block transmission time such that the channel gain and the phase delay can be estimated by the receiver and sent back to the transmitter. This enables the transmitters to implement channel inversion based power control [19].

3.1 *Cooperative Physical Layer Network Coding*

Consider a scenario of three randomly placed nodes S, R and D, as shown in Fig. 1. Source S transmits a packet of successive symbols, drawn independently from a Binary Phase Shift Keying (BPSK) constellation, to D. During the transmission of the first symbol, depicted as m_1 in Fig. 1, R overhears and decodes the transmitted data. During the subsequent time slots, S and R get synchronized and while S is transmitting the remaining sequence of symbols, i.e., m_k, $k \geq 2$, with transmitting power P_s, R repeatedly transmits the first symbol m_1 with transmitting power P_r. Here, P_s and P_r are the reduced transmission powers to be derived later in this section. Synchronous transmission of S and R to D is such that when they transmit the same symbol, the received sum power at the destination satisfies the minimum SNR threshold of γ_{th}; whereas, when they transmit different symbols, the received

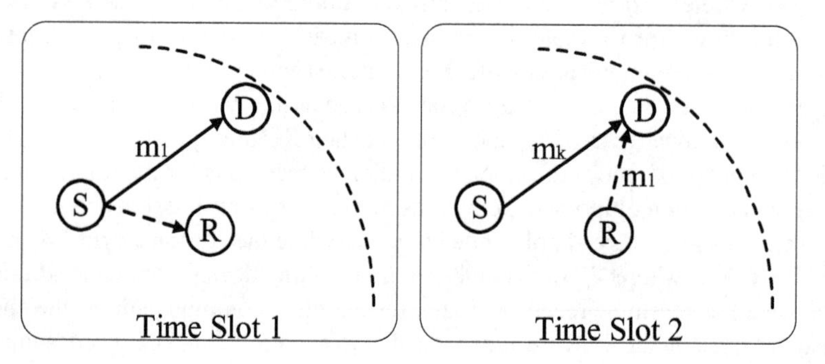

Fig. 1 A three node network and wireless broadcast advantage

signals cancel each other out. This implies that equal power must be received from the two transmitting nodes. Since D has already decoded one of the transmitted symbols, namely m_1, it can decode the remaining symbols, i.e., m_k, $k \geq 2$, from the combined received signal. When the same symbols are transmitted, the receiver SNR can be given as $\gamma_{\text{th}} = \frac{|\sqrt{P_s}h_{sd}+\sqrt{P_r}h_{rd}|^2}{P_\eta}$, where h_{sd} and h_{rd} represent the channel coefficients between $S - D$ and $R - D$, respectively, P_η is the receiver noise power while P_s and P_r represent the transmission power of S and R, respectively. Since the received power due to transmission from each transmitter node is the same, i.e., $\sqrt{P_s}h_{sd} = \sqrt{P_r}h_{rd}$, the aforementioned equation for receiver SNR can be written as $\gamma_{\text{th}} = \frac{|2\sqrt{P_s}h_{sd}|^2}{P_\eta}$, or $P_s = \frac{\gamma_{\text{th}}P_\eta}{4|h_{sd}|^2}$. Similarly, $P_r = \frac{\gamma_{\text{th}}P_n}{4|h_{rd}|^2}$. Thus, using CPLNC, the general form for power transmitted from node a_u to receiver r_v is given as

$$P_u = \frac{\gamma_{\text{th}}P_\eta}{4|h_{uv}|^2} \tag{3}$$

From the derived values for P_s and P_r, one can already see that each of the transmitter nodes only needs to transmit at 1/4 of its normal transmission power to achieve the required SNR at the destination. After down conversion at the destination, the received signal can be written as

$$\Omega'_d(t) = \frac{1}{2}\sqrt{\gamma_{\text{th}}P_\eta}(b_s + b_r) + \eta_d(t), \tag{4}$$

where b_s and b_r are the BPSK modulated bits transmitted by S and R, respectively, while η_d is the receiver noise. From (4), by focusing on the noiseless part of the received signal, it can easily be seen that the baseband equivalent signal is nonzero when the same symbols are transmitted and it is zero when the transmitted symbols are different. Thus, it becomes trivial to derive Table 1, which in effect, is the XNOR logic function. In Table 1, I_d represents the demodulator output at D.

Table 1 Modulation/demodulation mapping

Mapping at transmitter				Mapping at receiver	
Generated bits		Modulated bits		Received baseband	Output bit
S	R	S	R	D	D
I_s	I_r	b_s	b_r	Ω'_d	I_d
0	0	−1	−1	$-\sqrt{\gamma_{\text{th}}P_n}$	1
0	1	−1	1	0	0
1	0	1	−1	0	0
1	1	1	1	$\sqrt{\gamma_{\text{th}}P_n}$	1

Next, it is imperative to quantify the gains that can be achieved by using CPLNC. To this end, without loss of generality, it is assumed that the symbol period for each symbol is normalized to one second.

Definition 1 The energy saving gain of the cooperative transmission over a non-cooperative transmission, denoted by G, is defined as

$$G = \frac{\mathscr{E}_{nc} - \mathscr{E}_c}{\mathscr{E}_{nc}}, \tag{5}$$

where $\mathscr{E}_{nc}$ is the energy consumed by S in non-cooperatively transmitting the data towards D while $\mathscr{E}_c$ is the energy consumption if CPLNC is used to transmit the same data.

Assuming that the source S needs to transmit ℓ symbols towards the destination and by using the above definition, the energy saving gains that can be achieved by using CPLNC, as compared to non-cooperative point to point transmissions in a non-fading additive white Gaussian noise (AWGN) channel, can be written as [20]

$$G = \frac{3(\ell - 1)}{4\ell}\left(1 - \frac{d_{rd}^{\alpha}}{3d_{sd}^{\alpha}}\right), \tag{6}$$

where d_{sd} and d_{rd} are the distances between $S - D$ and $R - D$, respectively, while α is the pathloss exponent. Figures 2 and 3 present a three-dimensional view of the achievable energy saving gains. For Fig. 2, the distances between $S - D$ and $R - D$ were fixed at 100 m while the number of transmitted symbols was varied between 2 and 100. On the other hand, for Fig. 3, the number of transmitted symbols was fixed at 10, the distance between $S - D$ was fixed at 100 m while the distance between R and D was varied from 10 to 100 m. In both cases, it can be seen that gains of around 70 % are achievable. At this stage, it is worth mentioning that the above energy savings hold in the high SNR regime only since the performance changes when the bit error rate is taken into account [20].

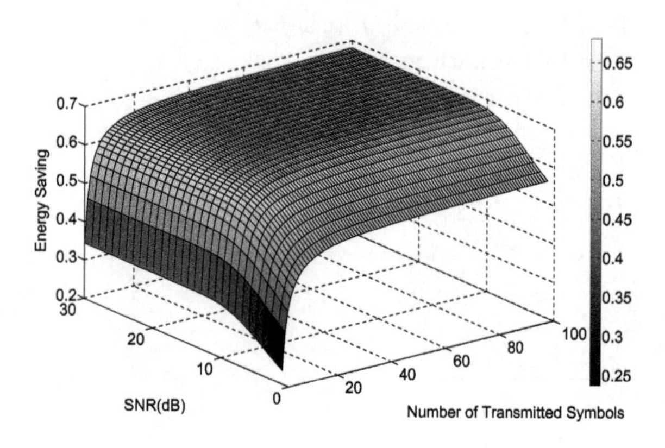

Fig. 2 Energy saving gain as a function of the received SNR and the number of transmitted symbols

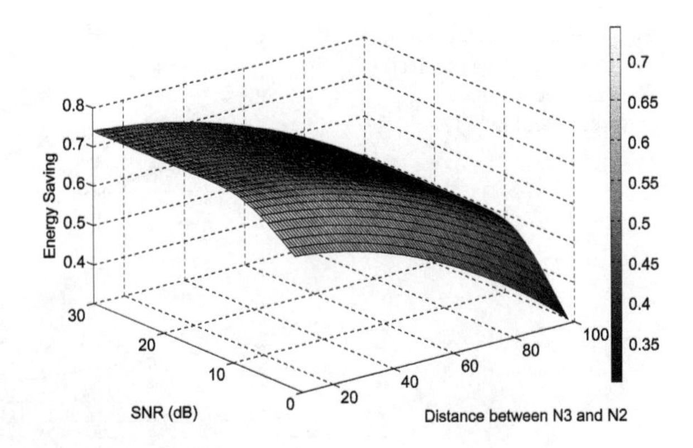

Fig. 3 Energy saving gain as a function of the received SNR and the distance between R and D

The performance of CPLNC in Rayleigh fading channels has also been analyzed in [21]. To this end, it is important to note that, as shown in (3), the minimum power required to satisfy the SNR requirement at the destination is inversely proportional to the channel power gain. However, for channel inversion with zero outage in a Rayleigh fading channel, infinite average power is required [19]. In order to transmit with a finite average power, a node is restricted to transmit only when the channel power gain is larger than a minimum threshold. Otherwise, an outage is declared. Let δ indicate the probability of this outage. With this restriction, CPLNC can be used only if R successfully overhears the first symbol and the channel gain between R and D is greater than a minimum threshold. Let ρ_s represent the probability that CPLNC is used. With these assumptions, the energy saving gain in a Rayleigh fading channel can be approximated as [21]

$$G \approx \frac{3(\ell-1)}{4\ell}\left(\frac{(1-\delta)^{2+\frac{d_{sr}^{\alpha}}{d_{sd}^{\alpha}}}}{1+\frac{d_{sd}^{\alpha}+d_{sr}^{\alpha}}{d_{sd}^{\alpha}}\delta}\frac{\Gamma}{\Theta}-\frac{d_{rd}^{\alpha}}{3d_{sd}^{\alpha}}\frac{\rho_s}{1+\delta}\right). \tag{7}$$

In the above equation, d_{sr} is the distance between the source and the relay, $\Gamma = \mathrm{Ei}\left(\frac{-(d_{sd}^{\alpha}+d_{sr}^{\alpha})\delta}{d_{sd}^{\alpha}+(d_{sd}^{\alpha}+d_{sr}^{\alpha})\delta}\right)$ and $\Theta = \mathrm{Ei}(-\delta)$, where Ei is the exponential function. Figure 4 plots the energy saving gains that are achieved by using CPLNC in a fading channel. The figure was obtained by implementing the scenario shown in Fig. 1. Similar to Fig. 3, the distance d_{sd} was fixed at 100 m. On the other hand, the distances d_{sr} and d_{rd} were set equal to each other and both were varied from 50 to 100 m. It can be seen that the Monte-Carlo simulations match nicely with the theoretical values approximated by (7). The figure shows that depending on the number of transmitted symbols and the inter-node distances, the achievable gains range between 25.3 and 65 %.

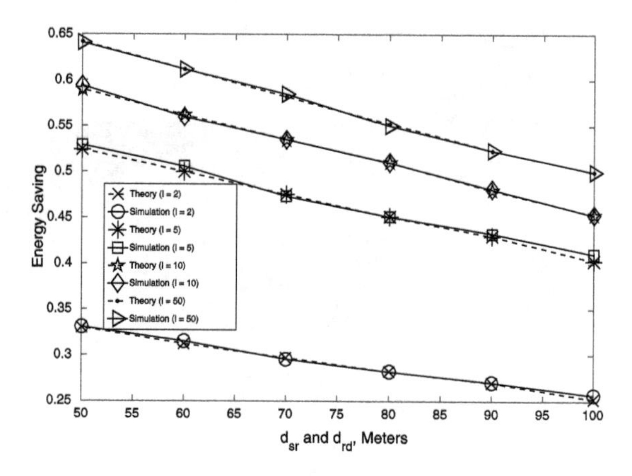

Fig. 4 Energy saving gain as a function of the received SNR and the distance between *R* and *D* [21]

3.2 CPLNC Based Energy Efficient Routing

The CPLNC scheme introduced in the previous subsection works purely at the PHY layer. As a next step, we modify higher layer protocols to achieve cross-layered designs for optimal energy savings. To this end, we focus on energy efficient routing at the network layer and incorporate CPLNC into two power saving routing algorithms, namely, power saving routing (PSR) and progressive power aware routing (PPAR), introduced originally in [22] and [23], respectively. PSR and PPAR work on the assumption that each node within the network knows its own location, the location of its immediate neighbors and the location of the destination. In the following, the modified versions of PSR and PPAR are denoted by MPSR and MPPAR, respectively.

CPLNC is utilized after the selection of the optimal next hop destination by PSR and PPAR. At each hop, the transmitting node *S* transmits the first symbol non-cooperatively towards the next hop destination *D*. Due to the broadcast nature of the wireless medium, this transmission is overheard by the neighboring nodes which are located between the source and the destination of each hop. These overhearing nodes report back to the sender about successful reception of the first symbol. The algorithm then selects the overhearing node located nearest to the hop destination to participate in the cooperation process. Following this selection, the algorithm decides whether or not it is feasible to use CPLNC for the given hop. This decision is based on threshold distances derived below:

Let the current, overhearing and destination nodes of each hop be denoted by *S*, *R* and *D*, respectively. Under the assumption that *S* needs to transmit ℓ symbols towards *D*, the energy consumption for non-cooperative point to point transmission is written as

$$\mathscr{E}_{woc} = \ell(\beta d_{sd}^{\alpha} + 2E), \tag{8}$$

where β is a constant scaling factor which reflects the effects of the power amplifier, receiver noise power and the transmit frequency. On the other hand, *E* represents the energy consumed by the transmitter or receiver circuitry and it is multiplied by 2 to account for both, the transmitter and receiver radio electronics power consumption.

With CPLNC, there will be an initial non-cooperative transmission, followed by $\ell - 1$ cooperative transmissions. The energy consumed by the non-cooperative transmission is the same as (8) above, except for the factor of ℓ and the fact that there is one transmitter and two receivers, i.e., E in (8) is multiplied by 3, instead of 2. On the other hand, for the cooperative transmissions, the energy consumption can be found by combining the $S - D$ and $R - D$ transmission powers, which can be calculated by utilizing (3) [21]. Let this combined energy consumption be denoted by $\mathscr{E}_{wc}$. Additionally, CPLNC also consumes extra energy for the multicast transmission when S informs R and D about the total number of symbols that are to be transmitted. Using this information, R and D can calculate the number of timeslots for which CPLNC is used and they can adjust their transmission and reception parameters accordingly. Let the multicast energy consumption be denoted by $\mathscr{E}_{mc}$, which is equal to the energy consumption of the first transmitted symbol, since there are two receivers and one transmitter. With all this information, the algorithm uses CPLNC if $\mathscr{E}_{wc} + \mathscr{E}_{mc} < \mathscr{E}_{woc}$. By plugging in the values for $\mathscr{E}_{woc}$, $\mathscr{E}_{wc}$ and $\mathscr{E}_{mc}$ in the aforementioned condition, the source S can obtain the distance thresholds which can be used to decide whether or not it is feasible to use CPLNC for a given transmission cycle.

The performance of the modified routing algorithms was evaluated by implementing a network of randomly distributed nodes within a square area of 500 m $\times$ 500 m. The source and destination nodes were located in the diagonal corners of the network. It was assumed that the transceiver devices consume 50 nJ/bit of energy while the power amplifier efficiency was set equal to 20 %. The noise power was fixed at -101 dBm while the minimum required SNR threshold was set equal to 10 dB. Frequency flat fading channel was considered which remained constant during a transmitted frame but varied from one frame to another. Lastly, all the computational results were verified by Monte-Carlo simulations using 10^8 samples.

Figure 5 compares the performance of MPPAR with the baseline algorithm, i.e., PPAR. For this comparison, the number of transmitted symbols was fixed to $\ell = 5$ while the pathloss exponent was varied between 2 and 4. It can be seen from the figure that, on average, MPPAR achieves energy saving of around 34 and 29 % for pathloss exponent values of 2 and 4, respectively. Interestingly, it is observed that the energy saving decreases with increasing α. This has to do with the inherent nature of PPAR, which transmits over longer distances when the channel gain is high and over smaller distances when the channel gain is low. With increasing values of pathloss exponents, the channel attenuation increases. Consequently, PPAR transmits over shorter distances, and this ultimately effects the performance of MPPAR because the probability of finding an overhearing node decreases when the distance between S and D is small. In other words, the probability of using CPLNC also decreases, thus, there is a drop in performance of MPPAR with increased pathloss exponent.

From the gain equations for CPLNC over a single hop, i.e., Eqs. (6) and (7), it is straightforward to see that the energy saving increases with increasing number of transmitted symbols. This observation also holds for the routing algorithms, as it is

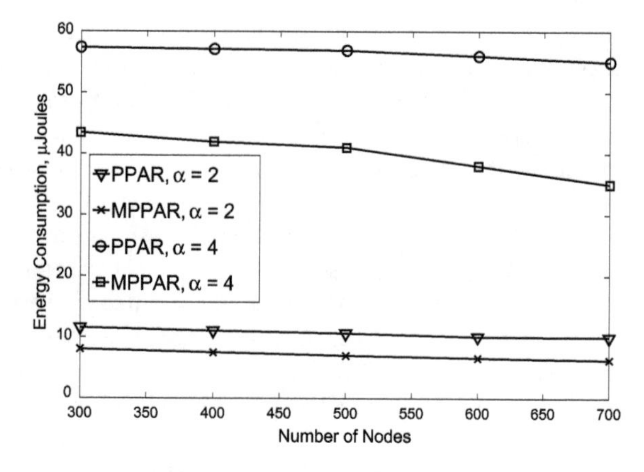

Fig. 5 Energy consumption as a function of the number of nodes in the network. $\ell = 5$ [21]

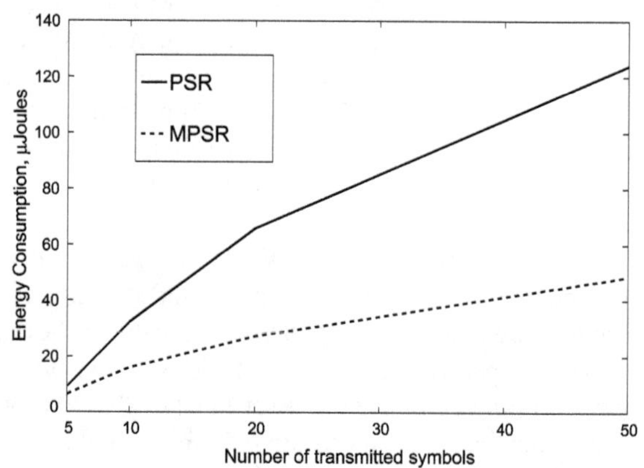

Fig. 6 Energy consumption as a function of the number of transmitted symbols. $\alpha = 2$ [21]

obvious from the plots shown in Fig. 6, where the energy consumptions of MPSR and PSR are compared for a fixed α and a variable number of transmitted symbols. It can be seen that for $\ell = 50$, MPSR minimizes the energy consumption by about 60 % as compared to PSR.

4 Cross-Layer Design Using Cooperative Type-I ARQ

The previous section focused on a cross-layer design which jointly optimized the performance of the PHY and network layers. This section goes a step further and presents a cross-layer algorithm which jointly optimizes the performance of the PHY, MAC and network layers. Specifically, the algorithm jointly optimizes the performance of PHY and MAC layers by finding a balance between the number of retransmission attempts required to successfully decode a transmitted frame and the energy consumption per each transmission attempt. To this end, it is assumed that the network employs coded type-I automatic repeat request (ARQ) at the MAC layer. With type-I ARQ, a frame is retransmitted if it is not decoded correctly at the receiver. Cooperative type-I ARQ is presented where a relay node assists the source node with its retransmission attempts. Following this, the energy efficiency performance of the network layer is optimized by designing energy-efficient routing algorithms which utilize cooperative ARQ as a basic building block. Unlike the previous subsection, where the channel coefficients were known to the transmitters, it is assumed that each transmitter only has partial channel state information (CSI), i.e., it knows the distance to the intended receivers but does not know the actual fading coefficients of the channel. Therefore, the source node uses an average transmission energy per symbol for transmitting its signal. In such situations, it is not always possible for the receiver to successfully decode the frame in a single transmission attempt. Consequently, coded type-I ARQ is employed to carry out the retransmission attempts until the frame is decoded successfully.

4.1 *Energy Consumption of Non-cooperative Type-I ARQ Systems*

This subsection presents a brief overview of the energy consumption in non-cooperative type-I ARQ systems. In such systems, the source node itself carries out all of the retransmission attempts, without any assistance from the nearby relays. The total average energy consumption per information bit for such a system can be found by deriving the average energy consumption per information bit for single transmission attempt and multiplying it with the expected number of retransmission attempts. The average energy consumption per information bit, for transmission between a transmitter a_u and receiver z_v, can be written as [17]

$$E_{uv}^{pb} = \frac{N + N_c}{N} A \gamma_v x_{uv}^{\alpha} + C, \tag{9}$$

where N and N_c are the number of information and overhead bits, respectively, γ_v is average received SNR, x_{uv} is the distance between the two nodes with α as the pathloss exponent. Finally, C represents the hardware energy consumption while A is a constant scaling factor which depends on the peak to average ratio of an M-ary modulation signal, transmit power amplifier efficiency, receiver noise power and the channel coding rate.

The expected number of retransmissions, including the original transmission, is a geometric random variable with a probability mass function given as $P_I(i) = (\text{FER})^{i-1}(1 - \text{FER})$, where FER is the frame error rate. For coded systems in a quasi-static Rayleigh fading channel, the frame error rate is approximately equal to the outage probability, where an outage occurs when the received SNR at the destination is less than a minimum required threshold γ_{th}. The minimum threshold γ_{th} depends on the channel code, the modulation scheme and the number of symbols per frame. It can be approximated as a linear function of $\log(N+N_c)$ and can modeled as $\gamma_{\text{th}} \simeq k_M \log(N+N_c)+b_M$, where k_M and b_M are constants which depend on the modulation scheme and the channel code [17]. Using the aforementioned SNR threshold and the average received SNR of γ_v, the outage probability, and consequently, the FER, can be approximated as [24]

$$\text{FER} \simeq \rho_{uv}^{o} = 1 - \exp\left(-\frac{\gamma_{\text{th}}}{\gamma_v}\right), \tag{10}$$

where ρ_{uv}^{o} represents the outage probability between a_u and z_v. Using FER from (10) in the probability mass function outlined above, the expected number of retransmission attempts can be shown to be equal to $1/\exp(-\gamma_{\text{th}}/\gamma_v)$. Multiplying this by (9) above, one obtains the total average energy consumption per information bit for non-cooperative type-I ARQ as

$$E_{uv}^{npb} = \frac{1}{\exp\left(-\frac{\gamma_{\text{th}}}{\gamma_v}\right)}\left(\frac{N+N_c}{N}A\gamma_v x_{uv}^{\alpha} + C\right). \tag{11}$$

It has been shown in [17] that (11) is convex in γ_v. Thus, the optimal target SNR is found by minimizing (11) with respect to γ_v. Using this optimal target SNR, the transmitter can adjust its transmission parameters to ensure a perfect balance between the energy consumption for each transmission attempt and the average number of transmission attempts required to successfully decode the source message at the receiver.

4.2 *Cooperative Type-I ARQ Systems*

In cooperative type-I ARQ systems, a relay node assists the source node with its retransmission attempts. This assistance is possible only if the relay is able to successfully decode the source signal while the destination is unable to do so. The scheme can be further illustrated with the scenario shown in Fig. 7, where the source, relay and destination nodes are denoted by S, R and D, respectively. Due to the broadcast nature of the wireless medium, R can overhear the $S-D$ transmissions. After transmitting the frame, the source awaits a positive or negative acknowledgement from the destination. If the destination is unable to successfully decode the frame, it sends out a negative acknowledgement (NACK), which is again intercepted by the relay R. If R had successfully overheard and decoded the frame in the previous transmission attempt, it retransmits this frame on behalf of S. Otherwise, the source retransmits the frame itself. Thus, a source only has to retransmit its frame until either R or D receives it successfully. In this setup, the energy saving is increased as compared to the non-cooperative case since R is located closer to D than S.

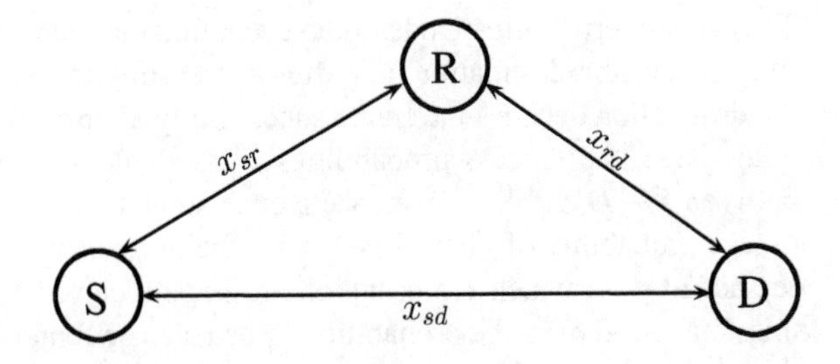

Fig. 7 A dual-hop network consisting of a source, relay and destination nodes [24]

In order to derive the total average energy consumption per bit for the aforementioned cooperative scheme, it is assumed that the distances between $S-D$, $S-R$ and $R-D$ are denoted by x_{sd}, x_{sr} and x_{rd}, respectively, as shown in Fig. 7. Furthermore, it is assumed that ρ^s_{sd}, ρ^s_{sr} and ρ^s_{rd} denote the probabilities of successful reception for the $S-D$, $S-R$ and $R-D$ links, respectively. Similar to the previous subsection on non-cooperative ARQ, the first step is to calculate the average energy consumption per information bit for each transmission that occurs within the cooperative setup. To this end, it can be seen that either the source or the relay transmits the data towards D. Let the energy consumption for these two transmissions be denoted by E^{pb}_{sd} and E^{pb}_{rd}, respectively. Since both E^{pb}_{sd} and E^{pb}_{rd} are point to point transmissions, their formulations remain similar to (9), except for the fact that target received SNR γ_v is replaced by γ_c, which is different since it is optimized for cooperative transmission, as discussed later in this subsection.

Next, in order to calculate the probability mass function for the expected number of retransmission attempts, it is noted that the following sequence of events occurs in cooperative transmissions:

1. S transmits its signal while R and D receive this signal.
2. If D successfully decodes the transmitted frame, the transmission cycle is completed and no retransmission attempts are required.
3. If D is unable to decode the frame while R is able to do so, R retransmits the frame on behalf of S.
4. If both D and R are unable to decode the frame, S retransmits the frame.

With these design considerations as well as the above mentioned probabilities of successful reception at R and D, the probability mass function of the expected number of retransmission attempts can be calculated as [24]

$$\begin{aligned} P(k_s,k_r) = &\left[(1-\rho^s_{sr})^{k_s-1}(1-\rho^s_{sd})^{k_s-1}\rho^s_{sd}\right]\delta[k_r] \\ &+\left[(1-\rho^s_{sr})^{k_s-1}(1-\rho^s_{sd})^{k_s}(1-\rho^s_{rd})^{k_r-1}\rho^s_{sr}\rho^s_{rd}\right], \end{aligned} \tag{12}$$

where k_s and k_r are the number of transmission attempts made by S and R, respectively, while $\delta[\cdot]$ is the dirac delta function. In the above equation, the first term represents the scenario where the destination decodes the frame successfully after k_s transmission attempts by the source, while the relay fails during these k_s attempts.

The second term corresponds to the event that the relay decodes the frame successfully before the destination and after k_s transmission attempts by S. Furthermore, the destination decodes the frame successfully after k_r transmission attempts by the relay. Next, the success probabilities ρ^s_{sd} and ρ^s_{rd} for point to point transmissions between $S - D$ and $R - D$, respectively, can be obtained simply by subtracting the outage probability of (10) from 1, with the only exception being the fact that γ_v is replaced by γ_c, which represents the average received SNR for cooperative transmissions. Moreover, the probability ρ^s_{sr} can also be found in a similar way except for the scaling factor which is induced due to the difference between x_{sd} and x_{sr} [24]. Having calculated all the aforementioned parameters, one can now calculate the total average energy consumption per information bit for cooperative type-I ARQ as

$$E^{cpb}_{sd} = \sum_{k_s=1}^{\infty} \sum_{k_r=0}^{\infty} P(k_s, k_r)(k_s E^{pb}_{sd} + k_r E^{pb}_{rd}). \tag{13}$$

By plugging in the values of the various parameters and after some rearrangement of terms, the above equation can be written as

$$E^{cpb}_{sd} = \frac{1}{\exp(-\frac{\gamma_{\text{th}}}{\gamma_c})} \left\{ \frac{A\gamma_c \left(\frac{N+N_c}{N}\right) \left[2\exp\left(\frac{\gamma_{\text{th}}}{\gamma_c}\left(\frac{1}{2} - \frac{x^\alpha_{sr}}{x^\alpha_{sd}}\right)\right) \sinh\left(\frac{\gamma_{\text{th}}}{2\gamma_c}\right) x^\alpha_{rd} + x^\alpha_{sd}\right]}{2\exp\left(\frac{\gamma_{\text{th}}}{\gamma_c}\left(\frac{1}{2} - \frac{x^\alpha_{sr}}{x^\alpha_{sd}}\right)\right) \sinh\left(\frac{\gamma_{\text{th}}}{2\gamma_c}\right) + 1} + C \right\} \tag{14}$$

In (14), the first term represents the expected number of retransmission attempts while the second term represents the average energy consumption per information bit for a single transmission attempt. Figure 8 shows that the theoretical results derived above, i.e., Eq. (14), match perfectly with the Monte-Carlo simulations. To plot this

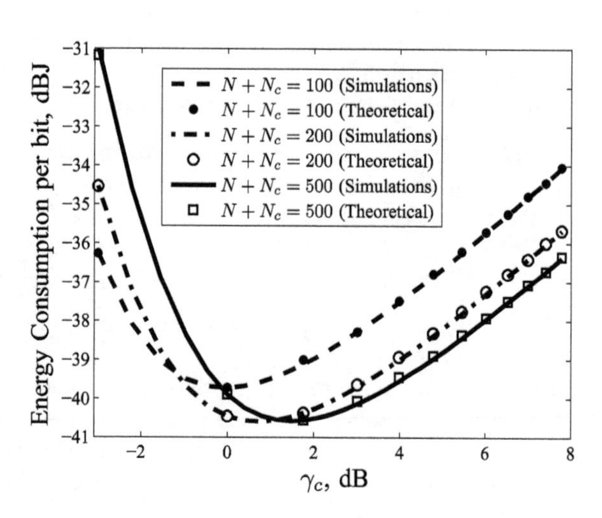

Fig. 8 Total average energy consumption per bit for cooperative transmission as a function of γ_c [24]

figure, the distance x_{sd} was fixed equal to 100 m while x_{sr} and x_{rd} were fixed at 50 meters. The rest of the simulation parameters were the same as those outlined in [24]. From the figure, it can also be seen that the energy consumption per bit is convex with respect to γ_c, therefore, the optimal SNR for cooperative transmission can be found by minimizing (14) over γ_c. Intuitively, it can also be seen from (14) that when $\gamma_c = \gamma_v$, the expected number of retransmission attempts with cooperative transmission becomes the same as that for non-cooperative transmissions.

Next, to quantify the energy saving gains achieved by cooperative type-I ARQ with respect to the non-cooperative ARQ, the gain equation defined earlier in Definition 1 is reused, with the only difference being the fact that $\mathscr{E}_{nc}$ and $\mathscr{E}_c$ are, respectively, replaced by E_{uv}^{npb} from (11) and E_{sd}^{cpb} from (14). With these replacements and by using x_{sd} for x_{uv} in (11), the energy saving gain can be written as,

$$G = \left(\frac{\Gamma}{\Gamma + \exp\left(\frac{\gamma_{th}}{\gamma_v}\frac{x_{sr}^{\alpha}}{x_{sd}^{\alpha}}\right)} \right) \left(\frac{x_{sd}^{\alpha} - x_{rd}^{\alpha}}{x_{sd}^{\alpha} + \frac{CN}{A\gamma_v(N+N_c)}} \right), \qquad (15)$$

where it is assumed that $\gamma_c = \gamma_v$ and $\Gamma = \exp\left(\frac{\gamma_{th}}{\gamma_v}\right) - 1$. Equation (15) provides some important insights about optimal relay selection in multihop wireless networks and it can be used as a design metric to that effect [24]. Figure 9 verifies the analysis of the achievable energy saving gains and also highlights the advantage of using cooperative ARQ with optimized γ_c. To implement this figure, the simulation parameters were set to be the same as those for Fig. 8, apart from the distances between $S - R$ and $R - D$. More specifically, x_{sr} was varied from 10 m to 90 m while x_{rd} was set to $100 - x_{sr}$. The figure provides some important insights that need further discussions. First of, it can be seen that when $\gamma_c = \gamma_v$, the theoretical results of (15) match perfectly with the simulation results. This verifies that (15) provides a very good

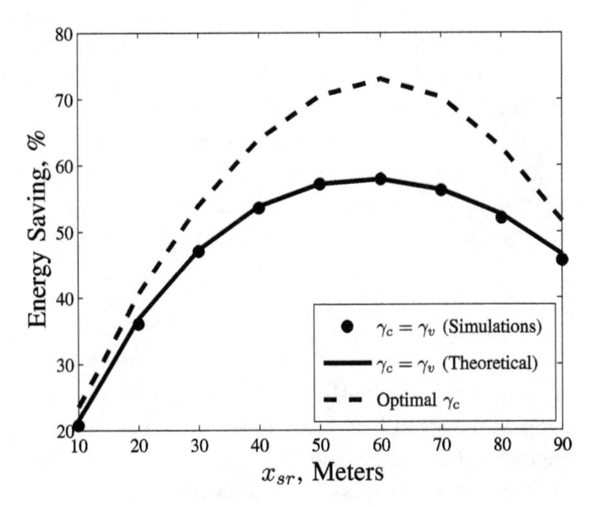

Fig. 9 Total average energy consumption per bit for cooperative transmission as a function of γ_c [24]

estimation of the achievable energy saving gains. Moreover, the figure also shows that the gain is maximized by using the optimal γ_c, which is found by minimizing (14) over γ_c. Overall, it can be seen that with the optimal γ_c, energy savings of more 70 % can be achieved, depending on the inter-node distances.

4.3 Cooperative ARQ Based Energy Efficient Routing

Continuing with our cross-layer design approach, this subsection integrates cooperative ARQ with the routing algorithms at the network layer. To this end, two energy-efficient routing algorithms are presented. The first algorithm, namely, cooperative cost-based shortest-path routing (CCB-SPR), utilizes cooperative ARQ as a basic building block and calculates the optimal energy-saving route based on the cooperative link costs. The second algorithm, called cooperation over non-cooperative shortest path (CONSP), first calculates a shortest-path route by using the non-cooperative routing (NCR) algorithm based on Distributed Bellmann-Ford algorithm. Then, cooperative transmission is used on top of this non-cooperative shortest path route.

4.3.1 CCB-SPR

With CCB-SPR, each node within the network initially calculates the cooperative link costs for each of its neighbors. To this end, a transmitter node a_u first finds the nearby nodes which are located between itself and a particular neighbor z_v, i.e., it finds the nodes that can successfully overhear the transmissions between a_u and z_v. Using each of these overhearing nodes as a relay, the node a_u calculates the cooperative link costs by using the formula presented in Eq. (14). Following these calculations, the overhearing node which gives the minimum cooperative link cost is selected to participate in cooperative transmission towards z_v. Lastly, a_u compares the selected cooperative link cost with the link cost for direct transmission towards z_v, calculated by using Eq. (11), and the minimum of the two is selected as the final link cost towards z_v. A similar procedure is adopted by all other nodes within the network. Once the link costs have been updated throughout the network, any shortest path algorithm, such as Bellmann-Ford or Dijkstra's algorithm, can be used to find the optimal energy saving route between a given pair of source and destination nodes.

4.3.2 CONSP

As mentioned previously, the CONSP algorithm initially calculates the optimal non-cooperative route between a given source-destination pair by using the distributed Bellmann-Ford algorithm. Then, this non-cooperative route is divided into groups of three consecutive nodes, as shown in Fig. 10. For each of these groups, the node

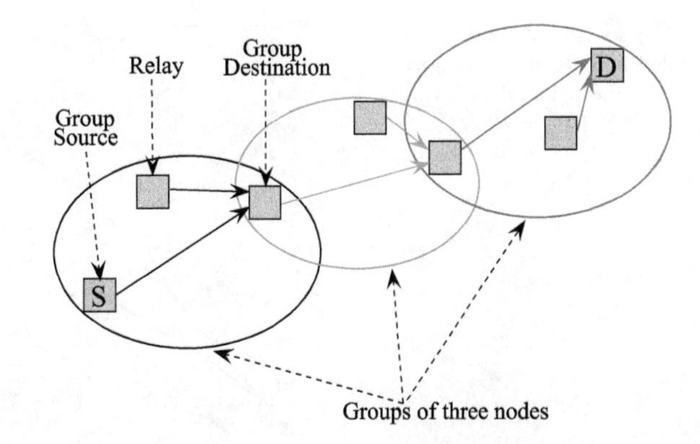

Fig. 10 Optimal non-cooperative route between the source and the destination. The figure shows the division of the path into groups of three nodes and cooperative transmission within each group [24]

located closest to the original source is designated as group source *S*, the one located closest to the destination is designated as group destination *D* while the one located in the middle of *S* and *D* is designated as group relay *R*. The group source *S* calculates the link cost for cooperative transmission towards *D* using *R*. If this cooperative link cost is lower than routing the data from *S* to *D* through *R*, then cooperative transmission is used within the given group. Otherwise, the message is route non-cooperatively within the group.

The performances of the aforementioned routing algorithms were evaluated through Monte-Carlo simulations. To this end, a square area of 100 m × 100 m was considered and between 10 and 100 nodes were distributed uniformly within this square area. Moreover, the source and the destination nodes were placed in the diagonal corners of the network. Figure 11 plots the energy savings achieved by using CCB-SPR and CONSP, as compared to NCR. As seen from this figure, CCB-SBR outperforms both CONSP and NCR. This happens because CCB-SPR utilizes the energy-efficient route which is optimized for cooperative transmissions. On the other hand, CONSP is only restricted to transmitting cooperatively over a non-cooperative shortest path route, which is not optimized for cooperative transmission. However, the performance improvement of CCB-SPR, as compared to CONSP, comes at the cost of increased route setup time since each node within the network is required to calculate new link costs for each of its neighbors. On the other hand, CONSP only has to transmit over a pre-selected non-cooperative route. Nevertheless, once the optimal route has been selected, CCB-SPR takes lesser time than CONSP to deliver the message to the destination since it requires lesser hops to reach the destination [24]. Figure 12 compares the route-setup time and the end-to-end latency of CCB-SPR and NCR. It can be seen that NCR performs better in terms of route-setup time while CCB-SPR has a lower end-to-end latency. Here, it must be emphasized that the route-setup for NCR and CONSP are the same since CONSP only transmits cooperatively over the route selected by NCR.

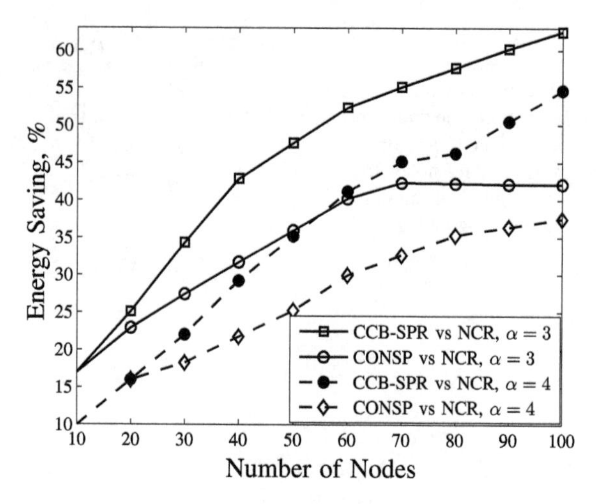

Fig. 11 Energy saving gain as a function of the number of nodes within the network [24]. $N + N_c = 1000$

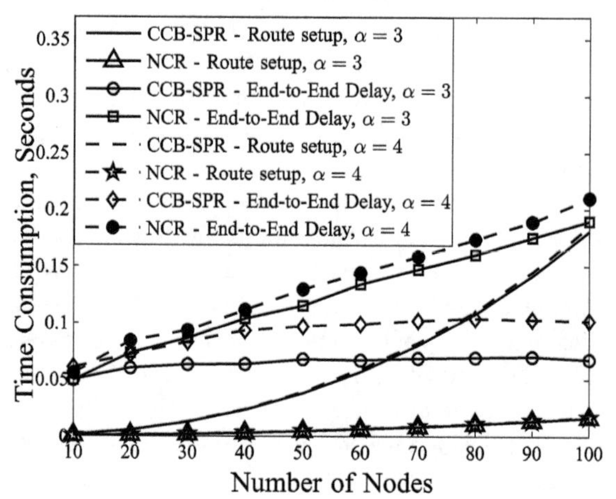

Fig. 12 Route setup time and end-to-end delay for CCB-SPR and NCR as a function of the number of nodes in the network [24]

5 Summary

The chapter focused on cross-layer designs for energy-efficient communications in wireless ad-hoc networks. Initially, the importance of cross-layered designs was highlighted. It was argued that treating a single layer at a time leads to sub-optimal solutions. Following this, the chapter dwelled into the various parameters that have to be taken into consideration when modeling energy consumption. A general energy consumption model, which took both the transmission and circuit energy consumption into consideration, was outlined and it was discussed how the general model could be tailored for specific scenarios under consideration. After outlining the methods for modeling energy consumption, the chapter presented two cross-layered design algorithms. For these designs, a bottom-up approach was adopted, whereby, energy efficient transmission strategies were introduced at the PHY layer and the protocols at the higher layers, i.e., MAC and network, were tailored to work in conjunction with the transmission technique at PHY. To this end, the first approach was based on CPLNC and it jointly optimized the performance of the PHY and network layers. On the other hand, the second approach jointly optimized the performance of PHY, MAC and network layers and it was based on a cooperative ARQ based transmission strategy. The performance of CPLNC and cooperative ARQ was analyzed in detail and the achievable energy saving gains were also derived. Lastly, computer simulations were used to evaluate the performance of the two cross-layered algorithms.

References

1. Cisco Inc.: Cisco visual networking index: global mobile data traffic forecast update, 2014-2019 white paper. http://www.cisco.com/c/en/us/solutions/collateral/service-provider/visual-networking-index-vni/white_paper_c11-520862.html (2015). Accessed 21 Apr 2015
2. Haardt, M.: Future mobile and wireless radio systems: challenges in European research. European Communication, Brussels, Belgium, Technical report (2008)
3. Han, et al.: Green radio: radio techniques to enable energy-efficient wireless networks. IEEE Commun. Mag. **49**(6), 46–64 (2011)
4. Cui, S., Goldsmith, A.J., Bahai, A.: Energy-constrained modulation optimization. IEEE Trans. Wireless Commun. **5**(5), 2349–2360 (2005)
5. Costa, F.M., Ochiai, H.: Energy-efficient physical layer design for wireless sensor network links. In: Proceedings of IEEE ICC, pp. 1–5 (2011)
6. Dong, L., Petropulu, A.: Weighted cross-layer cooperative beamforming for wireless networks. IEEE Trans. Signal. Proc. **57**(8), 3240–3252 (2009)
7. Ochiai, H., Mitran, P., Poor, H.V.: Collaborative beamforming for distributed wireless ad hoc sensor networks. IEEE Trans. Signal. Proc. **53**(11), 4110–4124 (2005)
8. Chatzipanagiotis, N., Liu, Y., Petropulu, A., Zavlanos, M.M.: Distributed cooperative beamforming in multi-source multi-destination clustered systems. IEEE Trans. Signal Proc. **62**(23), 6105–6117 (2014)
9. Kim, T., Kim, I.H., Sun, Y., Jin, Z.Y.: Physical layer and medium access control design in energy efficient sensor networks: an overview. IEEE Trans. Ind. Inform. **11**(1), 2–15 (2015)
10. Khan, J.A., Qureshi, H.K., Iqbal, A.: Energy management in wireless sensor networks: a survey. Comput. Electr. Eng. **41**, 159–176 (2015)
11. Ye, W., Heidemann, J., Estrin, D.: Medium access control with coordinated adaptive sleeping for wireless sensor networks. IEEE/ACM Trans. Netw. **12**(3), 493–506 (2004)
12. Fan, Z., Bai, S., Wang, S., He, T.: Delay-bounded transmission power control for low-duty-cycle sensor networks. IEEE Trans. Wirel. Commun. **14**(6), 3157–3170 (2015)
13. Steine M., Geilen M., Basten, T.: A distributed reconfiguration approach for quality-of-service provisioning in dynamic heterogeneous wireless sensor networks. ACM Trans. Sensor Netw. **11**(2), 34, 41 (2015)
14. Butt, M.M., Jorswieck, E.A., Ottersten, B.: Maximizing energy efficiency in multiple access channels by exploiting packet dropping and transmitter buffering. IEEE Trans. Wireless Commun. **14**(8), 4129–4141 (2015)
15. Zuo, J., Dong, C., Ng, S., Yang, L., Hanzo, L.: Cross-layer aided energy-efficient routing design for ad hoc networks. IEEE Commun. Surv. Tutor. **99**, 1–26 (2015)
16. Mansourkiaie, F., Ahmed, M.H.: Cooperative routing in wireless networks: a comprehensive survey. IEEE Commun. Surv. Tutor. **17**(2), 604–626 (2015)
17. Wu, J., Wang, G., Zheng, Y.R.: Energy efficiency and spectral efficiency tradeoff in type-i ARQ systems. IEEE J. Sel. Areas Commun. **32**(2), 356–366 (2014)

18. Akhtar, A.M., Nakhai, M.R., Aghvami, A.H.: Power aware cooperative routing in wireless mesh networks. IEEE Commun. Lett. **16**(5), 670–673 (2012)
19. Goldsmith, A.: Wireless Communications. Cambridge University Press, Cambridge (2005)
20. Akhtar, A.M., Nakhai, M.R., Aghvami, A.H.: On energy efficient routing using cooperative physical layer network coding. In: Proceedings of IEEE GLOBECOM, pp. 2791–2796 (2012)
21. Akhtar, A.M., Nakhai, M.R., Aghvami, A.H.: On the use of cooperative physical layer network coding for energy efficient routing. IEEE Trans. Commun. **61**(4), 1498–1509 (2013)
22. Stojmenovic, I., Lin, X.: Power-aware localized routing in wireless networks. IEEE Trans. Parallel Dist. Comput. **12**(3), 1122–1133 (2001)
23. Kuruvila, J., Nayak, A., Stojmenovic, I.: Progress and location based localized power aware routing for ad hoc and sensor wireless networks. Intl. J. Dist. Sens. Netw. **2**(2), 147–159 (2005)
24. Akhtar, A.M., Behnad, A., Wang, X.: Cooperative arq based energy efficient routing in multihop wireless networks. IEEE Trans. Veh. Technol. **64**(11), 5187–5197 (2015)

Energy Efficiency of Nonbinary Network-Coded Cooperation

Ohara Kerusauskas Rayel, João Luiz Rebelatto, Richard Demo Souza, Bartolomeu F. Uchôa-Filho and Yonghui Li

Abstract In this chapter, an energy efficiency analysis is performed in a wireless sensor network setup considering different communication schemes, including direct non-cooperative transmission, decoded-and-forward cooperation, and network-coded cooperation. The analysis is performed considering Nakagami-m block fading, so that the influence of some line-of-sight is taken into account, while the effect of the circuitry power consumption is also considered. The theoretical and numerical results show that the use of network coding can be considerably beneficial in terms of energy efficiency and that there exists an optimal number of cooperating nodes that minimizes the energy consumption for a given distance. With network coding and with an appropriate organization of cooperating nodes into clusters, the energy efficiency is maximized, leading to energy savings of an order of magnitude with respect to the direct non-cooperative transmission.

O.K. Rayel · J.L. Rebelatto · R.D. Souza (✉)
CPGEI, UTFPR, Av. Sete de Setembro, 3165, Room A-307, Curitiba, PR 80230-901, Brazil
e-mail: richard@utfpr.edu.br

O.K. Rayel
e-mail: oharakr@utfpr.edu.br

J.L. Rebelatto
e-mail: jlrebelatto@utfpr.edu.br

B.F. Uchôa-Filho
GPqCom/LCS, UFSC, Florianópolis, SC 88040-90, Brazil
e-mail: uchoa@eel.ufsc.br

Y. Li
SEIE, University of Sydney, Sydney, NSW 2006, Australia
e-mail: lyh@ee.usyd.edu.au

© Springer International Publishing Switzerland 2016
M.Z. Shakir et al. (eds.), *Energy Management in Wireless Cellular and Ad-hoc Networks*, Studies in Systems, Decision and Control 50,
DOI 10.1007/978-3-319-27568-0_8

1 Introduction

Recently, there has been a growing interest in ad hoc wireless networks composed of low cost and small sized devices such as wireless sensor networks (WSNs) and even machine-to-machine (M2M) networks [2, 29]. The typical devices in this case are most likely powered by batteries, which are often not easy and cheap to replace or recharge. Therefore, it is of paramount importance to increase the lifetime of these devices by efficiently utilizing their limited power sources. Moreover, from an environmental point of view, with the widespread deployment of WSNs and M2M communications, it is highly desirable that batteries used by these devices last for very long time, decreasing the impact of battery disposal.

A great part of the energy consumption of a wireless device is related to the data transmission, with the transmit power being a function of the channel conditions, range, target error rate, among other factors. In order to decrease the required transmit power the most common alternative is to use diversity techniques, such as channel coding [15] and multiple antennas [3, 25]. Alternatively, there is a recent trend to exploit the broadcast nature of the wireless channel to improve the overall performance by means of cooperation [13, 23]. The decode-and-forward (DF) cooperative protocol is one of the most adopted cooperative techniques [11, 13, 24, 27], in which the nodes often cooperate in pairs. In the first phase of the DF protocol (usually termed as broadcast phase) each user broadcasts its own information through orthogonal channels, while in the second phase (commonly referred to as cooperative phase) each user transmits the re-encoded information of its partner if it could be correctly decoded in the first phase. By doing so the cooperating nodes form a virtual antenna array that considerably improves the error performance seen at the common destination.

Several recent works have also demonstrated the benefits of the classical cooperative protocols (as the DF protocol) towards increasing the energy efficiency [7, 11, 24, 27]. However, as pointed out in [4, 17], in a more realistic scenario where the energy consumption of the transmitter and receiver circuitry is taken into account, there are situations where cooperation may be less energy efficient when compared to the direct transmission. This is due to the fact that, as presented in [6], the energy consumption of wireless networks is dominated by the transmit power only when the nodes are relatively far from each other. When the distance between the nodes decreases, the circuitry consumption becomes considerably more relevant in the energy efficiency analysis.

The *network coding technique* was proposed as a new strategy to attain maximum throughput in wired networks [1, 14], as well as to improve robustness [12]. Recently, this technique has been applied to wireless cooperative networks in order to improve their error performance [20, 22, 30, 32], by increasing the system's overall diversity order beyond the cooperative protocols limits. In cooperative systems with network coding, instead of just retransmitting the information of its partners, the cooperating users are able to transmit linear combinations (over a finite field GF(q)) of not only their partners' but also their own information. In [32], it was shown that, if the linear combinations are performed over a large enough field, the system diversity order is increased, improving the error performance. It should be mentioned that maximum diversity order could only be obtained with high probability through random network coding [9], by assuming infinitely large Galois Field, or through deterministic network coding under the assumption that the cooperating users also hear their partners' information at the cooperative phase [21].

In the scheme proposed in [32], called diversity network coding (DNC), each node transmits in the cooperative phase linear combinations over GF(q) of all the information received during the broadcast phase. A generalization of the DNC scheme, called generalized DNC (GDNC), was proposed in [22], which may achieve simultaneously both code rate and diversity order higher than in the DNC scheme. A feedback-assisted version of the GDNC scheme, called FA-GDNC, was proposed in [20], aiming at increasing the average code rate without degrading the system error performance. It is clear thus that network coding can improve the error rate performance of cooperative networks, and therefore reduce the required transmit power for achieving a given performance, potentially improving the energy efficiency, which was also pointed out in [7]. However, on the other hand, it is not clear if such savings in the required transmit power are not surpassed by the additional circuitry consumption, since in network-coded cooperative networks many nodes may collaborate, increasing the overall number of transmissions/receptions performed throughout the network.

The contributions of this chapter can be summarized as follows:

- We extend the preliminary results from [18, 19] and provide an extended analysis and results comparing the performance of several wireless transmission schemes in terms of energy efficiency (measured in J/bit), when taking into account, besides the energy consumed by the transmission, the consumption of the transmitter and receiver circuits;
- More specifically, we evaluate the energy efficiency of direct non-cooperative transmission, classical DF cooperation and network-coded cooperation (GDNC and FA-GDNC protocols);
- In the case of network coding cooperation, which may involve several nodes, we discuss the optimal number of cooperating nodes that maximizes the energy efficiency;
- The optimal number of cooperating nodes is then utilized to organize the network-coded cooperative network into clusters for the sake of designing more energy-efficient cooperative systems.

The rest of this chapter is organized as follows. Section 2 presents the system model and some fundamental concepts utilized for evaluating the energy efficiency of network-coded cooperative wireless networks. Section 3 discusses the energy efficiency definition, the power consumption model and the efficiency of the direct transmission and all the aforementioned cooperative protocols. Moreover, the optimal number of cooperating nodes is also discussed and its impact on the energy efficiency is demonstrated in Sect. 3. Section 4 presents some numerical results and Sect. 5 concludes the chapter.

Notation: We use lowercase bold letters to represent vectors. Operators + and − represent real number operations, while ⊕ is the binary sum (XOR), ⊞ and ⊟ are the sum and subtraction operations over a non-binary field, respectively. P denotes transmit power, while $\mathscr{P}_o$ stands for outage probability.

2 Preliminaries

2.1 System Model

This work considers the multiple access part of a cooperative wireless network, where M users have independent information to transmit to a common destination (D). A *frame* is defined as a vector of length N symbols. The time period corresponding to the transmission of one frame is referred as *subslot*. A time slot (TS) is defined as the time period that comprehends M subslots. In other words, a TS corresponds to M individual transmissions, performed in a round-robin fashion between the users and through time-orthogonal channels.

Omitting the time index, the baseband codeword received by user j after a transmission performed by the ith user can be written as

$$\mathbf{y}_{i,j} = \sqrt{P_i \gamma_{i,j}} h_{i,j} \mathbf{x}_i + \mathbf{n}_{i,j}, \tag{1}$$

where $\mathbf{x}_i \in \mathbb{C}^N$ and $\mathbf{y}_{i,j} \in \mathbb{C}^N$ are the transmitted and received packets, respectively, both of length N, with $i \in \{0, \ldots, M-1\}$ representing the transmitter index (and also the subslot index) and $j \in \{0, 1, \ldots, M\}$ standing for the received index (M corresponds to the destination). In (1), P_i is the transmission power, $h_{i,j}$ is the channel gain due to multipath fading between users i and j, and $\mathbf{n}_{i,j} \in \mathbb{C}^N$ is the zero-mean additive white Gaussian noise with variance $N_0/2$ per dimension, N_0 being the noise power spectral density per Hertz. We consider that the pathloss $\gamma_{i,j}$ between users i and j is [8]

$$\gamma_{i,j} = \frac{G\lambda^2}{(4\pi)^2 (d_{i,j})^\alpha M_t N_f}, \tag{2}$$

where G is the total gain of the transmit and receive antennas, λ is the wavelength, $d_{i,j}$ is the distance between the referred users, α is the pathloss exponent, M_t is the link margin and and N_f is the noise figure at the receiver [8]. The average signal-to-noise ratio (SNR) can be written as

$$\mathsf{SNR}_{i,j} = \frac{P_i \gamma_{i,j}}{N_0 B}, \tag{3}$$

where B is the bandwidth (in Hertz).

Moreover, $|h_{i,j}|$ is assumed to follow a block-fading model with Nakagami-m distribution and unitary energy, remaining constant within one subslot but changing in an independent and identically distributed (i.i.d.) fashion (in both space and time) between subslots. The choice for a Nakagami-m distribution is due to the fact that it is capable of representing the channel behavior in several scenarios, ranging from the total absence of line-of-sight (LOS) to scenarios with the strong presence of LOS. We also consider that all the receivers have perfect channel state information (CSI), but the transmitters do not have any CSI.

2.1.1 Outage Probability and Diversity Order

If we assume that the elements of $\mathbf{x}_i$ are Gaussian distributed and that all channels throughout the network experience the same average SNR, then the mutual information $\mathsf{MI}_{i,j}$ between $\mathbf{x}_i$ and $\mathbf{y}_{i,j}$, for a band of 1 Hz, is

$$\mathsf{MI}_{i,j} = \log_2(1 + |h_{i,j}|^2 \mathsf{SNR}). \tag{4}$$

An outage event occurs when $\mathsf{MI}_{i,j} < r$, where r corresponds to the attempted information rate (in bits/s/Hz). The probability of such event is called outage probability, which, for Nakagami-m fading is [4, 8, 28]:

$$\mathscr{P}_o = \frac{\Gamma(m, mg)}{\Gamma(m)}, \tag{5}$$

where $g = \frac{2^r - 1}{\mathsf{SNR}}$ and $\Gamma(a, b) = \int_0^b y^{a-1} e^{-y} dy$ corresponds to the lower incomplete Gamma function. For small values of b, it was shown in [28] that $\Gamma(a, b) \simeq (1/a) b^a$, which allows us to approximate (5) for the high-SNR region as

$$\mathscr{P}_o \simeq \frac{(mg)^m}{\Gamma(m+1)}. \tag{6}$$

The diversity order is another performance metric of interest, representing the derivative of the outage probability with respect to the SNR, being mathematically represented as [26]

$$\mathscr{D} \triangleq \lim_{\mathsf{SNR} \to \infty} \frac{-\log \mathscr{P}_{o,\mathsf{X}}}{\log \mathsf{SNR}}, \tag{7}$$

where $\mathscr{P}_{o,\mathsf{X}}$ is the overall outage probability of transmission scheme X.

2.2 *Decode-and-Forward (DF)*

In a wireless cooperative network, the users make use of the broadcast nature of the wireless channel in order to act as relays to their partners, providing spatial diversity and helping towards combating the fading inherent to this kind of channel [13, 22, 23, 30, 32]. In a cooperative scenario, the transmission round is usually divided in two phases: The *broadcast phase*, where the nodes broadcast their own information frames (IFs); and the *cooperative phase*, where the nodes transmit parity frames (PFs) to the destination node, which are redundant frames composed of IFs from the other nodes that were correctly received during the broadcast phase.

In the decode-and-forward (DF) cooperative protocol [13, 23], after broadcasting a single IF in the broadcast phase, the nodes just retransmit the IF from its partner (after decoding and re-encoding it) in the cooperative phase. In case of outage in the interuser channel, each user just retransmits its own IF in the cooperative phase. The outage probability achieved by such protocol can be shown to be [13]

$$\mathscr{P}_{o,\mathsf{DF}} \approx 0.5\mathscr{P}_o^2, \tag{8}$$

which, when substituted in (7), leads to diversity order $2m$, providing spatial diversity.

In the DF scheme, each node transmits twice per transmission round, one in the broadcast phase and one in the cooperative phase, so that the total number of transmissions is given by $2M$. In the broadcast phase, during each of the M broadcasts, all the other M remaining nodes ($M-1$ transmitting nodes plus the destination) try to recover the transmitted IF, leading to an amount of M^2 receptions. In the cooperative phase, all the transmissions are addressed only to the destination, yielding M receptions. Thus, the total number of receptions in the DF cooperative protocol is M^2+M, as summarized in Table 1 at the end of this section.

In order to perform a fair comparison between the direct non-cooperative transmission and the cooperative protocols in terms of spectral efficiency, one must compensate the half-duplexing loss inherent to the transmission of parities, by adjusting the attempted rate in (5) according to the code rate of the cooperative protocol. More specifically, in the DF scheme one has that the ratio between the number of transmitted IFs and the total number of transmitted frames (referred to as code rate) is

$$R_{\mathsf{DF}} = \frac{\mathsf{Number\ of\ IFs}}{\mathsf{Number\ of\ IFs} + \mathsf{Number\ of\ PFs}} = \frac{M}{2M} = \frac{1}{2}. \tag{9}$$

Thus, one must transmit with rate r/R_{DF} in order to perform a fair comparison to the direct transmission.

2.3 Network-Coded Cooperative Communication

The network coding technique [1, 12], initially proposed to increase the throughput in wired networks, has recently been applied to cooperative networks in order to improve their performance against errors [22, 30–32]. In network-coded cooperative networks, instead of just relaying individually the IFs of their partners in the cooperative phase, the users are able to transmit linear combinations of all the available IFs, possibly performed over a non-binary field GF(q). The transmission of linear combinations enables each IF to be transmitted through a larger number of independent channels, increasing the diversity order, provided that such linear combinations are performed in a way that the destination is able to individually recover each IF from the set of all received frames.

Several works consider scenarios where the coefficients from the linear combinations are randomly chosen [1, 12]. In this kind of approach, it can be shown that the probability of generating independent linear combinations tends to one for a sufficient large finite field q, leading to a full rank transfer matrix. However, since the complexity of the system as well as the amount of overhead increase with q [32], in this work we focus on deterministic network coding [22, 30, 32].

2.3.1 Dynamic Network Coding (DNC)

In [31], the authors showed that the use of non-binary network coding is necessary in order to achieve a diversity order higher than the binary network-coded (BNC) cooperative protocol from [30], and proposed the so called *Dynamic Network Coding* (DNC).[1]

Let us first consider a 2-user network where the interuser channel is not in outage (which happens with probability $1 - \mathscr{P}_o$) and considering that the IF and PF transmitted by User 1 (resp. User 2) are given respectively by I_1 and $I_1 \boxplus I_2$ (resp. I_2 and $I_1 \boxplus 2I_2$), the set of frames received at the destination node is equal to I_1, I_2, $I_1 \boxplus I_2$ and $I_1 \boxplus 2I_2$. It can be seen that the destination is able to decode the IFs I_1 and I_2 from any two out of the four received frames. Focusing on User 1 (the same result is valid to User 2 due to the system symmetry), an outage event occurs when the direct transmission of the frame I_1 and at least two out of the three remaining frames are not correctly decoded, which happens with probability [31]

$$\mathscr{P}_{\mathsf{A}} = \mathscr{P}_o \left[\binom{3}{2} \mathscr{P}_o^2 (1 - \mathscr{P}_o) + \mathscr{P}_o^3 \right] \approx 3\mathscr{P}_o^3, \tag{10}$$

where the approximation is valid for the high-SNR regime.

However, with probability $\mathscr{P}_o$ the interuser channel may be in outage. Considering that the users retransmit their own IF in this situation, and that the destination performs maximum ratio combining (MRC) [13, 26] after receiving multiple copies of the same frame, the following outage probability is obtained [13, 26]

$$\mathscr{P}_{\mathsf{B}} \approx \mathscr{P}_o^2 / 2. \tag{11}$$

[1]Later called *Diversity Network Coding* in [32].

Thus, the overall outage probability for the IF of User 1 in this 2-user network operating under the DNC scheme is [31]

$$\mathscr{P}_{o,\mathsf{DNC}}(M=2) = \mathscr{P}_o \mathscr{P}_\mathsf{B} + (1-\mathscr{P}_o)\mathscr{P}_\mathsf{A} \approx 3.5\mathscr{P}_o^3. \quad (12)$$

It can be seen from (12) that the diversity order is $\mathscr{D}_{\mathsf{DNC}}(M=2) = 3m$, by inserting (12) into (7).

The authors of [31, 32] also proposed an expansion of the DNC scheme to a scenario with M cooperative users, which, operating under a fixed and low code rate equal to

$$R_{\mathsf{DNC}} = \frac{\text{Number of IFs}}{\text{Number of IFs} + \text{Number of PFs}} = \frac{M}{M^2} = \frac{1}{M}, \quad (13)$$

is capable of achieving a diversity order equal to

$$\mathscr{D}_{\mathsf{DNC}} = 2M - 1. \quad (14)$$

2.3.2 Generalized Dynamic Network Coding (GDNC)

In [22], a generalization of the DNC scheme introduced in [32] was proposed, through an association between network coding and classical error correcting codes. The scheme from [22], referred to as generalized DNC (GDNC) was shown to be more flexible in terms of code rate, and being capable of achieving simultaneously both code rate and diversity order higher than the DNC scheme.

More specifically, through an analogy between the network transfer matrix and the generator matrix of a linear block code, it was shown in [22] that there is an equivalence between the system's diversity order and the minimum Hamming distance of the aforementioned block code. Having in mind that the minimum distance of a block code is upper bounded by the Singleton bound [16], and that this bound increases as the code dimensions increases, the GDNC scheme considers that during the broadcast phase each network user is able to broadcast a given number of k_1 IFs, while transmitting an arbitrary number of k_2 PFs in the cooperative phase, all of them composed of linear combinations of all the available information, over a large enough finite field GF(q), that satisfies the condition $q \geq M(k_1 + k_2)$ [22]. Figure 1 illustrates the GDNC protocol.

In [22], it was shown that the outage probability of the GDNC scheme with parameters (k_1, k_2, M) is given by

$$\mathscr{P}_{o,\mathsf{GDNC}} \approx \binom{k_1 + k_2 - 1}{k_2} \mathscr{P}_o^{M+k_2} \qquad (\text{for } k_2 \geq 2), \quad (15)$$

where $\binom{k_1+k_2-1}{k_2}$ corresponds to the binomial coefficient. From (15) and (7), it can be seen that the diversity order of the GDNC scheme is

$$\mathscr{D}_{\mathsf{GDNC}} = m(M + k_2), \quad (16)$$

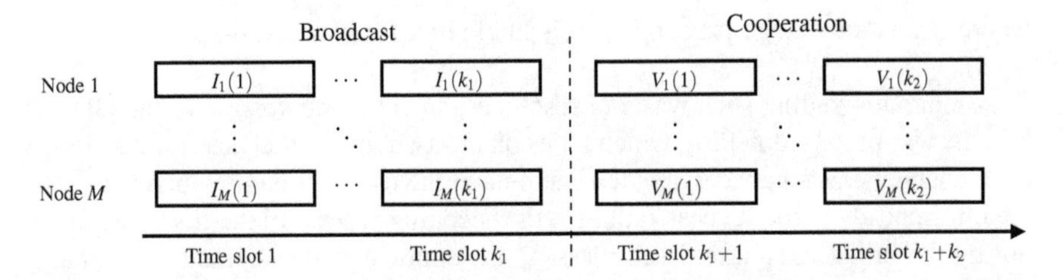

Fig. 1 GDNC scheme protocol. Each user initially broadcasts k_1 information frames $I_i(t)$ during the broadcast phase, and after that it transmits k_2 parity frames $V_i(t')$, composed of linear combinations over GF(q) of its own information and of all the frames from its partners that were correctly decoded during the broadcast phase. Each frame is transmitted at a different time subslot

which is achieved if a generator matrix from a maximum distance separable (MDS) code is used as the network transfer matrix (coefficients of the linear combinations), as for instance the well-known class of Reed-Solomon codes [22].

Since $M(k_1 + k_2)$ frames are transmitted in the GDNC scheme, among which only Mk_1 effectively carry new information, one has that its code rate is

$$R_{\mathsf{GDNC}} = \frac{Mk_1}{Mk_1 + Mk_2} = \frac{k_1}{k_1 + k_2}, \tag{17}$$

In the broadcast phase, during each of the Mk_1 transmissions by a given network user, all the other M network users ($M - 1$ cooperative users plus the destination) try to decode the broadcasted information, keeping its receiver circuits active. During the cooperative phase, all the transmissions are addressed only to the destination. Table 1 presents a summary of the numbers of transmissions and receptions performed in the GDNC scheme for each set of Mk_1 IFs effectively transmitted to the destination.

2.3.3 Feedback-Assisted GDNC (FA-GDNC)

In both DNC and GDNC schemes, once the network code is designed, it remains the same until the occurrence of a change in the network topology (a modification in the number of users, for instance) and a new code is required. However, this characteristic has a bad impact on the system rate: as the SNR increases, the probability that the destination can correctly decode all the IFs received in the broadcast phase also increases. In this case of successful decoding, the transmission of PFs would be no longer necessary.

Since the number of IFs transmitted during the broadcast phase in the GDNC scheme is Mk_1, the probability that all these frames are correctly decoded by the destination is given by

$$\Pr\{\mathsf{None\,in\,outage}\} = (1 - \mathscr{P}_o)^{Mk_1}, \tag{18}$$

where $\mathscr{P}_o$ is the outage probability of a single link for Nakagami-m fading, given in (5).

Aiming at avoiding such waste of resources, an enhanced version of the GDNC scheme was proposed in [20], which relies on the assumption that an error free feedback channel exists between the destination and the users, through which the users are informed about the success/failure in the decoding process of the IFs transmitted during the broadcast phase. Each message transmitted by the destination contains only one bit, named *outage bit* (OUT), which is sent back after each broadcast phase. $\mathsf{OUT} = 0$ means that the destination correctly decoded all the Mk_1 IFs, so new IFs can be generated and transmitted, and PFs transmissions are no longer necessary in the current cooperation round. According to (18), this event has probability $\Pr\{\mathsf{OUT}=0\} = (1 - \mathscr{P}_o)^{Mk_1}$. Otherwise, $\mathsf{OUT} = 1$ means that at least one out of Mk_1 information frames was not correctly decoded, which occurs with probability $\Pr\{\mathsf{OUT}=1\} = 1 - \Pr\{\mathsf{OUT}=0\}$. In this case, each user transmits k_2 parity frames, according to the original GDNC scheme in [22]. Each of these PFs contains all the IFs that the user could correctly decode during the broadcast phase.

Since all the users transmit k_2 PFs only when at least one of the IFs is not correctly decoded by the destination, the necessary condition for an outage event to occur in the FA-GDNC is the same as in the GDNC scheme, such that the outage probability and diversity order of the FA-GDNC scheme are also given respectively by (15) and (16).

The main difference from the original GDNC scheme lies on the average code rate. Since with the probability given in (18) there is no need to transmit parities, the average code rate of the FA-GDNC scheme is

$$R_{\mathsf{FA\text{-}GDNC}} = \frac{Mk_1}{Mk_1 + Mk_2\left(1 - \bar{\mathscr{P}}_o^{\,Mk_1}\right)} = \frac{k_1}{k_1 + k_2'}, \tag{19}$$

where $\bar{\mathscr{P}}_o = 1 - \mathscr{P}_o$ and $k_2' = k_2\left(1 - \bar{\mathscr{P}}_o^{\,Mk_1}\right)$. It can be seen from (19) that, since $0 \le \bar{\mathscr{P}}_o^{\,Mk_1} \le 1$, $R_{\mathsf{FA\text{-}GDNC}}$ is always greater than or equal to the GDNC scheme code rate from (17). More specifically, we can see from (19) that $R_{\mathsf{FA\text{-}GDNC}} \to 1$ as the SNR increases. When the SNR decreases, $R_{\mathsf{FA\text{-}GDNC}} \to \frac{k_1}{k_1+k_2}$. On the other hand, for a fixed SNR and with the number of users increasing, one can see from (19) that

$$\lim_{M\to\infty} R_{\mathsf{FA\text{-}GDNC}} = \frac{k_1}{k_1 + k_2}, \tag{20}$$

which means that, as the number of users increases, feedback becomes less and les necessary. The number of transmissions and receptions performed in the FA-GDNC scheme is presented in Table 1.

Table 1 Number of transmissions (# TX) and receptions (# RX) of the DF, DNC, GDNC and FA-GDNC schemes

	DF		DNC		GDNC		FA-GDNC	
	# TX	# RX	# TX	# RX	# TX	# RX	# TX	# RX
BP	M	M^2	M	M^2	Mk_1	M^2k_1	Mk_1	M^2k_1
CP	M	M	M^2-M	M^2-M	Mk_2	Mk_2	Mk_2'	Mk_2'
Total	$2M$	M^2+M	M^2	$2M^2-M$	$M(k_1+k_2)$	$M(Mk_1+k_2)$	$M(k_1+k_2')$	$M(Mk_1+k_2')$

BP and CP refer to the broadcast and cooperative phases, respectively

3 Energy Efficiency

After introducing the cooperative protocols, this work aims at evaluating their energy efficiency when taking into account, besides the energy spent with the transmission itself, the energy consumption of the transmitter and receiver's circuitry. In what follows we adopt as the performance metric the ratio between the overall energy consumed and the amount of information transmitted, measured in J/bit.

3.1 Power Consumption Model

We adopt the power consumption model from [6], where the overall consumed energy is given by

$$E = \frac{P_{\mathsf{amp}} + P_{\mathsf{tx}} + P_{\mathsf{rx}}}{R_b} \quad \text{(J/bit)}, \tag{21}$$

where $R_b = r \cdot B$ corresponds to the transmission rate (in bits/s), P_{tx} and P_{rx} are the power consumed respectively by the transmitter and receiver's circuitry, which, according to [6], depend on the power consumption of digital-to-analog converter (DAC) P_{DAC}, mixer P_{mix}, transmitter filters $P_{\mathsf{tx\_filt}}$, frequency synthesizer P_{syn}, low noise amplifier (LNA) P_{LNA}, intermediate frequency amplifier (IFA) P_{IFA}, receive filters $P_{\mathsf{rx\_filt}}$ and analog-to-digital converter (ADC) P_{ADC}, being given by

$$\begin{aligned} P_{\mathsf{tx}} &= P_{\mathsf{DAC}} + P_{\mathsf{mix}} + P_{\mathsf{tx\_filt}} + P_{\mathsf{syn}}, \\ P_{\mathsf{rx}} &= P_{\mathsf{syn}} + P_{\mathsf{LNA}} + P_{\mathsf{mix}} + P_{\mathsf{IFA}} + P_{\mathsf{rx\_filt}} + P_{\mathsf{ADC}}. \end{aligned} \tag{22}$$

In (21), $P_{\mathsf{amp}} = \frac{\xi}{\eta} P_i$ corresponds to the power employed by the amplifier in the transmission process, which depends on the transmission power P_i, as well as on the ratio between the drain efficiency η of the amplifier and the peak-to-average ratio ξ, which is $\xi = 3\left(\frac{\sqrt{\mathscr{M}}-1}{\sqrt{\mathscr{M}}+1}\right)$ for $\mathscr{M}$-QAM [6]. It is worth mentioning that, according to [10], in networks with relatively low transmission rate ($<$1 Mbps), like sensor networks, the energy consumption related to the channel encoding/decoding is very small compared to the overall consumption and can be neglected.

From (21), one can see that, in order to minimize the energy consumption E for a given rate R_b, one must minimize the power consumption of the amplifier (P_{amp}), since both P_{tx} and P_{rx} are fixed (hardware-dependent). In what follows we discuss how the cooperative protocols presented earlier can be used to reduce the overall energy consumption.

3.2 Energy Consumption of the Direct Transmission

Manipulating (6), one can obtain the minimum transmit power necessary to achieve a target outage probability $\mathcal{P}_o^*$ as

$$P_{i,\mathrm{DT}}^* = \frac{m\, N_0\, B(2^r - 1)}{\gamma_{i,j} \sqrt[m]{\mathcal{P}_o^*\, \Gamma(m+1)}}. \tag{23}$$

Thus, after placing (23) in (21), we have that the energy consumption of the direct transmission scheme can be written as

$$E_{\mathrm{DT}} = \frac{\frac{\xi}{\eta} P_{i,\mathrm{DT}}^* + P_{\mathrm{tx}} + P_{\mathrm{rx}}}{R_b} \quad \text{(J/bit)}. \tag{24}$$

3.3 Energy Consumption of the Cooperative Schemes

For the DF scheme, we find from (6) and (8) that the minimum power required to keep the outage probability below a certain threshold $\mathcal{P}_{o,\mathrm{DF}}^*$ is

$$P_{i,\mathrm{DF}}^* = \frac{m\, N_0\, B(2^{(r/R_{\mathrm{DF}})} - 1)}{\gamma_{i,j} \sqrt[2m]{2\mathcal{P}_{o,\mathrm{DF}}^*\, \Gamma(m+1)^2}}. \tag{25}$$

Recall that r is adjusted according to R_{DF} in (25) so that the effective spectral efficiency, in information bits/s/Hz, is made equal to that of the direct transmission, for a fair comparison among the two schemes.

Since the number of transmissions and receptions performed in the DF scheme are respectively $2M$ and $M^2 + M$ (see Table 1), the energy consumption (in J/bit) of the DF scheme is equal to

$$E_{\mathrm{DF}} = \frac{R_{\mathrm{DF}}}{R_b} \left[2 \left(\frac{\xi}{\eta} P_{i,\mathrm{DF}}^* + P_{\mathrm{tx}} \right) + (M+1) P_{\mathrm{rx}} \right]. \tag{26}$$

Note that the transmission rate R_b is also adjusted according to R_{DF} due to fair comparison purposes.

As presented earlier in this document, the diversity order of the DF scheme over Nakagami-m fading channel is equal to $2m$, reducing the required transmission power, as can be seen in (25). In Sect. 4, it is possible to evaluate if this energy savings at the amplifier is big enough to compensate for the additional transmissions and receptions needed by the cooperative scheme.

Through a similar procedure, after isolating the transmit power from the outage probability and taking into account the number of transmissions and receptions according to Table 1, it can be shown that the minimal transmit power and the energy efficiency of the GDNC and FA-GDNC schemes become respectively

$$P^*_{i,\mathrm{GDNC}} = \frac{m N_0 B(2^{(r/R_{\mathrm{GDNC}})} - 1)}{\gamma_{i,j} \Gamma(m+1)^{\frac{1}{m}} (\mathscr{P}^*_{o,\mathrm{GDNC}}/\mu)^{\frac{1}{m(M+k_2)}}}, \tag{27a}$$

$$E_{\mathrm{GDNC}} = \frac{R_{\mathrm{GDNC}}}{k_1\, R_b} \left[(k_1+k_2) \left(\frac{\xi}{\eta} P^*_{i,\mathrm{GDNC}} + P_{\mathrm{tx}} \right) + (M k_1 + k_2) P_{\mathrm{rx}} \right] \tag{27b}$$

and

$$P^*_{i,\mathrm{FA\text{-}GDNC}} = \frac{m N_0 B(2^{(r/R_{\mathrm{FA\text{-}GDNC}})} - 1)}{\gamma_{i,j} \Gamma(m+1)^{\frac{1}{m}} (\mathscr{P}^*_{o,\mathrm{FA\text{-}GDNC}}/\mu)^{\frac{1}{m(M+k_2)}}}, \tag{28a}$$

$$E_{\mathrm{FA\text{-}GDNC}} = \frac{R_{\mathrm{FA\text{-}GDNC}}}{k_1\, R_b} \left[(k_1+k'_2) \left(\frac{\xi}{\eta} P^*_{i,\mathrm{FA\text{-}GDNC}} + P_{\mathrm{tx}} \right) + (M k_1 + k'_2) P_{\mathrm{rx}} \right], \tag{28b}$$

where $\mu = \binom{k_1+k_2-1}{k_2}$. It is worthy mentioning that the energy consumption of the feedback channel in the FA-GDNC scheme has been neglected.[2]

From (27) and (28), one can notice that the energy consumption of both GDNC and FA-GDNC schemes depends on the number of cooperating nodes (M), besides the distance between them ($d_{i,j}$). In what follows we present some insights on the optimal number of users that minimizes the energy efficiency of the GDNC and FA-GDNC scheme for a given distance.

3.3.1 Optimal Number of Cooperating Nodes

By differentiating $E_{\mathrm{GDNC}}(M)$ with respect to M, equating it to zero and isolating M, it can be shown that the optimal (in the minimum energy consumption sense) number of cooperating users for the GDNC scheme is

$$M^*_{\mathrm{GDNC}} = \max\left\{ 2 \;;\; \frac{\ln(\Omega)}{2mW\left(\frac{1}{2}\sqrt{\frac{P_{\mathrm{rx}} R_{\mathrm{GDNC}} \ln(\Omega)}{m^2 \Psi}}\right)} - k_2 \right\}, \tag{29}$$

where $W(\cdot)$ corresponds to the Lambert-W function [5] and $\max\{2 \;;\; \cdot\}$ limits the minimum number of cooperative users to 2. Thus, from (29), it is possible to obtain the number of users that minimizes the energy consumption of the network for a given distance between them. This result can be used to create *cooperative clusters*, arranging the network users in smaller sets, where they cooperate with each other only within their associated clusters.

[2] Note that this is a reasonable assumption when the frame size is long enough, as discussed in [4].

For the DF scheme, by the same procedure, it can be shown that the optimal number of users that minimizes the energy consumption is $M^*_{\mathsf{DF}} = 2$, regardless the distance between the users. In the case of the FA-GDNC scheme, even though the energy consumption from (28b) still depends on the number of users M, due to the random nature of the code rate, it is hard (if possible) to obtain in a closed-form equation the optimal number of users that minimizes the energy consumption.

4 Numerical Results

In what follows we present some numerical results related to the previous analysis. The system parameters are in accordance to [6], and shown in Table 2. It is noteworthy that, for a target outage probability $\mathscr{P}_o^* = 10^{-4}$, the high SNR approximation of (6) can be shown to be tight.

In Fig. 2 the energy consumption (in J/bit) as a function of the distance (in meters) for a network with $M = 2$ cooperating users is presented, for the direct transmission, cooperation with the DF protocol and the GDNC scheme, the latter with $k_1 = k_2 = 2$ (values chosen so that the code rate of the GDNC scheme is equal to that of the DF scheme, and considering the restriction for k_2 in (15)). Each scheme was evaluated over Nakagami-m fading channels, with $m = 1$ (same as Rayleigh and referred to as NLOS, presented in Fig. 2a) and with $m = 2$ (with some line of sight, referred to as LOS, Fig. 2b). It can be seen that for short distances, the quadratic amount of receptions of the cooperative schemes compromises their energy efficiency, and thus the direct transmission becomes the most energy efficient scheme in this scenario. However, as the distance between users increases, the consumption of the transmit and receive circuits loses influence in (21). In this case, one can see that the GDNC and FA-GDNC schemes are the most energy efficient, due to their higher diversity order. More than that, feedback makes the FA-GDNC scheme slightly better than the GDNC scheme, mainly when the channel is in better conditions (LOS). It can also be noticed that the total absence of LOS between the users degrades more severely the performance of direct transmission and the DF scheme than that of the network-coded cooperative schemes.

Table 2 Parameters considered in the numerical results

M_l	40 dB	f_c	2.5 GHz	$\mathscr{M}$	4
N_f	10 dB	B	10 KHz	N_0	−174 dBm/Hz
G	5 dBi	r	1 b/s/Hz	α	4
P_{tx}	97.9 mW	η	0.35	$\mathscr{P}_o^*$	10^{-4}
P_{rx}	112.2 mW				

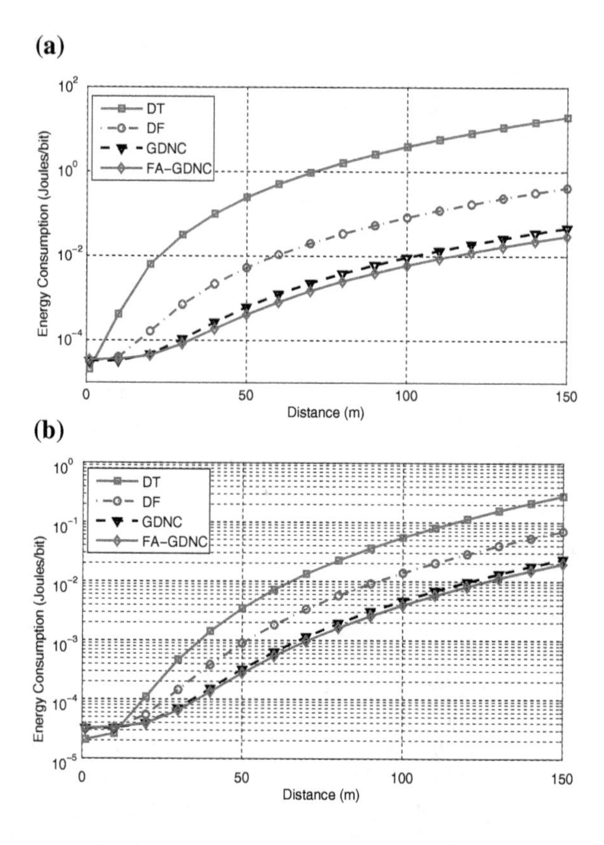

Fig. 2 Energy Consumption (in J/bit) as a function of the distance (in meters) for a network with $M=2$ cooperating users, considering direct transmission (DT), DF cooperation, the GDNC and FA-GDNC schemes, the latter two with $k_1=k_2=2$, all suffering Nakagami-m fading, with **a** $m=1$ (NLOS); **b** $m=2$ (LOS)

The relationship between the energy consumption (in J/bit) and the number of cooperative users M, while considering a fixed distance $d=50$ m is presented in Fig. 3, for the same schemes compared in Fig. 2.

One can see that, in the presence of LOS (Fig. 3b), the GDNC (the same holds for the FA-GDNC) scheme outperforms the direct transmission when $M<600$, but tends to be less efficient for a larger number of users. Again, this is due to the quadratic amount of receptions required in the GDNC scheme. It can also be noticed in Fig. 3 that the GDNC scheme has an optimal value for the number of users, where the energy consumption is minimum, depending on the presence or absence of LOS. Besides that, it is possible to see that there is a large range of M for which the energy consumption is not significantly affected. It can be seen that, despite being more energy efficient than GDNC for a small number of cooperating users, the FA-GDNC tends to the same energy consumption of the GDNC scheme as the number of users increases, LOS existing or not.

In Fig. 4, the optimal value M^* for the number of cooperative users is presented as a function of the distance for both the GDNC and FA-GDNC schemes. The results for the GDNC scheme were obtained both numerically from (27b) and analytically in accordance to (29). For the FA-GDNC scheme, only the result obtained numerically

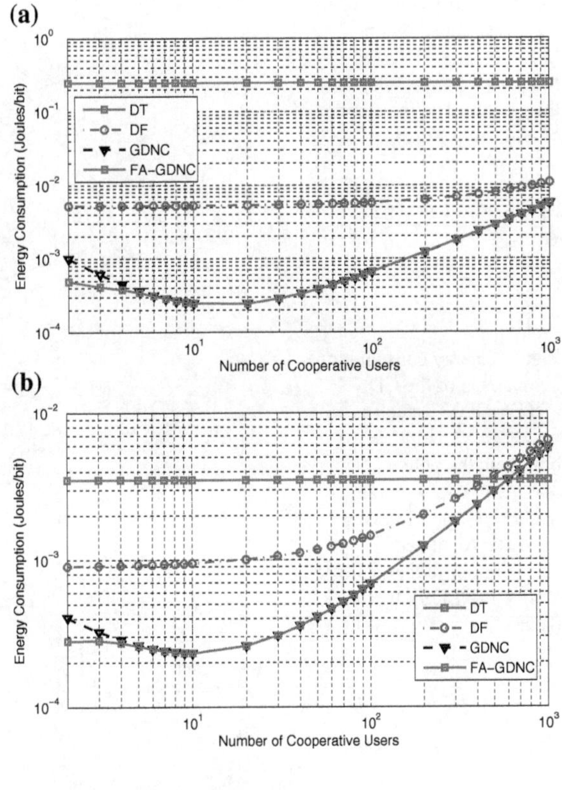

Fig. 3 Energy Consumption (in J/bit) as a function of the number of cooperative users, for $d = 50$ m, considering the direct transmission (DT), DF cooperation, the GDNC and FA-GDNC schemes, the latter two with $k_1 = k_2 = 2$, all suffering Nakagami-m fading, with **a** $m = 1$ (NLOS); **b** $m = 2$ (LOS)

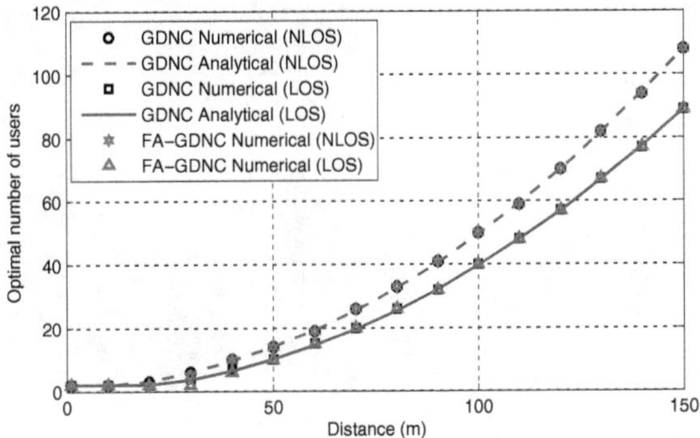

Fig. 4 Optimal value of the number of cooperative users as a function of the distance for the GDNC scheme, with $k_1 = k_2 = 2$, all suffering Nakagami-m fading, with m = 1 (NLOS) and m = 2 (LOS)

from (28b) is presented. It can be seen that, as the distance between the nodes increases, the number of users necessary to minimize the energy consumption also increases. However, for a large range of distances this number is small, being feasible in practice. It is also possible to see that the presence of LOS reduces the optimal value M^*. Figure 5 presents a 3D plot of the energy consumption as a function of both the distance and the number of cooperative users, for the NLOS scenario. As a general rule, we can conclude that the advantage of the GDNC scheme over the

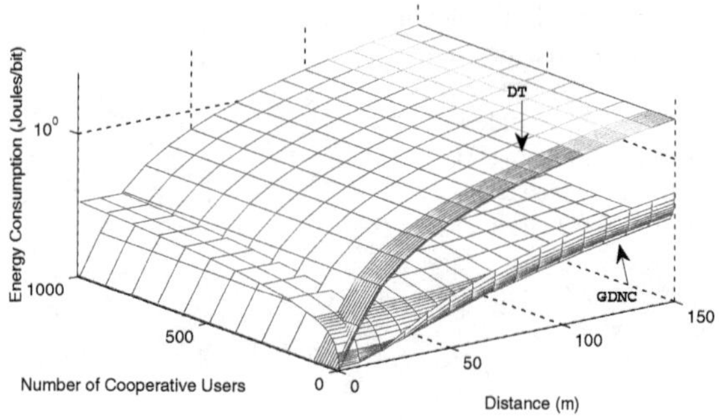

Fig. 5 Energy consumed by the direct transmission (NLOS) and GDNC (NLOS) schemes in relationship to the distance and the number of cooperative users

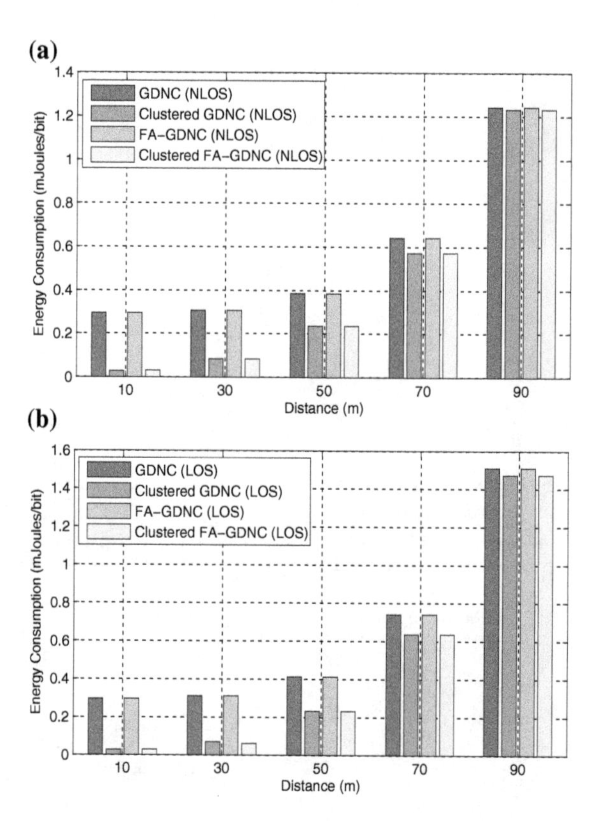

Fig. 6 Energy consumption comparison of the GDNC and FA-GDNC schemes with $k_1 = k_2 = 2$ and $M = 50$, both with the optimal number of users—arranged in clusters—as a function of the distance, under Nakagami-m fading, with **a** $m = 1$ (NLOS); **b** $m = 2$ (LOS)

direct transmission increases when the distance increases, and decreases with the number of users.

In Fig. 6 we present the advantage of operating with the optimal value M^* for the number of users in comparison to a system operating with a fixed number of users $M = 50$, for the GDNC and FA-GDNC schemes, both with $k_1 = k_2 = 2$. Thus, we assume that the total number of users M can be rearranged in a number of clusters with M^* users only, maximizing the energy efficiency. From the numerical results one can see that, for all the range of considered distances, the optimization in the number of cooperative users results in energy savings (the smaller the distance, the higher the savings). Nevertheless, one can see that the presence or not of LOS is not very significant under this metric.

5 Final Comments

In this chapter, we derived an energy efficiency analysis of a multiple access network considering different communication schemes, including direct non-cooperative transmission, decoded-and-forward cooperation, and network-coded cooperation, all of them subject to Nakagami-m block fading, so that the influence of some line-of-sight is taken into account. The effect of the circuitry power consumption was also considered in the analysis, showing that the use of network coding can be considerably beneficial in terms of energy efficiency and that there exists an optimal number of cooperating nodes that minimizes the energy consumption for a given distance. By the use of network coding and by the appropriate organization of cooperating nodes into clusters the energy efficiency is maximized and energy savings of an order of magnitude with respect to the direct non-cooperative transmission can be achieved.

References

1. Ahlswede, R., Cai, N., Li, S.Y., Yeung, R.: Network information flow. IEEE Trans. Inf. Theory **46**(4), 1204–1216 (2000)
2. Akyildiz, I., Su, W., Sankarasubramaniam, Y., Cayirci, E.: A survey on sensor networks. Commun. Mag. IEEE **40**(8), 102–114 (2002)
3. Alamouti, S.: IEEE J. Sel. Areas Commun. **16**(8), 1451–1458 (1998). doi:10.1109/49.730453
4. Brante, G.G.O., Kakitani, M.T., Souza, R.D.: Energy efficiency analysis of some cooperative and non-cooperative transmission schemes in wireless sensor networks. IEEE Trans. Commun. **59**(10), 2671–2677 (2011)
5. Corless, R.M., Gonnet, G.H., Hare, D.E.G., Jeffrey, D.J., Knuth, D.E.: On the Lambert W function. Adv. Comput. Math. **5**(1), 329–359 (1996)
6. Cui, S., Goldsmith, A., Bahai, A.: Energy-constrained modulation optimization. IEEE Trans. Wireless Commun. **4**(5), 2349–2360 (2005)
7. Feng, D., Jiang, C., Lim, G., Cimini, J.L.J., Feng, G., Li, G.: A survey of energy-efficient wireless communications. IEEE Commun. Surv. Tuts **15**(1), 167–178 (2013). doi:10.1109/SURV.2012.020212.00049
8. Goldsmith, A.: Wireless Communications. Cambridge University Press, Cambridge (2005)
9. Ho, T., Medard, M., Koetter, R., Karger, D., Effros, M., Shi, J., Leong, B.: A random linear network coding approach to multicast. IEEE Trans. Inf. Theory **52**(10), 4413–4430 (2006)
10. Howard, S.L., Schlegel, C., Iniewski, K., Iniewski, K.: Error control coding in low-power wireless sensor networks: when is ECC energy-efficient? EURASIP J. Wireless Commun. Netw. **2**, 1–14 (2006)
11. Hu, Y., Gross, J., Schmeink, A.: QoS-constrained energy efficiency of cooperative ARQ in multiple DF relay systems. IEEE Trans. Veh. Technol. **PP**(99), 1–1. doi:10.1109/TVT.2015.2399398 (2015)

12. Koetter, R., Médard, M.: An algebraic approach to network coding. IEEE/ACM Trans. Netw. **11**(5), 782–795 (2003)
13. Laneman, J.N., Tse, D.N.C., Wornell, G.W.: Cooperative diversity in wireless networks: efficient protocols and outage bahavior. IEEE Trans. Inf. Theory **50**(12), 3062–3080 (2004)
14. Li, S.Y., Yeung, R., Cai, N.: Linear network coding. IEEE Trans. Inf. Theory **49**(2), 371–381 (2003). doi:10.1109/TIT.2002.807285
15. Lin, S., Costello Jr, D.J.: Error Control Coding: Fundamentals and Applications. Prentice-Hall, Upper Saddle River (1983)
16. Macwilliams, F., Sloane, N.: The Theory of Error Correcting Codes. North Holland, Amsterdan (1977)
17. Quek, T., Dardari, D., Win, M.: Energy efficiency of dense wireless sensor networks: to cooperate or not to cooperate. IEEE J. Sel. Areas Commun. **25**(2), 459–470 (2007)
18. Rayel, O., Rebelatto, J., Souza, R., Uchoa-Filho, B., Li, Y.: Energy efficiency of network coded cooperative communications in nakagami-m fading. IEEE Signal Process. Lett. **20**(10), 960–963 (2013). doi:10.1109/LSP.2013.2276438
19. Rayel, O.K., Rebelatto, J.L., Souza, R.D., Uchôa-Filho B.F., Li, Y.: On the energy efficiency of feedback-assisted network coding in multiuser cooperative systems. In: Proceedings of the IEEE 23rd International Symposium on Personal, Indoor and Mobile Radio Communication (PIMRC' 12), Sydney, Australia (2012)
20. Rebelatto, J.L., Uchôa-Filho, B.F., Li, Y., Vucetic, B.: Adaptive distributed network-channel coding. IEEE Trans. Wireless Commun. **10**(9), 2818–2822 (2011)
21. Rebelatto, J.L., Uchôa-Filho B.F., Silva, D.: Full-diversity network coding for two-user cooperative communications. In: Proceedings of IEEE Information Theory Workshop (ITW' 11), Paraty, Brazil (2011)
22. Rebelatto, J.L., Uchôa-Filho, B.F., Li, Y., Vucetic, B.: Multiuser cooperative diversity through network coding based on classical coding theory. IEEE Trans. Signal Process. **60**(2), 916–926 (2012). doi:10.1109/TSP.2011.2174787
23. Sendonaris, A., Erkip, E., Aazhang, B.: User cooperation diversity: part I and part II. IEEE Trans. Commun. **51**(11), 1927–1948 (2003)
24. Sheng, Z., Fan, J., Liu, C., Leung, V., Liu, X., Leung, K.: Energy-efficient relay selection for cooperative relaying in wireless multimedia networks. IEEE Tran. Veh. Technol. **64**(3), 1156–1170 (2015). doi:10.1109/TVT.2014.2322653
25. Tarokh, V., Seshadri, N., Calderbank, A.: Space-time codes for high data rate wireless communication: performance criterion and code construction. IEEE Trans. Inf. Theory **44**(2), 744–765 (1998). doi:10.1109/18.661517
26. Tse, D., Viswanath, P.: Fundamentals of Wireless Communications. Cambridge University Press, Cambridge (2005)
27. Wang, X., Li, J.: Improving the network lifetime of MANETs through cooperative MAC protocol design. IEEE Trans. Parallel. Distrib. Syst. **26**(4), 1010–1020 (2015). doi:10.1109/TPDS.2013.110
28. Wang, Z., Giannakis, G.: A simple and general parameterization quantifying performance in fading channels. IEEE Trans. Commun. **51**(8), 1389–1398 (2003)
29. Wu, G., Talwar, S., Johnsson, K., Himayat, N., Johnson, K.: M2m: From mobile to embedded internet. Commun. Mag. IEEE **49**(4), 36–43 (2011)
30. Xiao, L., Fuja, T., Kliewer, J., Costello, D.: A network coding approach to cooperative diversity. IEEE Trans. Inf. Theory **53**(10), 3714–3722 (2007)
31. Xiao, M., Skoglund, M.: M-user cooperative wireless communications based on nonbinary network codes. In: Proceedings of IEEE Information Theory Workshop (ITW' 09), pp 316–320 (2009)
32. Xiao, M., Skoglund, M.: Multiple-user cooperative communications based on linear network coding. IEEE Trans. Commun. **58**(12), 3345–3351 (2010)

Energy Efficient Fundamental Theory and Technical Approach

Liqiang Zhao, Kun Yang and Guogang Zhao

Abstract With the goal of mitigating the environmental impact of information and communication technology (ICT) industry, "green" wireless communication technologies have drawn increasing attention from governments, academia and industry. Hence, energy management technique becomes one of the key considerations in the design of future wireless networks, especially for the inherently energy-constrained wireless ad-hoc networks (WANs) discussed in this chapter. Many schemes, which can be adopted at different layers of the protocol stack in order to accommodate the energy awareness in WANs, have been proposed in the recent past. However, it's absolutely impossible to expect the unilateral reduction of energy consumption as the only goal for the WANs since the diversiform energy-driven performance indicators have to be guaranteed for system's normal operation. Hence, recent research works focus on providing energy-efficient solutions with regarding to the QoS guarantee. In this chapter, we firstly carry out a comprehensive analysis of the relevant efficiency metrics as the fundamental evaluation, including the spectrum efficiency in b/s/Hz, energy efficiency in b/s/Hz/W (or b/Joule/Hz), area spectrum efficiency in b/s/Hz/km^2, and distance-related efficiency in (b m)/s/Hz/W, while we give out a comprehensive summary of optimization criteria for energy-efficient WANs. Secondly, we provide the taxonomy of various energy management schemes for the WANs covering all the layers of the protocol stack, specifically including the physical layer, medium access control (MAC) layer, network layer and cross-layer design. Importantly, we tend to discuss a range of energy-efficient MAC protocols proposed for the WANs, especially carrier sense multiple access with collision avoidance

L. Zhao (✉) · G. Zhao (✉)
State Key Laboratory of Integrated Service Networks, Xidian University,
2 Taibai South Road, Xi'an 710071, Shaanxi, China
e-mail: lqzhao@mail.xidian.edu.cn

G. Zhao
e-mail: ggzhao@s-an.org

K. Yang (✉)
School of Computer Science and Electronic Engineering (CSEE),
University of Essex, Wivenhoe Park, Colchester, Essex CO4 3SQ, UK
e-mail: kunyang@essex.ac.uk

© Springer International Publishing Switzerland 2016

M.Z. Shakir et al. (eds.), *Energy Management in Wireless Cellular and Ad-hoc Networks*, Studies in Systems, Decision and Control 50,
DOI 10.1007/978-3-319-27568-0_9

(CSMA/CA), the basic MAC protocol for WANs, and its potential energy management schemes. In each layer design, we note that the network energy consumption is always related with other performance indicators we can't ignore, for instance the network lifetime, connection reliability, network throughput, and etc. By exploiting the performance improvement potentials from the dependence among the original protocols layer, we finally pay attention to the cross layer design. In one word, the guiding study of fundamental theory goes together with the energy management with a specific system target. Meanwhile, we point out the future research perspectives in energy management for WANs.

1 Introduction

The development of high-throughput mobile communications systems is coming with a significant energy cost, which is economically unsustainable and unsuitable for future-proof communications. As one typical kind of network architecture tended to be adopted in future widely such as Internet of Things (IoT), sensors networks and so on, wireless ad-hoc networks (WANs) should also satisfy these rigorous energy consumption requirements. It will be even trickier that the WAN is an inherent energy-constrained system as formed by a collection of isolated nodes with limited energy resource, all of which have to achieve a complete wireless communication process by mutual cooperation without the aid of any established infrastructure. Despite of the transmission power (i.e., power consumed by radio transmitter dominating nodes' power consumption), signal processing portions for node's information reception, and even other correlative hardware-level network operations, also consume remarkable power [1, 2]. In one word, all the aspects associated with each network functionality in the WANs are driven by the limited energy resource distributed in the individual node.

Many schemes, which can be adopted at different layers of the protocol stack in order to accommodate energy awareness in WANs, have been proposed in the recent past [3]. However, it's absolutely impossible to expect unilateral energy consumption reduction as the only goal for the WANs since the diversiform performance indicators have to be guaranteed for system's normal operation. Furthermore, many exciting literatures are dedicated to the energy efficiency of WANs, e.g., adopting more efficient processor or optimization across protocol layers. However, a general evaluation performance criterion must be addressed first of all. Hence, the concept of "green" wireless communications have drawn increasing attention from governments, academia and industry.

Whilst a widely accepted definition of "green" wireless communications remains an open problem, there is a general consensus that it is synonymous with energy-efficient systems, leading to the open problem of green metrics. Indeed, the recent white paper from the Federal Communications Commission (FCC) Technological Advisory Council (TAC) alone refers to 25 different spectrum efficiency metrics, and this is not an exhaustive list. In many cases, researchers use the b/s/Hz spectrum efficiency, b/s/Hz/km^2 area spectrum efficiency [4] the b/TENU power efficiency [5], the b/s/Hz/W energy efficiency [6] and the (b m)/s/Hz/W distance-related efficiency [7] to evaluate the communications systems. In the WANs, the famous solid works by [8] and [9] give out an in-depth study to the WANs performance upper bound, namely transport and transmission capacity respectively.

By no means should the efficiency metrics above be classified as less efficient, since in the appropriate circumstances they are capable of considerably improving the overall performance of the entire network. However, the appropriate choice of the network optimization criteria based on different efficiency metrics can have a profound effect on the overall network performance. The most outstanding example happens in the design of WANs in network layer where various system targets are companied with abundant metrics, reflecting the selection importance. Therefore, above all, this chapter discusses the available efficiency metrics. To sum up, the fundamental theory along with appropriate choice of green metrics based on fickle physical scenario would offer the guiding for the future research on energy management schemes.

In the physical layer, energy management mainly refers to the structure optimization of battery systems, and furthermore how to enable all the hardware-level devices, e.g., processors-related parts, to make full use of the energy supplied by battery. The former tries to exploit the inherent batteries property to recover their charge when kept idle in order to increase the amount of energy provided by the power source. Recent research have illustrated that significant gain can be obtained in the total amount of energy supplied by them and the lifetime of WANs can be extended, e.g., [10, 11]. For the latter, we also note that the major consumers of power in the devices can lead to significant power saving potentials, e.g., by improving computational efficiency. Meanwhile, the energy can be consumed more efficiently by introducing advanced processors components, e.g., power amplifiers. Despite of the devices' self-improvement, this process may be related with higher layers information. For instance, we can switch off nodes duo to the in-existence of transmit routing path for saving energy based on routing information.

The performance of WANs depends highly on how the medium access control (MAC) protocol is designed, and so far, most WANs are implemented using wireless LANs (WLANs). IEEE 802.11x is one of the most influential WLAN standards, and its basic MAC protocol, distributed coordination function (DCF), is based on carrier sense multiple access with collision avoidance (CSMA/CA), as one of typical contention-based MAC protocols. Currently, CSMA/CA has been the facto MAC standard of WANs, and is widely used in almost all of the testbeds and simulations for WANs research. Up to now, since MAC protocols directly decide how the limited wireless resources are shared among existing nodes, the continuous investment in the MAC protocols of WANs has been aiming at improving the efficiency of CSMA/CA, and brings about a wealth of theoretical knowledge and practical engineering solutions. In this chapter, we will introduce some kinds of energy management schemes of ad hoc networks in MAC. mainly divided into three parts as follows: Firstly, we discuss how the nodes can save energy on idle-time in MAC layer [26, 36]; Secondly, the relationship between transmission power and data arrival rate of nodes is discussed [27, 28]; Finally, we introduce a heuristic MAC protocol to improve both the energy and spectrum efficiency of CSMA/CA in WANs.

In fact, the energy consumed by signals' transmission is always related to their routing paths. Meanwhile the routing paths are also influenced by distance between transmitter and receiver and geographic locations. Thus designing efficient energy management schemes in routing layer can offer a potential energy saving space. For example, a large amount of energy can be saved by dividing the original direct route path with long distance into several paths by means of signal relay. In comparison, some simple routing designs while ignoring energy consumption, e.g., only minimizing hop count, may result in energy dissipation, in particular when the nodes number in the network is small but the existence of traffic loads is heavy. Aimed at this, many research works have proposed the energy-aware routing protocols [17, 18]. On the other hand, we note that the state of node mode is also closely bound up with the energy consumption, e.g., four possible modes: transmit, receive, idle, and sleep (the least amount of energy) for radio transceiver. In other words, we also need to consider the nodes' mode change resulted from routing path selection. To sum up, the issue of designing appropriate and adjustable energy management solutions located in routing layer is urgent to addressed to reduce energy consumption.

In addition to the realization of energy savings at various layers of the protocol stack, some research, such as in [34] and [35], the authors suggest that instead of the independent consideration of these layers, a cross-layer design solution provides an efficient methodology to implement energy consumption in WANs. The author in [34] proposed an energy-efficient solution for addressing both transmit power control and scheduling where the solution is implemented by the interaction between the physical and MAC layers. In addition, the authors in [35] attempt to provide these solutions with end-to-end delay QoS guarantees for sessions.

In this chapter as illustrated in Fig. 1, we firstly focus on the fundamental theory of energy management in Sect. 2 by discussing diverse efficiency metrics which tends to guide the following scheme design and performance optimization. Secondly, we tend to introduce the design philosophy in each protocol stack severally from Sect. 3 to Sect. 5 and integrate them in the cross layer design in Sect. 6. Finally, we give out the summary while pointing out the future possible design perspectives in WANs.

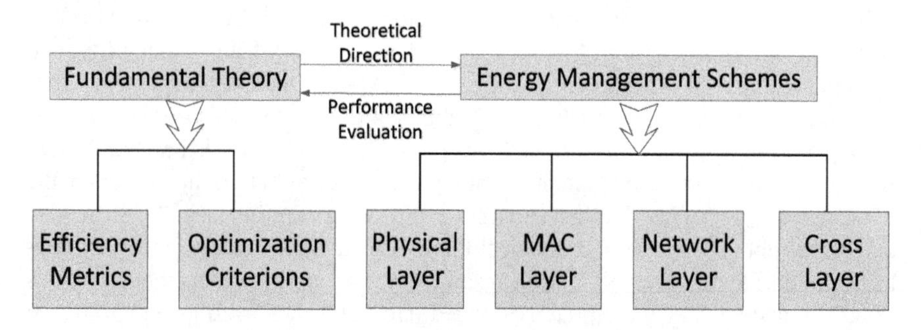

Fig. 1 Overall chapter structure and relationship for energy management in WANs

2 Fundamental Theory on Energy Management

In this section, we begin with the fundamental problems of energy management in the WANs: what's the meaning of energy-efficiency and what criterion we should obey when optimizing the corresponding target. Currently, a widely accepted definition of "green" wireless communications remains an open problem, there is a general consensus that it is synonymous with energy-efficient systems. As the importance of energy-efficient fundamental evaluation, many kinds of green efficiency metrics have been proposed in recent years as mentioned in last section. By no means should any efficiency metric be classified as less efficient, since in the appropriate circumstances they are capable of considerably improving the overall performance of the entire network. However, the appropriate choice of the network optimization criteria based on different efficiency metrics can have a profound effect on the overall network performance. Therefore, above all, this section considers the available efficiency metrics with analyzing their property mathematically in a comprehensive way to make preparations for the evaluation and measurement, and then propose a general optimization criterion with several system constraints.

2.1 Fundamental Evaluation: Efficiency Metrics

By the initial definition, efficiency is the ratio of the utility attained to the resources consumed. Clearly, the notion of efficiency is closely related to the specific definition of the utility and resources. In wireless communications, a user aims for successfully transmitting his packets over a certain distance to the distant receiver under specific QoS requirements, given the available resources. Hence, the radio utility metrics should include the ratio of successfully transmitted packets, QoS metrics (such as the throughput in b/s, delay and delay jitter in seconds, as well as the packet-loss-ratio) and the transmission distance in meter. The resource metrics should include all the radio resources consumed, which may be classified as time-, frequency-, space-, code-, power-, and diverse other resources in WANs.

There has been a lot of work on the definition of radio efficiency, but many researchers consider spectrum efficiency and energy efficiency as the principal efficiency metrics [7]. The spectrum efficiency of a PtP (Point-to-Point) link is defined as the number of bits per unit spectrum, which corresponds to bit-per-second-per-Hertz (b/s/Hz) in the case of single-input single-output (SISO) systems. By contrast, for multiple-input multiple-output (MIMO) systems it is equivalent to the bit-per-second-per-Hertz-per-antenna (b/s/Hz/antenna) metric. Spectrum efficiency should be carefully distinguished from the area spectrum efficiency (ASE) expressed in [b/s/MHz/km^2], because the latter takes into account the cellular frequency-reuse factor [4]. On the other hand, a host of spread-spectrum methods intentionally sacrifice the b/s/Hz spectrum efficiency for the sake of achieving a better bit-per-second-per-Hertz-per-Watt (b/s/Hz/W) energy efficiency. Moreover, [5] defines the power efficiency as the number of bits per thermal noise signal energy unit (TNEU), which may be viewed as an enhanced energy efficiency metric. Moreover, inspired by the initial definition of efficiency, we may define a distance-related efficiency metric, namely the successfully transmitted bits multiplied by the transmission distance per resource category in a specific domain, corresponding to (bit-meter)-per-second-per-Hertz-per-W ((b m)/s/Hz/W) [7].

Then we are interested in the mathematical property of the efficiency metrics above. Assuming P_t is the transmit power and d stands for the distance between the transmitter and receiver. If the pathloss is evaluated as αd^{β} where α and β are the pathloss factor, then we can get the signal to noise ratio (SNR) while B and n_0 stand for bandwidth and noise spectral density, i.e., S/n_0B, based on Shannon-Hartley theorem. Therefore, the spectrum efficiency of point-to-point (PtP) wireless link as illustrated in Fig. 2 is quantified as the throughput per spectrum unit and given as

$$\eta_s = \log_2\left(1 + \frac{S}{n_0 B}\right) = \log_2\left(1 + \frac{P_t \cdot \alpha d^{-\beta}}{n_0 B}\right), \text{b/s/Hz}. \tag{1}$$

Moreover, if the total amount of energy consumed at the transmitter is $aP_t + b$ including the part of radio frequency aP_t with b standing for the power consumed by

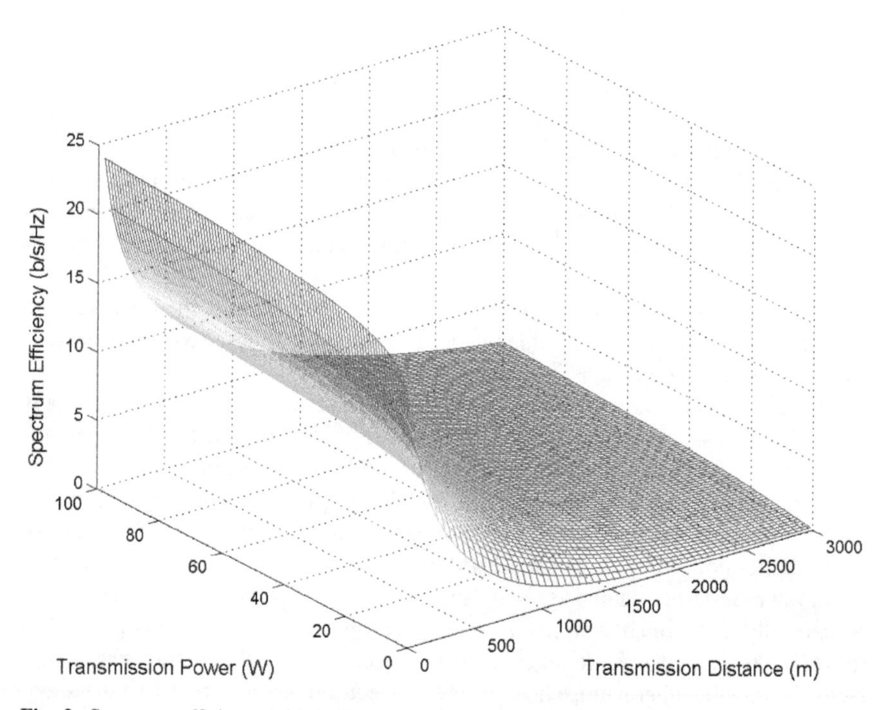

Fig. 2 Spectrum efficiency of PtP (Point-to-Point) link with respect to transmission power and transmission distance

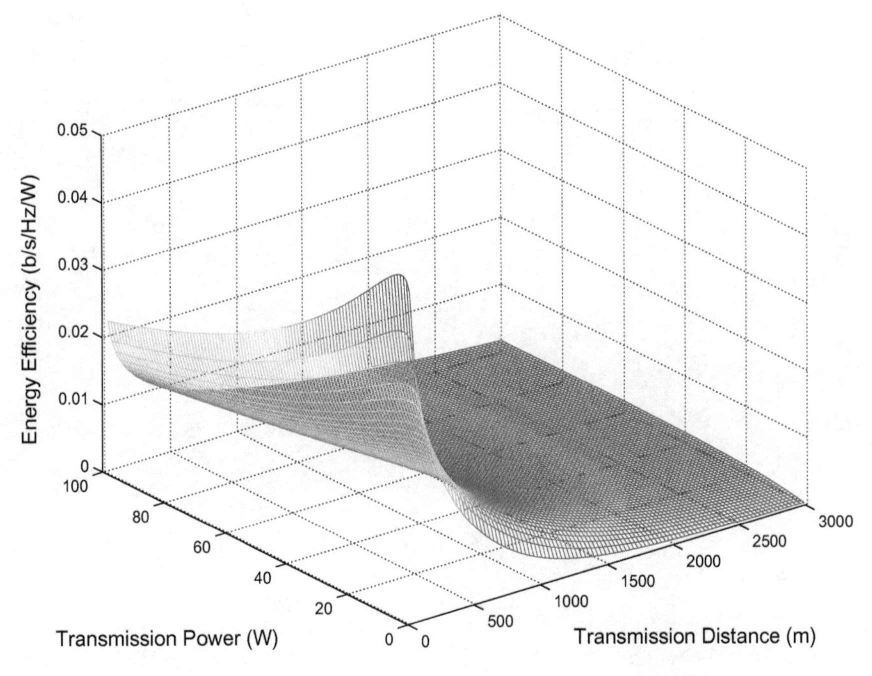

Fig. 3 Energy efficiency of PtP link with respect to transmission power and transmission distance

power amplifiers, signal processing, the cooling fans in eNBs, where a is the radio frequency efficiency factor, then we have energy efficiency in Fig. 3 expressed as

$$\eta_e = \frac{\eta_s}{aP_t + b} = \frac{\log_2\left(1 + \frac{\alpha P_t}{d^\beta n_0 B}\right)}{aP_t + b}, \text{b/s/Hz/W}. \quad (2)$$

We also illustrate the definition of our proposed distance-related efficiency in Fig. 5 as follows:

$$\eta_g = \eta_e \cdot d = \frac{d \cdot \log_2\left(1 + \frac{\alpha P_t}{d^\beta n_0 B}\right)}{aP_t + b}, (\text{b m})/\text{s/Hz/W}. \quad (3)$$

When γ_s denotes the average SNR recorded at the receiver, we have the b/TNEU power efficiency $\eta_{TENU} = \eta_s/\gamma_s$ as showed in Fig. 4. Naturally, all the above four efficiencies are dependent on both the transmission power and the transmission distance. We can know that the higher the transmission power and the shorter the transmission distance, the higher the receiver's SNR, which increases the bandwidth efficiency. The energy efficiency is a decreasing function of the transmission distance; but it is a convex function of the transmission power. The power efficiency is also a monotone function, hence the lower the transmission power and the longer the transmission distance, the higher the power efficiency. Finally, the distance-related efficiency is a

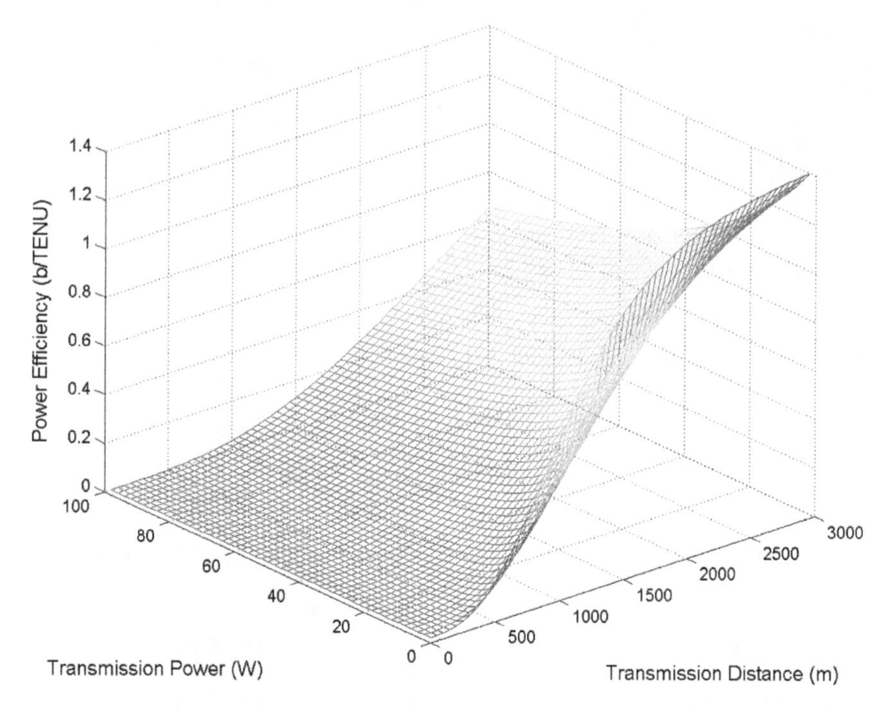

Fig. 4 Power efficiency of PtP link with respect to transmission power and transmission distance

two-dimensional convex function, which has a single peak value over the two arguments. In the appropriate circumstances, all the efficiency metrics above are relevant in terms of the overall performance of the entire network. However, the appropriate choice of the network optimization criteria based on different efficiency metrics can have a profound effect on system performance. Moreover, these definitions have different mathematical features, which are also closely related to their usage as a utility function in optimization problems. When furthermore extending efficiency metrics into system-level evaluation, much more system factors may be related with the resource utility while considering some factors resulting in the deterioration of performance, such as the interference. On the other hand, in most practical communication scenarios characterized by various performance-limiting factors including channel fading, interference as well as latency and complexity constraints, the actual attainable bandwidth-, energy-, power- and efficiency are considerably lower than the predicted values. Adaptive modulation and coding (AMC) is an appealing solution for a wireless link to approach the above efficiencies [6]. Moreover, the authors of [7] discusses the efficiencies of cellular networks, where all the detailed parameters are illustrated in [7].

There is no doubt that both spectrum and energy resource are precious and scarce. In recent years, many researchers are enthusiastic about optimizing both simultaneously or how to achieve tradeoff between them [6, 7]. Obviously, spectrum efficiency

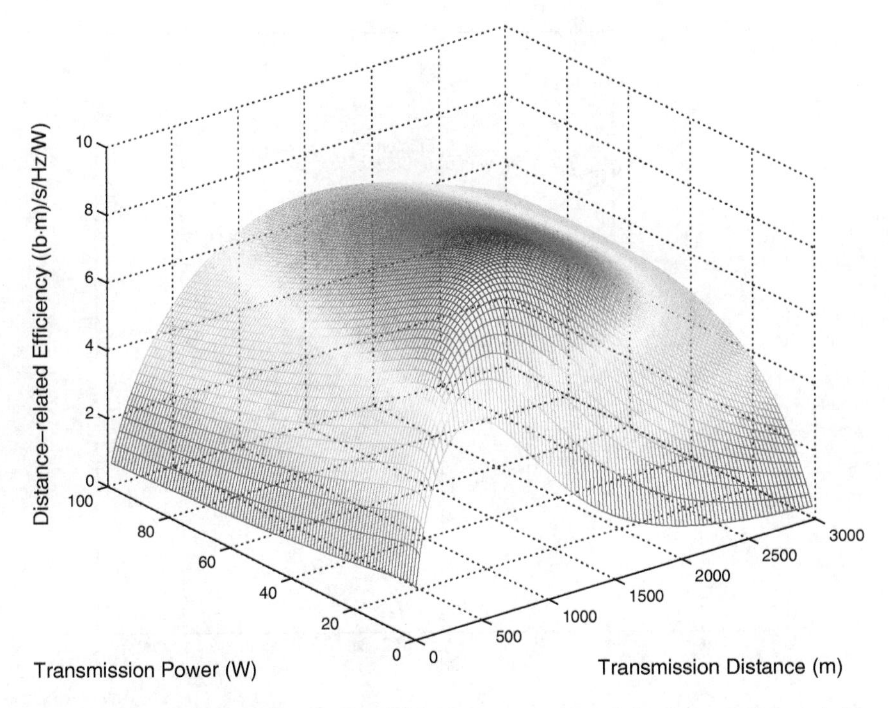

Fig. 5 Distance-related efficiency of PtP link with respect to transmission power and transmission distance

is increasing with respect to SNR increase. However, the mathematical property of energy efficiency is complex since it's also influenced by the energy consumption model at the end of radio transmitter, as showed in Fig. 6. When ignoring the impact of non-transmission power, i.e., $b = 0$, energy efficiency is a decreasing function with respect to SNR. In other words, it is completely impossible to achieve a synchronous optimization of spectrum and energy efficiency. In comparison, if we add the non-transmission power into our consideration ($b \neq 0$), we find that energy efficiency increases firstly and then decreases regarding to SNR. In other words, energy efficiency becomes a convex function with an unique optimality. Hence, we can optimize both of them in front of optimality and achieve a tradeoff behind it. In the next section, we continue to introduce a general fundamental optimization criterions with a specific efficiency metric and several system resource and users' QoS constraints.

2.2 *Fundamental Optimization Criterions*

For the practical WAN, there exist diversiform network forms duo to the usage in complex physical environment. As a matter of course, the standards do not explicitly specify the energy management in each protocols stack. Furthermore, this open

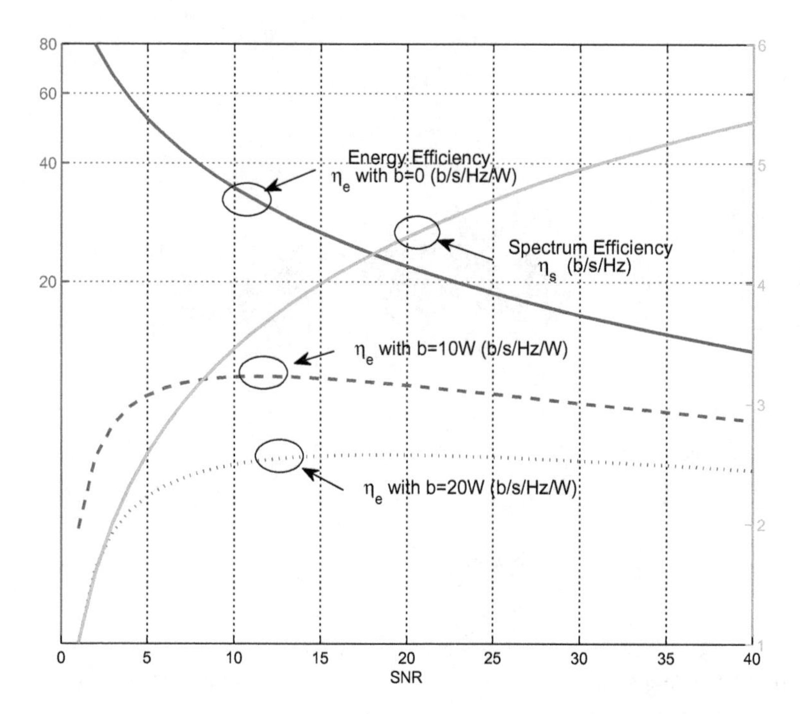

Fig. 6 Relationship between energy efficiency and spectrum efficiency with different b

structure facilities the creation of new innovative algorithms conceived for maximizing the system multiple performances. Most available energy management can be modelled as a convex optimization problem for the sake of maximizing the system's utility under two constraints, i.e.,

$$\arg\max \ \mathit{Efficiency} = \arg\max \ {}^{\text{Achieved utility}}\!/_{\text{Consumed resources}}$$

$$\begin{array}{l} \text{S.T.1(satisfying the minimal QoS requirements)} \\ \quad \begin{cases} \mathit{Bandwidth} > \mathit{Bandwidth}_{\min} \\ \mathit{Delay} < \mathit{Delay}_{\max} \\ \mathit{Jitter} < \mathit{Jitter}_{\max} \\ \mathit{PacketLossRate} < \mathit{PacketLossRate}_{\max} \end{cases} \\ \text{S.T.2(satisfying the constraints of available resources)} \\ \quad \begin{cases} \text{Allocated timeslots} \leq \text{Available timeslots} \\ \text{Allocated subcarriers} \leq \text{Available subcarriers} \\ \text{Allocated antennas} \leq \text{Available antennas} \\ \text{Transmitted power} \leq \text{Max transmission power} \end{cases} \end{array} . \tag{4}$$

As an original mathematical model above, we can introduce some related efficient methods to achieve the optimal solution. For example, the game theory can be regarded as an efficient method in Refs. [12, 13]. More specifically, the former solve an optimal energy efficient radio resource allocation problem in Low-Medium-Altitude aerial platform based TD-LTE networks against disaster. In the absence of constraints, the optimal system utility can be evaluated mathematically. However, the optima may be inapplicable to practical communication systems. For example, in [12, 13], the optimal transmission power values were obtained for achieving the maximal system utility based on game theory. However, the optimal power value may become negative or higher than the maximum value. Therefore, all the algorithms exploitation should carefully consider the above two constraints.

3 Physical Layer Design

In this section, we tend to address two aspects in physical layer design including processor management and battery system management. The former can make processors operating efficiently with lower energy consumption. In the contrary, the latter is aimed at increasing the power source supplied by the battery system.

3.1 Processor Management

The efficient processor design results in a significant improvement in the energy saving. Some design approach include adjusting clock speed CPUs, disk spin down, and flash memory. We give out some of the sources of power consumption in WANs and the corresponding solutions to reduce power consumption as stated in [15]. Major sources of power consumption in WANs are the transmitters and receivers of the communication module. The design of transceivers has a significant effect on power consumption. Hence, much attention must be taken while designing them. Meanwhile, switching off various units of the hardware while idling reduces energy consumption. Instead of switching off fully, different operation stages may be discussed, and each of them has a different power requirement level. Other techniques in this area are largely identical but with minor differences, paying attention the structure design of hardware and switching devices adaptively, hence we don't take much space here.

3.2 Battery System Management

Battery system management stands for the design with mainly taking into account of the battery and its internal characteristics. They try to maximize the amount of energy provided by the power source by exploiting the inherent property of batteries to recover their charge when kept idle. The lifetime of WANs covering each node is determined by the capacity of its energy source. Related work in [14] shows that node's lifetime can be extended by introducing techniques that make efficient utilization of the battery power. As summarized in [25], the stochastic model of the discharge pattern of batteries employs two key aspects affecting the node's lifetime: rate and the recovery capacity effect. To be specific, the authors of [14] illustrated that a pulsed current discharge applied for bursty stochastic transmissions improves battery lifetime. A battery subjected to pulsed current discharge possesses a higher lifetime than one with equivalent continuous current discharge. In [11] and [14], a model for battery pulsed discharge with recovery capacity effect is considered. The model proposed consists of a battery with a theoretical capacity and an initial nominal battery capacity. Battery behavior is considered as a discrete-time Markov process with the initial state and the fully discharged state 0. In [14] Chiasserini and Rao studied the battery behavior under two different modes of pulsed discharge: binary and generalized. Moreover, the battery system management can involve with other layers together including MAC layer, network layer and etc.

4 Energy Management Schemes in MAC Layer

In this section, we turn our attention to the energy management schemes in MAC layer and tend to discuss a range of energy-efficient MAC protocols proposed for the WANs, especially carrier sense multiple access with collision avoidance (CSMA/CA), the basic MAC protocol for WANs, and its potential energy management schemes. So far, most WANs are implemented using wireless LANs (WLANs). However, its performance is significantly limited by the use of an energy-consuming medium access control (MAC) protocol. IEEE 802.11x is one of the most influential WLANs standards, and its basic MAC protocol, distributed coordination function (DCF), is based on carrier sense multiple access with collision avoidance (CSMA/CA), one of typical contention-based MAC protocols. Currently, CSMA/CA has been the facto MAC standard of WANs, and is widely used in almost all of the testbeds and simulations for WANs research. So, we will discuss the varieties of energy management schemes in MAC layer. This section is a typical example to cover both efficiency metrics and the communication mechanism.

4.1 Spectrum and Energy Efficient MAC Protocol for WANs

As mentioned above, CSMA/CA is a probabilistic media access control (MAC) protocol in which a node verifies the absence of other traffic before transmitting on a shared transmission medium, such as a band of the electromagnetic spectrum. CSMA/CA uses a basic acknowledgment mechanism to verify successful transmissions, and an optional request-to-send/clear-to-send (RTS/CTS) handshaking mechanism to decrease collision overhead. In both cases, a binary exponential backoff mechanism is used. Before transmitting, a node generates a random backoff interval. The maximum contention window(CW_{max}) and minimum contention window(CW_{min}) are default values of CSMA/CA and value of contention window(CW) is counted in some algorithm between two above set values. The backoff time is slotted and the number of backoff slots is uniformly chosen in the range $[0, CW]$. At the first transmission attempt, the contention window, CW, is set equal to a value CW_{min}. After each unsuccessful transmission, CW is doubled up to the maximum value CW_{max}. Once CW reaches CW_{max}, it will remain at the value until the packet is transmitted successfully or the retransmission time reaches retry limit. While the limit is reached, retransmission attempts will cease and the packet will be discarded. There are three main kinds of energy management schemes' targets, including network lifetime, decreasing transmission power to save energy, another is enhancing energy efficiency (η_s) with consideration spectrum efficiency (η_e). Then we will introduce three kinds of energy management schemes in the followings.

It has been proved that η_e and η_s are inherently different and may not be maximized simultaneously. However, all these results are derived for a PtP wireless link and it does not consider any issue related with the MAC protocol. Hence, in the following we shall evaluate the two efficiencies of CSMA/CA, the basic MAC protocol of WANs. Starting from the model proposed in [29], we derive a formula that explicitly relates the η_e and η_s to the transmission probability, which is also related to the number of competing nodes and contention parameters (e.g., CW_{min}, m, and r).

We consider a scenario composed of a fixed number n of competing nodes, each operating in saturation conditions [29]. Let γ be the probability that a node transmits in a randomly chosen slot; W_{Tx}, W_{Rx} and W_{Lx} be the average transmitting, receiving and listening power respectively; P_{tr} be that there is at least one transmission in the considered slot and P_s be that exactly one node transmits on the channel upon at least one node's transmitting. Where $\bar{P}$ is the average payload size in bit, and the channel bandwidth B is fixed at 20 MHz in 802.11x. The analytical model given above is very convenient to determine the optimal transmission probability γ^* for the maximum efficiencies. Then, we get the following approximate solution of maximum η_e:

$$\gamma^* \approx \frac{1}{n}\sqrt[4]{\frac{4\mu W_{Lx}}{3T_c\,(3W_{Tx} - 2W_{Lx})}} \tag{5}$$

In the similar way, we can get an explicit formula of the optimal transmission probability to achieve the maximal η_S, as follows:

$$\gamma^* \approx \frac{1}{n}\sqrt{\frac{2\delta}{T_c}} \tag{6}$$

Furthermore, the η_e and η_s achieved by CSMA/CA in the case of the ACK access method are shown in Fig. 7. We can observe that: firstly, both efficiencies are a convex function over the transmission probability, but they cannot be maximized simultaneously; secondly, the optimal value γ for the maximum η_e is close to that

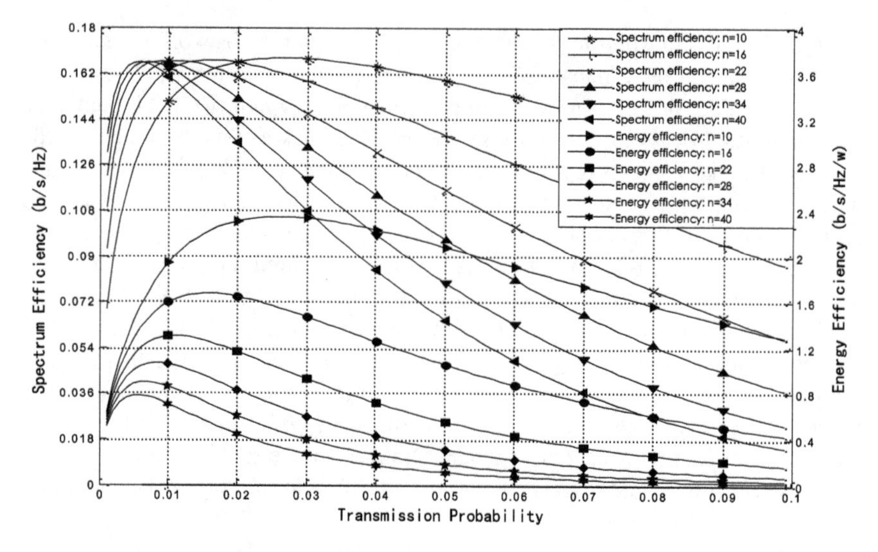

Fig. 7 Efficiencies versus the transmission probability

of the maximum η_s, so we can achieve the suboptimal, not the maximal, efficiencies simultaneously, i.e., we can get a good tradeoff between two efficiency metrics. It has been showed that the transmission probability (γ) depends on the network size and the contention parameters (CW_{min}, m, and r) in [30], as follows:

$$\gamma = \begin{cases} \frac{2(1-2P_{tr})(1-P_{tr}{}^{r+1})}{CW_{min}((1-(2P_{tr})^{m+1})(1-P_{tr})+(1-2P_{tr})(1-P_{tr}{}^{m+1}))} \\ \frac{2(1-2P_{tr})(1-P_{tr}{}^{r+1})}{CW_{min}((1-(2P_{tr})^{m+1})(1-P_{tr})+(1-2P_{tr})(1-P_{tr}{}^{r+1})+2^m P_{tr}{}^{m+1}(1-2P_{tr})(1-P_{tr}{}^{r-m}))} \end{cases} \tag{7}$$

As n is not a directly controlled variable, the only way to achieve optimal performance is to employ adaptive techniques to tune the values of contention parameters upon estimating the value of n [30]. So given the values of m and r, from Eqs. 5, 6, and 7, we can obtain the corresponding CW_{min} for the maximum BE and PE respectively, as shown in Fig. 8. We can observe that: firstly, both efficiencies of CSMA/CA are highly dependent on the number of competing nodes and the minimum contention window; secondly, the maximal efficiencies are very smooth, so even a nonnegligible difference in the estimate of the optimal value CW_{min} leads to similar energy and spectrum efficiency values.

Through the above theoretical analysis, an enhanced MAC protocol to approach both the suboptimal η_e and η_s by tuning the contention parameters upon estimating the number of competing nodes is proposed. In order to estimate the number of competing nodes in WANs precisely and timely, two estimation mechanisms are used to track the competing terminals [31], i.e., auto regressive moving average (ARMA) and Kalman Filters. A batch and sequential bayesian estimator is provided in [32], and we also proposed a frame-analytic estimation mechanism [33]. And

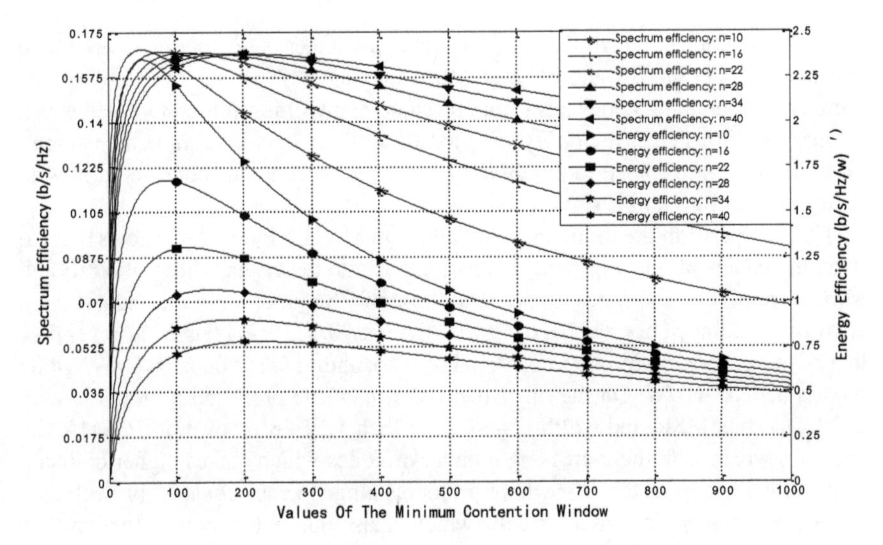

Fig. 8 Efficiencies versus the transmission probability

Table 1 Maximum efficiencies for the ACK method

Number of competing nodes	10		16		22		28		34		40	
CWmin	51	49	84	81	117	113	150	145	184	177	217	209
BE (b/s/Hz)		0.168		0.167		0.167		0.166		0.166		0.166
PE (b/s/Hz/W)	2.349		1.676		1.304		1.066		0.902		0.781	

then based on Eqs. 5, 6, and 7, we can explicitly compute the optimal CW_{min} that each node should adopt in order to achieve the maximum energy efficiency or the maximum spectrum efficiency within a considered network scenario, i.e., the number of competing nodes n, as shown in Table 1.

Therewith we choose a range of CW_{min} where the two efficiencies are not less than optimal value of maximum energy efficiency (MEE) and maximum spectrum efficiency (MSE) respectively. Here we choose optimal value as 99 % to simultaneously achieve high energy efficiency and spectrum efficiency. For instance, $[CW_{min,b1}, CW_{min,b2}]$ satisfies the required η_e and $[CW_{min,p1}, CW_{min,p2}]$ satisfies the required η_s. Then choose a value in the intersection of $[CW_{min,b1}, CW_{min,b2}]$ and $[CW_{min,p1}, CW_{min,p2}]$, i.e. satisfying the following:

$$CW^*_{min} \in \left[CW_{min,b1}, CW_{min,b2}\right] \bigcap \left[CW_{min,b1}, CW_{min,b2}\right] \tag{8}$$

In a word, firstly, each node estimates the number of competing nodes based on the proposed frame-analytic estimation mechanism in [10]. Secondly, each node adjusts its minimum contention window to the estimated number of competing nodes based on Table 1 to get the BE and PE tradeoff. Adjusting CW parameters based on the number of competing nodes has been presented many times in literature. However, to our best knowledge, the utility function of all the works is to maximize signal-to-interference-plus-noise ratio (SINR) or BE, and no one has considered how to optimize both BE and PE simultaneously.

Then we perform the following simulations in MATLAB. Figure 9 shows that the two efficiencies of our proposed protocol are a little lower than those of MEE and MSE respectively, and much higher than those of CSMA/CA. For instance, if there are 40 contending nodes, the BE of CSMA/CA (i.e., 0.132 b/s/Hz) is 23 % lower than that of our proposed protocol (i.e., 0.164 b/s/Hz), and its PE is 0.46 b/s/Hz/W while ours is 0.74 b/s/Hz/W. On the other hand, the η_s of the proposed protocol is only 2 % lower than MSE, and its PE is 4 % lower than MEE. Moreover, the CSMA/CA becomes worse with the increasing number of nodes which makes higher collision probability. The η_s of the other three protocols almost keeps constant or decreases slowly, but their η_e decreases sharply, which is also due to the more collisions. For instance, the η_e of CSMA/CA decreases 19 % from 0.162 to 0.132 b/s/Hz when the contending nodes increase from 10 to 40, while its PE drops 78 % from 2.11 to 0.46 b/s/Hz/W. In our proposed protocol, the η_S keeps constant at 0.164 b/s/Hz, and the η_e drops 66 % from 2.2 to 0.74 b/s/Hz/W. Furthermore, simulation results practically coincide with the analytical results in Table 1.

Then, we evaluate the proposed protocol in multi-hop environment. Figures 10 and 11 show he performance of energy and spectrum efficiency of the four protocol when the traffic is unsaturated. The performance of CSMA/CA is worse than all the other three protocols. For instance, Fig. 10 show that compared with CSMA/CA,

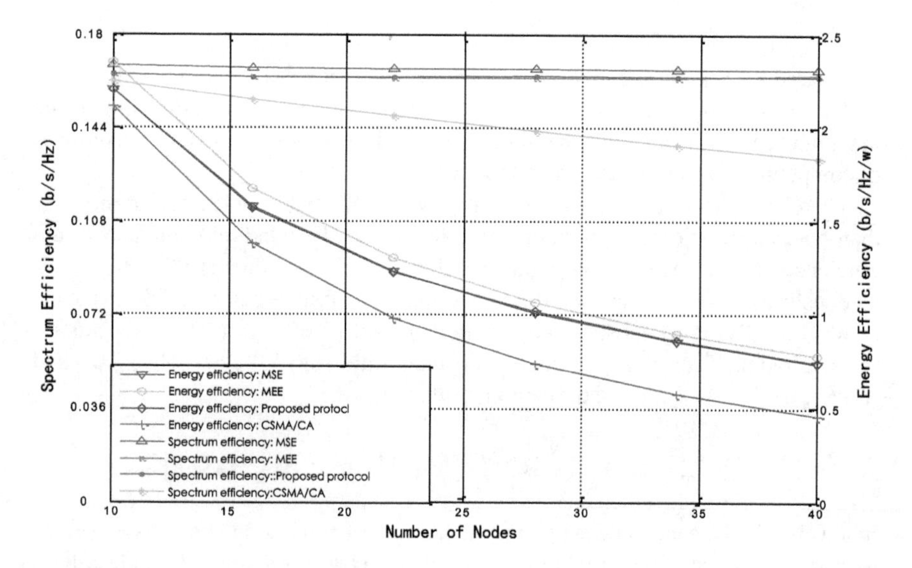

Fig. 9 Efficiencies of single hop WANs

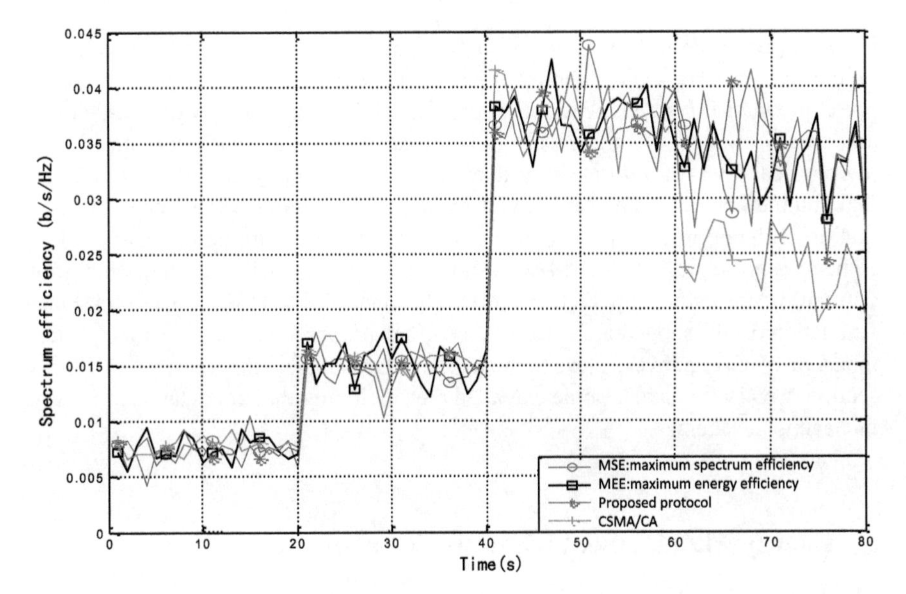

Fig. 10 Spectrum efficiency of multi-hop WANs

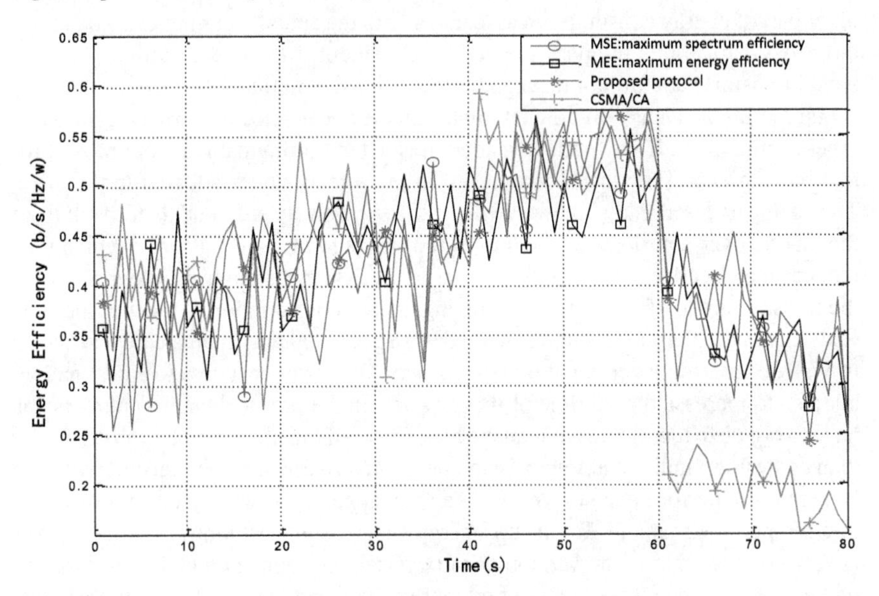

Fig. 11 Power efficiency of multi-hop WANs

our proposed protocol provides a 12 % larger η_s. However compared with MSE, thethe η_s of our proposed protocol is only 1 % lower than MSE. Figure 11 show that our proposed protocol is 20 % higher energy efficiency than traditional CSMA/CA. Moreover, its energy efficiency is 2 % lower than MEE.

For a greener WANs, we carries out a comprehensive analysis of the spectrum efficiency in b/s/Hz and the energy efficiency in b/s/Hz/W of CSMA/CA, the basic MAC protocol for WANs. Previous works concluded that the spectrum efficiency and energy efficiency are inversely proportional to each other if only considering the physical layer of a wireless system. However, our analysis also show that the maximum efficiencies are very smooth, even a non negligible difference in the estimation of the optimal value of contention parameters to a similar efficiency value. Hence we present a heuristic MAC protocol to approach both the suboptimal (not optimal) energy efficiency and spectrum efficiency of CSMA/CA in WANs. Final simulation results show that the BE and PE of the proposed protocol are much better than CSMA/CA, and very close to the maximal efficiencies respectively. In the next section, we also introduce some extended area including the processors and battery management schemes.

5 Energy Management in Network Layer

Despite of the development of energy management techniques in hardware-level, the other part of energy consumption associated with the actual communication through the WANs still limits the performance improvement if the energy utilized in each signal transmit path are not managed effectively. When aimed at a local scale, any signal routing path has a significant impact on the nodes selected to relay signals and other routing path. If we firstly consider an isolated routing path and optimize it for maximizing energy saving, it may result in more relay nodes adopted in this path. Then some nodes already involved with traffic excessively will tend to drop out from the WANs over a period of time duo to the shortage of battery energy, which thus reduces the effective connectivity number of WANs. In a similar way, switching off the transmit mode of nodes for minimizing energy consumption at most of the time is also not suitable. Therefore, the main design goal of energy management schemes in WANs's routing layer is not only to transmit data from a source to a destination, but also to increase the lifetime of the network. In the physical environment, nodes have to organize themselves to manage the energy all together, which is much harder than controlling individual transmission path. Moreover when the transmit paths are changed, the performance of WANs is also changed as it will result in a different interference environment. To sum up, energy management techniques in routing layer is very complex duo to the variability of physical environment and the in-existence of one "best", hence it is urgent to find more suitable and targeted system objective. In this section, we tend to illustrate the various research achievements of energy management schemes conducted in routing layer with classifying them in details, still considered as an ongoing and open research area up to now.

Recently, a number of energy management based on routing protocols have been discussed. In [16] the authors outline the key concepts of several proposed solutions and provide an analysis of them. Reference [17] proposes five power-aware metrics that can be used to classify routing protocols. In [18] the authors briefly summarize significant papers for each protocol layer and define several metrics for studying energy saving routing protocols. Reference [19] addresses design challenges of energy-efficient protocols in various layers and places especially on cross-layer design of these protocols. By reviewing a number of recent papers extensively, this section tends to classify them comprehensively.

5.1 Activity-Level Routing Protocols

To begin with, we note that the most straightforward method is reducing the energy consumption for each individual transmit routing path against the actual transmission of data between nodes in the network, i.e., activity-level routing protocols. To be specific, this method is mainly realized by source nodes' routing decisions before the delivery of a single data packet where it will calculate an optimized transmit path especially including the selection pattern of relay nodes and make the power consumption summation of each sub-path during the path minimized. Moreover, the physical channel quality is closely associated with power consumption. Up to now, there are two famous used power consumption models distinguished by the influence of channel fading [21]. Both have illustrated that in most cases communication between two nodes tends to consume less energy if relay nodes can be used for signal retransmission as showed in Fig. 12. Actually, this management is scoped in utilizing intermediate nodes to hold a packet instead of sending directly between nodes over large distances to reduce power consumption. Furthermore, it is worth noting that reliability of WANs should be considered as an important constraint when taking into account of energy consumption. For example, the goal of the energy management scheme in [20] is to route a packet along a path with minimum power consumption while also ensuring reliable communication when a high achievable successful data rate is obtained. Only the unicasting mode in the source node is discussed here, in comparison when broadcasting transmit mode is occupied, the ad hoc network layout may be transformed into a cellular system with centralized control mode.

For the practical WANs, minimizing the energy consumption of each individual transmit path should't be considered as the unique goal, since when the network traffic is not distributed uniformly, a small subset of nodes with limited energy charged may

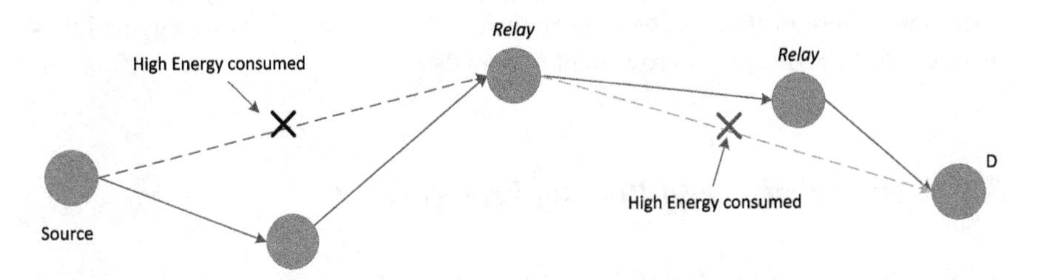

Fig. 12 Relay node for energy saving for a transmit path in WANs

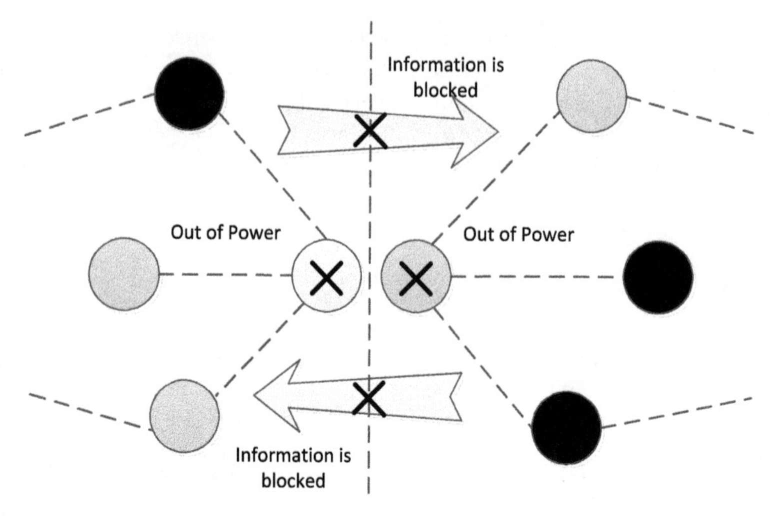

Fig. 13 Network separation duo to the dropout of nodes

drop out of networks in a short period of time, thus decreasing the effective connection ratio of networks, i.e., potential network partition in Fig. 13. In order to reduce power consumption while ensuring connection ratio, another category of routing path based scheme has drawn much attention with the issues of increasing network life. To be specific, this category tends to distribute the energy consumption among all nodes in a more balanced way. If the route with the maximal energy saving is always chosen for delivery, the subset of nodes along this route will be over-utilized and therefore drained in a short period of time, which may lead to network partitioning. During the realization process, this strategy is still based on each transmit path except for traffic should be routed through nodes that have sufficient remaining energy instead of selecting relay nodes just based on energy consumed. Currently, several hybrid schemes combining both goals have emerged and may be a more promising direction in the future.

To sum up, both of energy management schemes mentioned in this subsection are operated in routing-path manner with several system targets included. From the other side, we find that the definition of efficiency metrics is the most fundamental issue covering energy consumption, network lifetime and reliability. However, the operation pattern of transmit path-level always limit the scope of strategy and thus there exists performance improvement potentials.

5.2 Connectivity-Level Routing Protocols

In this subsection, we go forward one step by trying to reduce energy consumption while ensuring effective connectivity for the overall WANs, i.e., connectivity-level routing protocols, which is essential to almost any operation for a wireless network. However, there exist many complex relationships among the collection of transmit routing path. As a typical example, the connections in WANs are too dense, it may lead to severe interference at each receiver. In the comparison, when the density of transmit path is too sparse, the network is sensitive to broken link path.

Moreover, the physical factors in the network are always associated with its structure or performance, thus influencing the realization possibility of routing strategies. can be classified into topology control and passive energy saving [22, 23]. On one hand, the nodes' transmitting power (or transmitting range) can be adjusted to save energy while maintaining effective network connectivity. The network topology is formed by the links of each node in the network. In a wireless network, the number of links a node has is mainly determined by its transmission power. By managing its transmission power wisely, a node is able to not only maintain all necessary links, but also to reduce its power consumption. Controlling the topology of a wireless network by using adjustments to transmission power is presented in [22].

On the other hand, in WANs the radios utilized for communication consume power not only when operating (transmitting and receiving), but also when idle or listening [21]. This idle energy consumption is, over time, significant and cannot be ignored [24]. We tend to save energy by simply turning off some idle nodes, since energy consumption when a node's radio is idle is not negligible. The general goal of the protocols in this category is to turn off as many radios as possible while still maintaining the necessary network connectivity. Given this, only one node must be active in each cell, and all the other nodes can be put to sleep.

6 Cross-Layer Schemes of Energy Management

In this section, we tend to exploit the benefits of cross-layer information exchange, such as the knowledge of the frame error rate in the physical layer, the maximum number of retransmissions in the MAC layer and the number of relays in the network layer. Here, each layer must cooperate to manage energy of ad hoc networks, where we can build a module to process massages come from each layer as Fig. 14 shows. For example, we can use messages from two layer such as MAC and physical layer to reduce energy consuming [34].

In cross-layer design, detector need to collect parameters come from each layer for estimating the system state accurately. As Fig. 15 shows, in TCP (Transmission Control Protocol) layer, we need to know the event type and end-to-end QoS requirements so we can adopt the optimization strategy according to the prioritization of events. In network layer, we can collect the information of network topology (including nodes relative movement speed and acceleration, etc.) and in MAC layer channel contention status and queue status information are useful. In physical layer, the channel state and interference information are also can be used. Then, estimator processes all information come from physical, MAC, Network and TCP layer simultaneously and calculate system state. Each node of ad hoc networks can get

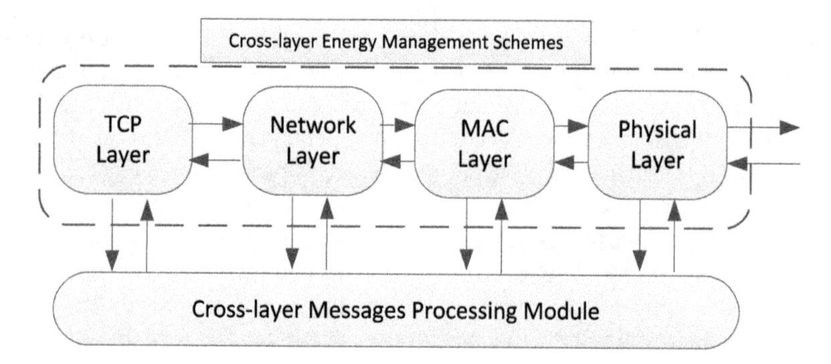

Fig. 14 Cross-layer information exchange structure

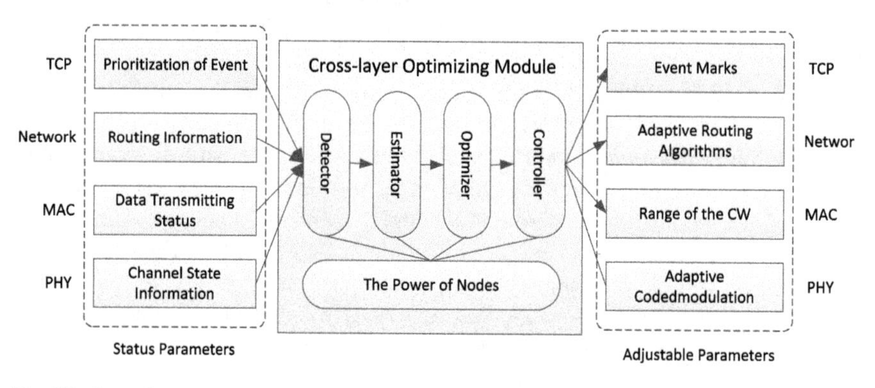

Fig. 15 Cross-layer energy management schemes work in WANs

system state through information exchanging between nodes. However, in a large-scale network such as vehicular ad hoc network (VANET), frequent communication is not realistic. Therefore, in estimator each node need to estimate relative distance and speed between nodes by some navigation devices or using incompletely cooperative game theory [35]. Next, optimizer, according to the system state, calculate the best energy management strategies. Finally, controller adapt parameters of each layer, such as event mark in TCP layer, routing algorithms in network layer, size of contention window in MAC, the adaptive modulation and coding in physical, to implement energy management schemes calculated by optimizer. We have provided a basic framework of cross-layer schemes, and we implement it in [33], which estimates the physical-specific and MAC-specific game state and adjusting the strategy to enhance throughput without additional energy consumption.

As a typical example, the author in [34] introduces a cross-layer protocol to increase throughput and decrease energy consumption during data transmissions. In the network layer, the proportion of successful data transmissions is considered, the number of channel contention events and the number of packets remaining in a node's queue in their proposed routing protocol. In the MAC layer, an adaptive contention window design that dynamically adjusted the range of the CW based on the proportion of successful data transmissions is proposed, which improved network throughput and energy consumption. The scheme considered the proportion of successful data transmissions, the traffic load of nodes, and channel contentions in the MAC layer were in the design of the energy-efficient routing protocol, which efficiently decreased energy consumption during data transmission and prolonged the network lifetime. Also, the routing path with a high proportion of successful data transmissions was given a higher probability of using the channel, and the range of the contention window is adjusted based on the proportion of successful data transmissions.

7 Summary

In this chapter, we illustrate the research status for the complete paradigm of energy management in WANs. First of all, the fundamental theory concentrated on the system resources utilization and further give out the comprehensive understanding of utility instead of reducing energy consumption unilaterally as the guiding role in the corresponding energy management scheme design. However, how to define an appropriate efficiency metric based on current system requirement and network type is always an open issue. During the design of energy management at all the protocols stack, the relationships between energy consumed and protocols decisions have been focused on by most researchers. We must realize reducing energy consumption can't be regarded as the unique goal since some other important performance indicators have to be guaranteed for system's normal operation. Moreover, it's very flexible to define efficiency metrics in different level, e.g., link-level, device-level and network level, etc., each of which can have a different target-level. Noteworthily an appropriate choice of the optimization target based on different efficiency metrics can have a profound effect on the overall network performance. Hence how to choice a better one is always is a challenging task. In physical layer, we are not only focusing on the energy management in battery system and hardware-level devices, but also the links to the protocols in other layers. How to integrate the physical techniques with other layers for a more promising design is still an urgent issue, besides the optimization in itself. Compared with the relatively isolated solutions in physical layer, in both MAC and network layer, we begin to fucus on some overall solutions to increase the system efficiency. For instance, both the spectrum efficiency and energy efficiency are considered in WANs and some solutions are given out to improve them at the same time. Meanwhile, the efficiency metric definition becomes more significant and we have to pay more attention to the user's QoS guarantee. It is worthy to note that we must consider many system performance indicators besides the energy consumption reduction or energy efficiency, e.g., network lifetime, network connections quality, reliability. For instance, there is an inherent tradeoff among many performance indicators, e.g., the tradeoff between lifetime and energy consumption. Finally, we are engaged with excavating the potentials from dependency between layers and illustrated some current techniques by cross-layer design. To sum up, it's hard to find a best energy management strategy in the complex WANs environments, how to achieve an efficient balance to make network operate with burdening more traffic is a permanent issue.

References

1. Chen, P., O'Dea, B., Callaway, E.: Energy efficient system design with optimum transmission range for wireless ad hoc networks. In: Proceedings of IEEE International Conference Communications vol. 2, pp. 945–952 (2002)
2. Deng, J., Han, Y.S., Chen, P.-N., Varshney, P.K.: Optimal transmission range for wireless ad hoc networks based on energy efficiency. IEEE Trans. Commun. **55**(9), 1772–1782 (2007)
3. Chen, Q., Gursoy, M.C.: Energy-efficient modulation design for reliable communication in wireless networks. In: Proceedings of 43rd Annual Conference on Information Sciences and Systems, Mar 2009, pp. 811–816
4. Mohammed-Slim, A., Goldsmith, A.J.: Area spectral efficiency of cellular mobile radio systems. IEEE Trans. Veh. Technol. **48**(4), 1047–1066 (1999)
5. Akhtman, J., Lajos, H.: Power versus bandwidth efficiency in wireless communications: the economic perspective. IEEE VTC Fall, Alaska, USA (2009)
6. Zhao, L., Cai, J., Zhang, H.: Radio-efficient adaptive modulation and coding: green communication perspective. In: 2011 IEEE 73rd Vehicular Technology Conference (VTC Spring), pp.1–5, 15–18 May 2011
7. Zhao, L., Zhao, G., O'Farrell, T.: Efficiency metrics for wireless communications. In: 2013 IEEE 24th International Symposium on Personal Indoor and Mobile Radio Communications (PIMRC), pp. 2825–2829, 8–11 Sept 2013
8. Gupta, P., Kumar, P.R.: The capacity of wireless networks. IEEE Trans. Inf. Theory **46**(2), 388–404 (2000)
9. Weber, S.P., Yang, X., Andrews, J.G., de Veciana, G.: Transmission capacity of wireless ad hoc networks with outage constraints. IEEE Trans. Inf. Theory **51**(12), 4091–4102 (2005)
10. Jayashree, S., Manoj, B.S., Siva Ram Murthy, C.: On using battery state for medium access control in ad hoc wireless networks. In: Proceedings of ACM MOBICOM'04, Sept 2004, pp. 360–373
11. Chiasserini, C.F., Rao, R.R.: Improving battery performance by using traffic-shaping techniques. IEEE JSAC **19**(7), 1385–1394 (2001)
12. Zhao, L., Yi, J., Adachi, F., Zhang, C., Zhang, H.: Radio resource allocation for low-medium-altitude aerial platform based TD-LTE networks against disaster. In: 2012 IEEE 75th Vehicular Technology Conference (VTC Spring), pp. 1–5, 6–9 May 2012
13. Zhao, L., Zhang, C., Zhang, H., Li, X., Lajos, H.: Power-efficient radio resource allocation for low-medium-altitude aerial platform based TD-LTE networks. In: 2012 IEEE Vehicular Technology Conference (VTC Fall), pp. 1–5, 3–6 Sept 2012
14. Chiasserini, C.F., Rao, R.R.: Pulsed battery discharge in communication devices. Proc MOBICOM **99**, 88–95 (1999)
15. Lahiri, K., et al.: Battery-driven system design: a new Frontier in low-power design. In: Proceedings of ASP-DAC/VLSI Design'02, Jan 2002, pp. 261–267
16. Lindsey, S., Sivalingam, K., Raghavendra, C.S.: Power optimization in routing protocols for wireless and mobile networks. In: Stojmenovic, I. (ed.) Handbook of Wireless Networks and Mobile Computing. Wiley, New York (2001)
17. Singh, S., Woo, M., Raghavendra, C.S.: Power aware routing in mobile ad hoc networks. In: Proceedings of 4th Annual International Conference on Mobile Computing and Networking, pp. 181–190, Oct 1998
18. Jones, C.E., et al.: A survey of energy efficient network protocols for wireless networks. Wireless Net. J. **7**(4), 343–358 (2001)

19. Goldsmith, A.J., Wicker, S.B.: Design challenges for energy-constrained ad hoc wireless networks. IEEE Wireless Commun. **9**(4), 8–27 (2002)
20. Subbarao, M.W.: Dynamic power-conscious routing for MANETs: an initial approach. In: Proceedings of 50th IEEE VTC, vol. 2, pp. 1232–1237 (1999)
21. Stojmenovic, I., Lin, X.: Power aware localized routing in wireless networks. IEEE Trans. Parallel Distrib. Sys. **12**(11), 33–1122 (2001)
22. Xu, Y., Heidemann, J., Estrin, D.: Geography informed energy conservation for ad hoc routing, In: Proceedings 7th Annual International Conference on Mobile Computing and Networking, July 2001, pp. 70–84
23. Ramanathan, R., Rosales-Hain, R.: Topology control of multihop wireless networks using transmit power adjustment. In: Proceedings of 19th Annual Joint Conference IEEE Computer and Communications Societies, Mar 2000, vol. 2, pp. 404–413
24. Xu, Y., Heidemann, J., Estrin, D.: Adaptive energy conserving routing for multihop ad hoc networks. Technical report 527, USC/Info. Science Institute, Oct 2000
25. Jayashree, S., Siva Ram Murthy, C.: A taxonomy of energy management protocols for ad hoc wireless networks. IEEE Commun. Mag. **45**(4), 104–110 (2007)
26. Jayashree, S., Manoj, B.S., Siva Ram Murthy, C.: On using battery state for medium access control in ad hoc wireless networks. In: Proceedings of ACM MOBICOM 04, pp. 360–373, Sept 2004
27. Agarwal, S., et al.: Route-Lifetime Assessment-Based Routing (RABR) protocol for mobile ad hoc networks. Proc. IEEE ICC **3**, 1697–1701 (2000)
28. Chen, L., Leneutre, J.: A game theoretic framework of distributed power and rate control in IEEE 802.11 WLANs. IEEE J. Sel. Areas Commun. **26**(7), 1128–1137 (2008)
29. Bianchi, G.: Performance analysis of the IEEE 802.11 distributed coordination function. IEEE J. Sel. Areas Commun. **18**(3), 535–547 (2000)
30. Zhao, L., Wu, J.Y., Zhang, H., Zhang, J.: Integrated QoS differentiation over IEEE 802.11 WLANs. IET Commun. **2**(2), 329–335 (2008)
31. Zheng, Y., Ning, F., Gao, F., Xu, Q., Gao, Z.: Kalman filter estimation of the number of competing terminals in IEEE 802.11 network based on the modified Markov model. In: 2010 3rd IEEE International Conference on Broadband Network and Multimedia Technology (IC-BNMT), pp. 438–442, 26–28 Oct 2010
32. Vercauteren, T., Toledo, A.L., Wang, X.: Batch and sequential Bayesian estimators of the number of active terminals in an IEEE 802.11 Network. IEEE Trans. Sig. Process. **55**(2), 437–450 (2007)
33. Zhao, L., Zou, X., Zhang, H., Ding, W., Zhang, J.: Game-theoretic cross-layer design in WLANs. In: International Wireless Communications and Mobile Computing Conference, 2008, IWCMC'08, pp. 570–575, 6–8 Aug 2008
34. Weng, C.-C., Chen, C.-W., Chen, P.-Y., Chang, K.-C.: Design of an energy-efficient cross-layer protocol for mobile ad hoc networks. IET Commun. **7**(3), 217–228 (2013)
35. Zhao, L., Zhang, J., Zhang, H.: Using incompletely cooperative game theory in wireless mesh networks. IEEE Netw. **22**(1), 39–44 (2008)
36. Zhao, L., Guo, L., Zhang, J., Zhang, H.: Game-theoretic medium access control protocol for wireless sensor networks. IET Commun. **3**(8), 1274–1283 (2009)

Part IV
Energy Management in Cognitive Radio Networks

Dynamic Spectrum Leasing for Cognitive Radio Networks—Modelling and Analysis

Maryam Hafeez and Jaafar Elmirghani

Abstract Incentive based dynamic spectrum leasing (DSL) has been suggested as a type of cognitive radio (CR) based communication in which the legacy network allows the cognitive radio nodes to utilize its spectrum for their communication in exchange for cooperative relaying services. The key objective of this chapter is to investigate the design space of a DSL empowered large scale CR network (CRN) collocated with a point-to-point primary communication link. The ultimate design objective is to improve both the network level energy efficiency and the spectral efficiency through the exploitation of cooperation gains rendered by the proposed optimally dimensioned DSL mechanism. This chapter presents a DSL scheme where the CRs cooperatively relay the data of the primary network for a duration of time. As a reward for the cooperation, the CRs are granted exclusive access to the primary spectrum for some time. To harness maximum gains in terms of energy efficiency (EE) for the primary network while maintaining its required quality of service and spectral efficiency (SE) of the CR network, a comprehensive model of DSL is presented. To this end, an accurate quantification of the random locations of the CR nodes and the optimal division of leasing time between the primary and secondary activities are two crucial factors. In this chapter, we consider a large scale cognitive random network. The spatial dynamics are modeled by using point process theory from stochastic geometry. Mutual agreement of the primary and secondary nodes on the leasing time division is studied using a game theoretic framework. The analysis indicates that DSL enables the primary to attain its required transmission rate and from 20 up to 50 % of the total leasing time is also reserved for the secondary activity. It is shown that the bargaining powers of the primary and secondary networks strongly dictate the proportion of cooperation and leasing time. Further, the EE of DSL based on the network geometry and optimal leasing time is analytically characterized. The simulation results reveal that DSL operation under such considerations can be significantly more energy efficient as compared to direct communication. A closer look helps to ascertain that DSL with a sparse secondary network can serve to be more than 10 times energy efficient while maintaining the

M. Hafeez (✉) · J. Elmirghani
University of Leeds, Leeds LS2 9JT, UK
e-mail: elmh@leeds.ac.uk

© Springer International Publishing Switzerland 2016

M.Z. Shakir et al. (eds.), *Energy Management in Wireless Cellular and Ad-hoc Networks*, Studies in Systems, Decision and Control 50,
DOI 10.1007/978-3-319-27568-0_10

same time-rate product as compared to direct communication for low CR densities. Hence DSL based communication enables the primary to communicate at its desired transmission rate and quality in an energy efficient manner and also enables the CR network to exploit the licensed spectrum for its own communication. In short, DSL is a useful technique for improving the efficiency of wireless communication with direct application to future networks.

1 Introduction

Over the last few decades, wireless communication has witnessed an immense growth in its technological sophistication and widespread deployment. It is been estimated that a capacity expansion by a factor of 1000 is needed in the next generation (5G) mobile networks [1]. In order to satisfy the sustained growth of mobile traffic, the development of more sophisticated and flexible radio networks is fundamental. This calls for additional spectral resources, planning/infrastructure deployment costs and energy requirements for the network operations. In the recent past, significant rise in the energy consumption of the communication networks has been recorded. Around 7.95 % rise in the energy demand of Telecom Italia network was observed in 2007. At the same time, British Telecom contributed to about 0.7 % of the total UK's energy consumption [2, 3]. It is predicted that in comparison to 2007, CO2 equivalent emissions of the communication network will increase by a factor of three until 2020. This corresponds to more than one third of the overall emissions in the UK [4, 5]. These alarming statistics and the rising costs have motivated the research and development community to target improving the energy efficiency of the mobile communication network by a factor of 1000 per transported bit for the emerging 5G networks [1].

A prime goal in the design of any wireless communication network is to maximize the spectral utilization while attaining the highest quality of communication. Increasing bandwidth and/or power are the two main approaches that directly follow from the Shannon's capacity of a wireless channel to enhance the communication rate [6]. Spectrum scarcity has already been recognized as one of the major problems faced in the deployment of new technologies and in the enhancement of the capacity of the existing ones. Moreover, the current energy consumption trends indicate that if the communication systems continue to develop and spread at the same pace, a significant portion of the total energy production of any country would be needed to meet the requirements of future communication systems [7, 8]. At this juncture, an ideal future wireless system would; (1) maximize the utilization of the existing bandwidth, (2) minimize the power consumption while supporting a high quality of communication.

Under-utilization of the electromagnetic spectrum due to the stringent spectrum allocation schemes has become a well established fact in a very short time [9]. This inefficient utilization of bandwidth is one of the main causes of the apparent free spectrum extinction. CRs are envisioned to be a possible solution to this problem.

They co-exist with licensed networks and enable optimum utilization of spectrum across both geographical and temporal domains. CRs dynamically exploit the spectral resources of the legacy (primary) network without causing any intervention in the primary network (PN) operations. Many different approaches to realize CRs have been suggested in [10, 11]. However, these approaches only provide intermittent/sporadic connectivity for the CRs without any QoS guarantees. Our goal here is to analyze a CR based architecture that exploits the licensed spectrum for its own utility and maintains the performance of the legacy network in terms of its communication rate while reducing the overall power consumption in the network. We aim to study that how the relaying services of a few geographically suitable CRs procures an exclusive spectral access to the entire CR network.

In contrast to passive spectrum sharing between a PN and a CRN (provisioned through hierarchical access mechanisms), DSL employs an active approach to improve the overall spectrum utilization through DSA [12]. This chapter is based on studying DSL where the PN leases a part of its spectrum to another spectrum-less network when the latter helps to improve the performance of the incumbent network. The aim of the chapter is to study DSL as a unified model for the mutual benefit of the incumbent spectrum users (primary users (PUs)) and non incumbent networks (secondary users (SUs) e.g., CRs). The primary metrics of quantifying the benefits of DSL are the spectral and energy efficiency of the network. This chapter explores how a network without having a pre-owned license to access the spectrum can help to improve the performance of a planned and deployed PN. It introduces a spectrum leasing framework that reconsiders spectrum allocation rights and policies, improves the current characterization by rigorously studying various degrees-of-freedom of the network, and also incorporates tools to enable the modelling of the intelligent and adaptive behaviour of such networks. It judiciously quantifies the service that a secondary network (SN) offers to get spectral access in return such that both primary and secondary network meet their own objectives of improved spectral utilization. It also studies how the proposed DSL schemes can help improve the energy efficiency of the primary network. More specifically, under DSL enabled DSA:

1. The PN has a certain incentive for rewarding the CRN with access to its licensed spectrum. Incentives can be either monetary or non-monetary in nature.
2. The PN can dynamically adapt the rewarding mechanism by observing changes in its incentive. In other words, the PN can actively control the amount of spectral resources it is willing to share across various dimensions of the Hertzian medium. Note that the radio spectrum has a multi-dimensional nature, i.e., variations across time, frequency, polarity, space, etc., all determine the available spectral resources.
3. The PN can ensure that the required quality of service (QoS) constraint for its own users is guaranteed. Thus transparency in terms of the performance of the PN is an intrinsic feature of DSL.

While it is easy to argue that DSL enabled CRNs have the potential to maximize the spectral utilisation, it is not clear if the potential gains are harnessed at a cost of increased energy consumption. This leads to the following design question:

> Is it possible to develop a DSL mechanism which maximizes the network level spectral efficiency (which is a function of individual spectral efficiencies of the PN and the CRN) while also ensuring an increase in the network wide energy efficiency?

Additionally, another related design issue stems from the fact that the existing literature on DSL, refrains from considering the impact of the network topology and propagation uncertainties on the promised potential gains. Specifically:

> Does DSL successfully deliver its promised spectral/energy gains under realistic channel propagation conditions while considering the topological uncertainties due to varying spatial dynamics of the CRN?

A practical DSL scheme that maximizes the spectral and energy efficiency of the network over a wide dynamic range of signal propagation conditions and node locations can be directly integrated in the design of future wireless networks. DSL provides a framework to caste the mutual interest of a variety of entities in the network to improve it overall performance by optimal division/allocation of available resources. The DSL scheme presented in this chapter finds direct application in efficient resource division in the 5G key concepts of carrier aggregation, license shared access (LSA), device-to-device (D2D) communication and offloading in heterogeneous networks etc. [13–15].

2 Research Objectives and Contributions

In this chapter, answers to the above-mentioned design issues/questions are investigated by developing a framework to quantify the performance of both the PN and the CRN under a proposed DSL mechanism. The aim of this chapter is to study how DSL can be used as an energy efficient alternative for PN communication while improving the spectral efficiency of the SN. The proposed DSL mechanism considers that in a dense CRN deployment, cooperation of the CRs with the PN can be traded for spectrum access opportunities. More specifically, the intrinsic distributed diversity gain provided by a cooperative relaying protocol and reduced propagation loss due to dense deployment can be treated as a resource which a CRN can offer to a PN to sustain its operations i.e., maintaining its QoS while reducing its energy expenditure. However, the improvement in the performance of the PN through cooperation comes at a cost paid by the CRs in terms of their energy consumption. Consequently, the CRs wish to trade the incurred cost for a spectrum access opportunity. Thus in a nutshell, the proposed DSL mechanism provides transmission opportunities to the CRs if they in return help in improving the energy utility of the PN through inter-network cooperation. Consequently by shrinking the transmission window of the PUs, during the remaining time the spectral resources are reserved to provide access to the cooperating CRs. Such a DSL approach for a CRN communication where the services of cooperative relaying by the CRs serve as an incentive for the PN spectrum leasing resulting in improved EE of the PN and better SE of the CRN is the focus in this chapter.

In the proposed DSL mechanism, the PN leases its spectrum to the SN, which in this chapter is a CRN, and forwards its data to the CRs for cooperative transmission. The SUs/CRs relay the PN data during some fraction of the leasing time. For the remaining leased time, the CRN exclusively uses the spectrum and carries out its own communication. The share in time and bandwidth for the secondary communication is the motivation for the CRs to cooperate with the PN. A PU is interested in maximizing the time for which the CR nodes relay its data. Greater negotiation power can help the PN to ensure that for most of the time the CRs relay its data. On the other hand, the CRs intend to schedule their own transmissions for most of the leasing time. The reward for the cooperative relaying of a few CRs can ripple across the entire CRN, enabling spectral access. However, the cooperating CRs need to negotiate with the PN to get a certain duration of leasing time so that the entire CR network can benefit from it. Such selfish yet rational behaviour of the PUs and SUs makes the appropriate division of the leasing time very important for successful DSL operation. An optimal division is described as a division which is mutually agreed upon and satisfies the demand of both networks. A greater negotiation power can help each network to procure more time for itself.

In this chapter, fundamental mathematical modelling and analysis of the proposed DSL mechanism for CRN is pursued. Despite the wide scale applicability and potential benefits of service based DSL, contributions in the existing literature are very limited. The following aspects of DSL for the CRNs need to be addressed:

1. In order to analyse a DSL empowered CRN, it is important to consider a realistic network topology for both the PN and the CRN. The necessity of a realistic geometry based network model manifests itself not only in topological considerations but also in terms of the efficient selection of the cooperation areas under the DSL mechanism. Unfortunately, it is common practice to ignore the network geometry in order to simplify the analytical model. However, such simplifications come at the cost of limited insights.
2. In order to ensure fairness and mutual satisfaction, it is important to divide the leasing time in a way that both the PUs and the CRs agree to their share of time. In previous studies, this division has been influenced more by the decision of the primary network which needs to possess cross-network channel state information (CSI) (i.e., CSI of the secondary network) to make the bargaining decisions. The CRN needs to observe the primary action and only decides in reaction to primary decision.
3. It is important to quantify how the division between cooperating and leasing time is dictated by the negotiating power of the PN and the CRN. Unlike existing studies, it is important to develop a comprehensive model to capture the scenarios where one network exercises greater influence on the decision, yet attains mutual agreement over the division of the leasing time and vice versa.
4. As mentioned earlier, the energy requirements of the design of any communication system has become a key concern due to the rapid growth in energy consumption. This warrants a formal analysis of the energy efficiency of leasing to measure its viability.

In summary, the main contribution of this chapter is to address the above mentioned design issues for enabling DSL based spectrum sharing in large scale wireless networks.

3 Key Findings

In this chapter, tools from stochastic geometry and game theory are used to build a quantitative framework for investigating the introduced design issues. The developed framework explicitly incorporates the impact of randomness rendered by the channel impairment process and geometry of the cooperation region on the DSL mechanism. In turn, these considerations demonstrate that a desired data transmission rate with a certain reliability can be provisioned for the PU links by leasing the spectrum to the CR nodes occupying spatially suitable locations. As a reward for cooperation, the entire CRN obtains access to the spectrum and thus the CRs can schedule their transmissions at a reasonable rate among themselves. The provision of negotiation between the PN and the CRN, over the division of leased time (reserved for cooperation with the PN and for the CRN communication) is ensured using the Nash bargaining framework. Unlike existing literature, a mutual agreement based division is attained that ensures proportional fairness for both networks. Also, the PN is not required to have CSI knowledge of the CRN. The work quantifies how the individual bargaining powers of the PN and the CRN can influence the division of cooperation and leasing time. Furthermore, it is shown that for equal bargaining powers, out of the total DSL operational time, 20–50 % of the time is reserved exclusively for the CRN which otherwise is dormant. It is demonstrated that the entire CRN can benefit from the leasing time that is procured by the cooperation of a few CRs with the PN. The variety of possible divisions of the leasing time ensures the flexibility and wide scale applicability of the considered DSL model. Moreover, the quantification of the energy requirements of the legacy and the DSL networks are established which to date was an open issue. The results indicate that DSL empowered networks can be more than 10x energy efficient as compared to traditional networks. It is shown that choosing a smaller cooperation area is more energy efficient for the PN. It is also shown that DSL is spectrally efficient at the network level where the CRN improves its spectral access considerably. At the same time, the QoS requirements for both the PN and CRN can be guaranteed.

4 Previous Work

Our work addresses three research areas in wireless communications (specifically CRNs); exploiting cooperative diversity, characterization of spectral leasing models and energy efficiency of the architecture. Energy efficiency has been explored in the context of cognitive radios by using adaptive modulation techniques [16] and optimal transmission duration estimation [17] in order to achieve power/bandwidth efficiency. Recently, cooperative diversity in cognitive radio networks has gained some attention. An overview of various possible ways of exploiting this diversity has been suggested in [18]. The existing literature on dynamic spectrum leasing can be characterized into three main types; (i) in which the incentive for leasing is based on monetary rewards, [19, 20], (ii) where leasing is allowed as long as the interference from the CRs is below an 'interference cap' [21, 22], (iii) where the incentive for leasing is based on service rewards [23–25], which is the model on which this study is based. For the first two types, numerous literary contributions exist, however, its survey is out of the scope of this chapter. Our focus is based on the third framework which was first explored by [23] where an analytical study of service based DSL is provided and cooperative diversity of the secondary relays has been exploited. In [24] the same framework is carried forward and applied in an ARQ based model where a portion of the retransmission slot is leased by the legacy network to the relays for their traffic in exchange for cooperative retransmission by the relays. In [25], the authors consider an infrastructured hierarchical spectrum leasing approach. In their work, they consider multiple primary nodes that select their respective individual relays for cooperation.

Game theoretic tools have been widely used to determine the amount and time for spectrum sharing. A comprehensive survey in [26] addresses the application of different games to model dynamic spectrum sharing. Previously, in [23, 25], a linear search based algorithm followed by a Stackelberg game was proposed to divide the leasing time between the primary and secondary activities. However, it does not cater for mutual agreement on leasing time division if (1) primary chooses a selfish time distribution as the leader and (2) the secondary in turn plays suboptimal strategy to hurt the interest of the primary in successive realizations of the game. The studies regarding the energy efficiency of CRNs mostly consider a generic scenario where spectrum sensing is employed. The study of the EE of DSL where nearest neighbor based communication is employed for the CR network is not studied in the literature.

Our work differs from the above in the following ways. Firstly, these studies abstract out the spatial geometry of the network. The impact of network geometry and provision of negotiation over the leasing time is not studied in these papers. Here, however, we investigate DSL in a geometric framework where the capacity of direct and DSL based communication is studied in terms of the spatial characteristics of the secondary network. Nash bargaining has been used for solving various problems of resource allocation in wireless networks [27, 28] and it is shown to attain a Pareto optimal solution that specifically discourages selfish behavior in the network. In [29] and its extension [30], DSL with spatial and bargaining based modeling was studied for spectral and energy efficiency gains for bidirectional communication. Physical layer techniques like network coding and beamforming were introduced to harness additional gains. However, in this work, the focus of the authors is to determine the fundamental behavior of DSL for unidirectional communication. The authors study the impact of bargaining powers of the two networks on the division of the leasing time. Also, unlike previous studies where a fixed geometric setup for CR receivers was considered, in this chapter a nearest neighbor based receiver model is considered. Nearest neighbor based receiver models find a direct application in future device to device (D2D) networks. The impact of the selection of the area of cooperation by the PN studied in this chapter is also a novel contribution. Finally, in this chapter, the entire CR network benefits from the leasing time which is a reward of the cooperative services of a few geographically suitable CR relays. To the best of our knowledge, the bargaining powers based modeling for DSL where the entire CR network communicates with nearest neighbors has not been carried out by any previous study. We use this framework to enable the primary and secondary users to reach a mutual agreement over the leasing time.

5 System and Network Model for DSL Empowered CRN

5.1 Network Geometric and Physical Layer Model

A primary link operating in the presence of a geographically co-located secondary network is considered. For simplicity, it is assumed that the primary communication link (P_{tx}, P_{rx}) is formed by a primary receiver (P_{rx}) located at the origin and a primary transmitter (P_{tx}) located at a distance $r_p > 1$ from P_{rx}. A region of 'exclusion' with radius ϵ is centred at the P_{rx}[1] to avoid excessive interference (see Fig. 1). Under the legacy operation of the PN, any transmission by the CRs is strictly forbidden within this exclusion area [31]. The secondary network is formed by the CR nodes, whose locations form a stationary Poisson point process (PPP) Φ of intensity λ. From the theory of PPP, the probability of finding $k \in \mathbb{N}$ CRs in an area $A \subset \mathbb{R}^2$ is given as

$$P_k = \Pr\{k \text{ nodes in } A\} = \frac{(\lambda\,|A|)^k}{k!}\exp\left(-\lambda|A|\right), \tag{1}$$

The average number of the CRs in an arbitrary region A with area $|A|$ is quantified as $\lambda|A|$. Each CR transmitter S_{tx} communicates with an associated CR receiver S_{rx} when spectral access is granted by the PN. In this chapter, it is considered that the CR receivers are associated with their nearest CR transmitter. In other words, 'nearest neighbour association' is adopted for the transmitter-receiver pairing in the CRN. Notice that such association mechanism indeed captures many emerging CR deployment paradigms. Specifically, it captures overlaid cellular CRN where the CR transmitters may be data aggregators for machine type communication or small cells associated with MUs based on the average path-loss, etc.

Based on the relative distances from the P_{tx} and the P_{rx}, the nodes lying within a radius r_p between the two primary nodes are expected to best serve as the potential relays for the PN in cooperation mode under DSL operation.[2] It is considered that

---

[1]Primary's exclusive region encapsulates those secondary nodes which are at such a small distance from the PU that any transmission from them directly interferes with the PN communication.

[2]Such a selection is inspired by the optimal forwarding area selection techniques [32, 33].

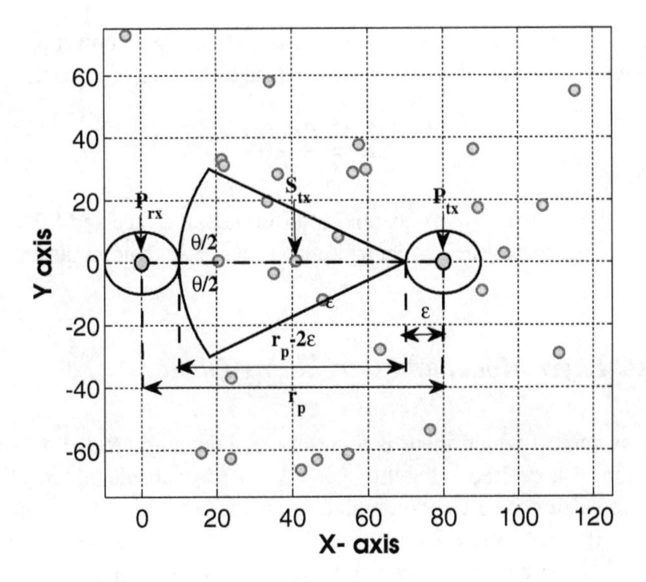

Fig. 1 Geometric model of the network

nodes within a radius ϵ from the P_{tx} or the P_{rx} are excluded from the cooperation phase of the DSL. This particular constraint reflects that only the nodes lying in the proximity of the half-way mark between the primary nodes can become cooperative relays. The key motivation behind such selection is to minimize the energy penalty, while balancing the average channel gain for the two hop communication. In other words, the condition where the average channel gain for the first hop is significantly larger than the second hop and vice versa are excluded. It is well known that an optimal relaying strategy can be devised by selecting relays which balance the average gains for both hops [32]. Consequently, the cooperation region, bounded by a sector, i.e., $\sec(\theta, r)$ of radius $r_p - 2\epsilon$ and an angle θ in radians, is considered to be the effective area of cooperation in DSL operation mode. Formally, it can be denoted as,

$$A_c\left(\theta, r_p, \epsilon\right) = \left\{(r, \theta) \in \mathbb{R}^2 : \epsilon < r \le r_p - \epsilon \text{ and } \theta \in [0, 2\pi]\right\},$$

where $\epsilon \ge 1$. The selected relays also form a PPP $\Phi_r \subset \Phi$ with an average number of CR relay nodes $k = \lambda \left|A_c\left(\theta, r_p, \epsilon\right)\right|$ in the region $A\left(\theta, r_p, \epsilon\right) \subset \mathbb{R}^2$.

It is assumed that the wireless channel suffers from path-loss and small-scale fading. For a distance r between any arbitrary pair of nodes, the channel between them can be expressed as $ahl\,(r)$ [34] where the fading power gain h is an independent and identically distributed (i.i.d.) exponential random variable with a unit mean, a is a frequency dependent constant and $l(r)$ is the distance dependent path-loss function. For the sake of simplicity, a is considered to be unity throughout the rest of the discussion. The power-law path-loss function $l(r) = \min(1, r^{-\alpha})$ is upper bounded by unity which corresponds to the reference distance. Also, $\alpha > 2$ is the operational environment dependent path-loss exponent. The noise at the receiver front end, is considered to be additive white Gaussian noise (AWGN) with power σ^2. For a given transmit power P and link distance r, the SNR at a receiver is given as

$$\text{SNR} = \frac{Phl\,(r)}{\sigma^2}. \tag{2}$$

Similarly, in the presence of co-channel interference, the received SINR is defined by adding the aggregate received interference power I in denominator of Eq. 2.

5.2 *MAC Layer Model and Bargaining Game*

A primary system in which there is a certain rate demand (R_{dir}) for a sustainable link operation at a desired reliability ($\tilde{\rho} = 1 - \rho$) is considered. In other words, the QoS demand for the PUs is completely characterized by the desired rate R_{dir} and the percentage of time $\tilde{\rho}$ over which this rate can be guaranteed. To meet this demand, the PU has a choice between continuing its communication in the legacy mode through direct communication or through the cooperative relaying of CRs via spectrum leasing mechanism. In direct communication, the primary transmitter communicates with its corresponding primary receiver at a rate R_{dir} for a duration T. The duration T corresponds to the duration of a temporal spectral resource such as the length of a transmission frame. Under DSL operational mode, the primary transmitter indicates its willingness to lease the spectrum for the same time duration T to the CR nodes inside a certain cooperation region $A_c\left(\theta, r_p, \epsilon\right)$. The choice of θ and willingness to lease are indicated over a dedicated control channel.

5.2.1 Phases of DSL

The process of dynamic spectrum leasing can be divided into three sub intervals:

BROADCAST The primary broadcasts its data to be relayed to the CR transmitters for a time $t_{ps} < T$.

COOPERATE During the second sub-interval, called the cooperation phase, k secondary nodes that are best suited for relaying on the basis of their geographical location, cooperatively relay the data of the P_{tx} to the P_{rx} for a time $t_{sp} < T$ by forming a distributed k-antenna array through ideal orthogonal distributed space time coding (DSTC) [35]. The details of DSTC codebook and operational parameters can be found in [35] and [23].

REIMBURSE Out of the total leased time T, the last sub-interval is reserved for the S_{tx_i} to carry out their own transmission to their respective receivers, S_{rx_i}. In other words, it is a fare that the primary has to pay in return for the relaying services of the secondary. In this duration $t_{ss} = T - t_{sp} - t_{ps}$, the primary refrains from transmission and grants exclusive access to the secondary network.

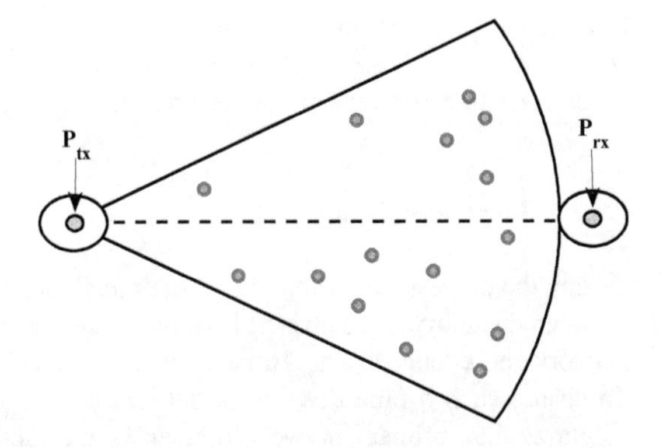

Fig. 2 Secondary nodes in the cooperation region

If the PN decides to seek the help of the CRs, it broadcasts a leasing beacon over the control channel. This beacon contains the information of cooperation and exclusion region θ, ϵ and the demand of relaying co-operation duration t_{sp}.[3] The concept of an exclusion region is exploited for minimizing the interference to the PU and also enhancing the cooperative transmission rate by selecting nodes within the exclusion region that lie between the primary transmitter and its corresponding receiver as shown in Fig. 2. The CR nodes employ listening mechanism over control channel. Beacon enabled signalling is adopted for DSL to initiate and agree on the leasing parameters. Listening only on the control channel is an energy efficient way for the CRs to monitor the primary activity. In this approach, the CRs only listen to short control messages, whereas, if the control channel is not used, then the CRs have to monitor the entire PU activity to learn about possible spectrum availabilities. The CRs are assumed to be aware of their location with respect to the primary transmitter and receiver. Upon the reception of the leasing beacon, only those CRs that lie within the desired cooperation region participate in cooperation. Based on the leasing information and the potential cooperation cost, the CRs also establish their reimbursement duration demand t_{ss}.[4]

The process of bargaining over the demand of t_{sp}, and t_{ss} is executed and if the negotiations are successful, a leasing agreement is reached. In DSL, during the first interval t_{ps}, the primary transmitter broadcasts its data to be relayed to the CRs at a low power, $\bar{P}_p < P_p$, since only geographically close CR relays need to receive and relay the data. In the second interval, the CRs cooperatively relay the data to the P_{rx} using DSTC for a duration t_{sp}. As a result of the leasing agreement, the entire CRN gets access to the spectrum for a duration t_{ss}. All the secondary nodes transmit with the same power P_s during the cooperation and reimbursement phase. P_s is significantly lower than the transmit power of the primary $P_s \ll P_t$. This maintains

[3]The PN is assumed to be aware of the average fading characteristics of its link with the secondary transmitters.

[4]The primary is assumed to be aware of the average fading characteristics of its link with the secondary transmitters.

low energy consumption in DSL and also ensures that in the last phase of secondary communication, the aggregate power of all the selected relay nodes does not increase excessively to avoid very high interference.

5.2.2 Bargaining Game

During the process of leasing, the most crucial factor is the division of leasing time between the above three phases. It is important that each operational element of the network gets enough share of time to meet its transmission throughput requirements. To ensure such a time division, a network level game is formulated where each of player, i.e., primary network (player 1) and the secondary network (player 2) engages itself in an arbitration for the time division over a control channel. As stated, the primary user initiates the leasing process. In response, the secondary users determine their demand and adopt a strategy according to the primary offer. If the offer is acceptable, the game is concluded and leasing is successful. If the CRs want to bargain further, another round of offer and respective response is played. In case the negotiations are unsuccessful, the game ends and the leasing is not done. It is further assumed that the CRs form a homogeneous network in terms of the hardware platform, leasing time demands and they do not show malicious or selfish behaviour.

During the process of leasing, the primary has a bargaining power Δp. The bargaining power of the primary determines the bias of the division of time in favour of the primary's demand. Similarly, the secondary CR network has a bargaining power Δs. The provision of variable bargaining powers in the model makes it flexible and adaptable to various real network settings. These include scenarios where the primary network has greater inherent power to determine the division of leasing time. For example, when the data traffic of the primary link is low or the channel conditions are favourable, the primary might have a greater bargaining power. Similarly scenarios where the CRs have a greater power can also be well studied using this model.

5.2.3 Assumptions

For simplicity and tractability of the analysis, it is assumed that the PN and the CRN are aware of the CSI within their respective networks. A practical implementation of such information exchange can be found in [36]. The CRs are aware of their location with respect to the primary transmitter and receiver. Moreover, the PU and the SUs are considered to be in perfect time synchronization with each other. Cost effective methodologies for implementing time synchronization in ad hoc networks have been suggested in [37], hence encouraging the proposal of the time sharing based communication scheme introduced here. The control beacon signal by the PU to initiate spectrum leasing can also be used for synchronization between the primary and the secondary nodes.

6 Analysis of DSL

6.1 Average Link Capacities R_{dir}, $\overline{R}_{ps}$, $\overline{R}_{sp}$ and $\overline{R}_{ss}$

Under conventional operation, the legacy network continues its communication over the direct link with its respective receiver at a certain rate R_{QoS}. Due to the small scale fading, the communication link is subject to outage. Thus, enforcing a certain reliability constraint restricts the operational rates to a limited regime. In other words, if $\tilde{\rho} = 1 - \rho$ is the reliability constraint, then the maximum rate which can be sustained is given as

$$R_{\text{dir}} = \sup \left\{ R_{\text{QoS}} : \text{pout}(R_{\text{QoS}}) \leq \rho \right\}, \tag{3}$$

where $\text{pout}(R_{\text{QoS}})$ is the link outage probability at a particular desired rate R_{QoS}. The performance of the direct link (P_{tx}, P_{rx}) pre-dominantly is noise limited, since it is assumed that there is no interference caused by the CRN to the primary transmission. The instantaneous capacity R_{QoS} of this link can be defined as,

$$R_{\text{QoS}} = \log_2(1 + \text{SNR}), \ \text{(bits/s)} \tag{4}$$

where SNR is as defined previously and h_p is the channel power gain between the source and the destination, P_p is the transmit power and $l(r_p)$ is the distance dependent path loss between the nodes. The ρ-outage rate R_{dir} is defined as the largest rate of transmission R such that the outage probability of the direct primary link is less than ρ. For a conventional operation mode it can be quantified as follows:

Lemma 1 *The ρ-outage rate, R_{dir}, for the link (P_{tx}, P_{rx}) is given as,*

$$R_{dir} = \log_2\left(1 - \left(\frac{P_p h_p l\left(r_p\right)}{\sigma^2}\right) \ln\left(1 - \rho\right)\right), \ (bits/s) \tag{5}$$

Proof The result can be derived following the same lines as in [29] Sect. IV Lemma 1. □

When the spectrum is leased to the SUs, the cooperative link performance is dictated by the attainable rate over the relay link, i.e., the cooperative channel capacity. The cooperative channel capacity depends upon both (i) the transmission rate R_{ps} achieved between the primary transmitter and any selected relay during the first leasing sub-interval and (ii) the rate R_{sp} between the selected relay nodes and the primary receiver assuming that DSTC cooperation is employed. Also, as mentioned earlier, nodes centred only in the effective area of communication, $A_c(\theta, r_p, \epsilon)$, are considered for cooperation.

Lemma 2 *The average transmission rate from the primary transmitter to secondary relay, $\overline{R}_{ps}$, is upper-bounded as,*

$$\overline{R}_{ps} = \log_2\left(1 + \left(\frac{\exp\left(-\lambda\frac{\theta}{2}\left(r_p - \epsilon\right)^2 - 1\right)}{(r_p - \epsilon) - C}\right)^{\alpha} \frac{\bar{P}_p}{\sigma^2}\right), \; (bits/s) \tag{6}$$

where $C = \sqrt{\frac{\pi}{2\lambda\theta}} \exp\left(-\lambda\frac{\theta}{2}\left(r_p - \epsilon\right)^2\right) erfi\left(\sqrt{\frac{2\lambda\theta}{\pi}}\right)$ and $erfi(x)$ is the imaginary error function such that $erfi(x) = 2\sqrt{\pi}\int_{t=0}^{x} \exp\left(-t^2\right) dt$.

Proof The result can be derived following the same lines as in [29] Sect. IV Lemma 2. □

In the second phase of cooperation, the selected secondary relays form a $k = \lambda A_c(\theta, r_p, \epsilon)$ antenna array and perform DSTC to send the data to the receiver with a rate R_{sp}. The rate of communication when DSTC is employed for multiple relay transmission to a common destination has been evaluated in [23, 35, 38, 39]. In the context of the geometric modelling of dynamic spectrum leasing, the DSTC communication rate is used and its mean value is determined considering the geometric parameters.

Lemma 3 *The average transmission rate, $\overline{R}_{sp}$, when k secondary relays, i.e., $k \in |\Phi_r|$ form an antenna array, where secondary relay i is located at a distance r_i from P_{rx} is given by*

$$\overline{R}_{sp} = \log_2\left(1 + \frac{\lambda\theta P_s}{\sigma^2}\left(\frac{(r_p - \epsilon)^{2-\alpha} - \epsilon^{2-\alpha}}{2-\alpha}\right)\right), \; (bits/s) \tag{7}$$

where, the secondary transmits with a power P_s, the channel gain between S_{tx} and P_{rx} is h_{sp_i}.

Proof The result can be derived following the same lines as in [29] Sect. IV Lemma 3. □

In the last phase of spectrum leasing, all the secondary transmitters communicate with their respective receivers. A nearest neighbour model of the CR source destination pairs is considered in this chapter where each transmitter only communicates with its nearest receiver [40] as shown in Fig. 2. It is of interest to know the average transmission capacity of the (S_{tx}, S_{rx}) link, $\overline{R}_{ss}$. In this case, all the secondary transmitters in the CR network simultaneously communicate with their receivers in order to utilize the leased bandwidth for their own transmission. In this phase, similar to the direct communication, a realistic situation is considered under which the secondary network also operates under a fixed QoS constraint R_{QoS_s}.

Lemma 4 *The average rate, $\overline{R}_{ss}$, for the link (S_{tx_i}, S_{rx_i}) where the channel power gain between the source i and its destination (nearest neighbour) is exponential h_{ss_i}, the transmit power P_s is given as,*

$$\overline{R}_{ss} = \frac{\pi^{\frac{3}{2}}\lambda}{\sqrt{\left(2^{R_{QoS_s}}-1\right)/\frac{P_s}{\sigma^2}}} \exp\left(\frac{\left(\pi\lambda\left(\tau\left(R_{QoS_s}\right)+1\right)\right)^2}{4\left(2^{R_{QoS_s}}-1\right)/\frac{P_s}{\sigma^2}}\right) \tag{8}$$

$$\times Q\left(\frac{\pi\lambda\left(\tau\left(R_{QoS_s}\right)+1\right)}{\sqrt{\left(2^{R_{QoS_s}}-1\right)/\frac{P_s}{\sigma^2}}}\right)\bar{R}_{th}.\,(bits/s) \tag{9}$$

where R_{QoS_s} is the desired threshold rate for secondary communication.

Proof The proof follows the same steps as in [41] in Sect. V. □

After computing the individual link transmission rates, the aim is to know the overall transmission rate achieved in the DSL operational mode. It is assumed that a decode and forward type single hop relaying mechanism is used in the cooperation phase. The effective DSL capacity R_{DSL} is then given as,

$$R_{\mathrm{DSL}} = \min\{\overline{R}_{ps}, \overline{R}_{sp}\}.\,(\text{bits/s}) \tag{10}$$

6.2 Optimal Division of Leased Time for Cooperation and Secondary Activity

The most critical factor in the operation of spectrum leasing is the optimal division of the total leased time T between the time t_{sp} reserved for cooperation with the primary at a cooperative rate R_{DSL} and the remaining time t_{ss} for the secondary activity at a rate $\overline{R}_{ss}$. The goal of the primary node is to ensure that its rate and quality of communication, R_{dir} and ρ-outage probability respectively, are maintained by maximizing the time t_{ps} and t_{sp}. The primary node can ensure that $\overline{R}_{ps}$ attains the QoS rate R_{dir} by a proper choice of t_{ps} such that $t_{ps}\overline{R}_{ps} = T R_{\mathrm{dir}}$. However, the remaining time $(T' = T - t_{ps})$ needs to be divided between phase two and three to get t_{sp} and t_{ss}.

The CR nodes intend to increase their benefits in terms of their spectrum utility and throughput by having spectrum access for maximum time and compensating for the cost of cooperation in relaying primary data. A very small fraction of t_{ss} will discourage the secondary, impacting cooperation and the overall throughput of the system suffers. On the other hand, prolonged t_{ss} will degrade the performance of the legacy network in terms of its bandwidth efficiency which is not acceptable in any case. Hence an intelligent division of time is very crucial for the operation of the network. Also, the secondary network must cooperate in relaying primary data for a time t_{sp} long enough so that the primary network maintains its communication standards. Hence the problem boils down to an optimal division of leasing time T' between phases two and three of DSL.

An optimal time division can be conveniently casted in the framework of Nash Bargaining: a game theoretic tool to model the situations of bargaining interactions. The situation can be modelled as a two player game using the Nash bargaining framework from cooperative game theory [42]. In this case, the primary transmitter is the first player whose utility is directly dependent upon the cooperation time t_{sp} and increases as it increases. For simplicity, we define the utility of the primary and the secondary node as;

$$\mathcal{U}_1(t) = t_{sp}, \tag{11}$$

and

$$\mathcal{U}_2(t) = t_{ss}, \tag{12}$$

respectively, where $t_{sp} + t_{ss} = T'$.

Bargaining as a two player game is considered because every single secondary node is representative of the utility of all the remaining secondary nodes as only the average rate values and equal transmit powers for all CRs are considered. The Nash bargaining framework is employed to model a situation in which the players negotiate for their agreement on a particular point out of a set of joint feasible payoffs $\mathcal{G}$. In this particular case, $\mathcal{G} \equiv \{g = (g_1, g_2) : g_i = \mathcal{U}_i(\mathbf{S}),\ i = 1, 2;\ \mathbf{S} \in S1 \times S2\}$, where the functions $\mathcal{U}_i(.)$ in this case of DSL are given in Eqs. 11 and 12. $\mathbf{S}$ is the strategy of the ith player in terms of the time it demands i.e., t_{sp}/t_{ss} from the strategy profile S_i. In Nash Bargaining, in case the negotiations render unsuccessful, the outcome of the game becomes $\mathcal{G} = (g_{01}, g_{02})$. It is a fixed vector known as the disagreement vector. The whole bargaining problem can be described conveniently by the pair $(\mathcal{G}, g_0)$. A pair of payoffs $\left(g_1^*, g_2^*\right)$ is a Nash Bargaining solution if it solves the following optimization problem

$$\max_{g_1, g_2} (g_1 - g_{01})^{\Delta_p} (g_2 - g_{02})^{\Delta_s} \tag{13}$$

$$\text{subject to} \begin{array}{l} (g_1, g_2) \in \mathcal{G} \\ (g_1, g_2) \geq \mathcal{G}_0 \end{array}.$$

If the set $\mathcal{G}$ is compact and convex, and there exists at least one $g \in \mathcal{G}$ such that $g > g_0$, then a unique solution to the bargaining problem $(\mathcal{G}, g_0)$ corresponds to the unique solution of the optimization problem [27, 42].[5] Here Δ_p and Δ_s defined as $\left\{\Delta_p, \Delta_s \in [0, 1]\,|\,\Delta_p = 1 - \Delta_s\right\}$ correspond to the bargaining powers of the primary and secondary network. Greater values of Δ_p and Δ_s correspond to higher bargaining powers. Increasing the bargaining power of a player corresponds to greater weightage

[5]From Eqs. 11 and 12, the compactness and convexity of $\mathcal{G}$ can be seen.

of its preferences over the preferences of the other player. Increasing the power of one player implies decreasing power of the other.

In this case, the fraction of leased time should be large enough to ensure that the time-rate product of cooperation time t_{sp} and cooperative rate $\overline{R}_{sp}$ is at least equal or greater than the direct communication time T and rate R_{dir} product. During the second sub-interval, a secondary node must have enough time to at least overcome its cooperation cost cP_s given its average transmission rate $\overline{R}_{ss}$. Here c is measures the bits transmitted per unit of power consumed.

Theorem 5 *The optimal proportion of time for cooperative relaying is*

$$t_{sp} = \frac{\Delta_p T' + \Delta_s\left(\frac{R_{dir}}{\overline{R}_{sp}}\right) - \Delta_p\left(\frac{cP_s}{\overline{R}_{ss}}\right)}{\Delta_p + \Delta_s}, \tag{14}$$

where the disagreement vector is $\left(t_{0p}, t_{0s}\right) = \left(\frac{TR_{dir}}{\overline{R}_{sp}}, \frac{cP_s}{\overline{R}_{ss}}\right)$ *and secondary activity time* $t_{ss} = T' - t_{sp}$.

Proof From the definition of Nash Bargaining solution, the time division problem for a 2-player game can be written as

$$\max\left(\Delta_p \log\left(\mathcal{U}_1(t) - g_{01}\right) + \Delta_s \log\left(\mathcal{U}_2(t) - g_{02}\right)\right), \tag{15}$$

$$\text{subject to}\, T' = t_{sp} + t_{ss}.$$

From the definition of Nash Bargaining solution, the time division problem for a 2-player game can be written in a logarithmic form as above. Such representation of the maximization problem ensures proportional fairness of the solution for both the players. Here the minimum required time for both primary and secondary is given as $(g_{01}, g_{02}) = \left(t_{0sp}, t_{0ss}\right) = \left(\frac{TR_{\text{dir}}}{\overline{R}_{sp}}, \frac{cP_s}{\overline{R}_{ss}}\right)$, which is the least time required to meet the respective objectives of QoS and cooperation cost compensation.The corresponding Lagrangian for the above optimization problem can be written as,

$$\begin{aligned} L(t_{sp}, \lambda_1, \lambda_2) = {} & \Delta_p \log\left(t_{sp} - t_{0sp}\right) + \Delta_s \log\left(t_{ss} - t_{0ss}\right) \\ & - \lambda_1\left(T' - t_{sp} + - t_{ss}\right). \end{aligned}$$

The original maximization problem can be solved by replacing t_{ss} by $T' - t_{sp}$ and using the first order necessary conditions,

$$\frac{\delta L}{\delta t_{sp}} = \frac{\Delta_p}{t_{sp} - t_{0sp}} + \frac{\Delta_s}{t_{sp} - T' - t_{0ss}} = 0, \tag{16}$$

This follows from the definition of the Nash Bargaining problem that there exists a vector **S** such that the optimal value of the optimization problem is strictly positive. Solving for Eq. 16 by using simple algebra, the result can be obtained as,

$$t_{sp} = \frac{\Delta_p T' + \Delta_s \left(\frac{R_{\text{dir}}}{R_{sp}}\right) - \Delta_p \left(\frac{cP_s}{R_{ss}}\right)}{\Delta_p + \Delta_s},$$

where the above equilibrium solution gives the optimal share of cooperation time t_{sp} out of the total leased time T that ensures a cooperative data transmission rate $R_{sp} \geq R_{\text{dir}}$. It reserves the rest of the time for secondary user that at least allows the secondary to utilize the spectrum to compensate for their transmission cost during the cooperation phase. □

7 Performance Evaluation of DSL

In this section, the design space of the DSL enabled CRN is investigated by employing the analytical model developed in the previous section. In order to verify the analysis and establish the validity of the assumptions made throughout, Monte Carlo simulations for the large scale DSL based CRN are performed. In order to simulate, a network radius of 200 m in which secondary nodes are Poisson distributed with mean λ is considered. Direct communication under an outage constraint ρ at a transmit power P_p is simulated. Similarly, the operational phases of DSL are simulated. For each realization of the Poisson network, a Rayleigh distributed channel coefficient is generated. The transmission rate at the receiver for each spatial instance of the network is averaged for 10^4 different channel coefficients. This process is in turn repeated for 10^4 realizations of Poisson distributed CR network with intensity λ and the transmission rate is averaged. Secondary network communication under interference considerations is also studied in a similar fashion. All the simulations are carried out in MATLAB. Normalized values for transmit powers P_p and P_s are used. It is assumed that the secondary network operates at a low power profile i.e., $\sim\frac{1}{10^{\text{th}}}$ of P_p. Similar power profiles can be found for devices like HeNB in LTE rel. 12 and other examples in heterogeneous networks [43, 44].

Firstly, the average achievable transmission rates under both the normal and leasing mode of network operation are studied as shown in Fig. 3a. The rate under normal primary communication at a transmit power P_p increases with improving channel conditions. Here, the reliability in terms of the probability of success ($p_{\text{suc}} = 1 - \rho$) of direct communication is assumed to be 90 %. The outage capacity, R_{dir}, defines the target capacity for communication in the primary network R_{th} for all operational modes i.e., direct and DSL. Under identical channel realizations, a demand for higher service quality (smaller ρ) straightforwardly results in lower R_{dir}.

For the capacity analysis of DSL, the average achievable transmission rates in the three phases of leasing are studied. The capacity of the primary to secondary communication in the first phase is strongly dependent upon the number of secondary nodes present in the area of cooperation. As mentioned earlier, in this analysis, the lower bound to this rate is studied by considering the average transmission rate between

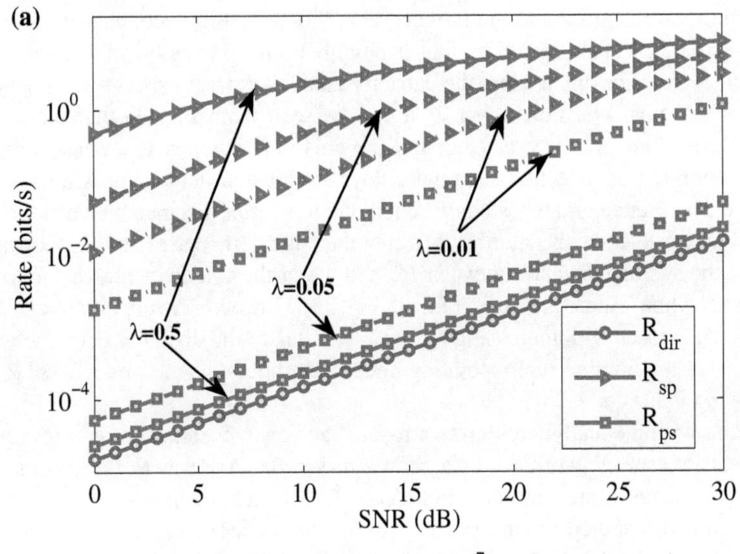

Rates vs. SNR. $P_p = 1, r_p = 10, \bar{P}_p = P_s = 0.1, \epsilon = 1$.

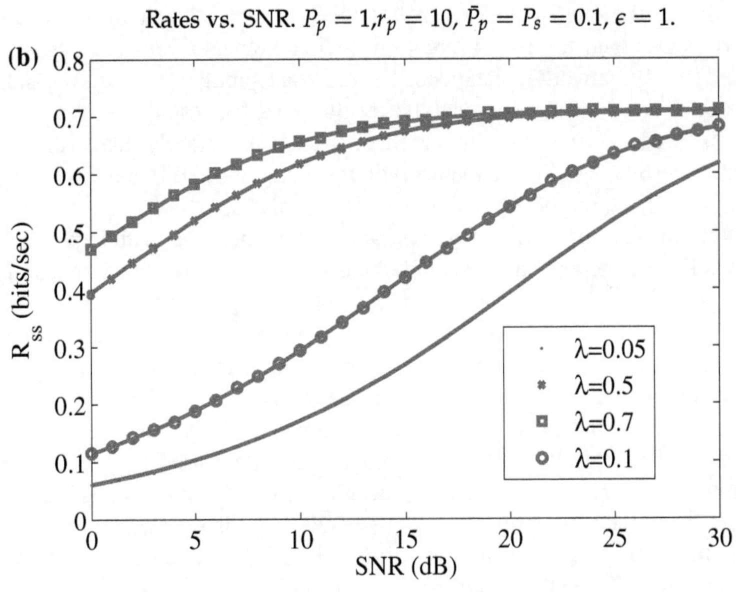

R_{ss} vs. SNR, $P_s = 0.1$.

Fig. 3 Achievable data rate of direct and DSL communication. **a** Rates versus SNR. $P_p = 1$, $r_p = 10$, $\bar{P}_p = P_s = 0.1$, $\epsilon = 1$. **b** R_{ss} versus SNR, $P_s = 0.1$

the primary transmitter and the farthest relay. For very low secondary density, e.g., $\lambda < 0.01$, the probability of finding a neighbour in the region of cooperation is extremely low. For this reason, the capacity analysis for very sparse secondary network is not possible. For higher λ, it can be seen from Fig. 3a that the average transmission rate $\overline{R}_{ps}$ is greater than R_{dir}. This phenomenon is a consequence of cooperation region selection such that relays are located in close proximity to both P_{tx} and P_{rx}. Hence greater rate is attained due to shorter distance between the relay and P_{tx}. However, if the number of secondary users increases in the cooperation region, the average distance between P_{tx} and the farthest node increases. Hence $\overline{R}_{ps}$ decreases when λ increases (lower line in Fig. 3a). However, the cooperative relaying rate $\overline{R}_{sp}$ increases with increasing relay density due to the diversity gain. Increasing λ increases the number of cooperating nodes, consequently, the rate $\overline{R}_{sp} \gg R_{\text{dir}}$ for increasing values of λ.

Along with the analytically drawn results, achievable transmission rates under a practical Poisson network are also shown in Fig. 3a. A PN with two nodes and a CRN for various λ. are simulated in MATLAB. For each realization of the network, exponential distributed channel power gain is generated. The successful transmission probability at the rate $R_{\text{QoS}}/R_{\text{QoS}_s}$ at the receiver for each spatial instance of the network is averaged for 10^4 different channel coefficients. This process is in turn repeated for 10^4 network realizations. The practical simulation results are indicated by the lines running over the analytic results (analytic results are indicated with markers). It can be seen that the practical simulations closely match the analytic evaluation results. It validates the analytic formulation of DSL and the simplifying assumptions made for the simplification of the analysis.

It is shown that the communication rate $\overline{R}_{ss}$ also increases with improving SNR values in Fig. 3b (here the desired QoS of the secondary network in terms of desired rate $\bar{R}_{\text{th}}$ is 0.5 bits/s). This is a consequence of the improved signal strength at the receiver. As the density of the secondary nodes increases, the average transmission rate increases. However, $\overline{R}_{ss}$ tends to saturate with increasing SNR at higher values of λ. It is a consequence of the interference limited behaviour of the channel. Increasing interference due to increasing λ limits the increase in $\overline{R}_{ss}$. It is clear that increasing the desired threshold rate $\bar{R}_{\text{th}}$ causes the average rate to decrease because the decoding threshold at S_{rx} is raised. Hence, a graphical illustration of this result is intentionally skipped. The practical simulation results of $\overline{R}_{ss}$ are also shown in Fig. 3b which verifies the analytical derivations. It is to be noted that the rest of the results are based upon the communication rates of direct and DSL communication, which have been shown to be in a close agreement with each other. Therefore, the practical simulations of the remaining results can safely be assumed to be accurate and hence are skipped for the sake of brevity.

In order to intelligently exploit the diversity gains of DSL at low power, it is important to determine the appropriate operational time of each phase of DSL. The primary itself determines and communicates for time t_{ps} during the first phase such that $t_{ps}\overline{R}_{ps} = T R_{\text{dir}}$. In Fig. 4a, the time t_{ps} reserved for R_{ps} is shown. It can be seen that it increases with increase in the secondary network density. This behaviour follows from the lower transmission rate achieved with increasing λ as discussed

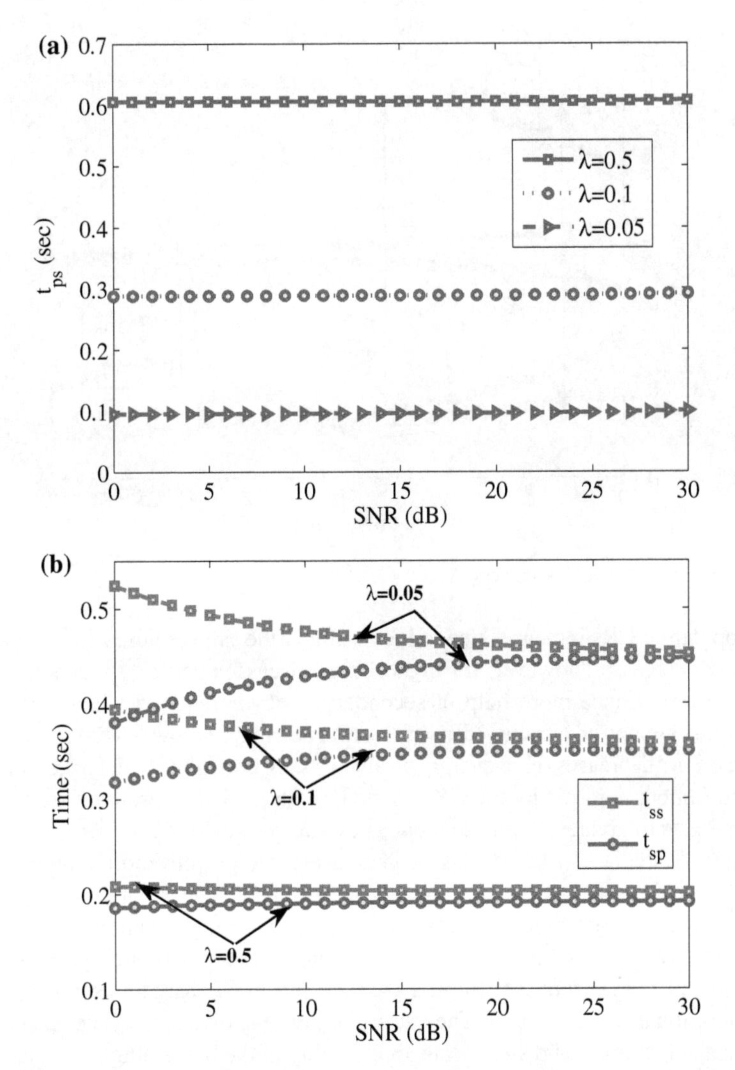

Fig. 4 Nash bargaining solution. **a** Time for first phase transmission R_{ps}. **b** Time bargain for primary and secondary transmitters. $T = 1, c = 0.05, \lambda = 0.5$

earlier. Correspondingly, the time share of t_{sp} and t_{ss} i.e., T' decreases with increasing λ since a major portion of the time is reserved for primary to secondary transmission in the first slot.

The optimal relation for the division of the remaining leased time T' is found in Eq. 14 and shown in Fig. 4b over a range of SNR values. At low CR densities, more time t_{sp} is required to harvest the gains from cooperative relaying. As the number

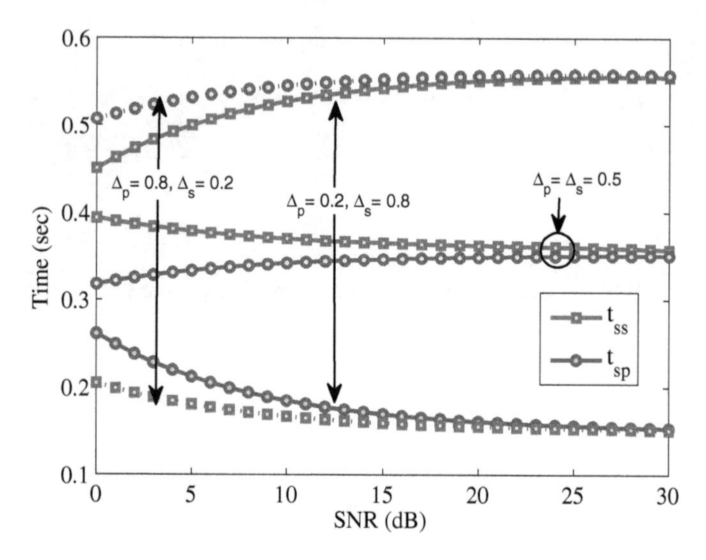

Fig. 5 Impact of bargaining powers

of cooperating CRs increases due to increasing λ, the time required for cooperative relaying decreases. However, for higher λ, as discussed earlier, t_{ps} gets the major share of time. Since more help of secondary nodes is required when the channel conditions are not favourable, CRs are reimbursed more at low SNRs. As the SNRs increases, t_{ss} decreases. However, t_{ss} is always long enough to satisfy the minimum reimbursement required by the CRs. Overall, both t_{sp} and t_{ss} assume low values at higher λ due to greater t_{ps} requirement as explained previously.

Figure 5, studies the bargaining powers of the two players and its impact on the division of time. It can be seen that the player with higher bargaining power is able to procure more time to increase its utility. The primary can get up to ∼20 % more time reserved for the cooperative relaying phase when its bargaining power is improved to 0.8 from 0.5. Similar increase in the CR bargaining power results in proportional increase in t_{ss}. The variety of possible divisions of the leasing time depicts the flexibility and wide scale applicability of the bargaining solutions. It can capture the scenarios where one player exercises greater influence on the decision.

8 Energy Efficiency of Spectrum Leasing Model

8.1 *Analytical Quantification*

In this section, the energy efficiency (EE) of the spectrum leasing model for cognitive radio networks is defined and quantified. The energy efficiency is the number of bits transmitted successfully across the channel per unit of energy consumed, given as,

$$EE = \frac{n_B}{J}, \text{ (bits/J)} \tag{17}$$

where n_B is the number of bits transmitted successfully and J is the energy consumed in Joules.

Theorem 6 *The energy efficiency of a licensed primary network employing direct communication EE_{dir} and while employing DSL, EE_{DSL} in terms of the number of successfully transmitted bits per unit energy can be given as*

$$EE_{\text{dir}} = \frac{n_{\text{dir}}}{TP_p}, \quad \text{and} \quad EE_{\text{DSL}} = \frac{n_{\text{DSL}}}{t_{ps}\bar{P}_p + t_{sp}P_s k}, \tag{18}$$

respectively, where n_{dir} is the number of successfully transmitted bits in direct communication, n_{DSL} are the successfully transmitted bits over the cooperative link.

Proof The number of bits successfully transmitted in the transmission duration of the direct link n_{dir} is given as [34];

$$n_{\text{dir}} = R_{\text{dir}}T, \tag{19}$$

where R_{dir} follows from the result in Lemma 1. In case the primary decides to lease the spectrum, the number of bits successfully transmitted in spectrum leasing is given as

$$n_{\text{DSL}} = \min\left(t_{ps}\overline{R}_{ps}, t_{sp}\overline{R}_{sp}\right), \tag{20}$$

where, $\overline{R}_{ps}$ and $\overline{R}_{sp}$ have been determined in Eqs. 6 and 7, respectively. The total energy consumed during direct communication is TP_p and that during DSL based cooperation is $t_{ps}\bar{P}_p + t_{sp}P_s k$ where the first term accounts for the energy consumed in P_{tx} to S_{tx} communication and the later for the energy consumption when k secondary transmitters cooperatively relay the data to P_{rx} for a duration equal to the leased time t_{sp} and then transmit their own traffic for a time t_{ss}. Similarly, we also quantify the energy efficiency of secondary communication phase as

$$EE_{\text{sec}} = \frac{n_{\text{sec}}}{P_s t_{ss}}, \text{ (bits/J)} \tag{21}$$

where $n_{\text{sec}} = t_{ss}\overline{R}_{ss}$. The total energy consumed when the secondary network communicates for a duration t_{ss} is $P_s t_{ss}$. □

8.2 Analytic Results

Following the analytical results for R_{dir}, $\overline{R}_{ps}$, $\overline{R}_{sp}$ and $\overline{R}_{ss}$, the energy efficiency is studied by observing both the direct link and DSL based communication over a variety of SNR values as shown in Fig. 6a. As in the discussion on the achievable

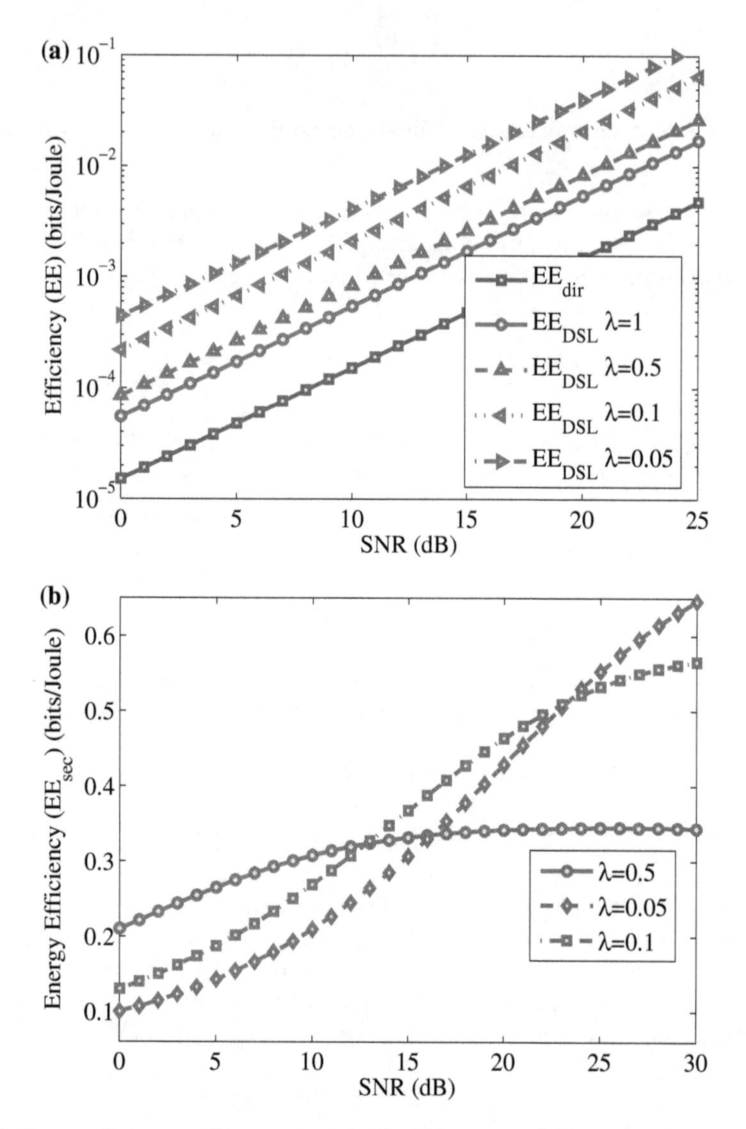

Fig. 6 Energy efficiency. **a** Direct communication EE_{dir} versus DSL communication EE_{DSL}, $P_p = 1,\ P_s = 0.1,\ T = 1, \theta = \frac{\pi}{4}$. **b** Secondary Network EE_{sec}, $P_s = 0.1$, time = t_{ss}

capacity and time division, the EE of direct and DSL communication is studied on the basis of secondary network density. It is clearly evident that the energy efficiency of DSL is significantly greater than that of the direct communication for smaller values of λ. This is because the transmit power of the primary and secondary in DSL mode is low. The selection of relays which are geographically closer to both P_{tx} and P_{rx} help in achieving the same transmission rate in shorter time and hence lower power. Also, the cooperative relaying based diversity benefits significantly increase the throughput at the primary receiver while maintaining a low transmit power. As λ increases, the EE of DSL decreases mainly due to two reasons;

1. The throughput of the cooperative DSL communication decreases as the average primary to secondary rate $\overline{R}_{ps}$ decreases with increasing λ (see Fig. 3a). The energy consumed in the first phase of DSL grows as the primary to secondary link operation time t_{ps} increases.
2. Also, in the second phase of DSL, aggregate transmit energy is higher due to increased number of relays.

It can be seen that the bargaining based leasing time division results in significantly more energy efficient communication via DSL as compared to direct communication when the secondary network is relatively sparse (i.e., $\lambda \leq 0.05$).

In Fig. 6b, we study the EE of the secondary network in the third phase of leasing. During this phase, the energy efficiency of the network improves with increasing SNR. It attains a maximum value as R_{ss} converges to a constant rate. Moreover, if the number of secondary transmitters is increased, the aggregate energy consumption is increased by the presence of greater number of interferes. Hence, the EE of the secondary network EE_{sec} decreases for high λ when DSL is operational in the interference limited regime. Hence, DSL for sparse secondary network is the most energy efficient solution for both primary and secondary networks.

Further, the effect of the angle θ of the sector of cooperation on the EE of DSL is investigated. From Fig. 7, it can be seen that the EE of DSL degrades with increasing the area of cooperation. This happens because increasing θ increases the number of cooperators which in turn increases the aggregate transmit power used for cooperation. Also, the probability of finding a farther neighbour increases as the area of

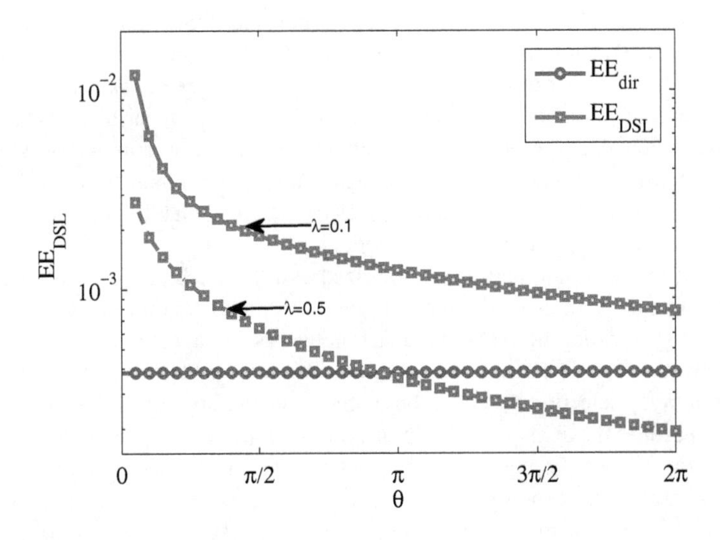

Fig. 7 Impact of θ on EE of DSL

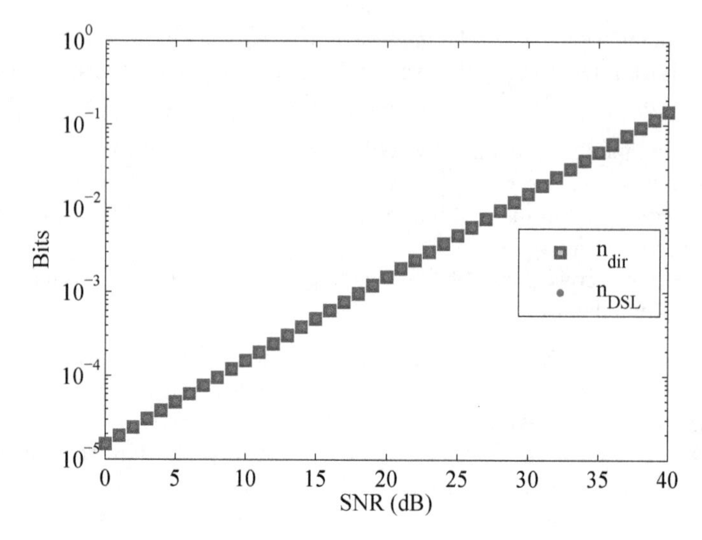

Fig. 8 Time-rate product: direct communication bits_{dir} versus DSL bits_{DSL} $\lambda = 0.5,\ T = 1$

cooperation increases and hence limits $\overline{R}_{ps}$ in the first phase. For low values of θ, $\overline{R}_{sp}$ is low due to limited number of cooperators. However, low θ results in high $\overline{R}_{ps}$, thus, an improved energy efficiency is observed. For very low values of θ, DSL is not viable because the probability of finding even a single relay is infinitely low. As soon as the cooperation area is wide enough to find a few relays in it, DSL becomes viable and most energy efficient.

Finally, the time rate product is analysed which determines the total number of bits that can be transmitted during both modes for a time T. The simulation results in Fig. 8 demonstrate the time-rate product of direct versus DSL communication. It can be seen that DSL communication achieves exactly the same performance in terms of the effective number of bits delivered to the primary receiver as compared to the direct communication. It simply implies that by using DSL, the primary can transmit the same amount of data as with direct communication. However, as discussed earlier, this transmission is more energy efficient than direct communication when the CR network is sparse. This result further verifies the practical viability and attraction for the legacy network to operate in DSL mode.

The entire discussion can be summarized as follows. DSL based transmission serves as an energy efficient alternative to direct communication when the secondary network is sparse. For these low populated networks, the aggregate energy and time requirements for cooperation and secondary network activity are low. Hence an intelligent relay selection based on the spatial characteristics of the network and the optimal leasing time division can help in exploiting the diversity gain of cooperative relaying to enhance the performance of legacy communication. It also allows the otherwise deprived secondary network to utilize its share in the bandwidth therefore improving the overall spectral utilization of the network.

9 Summary

In this chapter, a DSL scheme is presented that provides an elaborated implementation mechanism for dynamic resource sharing was presented. An analytical study of dynamic spectrum leasing based on a geometrical framework was presented and the relative link performances in terms of the achieved capacities in DSL and direct communication were evaluated. A Nash bargaining based approach for the determination of the appropriate leasing time was introduced. It was demonstrated that the proposed algorithm results in a division of time that satisfies the requirements of both primary and secondary networks. Based on these operational features, the energy efficiency was quantified and investigated through simulations. The results indicate that DSL is more energy efficient in most of the practical SNR regimes, hence making DSL a viable option for energy efficient communication. Such energy efficient solution can be achieved only if a sparse CR network is considered with DSL operation at a low transmit power as compared to that of the transmit power of the direct communication. DSL is shown to be more than 10x more energy efficient than direct communication when the CR density is low and/or the cooperation region is small. With only a few cooperating CR nodes, the entire CR network gets exclusive access to the spectrum. Hence DSL based communication enables the primary to communicate at its desired transmission rate and quality in an energy efficient manner and also enables the CR network to exploit the licensed spectrum for its own communication.

The scheme presented in this chapter can be further extended to study the possible delays incurred in the DSL communication. Also, the impact of the presence of any greedy CRs in the network is also an open issue. It is interesting to also consider individually autonomous entities in contrast to network level players in the game formulation of DSL. The energy and spectral efficiency of DSL under the above mentioned considerations is an important research question. In short, DSL is a useful technique for improving the efficiency of wireless communication with direct application to future networks.

References

1. Ericsson, N.: Ericsson mobility report. Stockholm, Sweden, Technical report, vol. 198, p. 287 (2014)
2. Proprietry, Q.: BT announces major wind power plans. http://stakeholders.ofcom.org.uk/binaries/research/technology-research/cograd_main.pdf (2007)
3. Athineos, M.: eSure: realize the goal of reducing energy consumption and CO_2 emissions. Global DC Power Engineering and Marketing. http://commsbusiness.co.uk/news/bt-announces-major-wind-power-plans/
4. Webb, M., et al.: Smart 2020: enabling the low carbon economy in the information age. The Climate Group, London vol. 1, no. 1, 1–1 (2008)
5. Fehske, A., Fettweis, G., Malmodin, J., Biczók, G.: The global footprint of mobile communications: the ecological and economic perspective. IEEE Commun. Mag. **49**(8), 55–62 (2011)

6. Shannon, C., Weaver, W.: The Mathematical Theory of Communication. University of Illinois Press, Urbana, Ill (1949)
7. Feeney, L., Nilsson, M.: Investigating the energy consumption of a wireless network interface in an ad hoc networking environment. In: Twentieth Annual Joint Conference of the IEEE Computer and Communications Societies (INFOCOM), vol. 3. IEEE, pp. 1548–1557 (2002)
8. Christensen, K., Gunaratne, C., Nordman, B., George, A.: The next frontier for communications networks: power management. Comput. Commun. **27**(18), 1758–1770 (2004)
9. E. FCC: Docket No 03-322 Notice of proposed rule making and order (2003)
10. Zhao, Q., Swami, A.: A survey of dynamic spectrum access: signal processing and networking perspectives. In: IEEE International Conference on Acoustics, Speech and Signal Processing (ICASSP), vol. 4, Apr 2007, pp. IV-1349–IV-1352
11. Goldsmith, A., Jafar, S.A., Maric, I., Srinivasa, S.: Breaking spectrum gridlock with cognitive radios: an information theoretic perspective. Proc. IEEE **97**(5), 894–914 (2009)
12. Jayaweera, S., Li, T.: Dynamic spectrum leasing in cognitive radio networks via primary-secondary user power control games. IEEE Trans. Wirel. Commun. **8**(6), 3300–3310 (2009)
13. R.S.P. Group: Report on collective use of spectrum and other sharing approaches, Technical report, vol. RSPG11-392 (2011)
14. Mueck, M., Jiang, W., Sun, G., Cao, H., Dutkiewicz, E., Choi, S.: Novel spectrum usage paradigms for 5G. http://cms.comsoc.org/SiteGen/Uploads/Public/Docs_TC_CN/WhitePapers/2014_11_White_Paper_Spectrum_Paradigms_v1.0.pdf (2014)
15. C. ECC: Licensed Shared Access (LSA), Report 205 (2014)
16. Grace, D., Chen, J., Jiang, T., Mitchell, P.: Using cognitive radio to deliver green communications. In: 4th International Conference on Cognitive Radio Oriented Wireless Networks and Communications (CROWNCOM), pp. 1–6, June 2009
17. Li, L., Zhou, X., Xu, H., Li, G., Wang, D., Soong, A.: Energy-efficient transmission in cognitive radio networks. In: 7th IEEE Consumer Communications and Networking Conference (CCNC), 2010, pp. 1–5, Jan 2010
18. Zhang, Q., Jia, J., Zhang, J.: Cooperative relay to improve diversity in cognitive radio networks. IEEE Commun. Mag. **47**(2), 111–117 (2009)
19. Duan, L., Huang, J., Shou, B.: Competition with dynamic spectrum leasing. In: IEEE Symposium on New Frontiers in Dynamic Spectrum, Apr 2010, pp. 1–11
20. Jia, J., Zhang, Q.: Competitions and dynamics of duopoly wireless service providers in dynamic spectrum market. In: Proceedings of the 9th ACM International Symposium on Mobile Ad Hoc Networking and Computing MobiHoc, 2008, pp. 313–322
21. Huang, J., Berry, R., Honig, M.: Auction-based spectrum sharing. Mobile Netw. Appl. **11**(3), 418 (2006)
22. Jayaweera, S., Vazquez-Vilar, G., Mosquera, C.: Dynamic spectrum leasing: a new paradigm for spectrum sharing in cognitive radio networks. IEEE Trans. Veh. Technol. **59**(5), 2328–2339 (2010)
23. Simeone, O., Stanojev, I., Savazzi, S., Bar-Ness, Y., Spagnolini, U., Pickholtz, R.: Spectrum leasing to cooperating secondary ad hoc networks. IEEE J. Sel. Areas Commun. **26**(1), 203–213 (2008)
24. Stanojev, I., Simeone, O., Spagnolini, U., Bar-Ness, Y., Pickholtz, R.: An auctionbased incentive mechanism for non-altruistic cooperative arq via spectrum-leasing. In: IEEE Global Telecommunications Conference, GLOBECOM 2009, Nov 2009, pp. 1–6
25. Yi, Y., Zhang, J., Zhang, Q., Jiang, T., Zhang, J.: Cooperative communication-aware spectrum leasing in cognitive radio networks. In: IEEE Symposium on New Frontiers in Dynamic Spectrum, 2010. IEEE, pp. 1–11
26. Ji, Z., Liu, K.: Cognitive radios for dynamic spectrum access–dynamic spectrum sharing: a game theoretical overview. IEEE Commun. Mag. **45**(5), 88–94 (2007)
27. Yaiche, H., Mazumdar, R., Rosenberg, C.: A game theoretic framework for bandwidth allocation and pricing in broadband networks. IEEE/ACM Trans. Netw. **8**(5), 667–678 (2000)
28. Cagalj, M., Ganeriwal, S., Aad, I., Hubaux, J.: On selfish behavior in CSMA/CA networks. In: Proceedings IEEE INFOCOM 2005. 24th Annual Joint Conference of the IEEE Computer and Communications Societies, pp. 2513–2524 (2005)

29. Hafeez, M., Elmirghani, J.: Analysis of dynamic spectrum leasing for coded bidirectional communication. IEEE J. Sel. Areas Commun. **30**(8), 1500–1512 (2012)
30. Hafeez, M., Elmirghani, J.: Dynamic specrum leasing for beamforming cogniive radio neworks using nework coding. In: 2013 IEEE International Conference on Communications (ICC), June 2013, pp. 2840–2845
31. Vu, M., Devroye, N., Tarokh, V.: The primary exclusive region in cognitive networks. In: 5th IEEE Consumer Communications and Networking Conference (CCNC), Jan 2008, pp. 1014–1019
32. Zorzi, M., Rao, R.R.: Geographic random forwarding (geraf) for ad hoc and sensor networks: multihop performance. IEEE Trans. Mobile Comput. **2**(4), 337–348 (2003)
33. Chen, D., Deng, J., Varshney, P.: Selection of a forwarding area for contentionbased geographic forwarding in wireless multi-hop networks. IEEE Trans. Veh. Technol. **56**(5), 3111 (2007)
34. Goldsmith, A.: Wireless Communications. Cambridge University Press, Cambridge (2005)
35. Laneman, J.N., Wornell, G.W.: Distributed space-time-coded protocols for exploiting cooperative diversity in wireless networks. IEEE Trans. Inf. Theory **49**(10), 2415–2425 (2003)
36. Etkin, R., Parekh, A., Tse, D.: Spectrum sharing for unlicensed bands. IEEE J. Sel. Areas Commun. **25**(3), 517–528 (2007)
37. Goldsmith, A., Wicker, S.: Design challenges for energy-constrained ad hoc wireless networks. IEEE Wirel. Commun. **9**(4), 8–27 (2002)
38. Elkourdi, T., Simeone, O.: An information-theoretic view of spectrum leasing via secondary cooperation. In: IEEE International Conference on Communications (ICC), pp. 1–6, May 2010
39. Elkourdi, T., Simeone, O.: Spectrum leasing via cooperative interference forwarding. IEEE Trans. Veh. Technol. **62**(3), 1367–1372 (2013)
40. Baccelli, F., Błaszczyszyn, B.: Stochastic geometry and wireless networks, Volume II-Applications. In: Foundations and Trends in Networking. NoW Publishers. Citeseer (2009)
41. Hafeez, M., Elmirghani, J.: Green Licensed Shared Access (LSA). IEEE J. Sel Areas Commun. (2015)
42. Nash, J. Jr.: The bargaining problem. Econometrica: J. Econometric Soc. **18**(2), 155–162 (1950)
43. Arnold, O., Richter, F., Fettweis, G., Blume, O.: Power consumption modeling of different base station types in heterogeneous cellular networks. In: Future Network and Mobile Summit. IEEE, 2010, pp. 1–8
44. G.T. version 12.0.0 Release 12: LTE; evolved universal terrestrial radio access (E-UTRA); TDD home eNode B (HeNB) radio frequency (RF) requirements analysis. http://www.etsi.org/deliver/etsi_tr/136900_136999/136922/12.00.00_60/tr_136922v120000p.pdf (2014)

Energy-Aware Cognitive Radio Systems

Ebrahim Bedeer, Osama Amin, Octavia A. Dobre and Mohamed H. Ahmed

Abstract The concept of energy-aware communications has spurred the interest of the research community in the most recent years due to various environmental and economical reasons. It becomes indispensable for wireless communication systems to shift their resource allocation problems from optimizing traditional metrics, such as throughput and latency, to an environmental-friendly energy metric. Although cognitive radio systems introduce spectrum efficient usage techniques, they employ new complex technologies for spectrum sensing and sharing that consume extra energy to compensate for overhead and feedback costs. Considering an adequate energy efficiency metric—that takes into account the transmit power consumption, circuitry power, and signaling overhead—is of momentous importance such that optimal resource allocations in cognitive radio systems reduce the energy consumption. A literature survey of recent energy-efficient based resource allocations schemes is presented for cognitive radio systems. The energy efficiency performances of these schemes are analyzed and evaluated under power budget, co-channel and adjacent-channel interferences, channel estimation errors, quality-of-service, and/or fairness constraints. Finally, the opportunities and challenges of energy-aware design for cognitive radio systems are discussed.

E. Bedeer (✉)
University of British Columbia, 3333 University Way, Kelowna, BC V1V 1V7, Canada
e-mail: ebrahim.bedeer-mohamed@ubc.ca; e.bedeer@sce.carleton.ca

O. Amin
King Abdullah University of Science and Technology,
Thuwal, Makkah Province, Saudi Arabia
e-mail: osama.amin@kaust.edu.sa

O.A. Dobre · M.H. Ahmed
Memorial University of Newfoundland, St. John's, NL A1B 3X5, Canada
e-mail: odobre@mun.ca

M.H. Ahmed
e-mail: mhahmed@mun.ca

© Springer International Publishing Switzerland 2016

M.Z. Shakir et al. (eds.), *Energy Management in Wireless Cellular and Ad-hoc Networks*, Studies in Systems, Decision and Control 50,
DOI 10.1007/978-3-319-27568-0_11

1 Introduction

Cognitive radio (CR) can considerably enhance the spectrum utilization efficiency by dynamically sharing the spectrum between licensed/primary users (PUs) and unlicensed/secondary users (SUs) [29]. This is achieved by granting SUs opportunistic access to the white spaces within PUs spectrum, while controlling the interference to PUs, i.e., overlay approach [29]. Alternatively, the SUs and PUs may coexist in the same spectral band, i.e., underlay approach [29]. In other words, an overlay coexistence scenario holds if the SUs are allowed to only access the vacant PUs frequency bands. On the other hand, an underlay coexistence scenario holds if the SUs can access occupied PUs frequency bands while meeting the PUs interference thresholds. The performance of overlay approaches is limited by the mutual interference between SUs and PUs, while co-channel interference places stringent power transmission limitations on the SUs transmission in underlay approaches.

In the past few years, energy-aware communications receive a lot of attention from research and industrial communities due to the rising energy costs to operate future wireless networks, ecological, and environmental reasons [16, 24]. Hence, designing energy-aware CR systems is important to improve both the energy and spectrum efficiencies. The performance of the energy-aware CR systems can be improved by properly allocating the available resources while meeting the imposed constraints, e.g., transmit power, interference, quality-of-service (QoS), and/or fairness constraints. A major difference between CR and conventional systems is the capability of the CR to sense the surrounding wireless environment. In fact, the available resources include transmit power and/or subchannels as in conventional wireless systems in addition to other parameters related to the sensing process such as sensing, transmission, and idle durations; number of SUs sensing the PUs spectrum; and/or SUs assignment to sense the channel.

The available PUs spectrum can be sensed using SUs independently, i.e., single node sensing; however, the performance is degraded by the hidden node problem[1] [51]. One way to improve the performance of single node sensing is to use cooperative sensing where multiple SUs may cooperate to sense the spectrum [51]. Cooperative sensing shows better sensing performance at the cost of increasing the consumed energy in the sensing process.

The sensing process can be done periodically, i.e., at the beginning of each frame, then the SU decides about its transmission based on the result of the sensing process. For example, if the sensing outcome is that the PU is not using the channel, then the SU can transmit; otherwise, the SU has to wait until the next frame to repeat the process [32]. Another possible approach for spectrum sensing is the sequential approach where the SU senses the channel sequentially and each time after sensing the SU has to make a decision on whether to continue to sense the next channel or to start transmission [40]. This approach is efficient as the SU does not wait for the next

[1] PUs may not be detected correctly due to many reasons including shadowing, deep fading, and/or location.

frame if the PU channel is not available; however, this is at the cost of consuming more energy in the sensing process.

In this chapter, we provide an overview of the resource allocation problems in energy-aware CR systems. In Sect. 2 we summarize the commonly used energy-aware metrics for CR systems. Section 3 discusses the parameters affecting the resource allocation in energy-aware CR systems. Finally, future trends in energy-aware design are discussed in Sect. 4.

2 Energy-Aware Metrics

Different energy-aware metrics are used to quantify the energy efficiency (EE) of CR systems through considering various parameters that affect EE. A presentation of such metrics is provided in the following sub-sections and summarized in Fig. 1. For the reader's convenance, all the symbols are defined when they first appear, as well as in Tables 1 and 2.

2.1 Energy Efficiency (EE) Metrics

The instantaneous EE is widely used when perfect channel-state-information (CSI) is available. This is defined as the instantaneous transmission rate (or throughput) divided by the instantaneous transmit and consumed circuitry power (or energy), and is expressed in bits/joule (bits/J) [10, 12, 26, 35, 36, 45, 46, 50].

For instance, for multicarrier communication systems, if a single SU employs an orthogonal frequency division multiplexing (OFDM) in order to access the available spectrum, the instantaneous EE is defined as [36, 46]

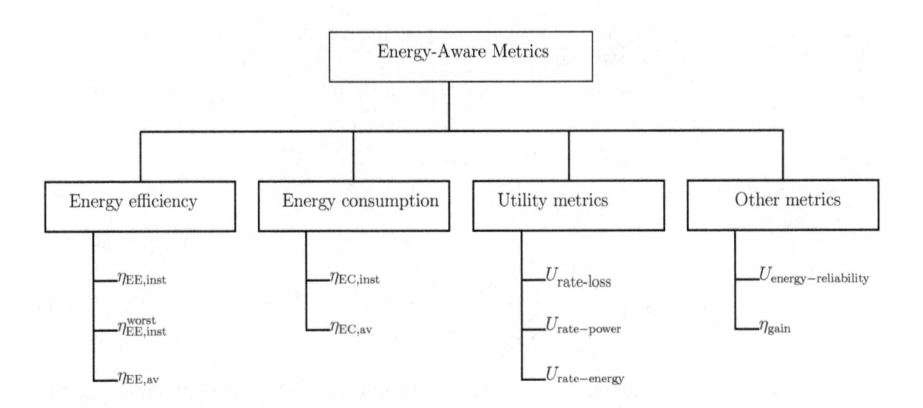

Fig. 1 Summary of EE-aware metrics

Table 1 List of symbols

α	Path loss exponent
β	Cooperative sensing gain
γ_n	Channel-to-interference-plus-noise ratio (CINR) of nth subcarrier
$\gamma_{k,n}$	CINR for the nth subcarrier used by the kth SU
Δ	Step of the power control
ζ	Power amplifier efficiency
$\eta_{\mathrm{EE,inst}}$	Instantaneous EE
$\eta_{\mathrm{EE,inst}}^{\mathrm{worst}}$	Worst EE
$\eta_{\mathrm{EC,inst}}$	Instantaneous energy consumption
$\eta_{\mathrm{EC,av}}$	Average energy consumption
η_{gain}	Cooperative sensing gain per joule per SU
$\eta_{\mathrm{EE,inst}}^{\mathrm{inv}}$	Inverse of instantaneous EE
$\eta_{\mathrm{EE,av}}^{\mathrm{inv}}$	Inverse of average EE
Λ	Bit error rate
μ	SU sleep rate
ν	Censoring rate
ϖ_{m}	Density of mobile base station of the 1st tier network
ϖ_{s}	Density of small cell access points of the 2nd tier network
$\rho_{k,n}$	Channel allocation indicator
$\sigma_{\Delta\mathrm{H}}^2$	Minimum mean square error
σ_n^2	Noise variance
$\phi_{\mathrm{PU}}^{\mathrm{idle}}$	Probability that PU is not using the channel
$\phi_{\mathrm{PU}}^{\mathrm{busy}}$	Probability that PU is using the channel
ϕ_{fa}	Probability of false-alarm
ϕ_{md}	Probability of mis-detection
$\phi_{\mathrm{PU}}^{\mathrm{re}}$	Probability that PU reoccupies the channel during the SU transmission
ϕ_n	Posterior probability that the PU channel n is identified to be idle while it is truly idle
ϕ_{UE}	Probability that there is a user equipment within the coverage of a small cell access points (SAP) unit
ϕ_{s}	Success probability of a typical user within the coverage area of SAP
$\phi_{n,\mathrm{fa}}$	Probability of false alarm of channel n
ϕ_n^{idle}	Probability that channel n is idle

$$\eta_{\mathrm{EE,inst}} = \frac{B \sum_{n=1}^{N} \log_2(1 + \gamma_n\, p_n)}{\frac{1}{\zeta} \sum_{n=1}^{N} p_n + p_{\mathrm{c}}} \quad \text{(bits/J)}, \tag{1}$$

where B is the subcarrier bandwidth, N is the total number of subcarriers, ζ is the power amplifier efficiency, γ_n is the channel-to-interference-plus-noise ratio (CINR) for subcarrier n, p_n is the power allocated to subcarrier n, and p_{c} is the circuit power

Table 2 List of symbols (cont'd)

B	bandwidth
d_{PS}	Distance from the PU transmitter to the SU receiver
d_{SS}	Distance from the SU transmitter to the SU receiver
$E_{k,n}$	energy consumed by the kth user to employ the nth channel
E_k^{idle}	Energy consumed by the kth user while being in idle state
E^{se}	Energy consumed by all the cooperative sensing SUs
E^{tr}	Energy consumed during the transmission
E_{c}	Consumed energy
G	path loss
$\hat{H}_n$	Estimate of the complex channel gain on subcarrier n
$\mathscr{I}_n$	Interference from all PUs to subcarrier n
K	total number of SUs
K^{relay}	Number of relays
K^{se}	Number of sensing users
$\mathscr{L}(p_n)$	Loss function of power consumed in subcarrier n
ℓ	Packet length
n	Subcarrier index
N	Total number of subcarriers
N_{samples}	Total number of samples
p_n	Power allocated to subcarrier n
p_c	Circuitry power consumption
$p_{k,n}$	Power allocated to the nth subcarrier employed by the kth SU
p^{se}	Sensing power
p^{tr}	Transmission power
p_{SU}	SU transmit power
p^{PA}	Power consumed in the power amplifier
p^{LNA}	Power consumed in low noise amplifier
p^{ADC}	Power consumed in the ADC
p^{syn}	Synchronization power
p^{proc}	Processing power
p^{relay}	Average transmit power of a given relay
$\mathbf{Q}_k$	Covariance matrix of kth SU
$r_{k,n}$	Rate of the kth user employing the nth channel
r	Transmission rate
r_k	Rate of kth SU
t	Total frame duration
$t_{k,n}^{\mathrm{cs}}$	Time required by the kth user to tune to the nth channel
t^{se}	Sensing duration
t^{tr}	Transmission duration
T	is the total slot length

(continued)

Table 2 (continued)

B	bandwidth
T_{samples}	Sampling period
$U_{\text{rate-loss}}$	Rate-loss utility metric
$U_{\text{rate-power}}$	Rate-power utility metric
$U_{\text{rate-energy}}$	Rate-energy utility metric
$U_{\text{energy-reliability}}$	Energy-reliability utility metric
w_k	relative weight of the EE of the kth SU

consumption. In case of imperfect CSI on the link between the SU transmitter and receiver, the instantaneous EE is defined as [12]

$$\eta_{\text{EE,inst}} = \frac{B\sum_{n=1}^{N}\log_2(1+(|\hat{H}_n|^2 G\, p_n)/(\sigma_{\Delta\text{H}}^2 G\, p_n + \sigma_n^2 + \mathcal{J}_n))}{\frac{1}{\zeta}\sum_{n=1}^{N} p_n + p_{\text{c}}} \quad \text{(bits/J)}, \quad (2)$$

where $\hat{H}_n$ is the estimate of the complex channel gain on subcarrier n, $\sigma_{\Delta\text{H}}^2$ is the minimum mean square error of the channel estimate, G is the path loss between the SU transmitter and receiver, σ_n^2 is the noise variance, and $\mathcal{J}_n$ is the interference from all the PUs to subcarrier n. On the other hand, if multiple SUs employ orthogonal frequency division multiple access (OFDMA) as the access technique to the available spectrum, then the instantaneous EE is defined as [45]

$$\eta_{\text{EE,inst}} = \frac{B\sum_{k=1}^{K}\sum_{n=1}^{N}\rho_{k,n}\log_2(1+\gamma_{k,n}\, p_{k,n})}{\frac{1}{\zeta}\sum_{k=1}^{K}\sum_{n=1}^{N}\rho_{k,n} p_{k,n} + p_{\text{c}}} \quad \text{(bits/J)}, \quad (3)$$

where K is the total number of SUs, $\rho_{k,n}$ is an integer variable that can be either 1 (if the nth subcarrier is occupied by the kth SU) or 0 (otherwise), $\gamma_{k,n}$ is the CINR for the nth subcarrier used by the kth SU, and $p_{k,n}$ is the power allocated to the nth subcarrier employed by the kth SU. The aforementioned metrics improve the EE of the whole system without guarantee on the achieved EE of individual users. The following worst EE metric improves fairness (in terms of EE), as it maximizes the EE of the user with the lower EE value. It can be defined as [50]

$$\eta_{\text{EE,inst}}^{\text{worst}} = \min_{k}\; w_k \frac{B\sum_{n=1}^{N}(1-\phi_{\text{md}})\,\rho_{k,n}\log_2(1+\gamma_{k,n} p_{k,n})}{\frac{1}{\zeta}\sum_{n=1}^{N} p_{k,n} + p_{\text{c}}} \quad \text{(bits/J)}, \quad (4)$$

where w_k is the relative weight of the EE of the kth SU, being used to reflect certain fairness between SUs and ϕ_{md} is the posterior probability of miss-detection (i.e., when certain PUs bands are identified to be vacant while they are truly occupied). Maximizing $\eta_{\text{EE,inst}}^{\text{worst}}$ can be viewed as the maximization of the minimum (worst) EE for all SUs to guarantee fairness between users. However, this comes at the expense of deterioration of the average EE of the system. To strike a balance between performance and fairness, the following weighted average EE is additionally defined as

$$\eta_{\mathrm{EE,inst}} = \sum_{k=1}^{K} w_k \frac{B \sum_{n=1}^{N} (1 - \phi_{\mathrm{md}})\, \rho_{k,n} \log_2(1 + \gamma_{k,n} p_{k,n})}{\frac{1}{\zeta} \sum_{n=1}^{N} p_{k,n} + p_{\mathrm{c}}} \quad \text{(bits/J)}. \qquad (5)$$

For the single carrier (SC) single-input single-output (SISO) systems, the definition of the instantaneous EE may vary depending on the considered scenario. For example, for a CR base station serving K users in a time synchronized manner on a set of N channels, the instantaneous EE is defined to consider the energy consumed in switching the SU RF chain from a frequency band in a certain frame to another frequency band in the following frame as [10]

$$\eta_{\mathrm{EE,inst}} = \frac{\sum_{k=1}^{K} \sum_{n=1}^{N} \rho_{k,n}\, r_{k,n}\, (t - t_{k,n}^{\mathrm{cs}})}{\sum_{k=1}^{K} \sum_{n=1}^{N} \rho_{k,n} E_{k,n} + \sum_{k=1}^{K} (1 - \sum_{n=1}^{N} \rho_{k,n}) E_k^{\mathrm{idle}}} \quad \text{(bits/J)}, \qquad (6)$$

where $r_{k,n}$ is the rate of the kth user employing the nth channel, t is the total frame duration, $t_{k,n}^{\mathrm{cs}}$ is the time required by the kth user to tune to the nth channel, $E_{k,n}$ is the energy consumed by the kth user to employ the nth channel (it is function of transmission duration, channel switching duration, idling duration, transmit power, channel switching power, idling power, and circuitry power), and E_k^{idle} is the energy consumed by the kth user while being in idle state (that is a function of idling power and idling duration). For an SC multiple-input multiple-output (MIMO) broadcast system, the instantaneous EE is defined as [35]

$$\eta_{\mathrm{EE,inst}} = \frac{\sum_{k=1}^{K} r_k}{\frac{1}{\zeta} \sum_{k=1}^{K} \mathrm{Tr}(\mathbf{Q}_k) + p_{\mathrm{c}}} \quad \text{(bits/J)}, \qquad (7)$$

where r_k is the rate of the kth SU, $\mathbf{Q}_k$ is the covariance matrix of the kth SU signal, and $\mathrm{Tr}(\mathbf{X})$ denotes the trace of matrix $\mathbf{X}$.

When the instantaneous channel coefficients are not available, an average EE is defined as the ratio of the average transmission rate (or throughput) to the average transmit and consumed circuitry power (or energy) [2, 7, 40–42, 48]. For example, for an SC SISO scenario where an SU overlays a PU and concurrent transmission is required, the average EE is defined as [41]

$$\eta_{\mathrm{EE,av}} = \frac{B \log_2(1 + \Delta (d_{\mathrm{PS}}/d_{\mathrm{SS}})^{\alpha})}{p_{\mathrm{SU}}} \quad \text{(bits/J)}, \qquad (8)$$

where Δ is the step of the power control, d_{PS} and d_{SS} are the distances from the PU transmitter to the SU receiver and from the SU transmitter to the SU receiver, respectively, α is the path loss exponent, and p_{SU} is the SU transmit power. The transmission rate in the numerator of (8) considers the signal-to-interference ratio of the SU link as $p_{SU}(1/d_{SS})^{\alpha}/p_{PU}(1/d_{PS})^{\alpha}$, where $\Delta = p_{SU}/p_{PU}$ and p_{PU} is the PU transmit power. For a single node periodic sensing, the average EE can be defined to consider both the time duration and consumed power in the sensing, transmission, and idling as [48]

$$\eta_{EE,av} = \frac{\phi_{PU}^{idle}\,(1-(1+w)\,\phi_{fa})\,t^{tr}\,r}{(p^{se}-p^{idle})\,t^{se}+T\,p^{idle}+(p^{tr}-p^{idle})\,t^{tr}(\phi_{PU}^{busy}\,\phi_{md}+\phi_{PU}^{idle}\,(1-\phi_{fa}))} \quad \text{(bits/J)}, \quad (9)$$

where ϕ_{PU}^{idle} and ϕ_{PU}^{busy} are the probabilities that the PU is not using the channel and occupying the channel, respectively, and ϕ_{fa} and ϕ_{md} are the probabilities of false-alarm and mis-detection, respectively. t^{se} and t^{tr} are the sensing and transmission durations, respectively, and p^{se} and p^{tr} are the sensing and transmission powers, respectively. r is the transmission rate, the total slot length T (that includes idling, sensing, and transmission durations) and w is a weight that reflects the loss of the potential reward (i.e., number of transmitted bits). The transmission rate in (9) can be expressed as the difference between the transmission reward $\phi_{PU}^{idle}\,t^{tr}\,r$ (i.e., when the SU correctly detects the PU's absence, and hence, the SU transmits) and the waste of potential reward $\phi_{PU}^{idle}\,(1+w)\,\phi_{fa}\,t^{tr}\,r$ (i.e., when the SU falsely reports the PU's presence, and hence, the SU does not transmit). Additionally, the average consumed energy consists of the following parts: (1) the energy when the SU successfully detects the PU's presence, and hence, the SU does not transmit with a probability $\phi_{PU}^{busy}\,(1-\phi_{md})$, (2) the energy when the SU mis-detects the PU's existence, and hence, the SU transmits with probability $\phi_{PU}^{busy}\,\phi_{md}$, (3) the energy when the SU falsely detects the PU's existence, and hence, the SU does not transmit with probability $\phi_{PU}^{idle}\,\phi_{fa}$ and 4) the energy when the SU correctly the PU's presence, and hence, the SU transmits with probability $\phi_{PU}^{idle}\,(1-\phi_{fa})$. Similarly, the average EE is defined to consider the SU circuity power and the PU activity during the SU transmission as [42]

$$\eta_{EE,av} = \frac{\phi_{PU}^{idle}\,(1-\phi_{fa})\,t^{tr}(1-\phi_{PU}^{re})\,r}{(p^{se}+p_c)\,t^{se}+(\frac{1}{\zeta}\,p^{tr}+p_c)\,t^{tr}(\phi_{PU}^{busy}\,\phi_{md}+\phi_{PU}^{idle}\,(1-\phi_{fa}))} \quad \text{(bits/J)}, \quad (10)$$

where ϕ_{PU}^{re} is the probability that the PU reoccupies the channel during the SU transmission.

For a single node sequentially sensing $n = 1, \ldots, N$ channels, the average EE is defined as [40]

$$\eta_{EE,av} = \frac{\mathbb{E}\{\sum_{n=1}^{N}\phi_n\,t^{tr}\,r_n\}}{\mathbb{E}\{\sum_{n=1}^{N}E_n^{se}+E_n^{tr}\}} \quad \text{(bits/J)}, \quad (11)$$

where ϕ_n is the posterior probability that the PU channel n is identified to be idle while it is truly idle and $\mathbb{E}$ is the expectation operator. As can be seen, the numerator of (11) accounts for the average number of transmit bits, while the denominator accounts for the total average consumed energy. In a cooperative sensing scheme, the average EE can be defined to account for the total energy consumed by all SUs in the sensing process as well as the energy required for transmission as [2]

$$\eta_{\mathrm{EE,av}} = \frac{\phi_{\mathrm{PU}}^{\mathrm{idle}}\,(1-\phi_{\mathrm{fa}})\,r\,t^{\mathrm{tr}}}{E^{\mathrm{se}}+E^{\mathrm{tr}}} \quad \text{(bits/J)}, \tag{12}$$

where E^{se} and E^{tr} are the energies consumed by all the cooperative sensing SUs and during the transmission, respectively.

It is worthy to mention that the inverse of either the instantaneous EE, i.e., $\eta_{\mathrm{EE,inst}}^{\mathrm{inv}} = 1/\eta_{\mathrm{EE,inst}}$ [26] or the average EE, i.e., $\eta_{\mathrm{EE,av}}^{\mathrm{inv}} = 1/\eta_{\mathrm{EE,av}}$ [1] can be used as a measure of the EE.

2.2 *Energy Consumption (EC) Metrics*

The total instantaneous/average consumed energy in a given transmission can be also used as an energy-aware metric, and it is measured in joule (J) [19, 25, 27, 31, 34, 47, 49].

For example, an energy-aware metric that considers the transmit power in addition to the the power consumed in the analog circuit component, bit resolution of the analog-to-digital converter (ADC), and the input backoff of the power amplifier is expressed as [19]

$$\eta_{\mathrm{EC,inst}} = p^{\mathrm{PA}}\,t^{\mathrm{tr}} + (p^{\mathrm{LNA}}+p^{\mathrm{ADC}})(t^{\mathrm{tr}}+t^{\mathrm{se}}) \quad \text{(J)}, \tag{13}$$

where p^{PA}, p^{LNA}, and p^{ADC} are the powers consumed in the power amplifier, low noise amplifier, and ADC, respectively.

For a single node sensing, the instantaneous energy consumption of multiple SUs sharing the spectrum with multiple PUs is expressed as [25]

$$\eta_{\mathrm{EC,inst}} = \sum_{k=1}^{K} t_k^{\mathrm{tr}}\,||\mathbf{p}_k||_2^2 \quad \text{(J)}, \tag{14}$$

where $\mathbf{p}_k$ is the transmit beamforming vector of the kth user. The average energy consumption of a two-tier heterogenous network, where the 2nd tier network is equipped with SAP with cognitive capabilities, can be expressed as [47]

$$\begin{aligned}\eta_{\mathrm{EC,av}} = \frac{\varpi_{\mathrm{s}}}{\varpi_{\mathrm{m}}}(p^{\mathrm{syn}}(t^{\mathrm{se}}+t^{\mathrm{tr}}) + \phi_{\mathrm{UE}}\,\phi_{\mathrm{s}}(p^{\mathrm{se}}\,t^{\mathrm{se}}+\phi_{\mathrm{md}}\,p^{\mathrm{proc}}\,t^{\mathrm{tr}}) \\ +(1-\phi_{\mathrm{UE}})\,\phi_{\mathrm{s}}(p^{\mathrm{se}}\,t^{\mathrm{se}}+\phi_{\mathrm{fa}}\,p^{\mathrm{proc}}\,t^{\mathrm{tr}})) \quad \text{(J)},\end{aligned} \tag{15}$$

where ϖ_m and ϖ_s are the densities of mobile base station of the 1st tier network and the small cell access points of the 2nd tier network, respectively, p^{syn} and p^{proc} are the synchronization and processing powers, respectively, ϕ_{UE} is the probability that there is a user equipment within the coverage of an SAP unit, and ϕ_s is the success probability of a typical user within the coverage area of SAP.

For a cooperative sensing distributed CR sensor network, where the SU may turn off its sensing and transmission capabilities in order to save energy, the average energy consumption is expressed as [34]

$$\eta_{EC,av} = (1-\mu)\left(\sum_{k=1}^{K} E_k^{se} + E_k^{tr}\,(1-\nu)\right) \quad \text{(J)}, \tag{16}$$

where μ is the SU sleep rate, i.e., the probability that the SU is turned off and ν is the censoring rate, i.e., the probability that the SU has no output from the sensing process. The average energy consumption of another cooperative spectrum sensing scheme where SUs can be considered as relays can be defined as [31]

$$\eta_{EC,av} = K^{relay}\,N_{samples}\,T_{samples}\,p^{relay} \quad \text{(J)}, \tag{17}$$

where K^{relay} is the number of relays, $N_{samples}$ is the total number of samples, $T_{samples}$ is the sampling period, and p^{relay} is the average transmit power of a given relay.

2.3 *Utility Metrics*

Utility functions can be used to measure the EE. For example, a utility function that considers the subcarrier availability, and hence, the reliability of transmission is defined as the difference between the SU transmission rate and its prospective rate loss due to sensing errors or to the PU reoccupying the channel during the SU transmission [28]

$$U_{rate\text{-}loss} = B\sum_{n=1}^{N} \log_2(1+\gamma_n\,p_n) - \phi_n\,\mathcal{L}(p_n) \quad \text{(bits/sec)}, \tag{18}$$

where $\mathcal{L}(p_n)$ is a real-valued increasing concave and normalized average loss function of the power consumed in subcarrier n and ϕ_n denotes the probability that a PU reoccupies channel n during the SU transmission or the probability that sensing errors occur during the PU transmission. Hence, $\phi_n\,\mathcal{L}(p_n)$ represents the average rate loss due to sensing errors or collision with PU transmission.

Another utility function that uses concepts of multi-objective optimization to maximize the transmission rate with the least amount of transmit power is defined as [11, 23]

$$U_{\text{rate-power}} = w_1\, B \sum_{k=1}^{K} \sum_{m=1}^{M} \log_2(1 + \gamma_{k,m}\, p_{k,m}) - w_2 \sum_{k=1}^{K} \sum_{m=1}^{M} ||\mathbf{p}_{k,m}||_{\text{F}}^2 \quad \text{(dimensionless)}, \tag{19}$$

where w_1 and w_2 are the relative weighting coefficients associated with the competing objectives, $m = 1, ..., M$ denotes the index of the transmit stream, and $||.||_{\text{F}}^2$ is the Frobenius matrix norm. Here, the weighting coefficients w_1 and w_2 include normalization factors such that the competing objectives are within the same range, and hence, the metric in (19) is dimensionless [11, 13]. It is worthy to mention that if the circuity power is considered in this metric, the optimal EE solution can be achieved at certain values of w_1 and w_2 [4]. Similarly, a utility function that maximizes the transmission rate with the least amount of consumed energy in the sensing process is defined as [43]

$$U_{\text{rate-energy}} = w_1 \sum_{n=1}^{N} r_n\, (1 - \phi_{n,\text{fa}})\phi_n^{\text{idle}} - w_2\, (t^{\text{se}}\, p^{\text{se}}\, K^{\text{se}} + t^{\text{se}}\, p^{\text{idle}}\, (K - K^{\text{se}})) \quad \text{(dimensionless)}, \tag{20}$$

where $\phi_{n,\text{fa}}$ is the probability of false alarm of channel n, ϕ_n^{idle} is the probability that channel n is idle, and K^{se} is the number of sensing users.

2.4 Other Metrics

Other additional dimensionless energy-aware metrics can be used, which capture other aspects of the energy consumed in the transmission. A novel dimensionless energy-aware metric is defined that captures the actual (and total) energy consumed in transmitting (and receiving) one packet, average PU interference time, average PU transmission time, and reliability of transmission as [39]

$$U_{\text{energy-reliability}} = \frac{p^{\text{tr}} \ell}{p^{\text{tr}} \ell + E_{\text{c}} r}\, (1 - \Lambda)^{\ell} \quad \text{(dimensionless)}, \tag{21}$$

where ℓ is the packet length in bits, E_{c} is the consumed energy of a node before the actual transmission occurs, and Λ is the bit error rate. As can be seen, $p^{\text{tr}}\ell$ represents the actual energy required to transmit one packet, $p^{\text{tr}}\ell + E_{\text{c}}$ is the total energy spent to transmit and receive one packet, and $(1 - \Lambda)^{\ell}$ accounts for the reliability of transmission, i.e., all the ℓ bits are received correctly.

A novel metric that captures the sensing gain of cooperative spectrum sensing is defined as the cooperative sensing gain in dB per joule per SU [30]

$$\eta_{\text{gain}} = \frac{\beta}{K^{\text{se}}\, E^{\text{se}}} \quad \text{(dB/J per SU)}, \tag{22}$$

where β is the cooperative sensing gain in dB and it is defined as the difference between the mis-detection threshold and the probability of mis-detection of cooperative spectrum sensing.

3 Energy-Aware Resource Allocation

In order to improve the energy efficiency performance of CR systems, different parameters contributing to the total energy consumption can be optimized, e.g., transmit power, power consumed in the transmitter and receiver circuitry, allocation of subchannels, sensing duration, idle duration, transmission duration, sensing access strategy, number of sensing SUs, and/or SUs assignment to sense the channel [1, 2, 7, 10, 12, 19, 23, 25–28, 30, 31, 34–36, 39–43, 45–50]. When the energy consumed in the sensing process can be neglected, then transmit power, consumed circuitry power, and/or frequency bands/subcarriers are optimized to improve the energy efficiency performance of the CR networks [1, 10, 12, 23, 26, 28, 35, 36, 39, 41, 45, 46, 50, 52]. Otherwise, additional parameters have to be optimized such as sensing duration, idling duration, transmission duration, sensing access strategy, number of sensing SUs, and/or SUs assignment to sense the channel [2, 7, 19, 25, 27, 30, 31, 34, 40, 42, 43, 47–49], in addition to adjusting the previous parameters accordingly.

3.1 Sensing-Less Energy-Aware Resource Allocation

Energy-aware CR systems can be designed while neglecting the effect of the energy spent in the sensing process. In such a case, transmit power [1, 28, 35, 36, 41, 46], signal-to-interference-plus-noise ratio (SINR) threshold [23], frequency bands/subcarriers [9, 26, 45, 50], and/or packet length [39] are optimized separately or jointly in order to improve the CR system energy efficiency.

3.1.1 Power Allocation of Multicarrier Systems

OFDM is recognized as an attractive modulation technique for CR networks due to its spectrum shaping flexibility, adaptivity in allocating vacant radio resources, and capability in monitoring the spectral activities of PUs [33]. The available spectrum can be accessed by a single SU employing OFDM modulation [28, 35, 46] or by multiple SUs employing OFDMA [26, 45, 50]. In case of a single SU accessing the spectrum, the power allocated to each OFDM subcarrier should be optimally allocated in order to improve EE. On the other hand, if multiple SUs access the spectrum, the subcarriers assigned to different SUs should be additionally optimized. In either case, a side-by-side frequency spectrum model is adopted as shown in Fig. 2,

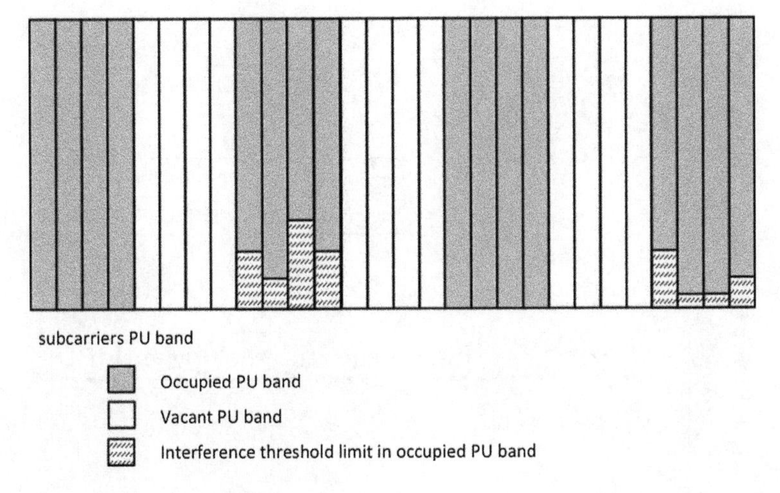

Fig. 2 Side-by-side frequency spectrum model

where the available frequency spectrum is divided and licensed to multiple PUs that do not necessarily use it all the time or at all the geographical locations [8].

In a simple overlay coexistence scenario, one SU is assumed to coexist adjacent in the frequency domain with one PU [46]. To further simplify the scenario, perfect CSI is assumed to be known between the SU transmitter and receiver pair, between the SU transmitter and the PU receiver, and between the PU transmitter and the SU receiver. This assumption is practically challenging; however, CSI can be obtained through cooperation/feedback between the PUs and SUs with negligible error, especially at high SNR [53]. The CR system uses the instantaneous knowledge of the CSI and optimizes the transmit power in order to maximize $\eta_{\text{EE,inst}}$ in (1) of the SU while guaranteeing total power constraint to reflect the SU power amplifier limitations and adjacent interference constraint to control the amount of the leaked interference to the PU. The formulated optimization problem is non-convex and it is solved using concepts of fractional programming, i.e., Dinkelbach method [22]. Results show that the EE can be maximized at the expense of deteriorating the SU rate.

Towards a more practical scenario where the SU coexists with multiple PUs in the overlay approach and guarantees certain QoS requirements, a minimum supported rate is added to the optimization problem [36]. A novel method, namely the generalized waterfilling factor aided search (WFAS), is proposed to solve the non-convex optimization problem [36]. A simplified version of the WFAS method is further presented. This version has simpler structure when compared to the Dinkelbach method in [46], i.e., the EE optimization problem is solved through the well-known rate maximization problem. Hence, the complexity order of the proposed simplified WFAS is much lower than its counterpart in [46].

A more practical scenario of an SU coexisting with multiple PUs is when the SU does not rely on perfect CSI knowledge [12]. Channel estimation errors on the links between the SU transmitter and receiver pair are considered. Additionally,

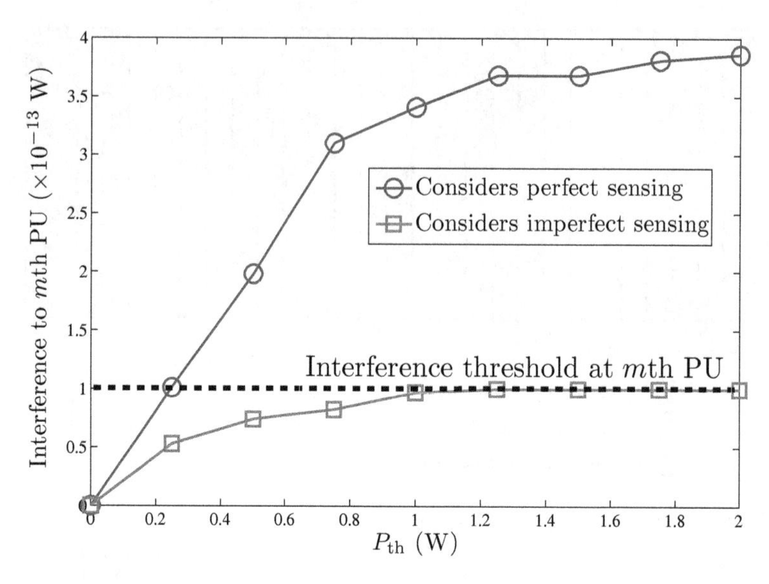

Fig. 3 Effect of perfect and imperfect sensing assumptions on the interference leaked to the mth PU. Simulation parameters are as in [12]

the SU does not have perfect sensing capabilities and only knows the statistics of the channels to the PUs receiver. The formulated optimization problem minimizes the SU instantaneous EE (the inverse of (2)) while guaranteeing certain power budget, minimum supported QoS, limited co-channel interference (CCI), and limited adjacent channel interference (ACI) constraints. The non-convex optimization problem is transformed to an equivalent one and solved using Dinkelbach algorithm [22]. Numerical results show that increasing the channel estimation errors deteriorates the EE. Additionally, the assumption that the SU has perfect sensing capabilities leads to violation of the interference constraints in practice as can be seen in Fig. 3. In other words, if the SU is assumed to sense the PUs activities perfectly, which is not necessarily true in practical scenarios, then the interference received at the PUs can exceed the predefined threshold. Hence, the practical case of the SU with limited sensing capabilities should be considered.

A generalization to the previous scenarios is when the SU accesses the spectrum in both overlay and underlay approaches [28]. In such a scenario, the SU is allowed to use the frequency band of a PU (underlay) that is located at a distant geographical location from the PU, and hence, it is required to transmit with lower transmit power such that no harmful interference occurs to the co-channel PU. This is in addition to the SU coexistence with PUs in unused adjacent frequency band (overlay). Due to practical considerations, only knowledge of the distance-based path loss is taken into account on the links between the SU transmitter and the PUs receivers. The SU considers a risk-return model to maximize $U_{\text{rate-loss}}$ (18), where the power allocated to a certain subcarrier is recognized as an investment, while loss of useful power when the PU reoccupies the band is recognized as a risk. The optimization problem additionally imposes total transmit power, CCI, and ACI constraints. Optimal and suboptimal algorithms are proposed to show the performance of the SU. It was shown that valuable resources can be saved, e.g., battery life, by selectively allocating higher power to underutilized subcarriers and lower power to subcarriers with high PUs activities.

3.1.2 Power and Subcarrier Allocation of Multicarrier Systems

When multiple SUs are allowed to access the spectrum of multiple PUs [26, 45, 50], the aforementioned models can be extended as follows. An underlay downlink scenario is considered where a cognitive base station and multiple SUs receivers share the licensed spectrum of PUs [45]. The SUs access technology is assumed to be OFDMA and perfect CSI is assumed between the base station and the SU receivers, between the CR base station and the PUs receivers, and between the PUs transmitters and the SUs receivers. The CR base station allocates the power and subcarriers of the SUs in order to maximize $\eta_{\mathrm{EE,inst}}$ in (3) of the whole CR network subject to total transmit power, ACI, and certain QoS of each SU (in terms of minimum supported rate) constraints. The formulated problem is a mixed-integer programming problem that is computationally hard to solve. In order to overcome the complexity burden, the concept of time-sharing is adopted, i.e., two different SUs can share the same subcarrier, and then a hypograph form is used to convert the resultant non-convex optimization problem into a convex one, where a barrier method [17] is used to find the solution.

Another possible scenario is when the available OFDMA resources (i.e., power and subcarriers) are optimally distributed to enable communications between the SU transmitter and receiver pairs [26]. An optimization problem is formulated for each SU transmitter and receiver pair in order to minimize $\eta^{\mathrm{inv}}_{\mathrm{EE,inst}}$ (the inverse of (3)) of each pair, while guaranteeing acceptable interference to existing users (PUs and SUs with already established connections) and minimum supported rate for each SU pair. The proposed solution is named energy-efficient waterfilling solution, where the optimality is obtained in the constraint interval. This is in contrast to the classical waterfilling solutions to maximize the rate or margin, where the optimality point is found at the constraint boundary. Numerical results show that for multiple transmitter and receiver pairs employing OFDMA, the optimal EE solution for a certain transmitter and receiver pair does not necessarily select the best subcarrier for transmission. This is in contrast to spectral efficiency maximization problems, where the optimal solution for a given pair selects the best subcarrier for transmission.

An upper bound on the performance is achieved when perfect CSI is assumed [26, 35, 45, 46]. In practical scenarios where perfect CSI is not available, the achieved performance is expected to degrade. In fact, only knowledge of average channel power gain on the links between the SUs transmitters to the PUs receivers is assumed in [50]. The worst $\eta^{\mathrm{worst}}_{\mathrm{EE,av}}$ and average $\eta_{\mathrm{EE,av}}$ EE metrics in (4) and (5), respectively, are to be optimized while guaranteeing a predefined fairness between SUs. In order to maintain fairness between users, the lowest SU EE is maximized and in order to improve the EE of the CR network as a whole, the summation of the weighted SUs EE is maximized, i.e., average EE. The formulated optimization problems are non-convex integer optimization problems that guarantee total transmit power and per subcarrier average interference[2] constraints and neglect ACI constraint. The optimization problems are relaxed to convex ones, where a general concave envelope function is used to find a near-optimal solution.

3.1.3 Power Allocation of Single Carrier Systems

Other systems that adapt the power may not employ OFDM(A) to access the available spectrum in CR networks [1, 23, 36, 41]. In such cases, single carrier for SISO systems [41] or MIMO systems [1, 23, 36] can be used. For example, an interference-limited spectrum sharing CR ad-hoc network is considered in [41], where simultaneous transmission between an SU pair and a PU pair is requested. A forbidden transmission region around the PU receiver is defined, where no SU transmission is allowed to guarantee the PU QoS. On the other hand, an effective cognitive region is defined, where the SU transmission is allowed if and only if the QoS of the SU can be achieved (in this case, a concurrent transmission between both the PU pair and the SU pair is possible while meeting both users QoS). The step size of the SU power control is adapted in order to maximize $\eta_{\mathrm{EE,av}}$ in (8) while neglecting the SU circuitry power consumption. It was shown through simulation examples that the concurrent transmission region can be expanded if power control is used.

Multiple SUs share the spectrum in an underlay fashion with a single PU in [23]. It is assumed that the SUs and the PU have MIMO capabilities. The target is to maximize the transmission rate and the number of admitted SUs to use the spectrum with the least possible transmit power. Accordingly, a MOOP problem is formulated to maximize $U_{\mathrm{rate-power}}$ (19) subject to peak power and interference constraints. The formulated problem considers the effect of imperfect CSI (including quantization and estimation errors) through adopting a normal-bounded channel imperfection model [44]. In addition to the power, the SINR threshold is considered as one of the optimization variables in order to have an efficient admission control without additional integer variables. In other words, a certain data stream of a certain SU is admitted for transmission if its received SINR is above a certain optimal threshold. The MOOP problem utility function is combined into a single objective function using the weighting sum method, where weighting coefficients are adopted to reflect the relative importance of the competing rate and power. The resulting problem is solved using concepts of semidefinite programming [17].

A MIMO boradcast scenario is considered in [35], where a single SU transmitter and multiple SUs receivers are assumed to share the spectrum in an underlay fashion with multiple PUs [35]. To maximize $\eta_{\mathrm{EE,inst}}$ in (7), an optimization problem is formulated subject to transmit power, interference, and minimum supported

[2]This is to guarantee PU protection in case of incorrect sensing of the channel.

rate constraints. The problem is non-convex and in order to reach the optimal solution, it was transformed into a one-dimensional optimization problem with a quasi-concave objective function that is solved using a golden search. Numerical results show that the EE can be significantly improved at high SNR regime or for a large number of SUs.

An uplink EE scheduling in CR networks is when an SU base station is serving a number of SUs in a time-slotted manner [10]. At the beginning of each frame, each SU gets a list of all the available frequency bands and estimates its capacity to each idle band. The SUs send this information to the base station along with the number of bits in its buffer. The CR base station assigns/schedules one SU to one frequency band in order to maximize $\eta_{\mathrm{EE,inst}}$ in (6) of all SUs. The CR base station considers the energy consumed in switching the SU RF chain from a frequency band in a certain frame to another frequency band in the following frame. The formulated scheduling optimization problem is solved by the Charnes-Cooper method [6] with high computational complexity. Then, two suboptimal solutions are provided to reduce the complexity. Results indicate that if there are many frequency bands to schedule, then a significant improvement in the EE is achieved when compared to other works in the literature that maximize the capacity. On the other hand, if there is a limited number of frequency bands, then the achieved EE is comparable to other works that maximize the capacity.

3.1.4 Packet Length Allocation

For a CR sensor network, increasing the packet size increases the network utilization; however, this is at the cost of increasing the packet loss probability if the PUs reoccupy the channel. On the other hand, reducing the packet size reduces the interference to PUs; however, this suffers from extensive overhead. Accordingly, it is important to find the optimal packet length that maximizes the EE of CR sensor networks. An overlay CR sensor network is considered to operate in sleep or active mode [39]. The sleep mode consists of ready, monitor, observe, and deep sleep periods, while the active mode consists of periods of sensing, decision, handoff, transmit, and receive. An optimization problem to maximize the $U_{\mathrm{energy-reliability}}$ in (21) is formulated subject to interference and maximum distortion level for PU reliable detection constraints, and it is solved using sequential quadratic programming [15]. Results show that changes in the PU behavior in the target BER can significantly change the optimal packet length.

3.1.5 Relays Placement

To enhance the EE of CR networks, a new architecture is introduced, namely, cognitive capacity harvesting networks [52]. In such an architecture, SUs are not equipped with cognitive capabilities in order to reduce their energy consumption, and the sensing capabilities are moved to a set of relay stations. The optimal placement of the relay stations significantly affects the EE of the CR network. An optimization problem is formulated to minimize the placement size subject to EE and spectrum efficiency constraints. Due to the NP-hardness of the problem, a two step heuristic algorithm is proposed to find a near-optimal solution. Results show that the proposed heuristic algorithm outperforms the random placement and the number of required relay stations are at most twice the number obtained from the optimal solution.

3.2 Sensing-Aware Energy-Aware Resource Allocation

The sensing process plays an important role in EE optimization in CR networks [7, 19, 25, 27, 30, 31, 34, 40, 42, 43, 47, 48]. Increasing the sensing time improves the detection probability at the expense of consuming more energy, whereas reducing the sensing time results in more collisions with existing PUs due to false detection. Furthermore, increasing the transmission time increases the CR network performance; however, data loss may happen as PUs may reoccupy the channel at any point in the SU transmission. For a fixed frame duration, reducing the transmission time directly limits the network performance and leads to longer sensing duration, and hence, wastes energy. A single SU can sense the available spectrum for possible opportunities for its transmission [19, 25, 27, 40, 42, 47–49] or multiple SUs can perform the sensing in a cooperative manner in order to improve the sensing performance at the expense of spending more energy [7, 30, 31, 34, 43].

3.2.1 Single Node Sensing

Periodic Sensing:

For a single SU sensing the channel, a general time-slotted CR system is considered where the frame is divided between sensing, transmission, and possible idling[3] durations [48]. The sensing and transmission durations are optimized to maximize $\eta_{\mathrm{EE,av}}$ in (9) subject to the probability of detection and maximum energy consumption constraints. Results show that shorter sensing durations can be achieved at the expense of increasing the sensing power, while larger idling power requires longer sensing durations. A more practical scenario is encountered when the PUs can reoccupy the channel during the SU transmission [42]. Stringent QoS of the SU is assumed, and hence, the successful transmission of the SU occurs if and only if the whole frame is sent correctly. Consequently, the sensing and transmission durations are optimized in order to maximize $\eta_{\mathrm{EE,inst}}$ in (10) subject to detection probability and interference to PUs constraints. The optimization problem is non-convex, and a suboptimal alternate search algorithm is proposed to reach an acceptable solution. It was

[3]Idle duration is important when the SU has to stop its transmission when a PU reoccupies the channel.

shown through numerical evaluations that significant improvements in the EE can be achieved by optimizing the sensing/transmission duration when compared to fixed sensing/transmission duration techniques [32].

Another scenario of single node sensing is discussed for heterogenous networks (HetNets) to improve their EE [47]. In HetNets, macro-cells are deployed to guarantee coverage, being overlayed by small cells (i.e., micro-cells, pico-cells or femto-cells) that offload traffic from the macro-cells in order to support local traffic demands [5]. However, deployment of small cells base stations is accompanied by excessive energy consumption [5]. Traffic demands may fluctuate over time, frequency, and space; hence, sleep mode techniques can be used to reduce the energy consumption [20]. To reduce the signalling overhead and hence improve the EE, a distributed sleep mode that does not require user location information is proposed in [47]. This is achieved by adopting small cell access points (SAPs) that have cognitive capabilities to periodically sense the channel if a user is active within its coverage area. The sensing time and probability are optimized to minimize $\eta_{\mathrm{EC,av}}$ in (15) in synchronization, sensing, processing, and transmission, while considering random locations of the users within the SAP coverage area. Results showed the tradeoff between the energy consumption and the amount of traffic that can be offloaded from macro-cells and it was shown that the consumed energy linearly increases with the density of the SAPs. Additionally, it was shown that knowledge of the interference environment can lead to significant reduction of the SAPs energy consumption.

The energy consumption of a CR sensor network can be reduced by selecting the operation mode, i.e., channel sensing, channel switching, and data transmission [27]. The CR sensor network is assumed to sense the channel continuously in order to identify vacant PU bands. In order to guarantee the PU protection, SUs need to switch the channel as fast as possible if a PU returns to use the channel. To reduce the switching time, the CR sensor network prepares a candidate channel called backup channel. An algorithm that considers errors in the sensing process is proposed to select the operation mode using the concepts from the partially observable Markov decision process [37].

In an underlay approach, multiple SUs are assumed to share the spectrum with multiple PUs [25] and both the SUs and PUs can be equipped with single or multiple antennas to access the spectrum. An uplink scenario is considered where the SUs transmitters communicate with a single SU base station using the time division multiple access protocol. The sensing time and the beamforming of the SUs is optimized in order to minimize $\eta_{\mathrm{EC,inst}}$ in (14) for the SUs while satisfying transmit power, interference to PUs, and QoS (in terms of minimum supported rate) for the SUs constraints. The SU is assumed to have either perfect or imperfect CSI on the links to the PUs receivers. In case of the availability of perfect CSI, the interference constraints have to be satisfied in a deterministic way, i.e., for every channel realization. In such case, closed form expression for the optimal solution can be reached if the SUs are under-utilized and a heuristic sub-optimal solution is proposed if the SUs are heavily-utilized. On the other hand, if perfect CSI is not available, the interference constraint has to be satisfied in a statistical way and the optimization problem is solved through decomposition [17]. Numerical results show that the energy consumption is significantly reduced compared to works that maximize the transmission rate, with less generated interference to PUs.

Another perspective of minimizing the energy consumption in delay-constrained CR systems is achieved when the SU adapts its transmission rate, i.e., reduces its transmission rate under deep fading channel and increases the rate for good channel conditions, in order to deliver a target payload [49]. The optimization problem is formulated as a discrete-time Markov decision process [37], where a low complexity algorithm is proposed to find the optimal policy. The results show that the impact of energy overheads is more significant for delay-insensitive traffic when compared to delay-sensitive scenarios.

For short range communications, the energy consumption in the transmitter and receiver circuitry is more significant than the energy required for transmitting the data [21]. Therefore, optimizing the sensing/transmission duration only is not sufficient and other parameters affecting the energy consumed in the circuitry should be optimized as well [19]. The optimization variables are the sensing duration, the input backoff of the power amplifier, the power consumed in the low noise amplifier, and the bit resolution of the ADC. During the sensing duration, only the energy consumed in the SU receiver is considered, i.e., the energy consumed in the low noise amplifier and ADC. On the other hand, during the transmission duration, the power consumed in the power amplifier of the SU transmitter is additionally considered [19]. The results show that for strong interference from the PU, less transmission energy and more circuit energy are required. Additionally, the SU receiver, especially ADC, consumes more energy when compared to traditional systems, and hence, it is preferable to operate the power amplifier of the SU transmitter at high input backoff values to compensate for such energy loss.

Sequential Sensing:

Unlike periodic sensing [19, 25, 27, 42, 47–49], a sequential sensing approach is optimized in order to maximize the average EE [40]. In sequential sensing, the SU has to make a decision on whether to continue to sense the next channel or to start its transmission, i.e., a sequence of decisions has to be made before the SU transmission. Sequential sensing is efficient as it allows the SU to sense another channel if the the current sensed channel is identified to be busy, instead of waiting until the beginning of the next frame. The sensing-access strategies to be optimized are the sensing strategy (i.e., when to stop sensing and start transmission), access strategy (i.e., determine the transmit power level during the transmission), and the sensing order (i.e., which channel to sense if the current channel is changed). The formulated optimization problem is a stochastic sequential decision-making problem, i.e., the decision maker has to decide at each state after observing the current system state [14], that is difficult to solve directly. A sub-optimal solution can be reached by transforming the problem into a parametric formulation that rewards the throughput and penalizes the transmit power. Numerical results show that the sensing strategy has a certain threshold, i.e., the SU transmits if the channel is good enough; otherwise, the

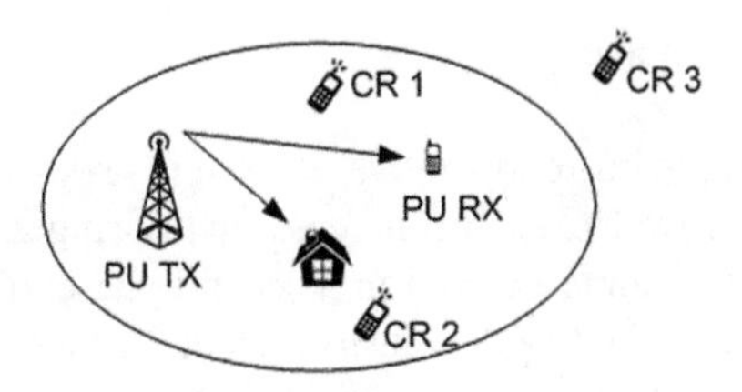

Fig. 4 Cooperative sensing

SU should keep sensing. Additionally, the optimal power allocation has a waterfilling structure and the optimal sensing order is to choose to sense a channel associated with the maximum expected future net reward, i.e., average EE.

3.2.2 Cooperative Sensing

Using a single SU to sense the spectrum has its own difficulties and limitations [19, 25, 27, 40, 42, 47–49], e.g., multi-path fading, shadowing, and hidden terminal problem [51]. For example, consider Fig. 4 where CR 2 cannot receive the PU transmission due to shadowing and CR 3 is located outside the PU transmission coverage, and hence, not aware of its transmission. One possible way to overcome these shortcomings is to use cooperative spectrum sensing, where more than one SU is allowed to collaboratively sense the channel [51]. For example, in Fig. 4 CR 1 can sense the PU signal and share the sensing results with CR 2 and CR 3. One of the main challenges of EE optimization in the cooperative sensing CR environment, when compared to single SU sensing strategies, is to find the optimal number and the assignment policy of SUs participating in the sensing process [7, 30, 31, 34, 43].

General Scheme:

A general cooperative sensing model can be considered as a two step process, where the first step is to find the number and assignment of SUs to do the sensing and the second step is to determine the sensing duration, detection threshold, etc. The most challenging part of cooperative sensing is related to the first step [30, 43]. The more SUs participating in the sensing process, the higher the sensing accuracy. However, this will be accompanied by an increase in the consumed energy in the sensing process. Some SUs are better candidates for sensing if they are subject to less noise or not in deep fading, and it is assumed that each SU can only be assigned to sense one frequency band in a given time slot [30]. In this case, the optimization problem reduces to finding the optimal number of SUs to sense the spectrum and to find which SU is assigned to sense a certain channel. A greedy algorithm is used to find the optimal solution and it was shown that significant performance gain is achieved especially with a large number of SUs [30].

Sensing and Relaying Scheme:

Another possible scenario for cooperative spectrum sensing is when SUs can be considered as relays as well [31]. In this case and during a first phase, all relays monitor the activity of a single PU and in a second phase, all relays amplify their PU received signal and send it to a data fusion center. In such a scenario, the optimal number of sensing/relays SUs, their amplifying gain, and the number of processed samples are to be optimized in order to reduce the energy consumption subject to the probabilities of false-alarm and detection constraints. In modeling the energy consumption, the circuit energy consumption in the relays, in the fusion center, and in the PUs are neglected under the condition that the transmission distance from the PU to the relays and from the relays to the fusion center is large. The formulated optimization problem is solved using an exhaustive search which is computationally complex. In order to reduce the complexity, equal power allocation of the relays is assumed and closed-form expressions for the optimal number of samples and the number of relays were obtained. It was shown that for higher PU transmit power, a reduced number of relays or samples is needed [31].

Sleep-Mode Based Cooperative Scheme:

Another scenario for distributed spectrum sensing is considered in [34], where the target is to minimize $\eta_{\mathrm{EC,av}}$ in (16) subject to a minimum probability of detection and a maximum false alarm probability constraints. In order to achieve this target, a sleep mode rate and censoring thresholds are to be optimized. In the sleep-mode, an SU turns off its sensing and transmission functions in order to save energy. On the other hand, censoring thresholds define the detection thresholds where an SU is confident of its sensing results. This can be explained as follows, if the output of the energy detector is greater than a certain threshold, then the SU reports its sensing decision as the PU exists; if the output of the energy detector is less than a certain threshold, then the SU reports its sensing decision as the PU does not exist; and if the output of the energy detector is between these two thresholds, the SU makes no decision and enters the sleep-mode. The optimization problem is solved using the alternating search algorithm [17] under two setups: the blind setup, when there is no prior information about the PU presence and the information-aided setup, when such information is available. The results indicate that when the energy consumed in transmission exceeds the energy consumed in sensing the sleep rate is higher than its counterpart when the the transmission and sensing energies are equal. Additionally, the minimized energy consumption is reduced significantly and becomes almost independent of the number of sensing users for a large number of users.

To conclude the sensing-aware energy-aware resource allocation section, different sensing parameters directly affect EE of CR systems and should be carefully selected. For single node sensing time-slotted systems, the sensing duration plays a crucial rule in maximizing the EE of the system. For instance, increasing the sensing time improves the detection probability at the expense of consuming additional energy, while decreasing the sensing time reduces the energy wasted in the sensing process but at the expense of more collisions with PUs. For cooperative sensing, the problem is even more challenging, as the system designer needs to select the optimal number of SUs to sense the PUs spectrum. For the short range communications when the energy consumption in the circuitry is more than the transmission energy, optimizing the sensing duration is not enough and optimizing the input backoff of the power amplifier becomes crucial.

4 Challenges and Opportunities

The performance of cooperative spectrum sensing can be severely degraded by misbehaving SUs, i.e., malicious users, that provide false information about sensing to the fusion center in order to have a wrong final decision and then use the spectrum exclusively [18]. This is achieved by reporting that the spectrum is occupied such that the fusion center produces a final decision that the spectrum is occupied and no legitimate SUs should access the spectrum; then, the malicious SUs become the only users to access the spectrum. A possible solution to this problem is that the fusion center accepts only reported results from authenticated SUs. This authentication can be achieved by transmitting extra bits for cryptographic purposes; however, additional energy has to be spent in these overhead bits. Then, the number of the security bits has to be optimized in order to maximize $\eta_{\mathrm{EE,av}}$ in (12) [2]. An average EE metric that captures the influence of malicious users is formulated, i.e., the malicious users tend to increase the false alarm probability, which decreases the successful amount of bits transmitted, and hence, reduces the EE. The optimal number of security bits depends on the fusion rate, fusion rule, number of malicious users, and number of legitimate users. The optimal number of bits increases as the number of malicious users increases, which decreases the EE.

Improving the SU connectivity comes mostly at a cost of increasing the energy consumption in CR systems. One of the most recent approaches to boost the SU performance is the use of improper Gaussian signaling, which creates more opportunity than the traditional proper signaling schemes, to minimize its outage while satisfying PU QoS [3]. Thus, investigating the EE-SE trade-off optimization problem in the underlying SU system is important in order to tune the SU power and circularity coefficient to accomplish an acceptable balance between EE and SE.

Additionally, employing full duplex in CR is considered an efficient way to improve the SE after the recent progress in self-interference cancelation. On the other hand, this may not be the case as discussed in a generic MIMO wireless communication system [38]. Since the CR has additional channel sensing cost, designing energy-efficient full-duplex CR system is a challenging issue. To tackle this issue, CRs need to operate with new adaptive full duplex modes and polices to accomplish a balance between efficient sensing, SE, and EE.

Another promising trend to improve the EE of future CR networks is through energy harvesting CRs This is achieved by tapping energy from readily available ambient sources, e.g., wind, solar, and radio frequency signals. However, knowing the instantaneous energy arrival is crucial in order to optimize the performance of energy harvesting CR networks. Energy arrivals can be assumed to be available in order to simplify the optimization framework. However, such an assumption may not be realistic; in this case, statistical knowledge about the energy arrival can be assumed. It is expected that energy harvesting will extend the lifetime of CR networks with improved EE performance.

References

1. Akin, S., Gursoy, M.C.: On the throughput and energy efficiency of cognitive MIMO transmissions. IEEE Trans. Wireless Commun. **62**(7), 3245–3260 (2013)
2. Althunibat, S., Sucasas, V., Marques, H., Rodriguez, J., Tafazolli, R., Granelli, F.: On the trade-off between security and energy efficiency in cooperative spectrum sensing for cognitive radio. IEEE Commun. Lett. **17**(8), 1564–1567 (2013)
3. Amin, O., Abediseid, W., Alouini, M.S.: Outage performance of cognitive radio systems with improper gaussain signaling. In: IEEE International Symposium on Information Theory, pp. 1851–1855 (2015)
4. Amin, O., Bedeer, E., Ahmed, M., Dobre, O.: Energy efficiency—spectral efficiency trade-off: a multiobjective optimization approach. IEEE Trans. Veh. Technol. (to appear)
5. Andrews, J.G., Claussen, H., Dohler, M., Rangan, S., Reed, M.C.: Femtocells: past, present, and future. IEEE J. Sel. Areas Commun. **30**(3), 497–508 (2012)
6. Bajalinov, E.B.: Linear-Fractional Programming: Theory, Methods, Applications and Software, vol 84. Springer, New York (2003)
7. Ban, T.W., Choi, W., Sung, D.K.: Capacity and energy efficiency of multi-user spectrum sharing systems with opportunistic scheduling. IEEE Trans. Wireless Commun. **8**(6), 2836–2841 (2009)
8. Bansal, G., Hossain, M., Bhargava, V.: Optimal and suboptimal power allocation schemes for OFDM-based cognitive radio systems. IEEE Trans. Wireless Commun. **7**(11), 4710–4718 (2008)
9. Bayhan, S., Alagoz, F.: Scheduling in centralized cognitive radio networks for energy efficiency. IEEE Trans. Veh. Technol. **62**(2), 582–595 (2013)
10. Bayhan, S., Eryigit, S., Alagoz, F., Tugcu, T.: Low complexity uplink schedulers for energy-efficient cognitive radio networks. IEEE Commun. Lett. **2**(3), 363–366 (2013)
11. Bedeer, E., Dobre, O.A., Ahmed, M.H., Baddour, K.: A multiobjective optimization approach for optimal link adaptation of OFDM-based cognitive radio systems with imperfect spectrum sensing. IEEE Trans. Wireless Commun. **13**(4), 2339–2351 (2014)
12. Bedeer, E., Amin, O., Dobre, O., Ahmed, M., Baddour, K.: Energy-efficient power loading for OFDM-based cognitive radio systems with channel uncertainties. IEEE Trans. Veh. Technol. **64**(6), 2672–2677 (2015)
13. Bedeer, E., Dobre, O., Ahmed, M., Baddour, K.: Rate-interference tradeoff in OFDM-based cognitive radio systems. IEEE Trans. Veh. Technol. (to appear)
14. Bertsekas, D.P., Bertsekas, D.P., Bertsekas, D.P., Bertsekas, D.P.: Dynamic Programming and Optimal Control, vol. 1. Athena Scientific Belmont, MA (1995)
15. Boggs, P.T., Tolle, J.W.: Sequential quadratic programming. Acta numerica **4**, 1–51 (1995)
16. Bolla, R., Bruschi, R., Davoli, F., Cucchietti, F.: Energy efficiency in the future internet: a survey of existing approaches and trends in energy-aware fixed network infrastructures. IEEE Commun. Surveys Tutor. **13**(2), 223–244 (2011) (Second Quarter)

17. Boyd, S., Vandenberghe, L.: Convex Optimization. Cambridge University Press, Cambridge (2004)
18. Chen, R., Park, J.M., Hou, Y.T., Reed, J.H.: Toward secure distributed spectrum sensing in cognitive radio networks. IEEE Commun. Mag. **46**(4), 50–55 (2008)
19. Chen, Y., Nossek, J., Mezghani, A.: Circuit-aware cognitive radios for energy-efficient communications. IEEE Wireless Commun. Lett. **2**(3), 323–326 (2013)
20. Cheung, W.C., Quek, T.Q., Kountouris, M.: Throughput optimization, spectrum allocation, and access control in two-tier femtocell networks. IEEE J. Sel. Areas Commun. **30**(3), 561–574 (2012)
21. Cui, S., Goldsmith, A.J., Bahai, A.: Energy-constrained modulation optimization. IEEE Trans. Wireless Commun. **4**(5), 2349–2360 (2005)
22. Dinkelbach, W.: On nonlinear fractional programming. Manage. Sci. **13**(7), 492–498 (1967)
23. Du, H., Ratnarajah, T.: Robust utility maximization and admission control for a MIMO cognitive radio network. IEEE Trans. Veh. Technol. **62**(4), 1707–1718 (2013)
24. Ericsson, A.B.: Sustainable energy use in mobile communications. (2007)
25. Fu, L., Zhang, Y.J.A., Huang, J.: Energy efficient transmissions in MIMO cognitive radio networks. IEEE J. Sel. Areas Commun. **31**(11), 2420–2431 (2013)
26. Gao, S., Qian, L., Vaman, D.R.: Distributed energy efficient spectrum access in cognitive radio wireless ad hoc networks. IEEE Trans. Wireless Commun. **8**(10), 5202–5213 (2009)
27. Han, J.A., Jeon, W.S., Jeong, D.G.: Energy-efficient channel management scheme for cognitive radio sensor networks. IEEE Trans. Veh. Technol. **60**(4), 1905–1910 (2011)
28. Hasan, Z., Bansal, G., Hossain, E., Bhargava, V.: Energy-efficient power allocation in OFDM-based cognitive radio systems: a risk-return model. IEEE Trans. Wireless Commun. **8**(12), 6078–6088 (2009)
29. Hossain, E., Bhargava, V.: Cognitive Wireless Communication Networks. Springer, New York (2007)
30. Huang, D., Kang, G., Wang, B., Tian, H.: Energy-efficient spectrum sensing strategy in cognitive radio networks. IEEE Commun. Lett. **17**(5), 928–931 (2013)
31. Huang, S., Chen, H., Zhang, Y., Zhao, F.: Energy-efficient cooperative spectrum sensing with amplify-and-forward relaying. IEEE Commun. Lett. **16**(4), 450–453 (2012)
32. Liang, Y.C., Zeng, Y., Peh, E.C., Hoang, A.T.: Sensing-throughput tradeoff for cognitive radio networks. IEEE Trans. Wireless Commun. **7**(4), 1326–1337 (2008)
33. Mahmoud, H., Yucek, T., Arslan, H.: OFDM for cognitive radio: merits and challenges. IEEE Wireless Commun. Mag. **16**(2), 6–15 (2009)
34. Maleki, S., Pandharipande, A., Leus, G.: Energy-efficient distributed spectrum sensing for cognitive sensor networks. IEEE Sensors J. **11**(3), 565–573 (2011)
35. Mao, J., Xie, G., Gao, J., Liu, Y.: Energy efficiency optimization for cognitive radio MIMO broadcast channels. IEEE Commun. Lett. **17**(2), 337–340 (2013)
36. Mao, J., Xie, G., Gao, J., Liu, Y.: Energy efficiency optimization for OFDM-based cognitive radio systems: a water-filling factor aided search method. IEEE Trans. Commun. **12**(5), 2366–2375 (2013)
37. Monahan, G.E.: State of the art-a survey of partially observable markov decision processes: theory, models, and algorithms. Manage. Sci. **28**(1), 1–16 (1982)
38. Nguyen, D., Tran, L.N., Pirinen, P., Latva-aho, M.: Precoding for full duplex multiuser MIMO systems: spectral and energy efficiency maximization. IEEE Trans. Signal Process **61**(16), 4038–4050 (2013)
39. Oto, M.C., Akan, O.B.: Energy-efficient packet size optimization for cognitive radio sensor networks. IEEE Trans. Wireless Commun. **11**(4), 1544–1553 (2012)
40. Pei, Y., Liang, Y.C., Teh, K.C., Li, K.H.: Energy-efficient design of sequential channel sensing in cognitive radio networks: optimal sensing strategy, power allocation, and sensing order. IEEE J. Sel. Areas Commun. **29**(8), 1648–1659 (2011)
41. Sanchez, S.M., Souza, R.D., Fernández, E.M.G., Reguera, V.A.: Rate and energy efficient power control in a cognitive radio ad hoc network. IEEE Signal Process Lett. **20**(5), 451–454 (2013)

42. Shi, Z., Teh, K., Li, K.: Energy-efficient joint design of sensing and transmission durations for protection of primary user in cognitive radio systems. IEEE Commun. Lett. **17**(3), 565–568 (2013)
43. Sun, X., Tsang, D.: Energy-efficient cooperative sensing scheduling for multi-band cognitive radio networks. IEEE Trans. Wireless Commun. **12**(10), 4943–4955 (2013)
44. Wang, J., Palomar, D.P.: Worst-case robust MIMO transmission with imperfect channel knowledge. IEEE Trans. Signal Process **57**(8), 3086–3100 (2009)
45. Wang, S., Ge, M., Zhao, W.: Energy-efficient resource allocation for OFDM-based cognitive radio networks. IEEE Trans. Commun. **61**(8), 3181–3191 (2013)
46. Wang, Y., Xu, W., Yang, K., Lin, J.: Optimal energy-efficient power allocation for OFDM-based cognitive radio networks. IEEE Commun. Lett. **16**(9), 1420–1423 (2012)
47. Wildemeersch, M., Quek, T., Slump, C., Rabbachin, A.: Cognitive small cell networks: energy efficiency and trade-offs. IEEE Trans. Commun. **61**(9), 4016–4029 (2013)
48. Wu, Y., Tsang, D.H.: Energy-efficient spectrum sensing and transmission for cognitive radio system. IEEE Commun. Lett. **15**(5), 545–547 (2011)
49. Wu, Y., Lau, V.K., Tsang, D.H., Qian, L.P.: Energy-efficient delay-constrained transmission and sensing for cognitive radio systems. IEEE Trans. Veh. Technol. **61**(7), 3100–3113 (2012)
50. Xiong, C., Lu, L., Li, G.: Energy-efficient spectrum access in cognitive radios. IEEE J. Sel. Areas Commun. **32**(3), 550–562 (2014)
51. Yucek, T., Arslan, H.: A survey of spectrum sensing algorithms for cognitive radio applications. IEEE Commun. Surveys Tutor. **11**(1), 116–130 (2009)
52. Yue, H., Pan, M., Fang, Y., Glisic, S.: Spectrum and energy efficient relay station placement in cognitive radio networks. IEEE J. Sel. Areas Commun. **31**(5), 883–893 (2013)
53. Zhang, L., Liang, Y.C., Xin, Y.: Joint beamforming and power allocation for multiple access channels in cognitive radio networks. IEEE J. Sel. Areas Commun. **26**(1), 38–51 (2008)

Cognitive Radio Energy Saving and Optimization

Yunfei Chen

Abstract In an ad hoc cognitive radio network, energy management is of paramount importance, as it directly determines the lifetime of the cognitive radio as well as the interferences to the licensed users for which the regulatory obligations of cognitive radios must be fulfilled. When the transmission power is fixed, this boils down to the management of the cognitive radio operation time consisting of a dedicated sensing period and a transmission period. In this chapter, different energy saving techniques that use non-coherent sensing, decision-feedback sensing, or censored sensing to reduce the amount of total energy consumption incurred by sensing will be investigated. We will also look into energy optimization techniques that minimize the energy use by taking the physical layer sensing and upper layer throughput into account. Extensive analysis and simulation will be provided to obtain useful guidance on energy management in ad hoc cognitive radio networks.

1 Cognitive Radio Energy Saving

Cognitive radio has attracted great research interest in recent years, owing to its promise in solving the so-called spectrum scarcity problem [1]. By sensing the radio environment from time to time, cognitive radio is able to determine the availability of the licensed bands and switch its operating frequency to the available bands automatically. This allows the re-use of unoccupied licensed spectrum to improve spectrum efficiency. Thus, current cognitive radios usually operate in two statuses: spectrum sensing and data transmission. For example, in the proposed IEEE 802.22 standard [2], a dedicated quiet period is allocated for spectrum sensing to avoid any interferences from other cognitive radios. Although this improves the sensing accuracy, it does lead to extra energy consumption due to the additional sensing period. To improve the energy efficiency, in this section, we consider two energy-saving techniques that perform sensing without dedicated sensing periods to remove the

Y. Chen (✉)
School of Engineering, University of Warwick, CV4 7AL Coventry, UK
e-mail: Yunfei.Chen@warwick.ac.uk

© Springer International Publishing Switzerland 2016

M.Z. Shakir et al. (eds.), *Energy Management in Wireless Cellular and Ad-hoc Networks*, Studies in Systems, Decision and Control 50,
DOI 10.1007/978-3-319-27568-0_12

additional sensing period for less energy consumption as well as one energy-saving technique that performs censoring before sensing to reduce the number of cognitive radios involved in sensing and therefore to reduce the total energy use.

The spectrum sensing problem can be formulated as a binary hypotheses testing problem as

$$\begin{aligned} H_0 &: r_i[k] = S_i[k] + n_i[k] \\ H_1 &: r_i[k] = a_i[k] + S_i[k] + n_i[k] \end{aligned} \tag{1}$$

where $r_i[k]$ is the kth time sample received at the ith cognitive radio, $S_i[k]$ is the kth sample of the interference from other cognitive radios received at the ith cognitive radio, $a_i[k]$ is the kth sample of the possible licensed user signal received at the ith cognitive radio, $n_i[k]$ is the kth noise sample received at the ith cognitive radio, $i = 1, 2, \ldots, I$ index the number of cognitive radios in sensing and $k = 1, 2, \ldots, K$ index the sample size used in each cognitive radio within a given period of time or frame. The task of spectrum sensing is to determine which of H_0 and H_1 is true based on the received samples $r_i[k]$. The interference from other cognitive radios $S_i[k]$ reduces the accuracy of spectrum sensing.

In traditional cognitive radios, spectrum sensing is performed in a dedicated sensing period where all cognitive radios stop transmission such that $S_i[k] = 0$, giving

$$\begin{aligned} H_0 &: r_i[k] = n_i[k] \\ H_1 &: r_i[k] = a_i[k] + n_i[k]. \end{aligned} \tag{2}$$

This can be considered as coherent sensing. The coherence is achieved at the expense of extra energy consumption during the dedicated sensing period. Hence, to improve energy efficiency, one can perform sensing using (1) directly without any dedicated sensing period. This can be considered as non-coherent sensing. Alternatively, one may also use the estimated interference from other cognitive radios as

$$\begin{aligned} H_0 &: r_i[k] = S_i[k] + n_i[k] - \hat{S}_i[k] \\ H_1 &: r_i[k] = a_i[k] + S_i[k] + n_i[k] - \hat{S}_i[k] \end{aligned} \tag{3}$$

where $\hat{S}_i[k]$ is the estimated value of the interference using data decisions. This can be considered as decision-feedback sensing that also performs sensing without any dedicated sensing period.

1.1 Non-coherent Sensing

In this case, the samples in (1) are used directly for spectrum sensing. The benefit of doing this is two-fold: first, it reduces the energy consumption by not having a dedicated sensing period and second, it can perform sensing at any time to reduce collision with the licensed user. The disadvantage is the reduced sensing accuracy

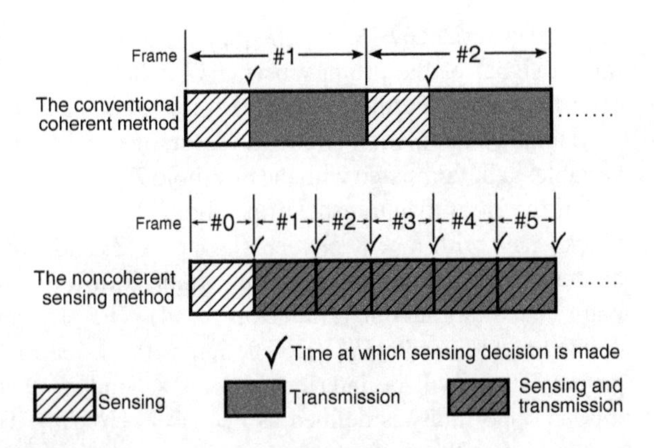

Fig. 1 Comparison of the coherent and non-coherent sensing schemes [3]

due to the interference, which can be alleviated by choosing the sensing parameters of K and I. Figure 1 compares the conventional coherent sensing with the new non-coherent sensing. The actual detector of non-coherent sensing depends on different assumptions of the licensed user signal.

1.1.1 Fast Fading Licensed User Signal

When the licensed user signal experiences fast fading, spectrum sensing can be described as

$$\begin{aligned} H_0 &: r_i[k] = S_i[k] + n_i[k] \\ H_1 &: r_i[k] = h_i[k]d[k] + S_i[k] + n_i[k] \end{aligned} \tag{4}$$

where $h_i[k]$ is the gain of the fast fading channel, $d[k]$ is the transmitted signal of the licensed user and other symbols are defined as before. Further, in a fast fading Rayleigh channel, $h_i[k]$ is complex Gaussian with mean zero and variance $2\sigma_h^2$. If isotropic scattering is assumed, one has the covariance between $h_i[k_1]$ and $h_i[k_2]$ at the ith cognitive radio as $E\{h_i[k_1]h_i[k_2]^*\} = 2\sigma_h^2 \cdot J_0(2\pi(k_1 - k_2)f_0T)$ [4], where $k_1, k_2 = 1, 2, \ldots, K$, f_0T is the normalized Doppler shift and $J_0(\cdot)$ is the zero-th order Bessel function of the first kind [5, Eq. (9.1.1)]. The fading gains at different cognitive radios are independent such that $E\{h_{i_1}[k_1]h_{i_2}[k_2]^*\} = 0$ for $i_1 \neq i_2$. Also, for additive white Gaussian noise, $n_i[k]$ is complex Gaussian with mean zero and variance $2\sigma_n^2$. The real and imaginary parts of $n_i[k]$ are circularly symmetric. The noise samples at different cognitive radios are also independent.

Furthermore, the kth sample of interference at the ith cognitive radio can be expressed as $S_i[k] = \sum_{j=1, j\neq i}^{I} S_j[k]$, where $S_j[k]$ is the kth sample of the interfering signal from the jth cognitive radio. Also, assume that the interfering signals are independent and Gaussian, and that they are identically distributed each with mean $E\{S_j[k]\} = 0$ and covariance $E\{S_j[k_1]S_j[k_2]^*\} = 2\varepsilon^2 \cdot sinc((k_1 - k_2)BT)$, where square spectrum is assumed for easy manipulation, B is the signal bandwidth, and T is the sampling period. Thus, one has $S_i[k]$ as complex Gaussian with mean zero and covariance $E\{S_i[k_1]S_i[k_2]^*\} = (I-1) \cdot 2\varepsilon^2 \cdot sinc((k_1-k_2)BT)$. The transmitted signal of the primary user $d[k]$ is deterministic, and it is the same for all cognitive radios. Further, assume that it is a constant such that $d[k] = d$.

Using (1) and the likelihood ratio test, one can derive $D = \mathbf{r}^H\mathbf{U}\mathbf{r}$ as the decision variable, to be compared with the threshold T_1, where $\mathbf{r}$ is the received sample vector, $(\cdot)^H$ is the Hermitian transpose operation, $\mathbf{U}$ is a $IK \times IK$ block matrix having $I \times I$ blocks each with $K \times K$ elements given in [3] and T_1 is the detection threshold to be determined. Note that D is a quadratic form of Gaussian random variables, as $\mathbf{r}$ are Gaussian random variables from (4). The probability of false opportunity is defined as $P_{f.o.} = Pr\{H_0|H_1\}$. The larger the value of $P_{f.o.}$ is, the more likely the primary users will be interfered by the secondary users. Also, the probability of missed opportunity is defined as $P_{m.o.} = Pr\{H_1|H_0\}$. The larger the value of $P_{m.o.}$ is, the poorer the performance of spectrum sensing will be. In spectrum sensing, $P_{f.o.}$ is often predetermined as $P_{f.o.} = \beta$ according to the primary users' tolerance to interferences.

In order to determine T_1, the probability density function (PDF) of D is required. However, the exact PDF of D is difficult to obtain, if not impossible. Motivated by the fact that D follows a Gamma distribution when $\mathbf{U}$ is an identity matrix, a Gamma approximation to D is considered. The probability of missed opportunity is $P_{m.o.} \approx 1 - gamcdf\left(gamcdf^{-1}(P_{f.o.}, k_1, \theta_1), k_0, \theta_0\right)$ using Gamma approximation, where $gamcdf(x, k, \theta)$ is the cumulative distribution function (CDF) of Gamma distribution with parameters k and θ, $gamcdf^{-1}(x, k, \theta)$ is the inverse of $gamcdf(x, k, \theta)$, and the parameters of k_0, θ_0, k_1 and θ_1 can be derived by matching the first- and second-order moments of D with those of a Gamma distribution. Figure 2 compares simulation results using the true distribution with that using the Gamma approximation. The simulation is performed using 10^6 runs for each point. The sample mean is calculated as $\bar{P}_{m.o.} = \sum_{i=1}^{10^6} P^i_{m.o.}/10^6$, where $P^i_{m.o.}$ is the probability of missed opportunity in the ith run. The upper limit (UL) and the lower limit

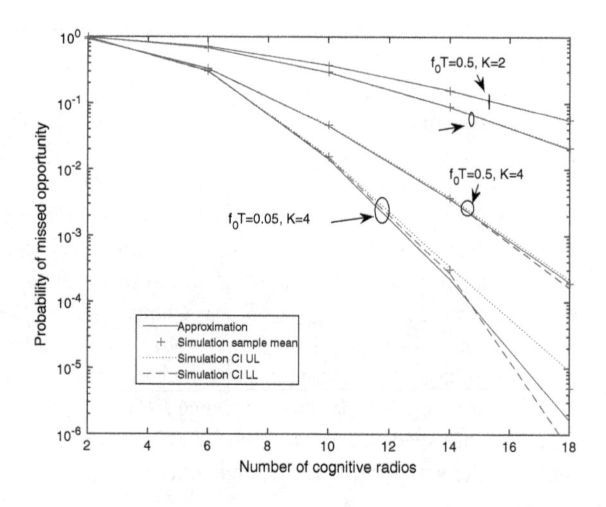

Fig. 2 Comparison of the simulation result and the Gamma approximation when $SNR = 0\ dB$, $SIR = 0\ dB$ and $BT = 1$

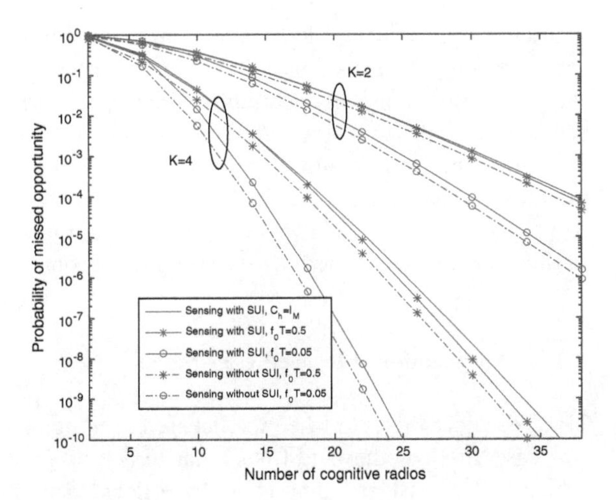

Fig. 3 Probability of missed opportunity for different values of K and f_0T when $SNR = 0\ dB$, $SIR = 0\ dB$ and $BT = 1$

(LL) of the confidence interval (CI) are calculated as $\bar{P}_{m.o.} + \frac{c\bar{\sigma}}{\sqrt{10^6}}$ and $\bar{P}_{m.o.} - \frac{c\bar{\sigma}}{\sqrt{10^6}}$, respectively, where $\bar{\sigma}^2 = \sum_{i=1}^{10^6}(P^i_{m.o.} - \bar{P}_{m.o.})^2/(10^6 - 1)$, $c = 1.96$ gives a confidence level of 95 %. One sees that the Gamma approximation matches well with the simulation in most cases considered. When $f_0T = 0.05$ and $K = 4$, they have noticeable difference.

Next, some numerical examples are shown for the new non-coherent sensing. In the examples, define the average signal-to-noise ratio (SNR) as $SNR = \frac{\sigma_h^2|d|^2}{\sigma_n^2}$ and the average signal-to-interference ratio (SIR) from a single interferer as $SIR = \frac{\sigma_h^2|d|^2}{\varepsilon^2}$, since $2\sigma_h^2 = E\{|h_i[k]|^2\}$. Also, the parameters are set as $\sigma_h^2 = 1/2$ and $d = 1$, while σ_n^2 and ε^2 change according to the average SNR and the average SIR, respectively. Also, $P_{f.o.} = 0.01$ is used. Figures 3 and 4 compare the performances of coherent sensing and non-coherent sensing under different conditions. As expected, there

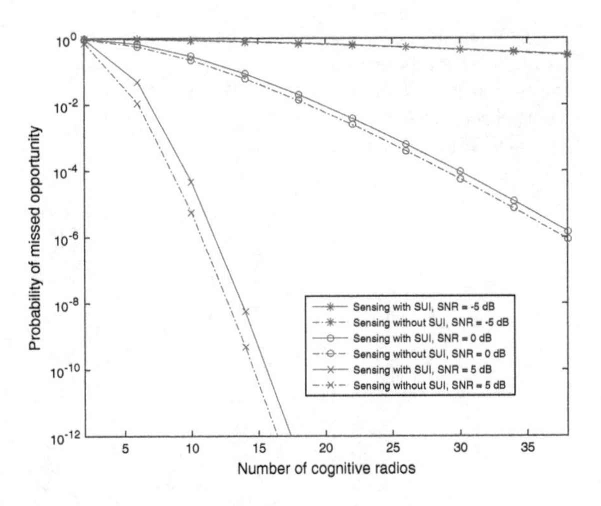

Fig. 4 Probability of missed opportunity for different values of SNR when $SIR = 0\ dB$, $K = 2$, $f_0T = 0.05$ and $BT = 1$

is always a performance degradation due to the interferences from other cognitive radios in non-coherent sensing. However, the diversity gain or the decreasing rate of the probability of missed opportunity with respect to the number of cognitive radios remains the same for coherent and non-coherent sensing schemes, which is desirable. Moreover, this performance degradation can be reduced by controlling the system parameters, either decreasing the values of *SNR* or K, or increasing the value of f_0T. Thus, one can save the energy of cognitive radio by not having a dedicated sensing period while achieving performances similar to coherent sensing.

1.1.2 Slow Fading Licensed User Signal

Next, consider the case when the licensed user signal experiences slow fading. In this case, the hypothesis test is similar to (4), except that $h_i[k]$ is replaced by h_i, which is the constant fading gain of the licensed user signal at the ith secondary user that does not change with time k. This is equivalent to setting $f_0T = 0$ in fast fading channels. Compared with the fast fading channel, the dependence of the channel gain on the time k does not exist in this case. Thus, using the likelihood test, one has the decision variable as $D = \frac{1}{2}\mathbf{r}^H\mathbf{G} + \frac{1}{2}\mathbf{G}^H\mathbf{r}$, to be compared with the threshold T_2, where the symbols can be found in [3]. One sees that D is a linear form of Gaussian random variables. Thus, D is also Gaussian. One has $P_{m.o.} = 1 - \int q(\frac{m_1}{\sigma_0} - q^{-1}(1 - P_{f.o.}, 0, 1), 0, 1)p(\mathbf{h})d\mathbf{h}$ as the probability of missed opportunity, where $q(x, 0, 1)$ is the CDF of a standard Gaussian random variable with mean 0 and variance 1, $q^{-1}(x, 0, 1)$ is its inverse, $\mathbf{h} = [h_1, h_2, \ldots, h_I]$, $p(\mathbf{h})$ is the joint PDF of $\mathbf{h}$, $d\mathbf{h} = dh_1dh_2 \cdots dh_I$, m_1 and σ_0 are calculated as the mean of D when H_1 is true and the variance of D when H_0 is true.

Figures 5 and 6 compare the performances of coherent sensing and non-coherent sensing under different conditions. Again, non-coherent sensing is inferior to

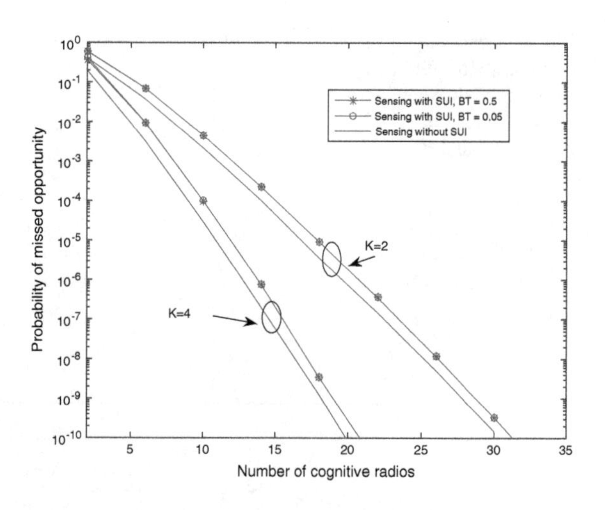

Fig. 5 Probability of missed opportunity for different values of M and BT in slow fading channels when $SNR = 0\ dB$ and $SIR = 0\ dB$

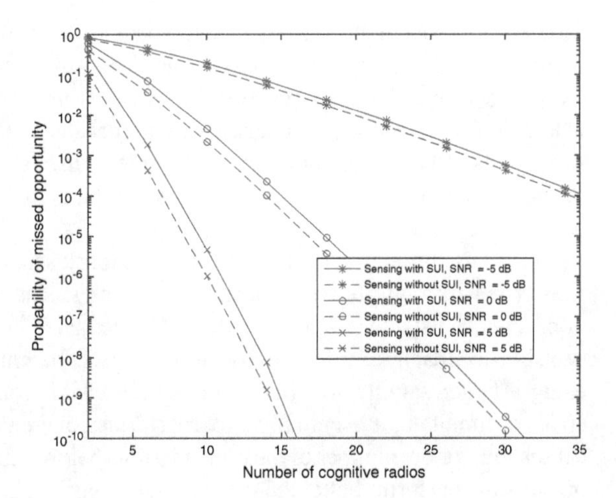

Fig. 6 Probability of missed opportunity for different values of SNR in slow fading channels when $SIR = 0\ dB$, $M = 2$ and $BT = 0.05$

coherent sensing in all the cases considered, due to the extra interferences from other cognitive radios after removing the dedicated sensing period. However, the diversity gain of non-coherent sensing remains the same as that of coherent sensing. Also, one may choose appropriate system parameters to achieve the same performance as coherent sensing.

In [3], the case when the interferences from other cognitive radios are independent has been investigated. Here, the case when the interferences from other cognitive radios are totally correlated has been investigated. Together they give the upper and lower bounds of the spectrum sensing performances for interferences with arbitrary correlation in non-coherent sensing. These performance bounds will be useful guidance in cognitive radio system designs.

1.2 Decision-Feedback Sensing

In the above subsection, the received samples containing interferences from other cognitive radios are used directly for energy saving, following the removal of the dedicated sensing period. Alternatively, one may "clean" these samples by removing interferences from other cognitive radios in the samples for higher accuracy while still saving energy, as in (3). This requires us to feedback the data decisions from relevant cognitive radios to the spectrum sensing operation and remove the interferences caused by these cognitive radios using the data decisions.

A previous work using this idea assumed that the data decisions are always correct such that the interferences from other cognitive radios can be completely removed [6]. This is too ideal. In reality, errors in the data decisions will occur such that the interferences from other cognitive radios cannot be completely removed. Then, the sensing accuracy using these slightly "contaminated" samples in (3) will be reduced.

This reduced accuracy will then degrade the cognitive radio data transmission performance in the next transmission due to possible undetected interferences from the licensed user. This causes even more errors in data decisions. Eventually, the whole performance may be degraded by accumulated errors over several transmissions. This effect occurs only when multiple transmissions are considered. During this process, the licensed user may change its status from presence to absence or the other way around.

Thus, this subsection studies decision-feedback sensing assuming practical decision errors in multiple data transmission frames. The effect of different modulation schemes used by cognitive radios is investigated. The investigation shows that decision-feedback sensing initially outperforms coherent sensing by benefiting from using a larger sample size but as the data transmission goes on and the decision errors accumulate, it eventually underperforms coherent sensing. Thus, there exists a threshold for the number of the transmissions below which decision-feedback sensing not only performs better than coherent sensing but also saves energy for operation without dedicated sensing period.

To start with, a few assumptions need to be made. In particular, the cognitive radios operate on the basis of L consecutive transmissions or frames. Each transmission lasts a duration of T_o. During these L transmissions, the occupancy status of the licensed user changes dynamically, defined by its channel holding time. The probability that the licensed user changes its occupancy status during the lth transmission is given by $p_{\lambda(l)} = F(lT_o) - F\big((l-1)\,T_o\big)$, using the probability mass function (PMF) of the channel holding time [7], where $F(t)$ is the cumulative distribution function (CDF) of the channel holding time, in this case, is assumed exponential such that $F(t) = 1 - e^{-\lambda t}$ [8], and $l = 1, 2, \ldots\ldots L$ index the transmissions. It is further assumed that the licensed user does not change its occupancy status during the data transmission. Also, for simplicity, one sensing cognitive radio and one interfering cognitive radio are assumed. Based on these assumptions, we can study the performance of decision-feedback sensing.

The cognitive radio decodes the received samples first. Since the decoding process is not ideal, interferences cannot be completely removed. The sensing accuracy using these samples will therefore be adversely affected. This further degrades the decoding performance. Following these steps, we study the decoding process first.

The kth received sample in the lth transmission at the sensing cognitive radio in (1) can be rewritten as $r[k + lK] = p \cdot a[k + lK] + S[k + lK] + n[k + lK]$, where $p = 0$ indicates H_0 when the licensed user is absent, $p = 1$ indicates H_1 when the licensed user is present, $a[k + lK]$ is the licensed user signal sampled at the kth time in the lth frame, $S[k + lK]$ is the interfering cognitive radio signal sample at the kth time in the lth frame with $s[k + lK] = -\sqrt{\varepsilon_b}$ or $s[k + lK] = +\sqrt{\varepsilon_b}$, ε_b is the bit energy of the cognitive radio signal and $n[k + lK]$ is the Gaussian noise sample with mean zero and variance σ_n^2, $k = 1, 2, \ldots, K$, $K = \frac{T_0}{T}$ is the total number of samples in one frame and T is the sampling interval. Compared with (1), the index of i is dropped for convenience because only one interfering cognitive radio is considered. In this section, consider only binary phase shift keying such that only the real parts of the samples are needed for sensing and decoding, giving

$R[k+lK] = p \cdot A[k+lK] + S[k+lK] + N[k+lK]$, where $R[k+lK] = Re\{r[k+lK]\}$, $A[k+lK] = Re\{a[k+lK]\}$ and $N[k+lK] = Re\{n[k+lK]\}$.

The signal is decoded. Denote the *a priori* probabilities of $-\sqrt{\varepsilon_b}$ and $+\sqrt{\varepsilon_b}$ as $P_{(-\sqrt{\varepsilon_b})}$ and $P_{(+\sqrt{\varepsilon_b})}$, respectively. When the licensed user is absent, the error rate of each sample can be derived as $E_{H_0} = Q\left(\sqrt{2\gamma_s}\right)$, where $\gamma_s = \frac{\varepsilon_b}{\sigma_n^2}$ is the cognitive radio SNR. Assume that out of the K transmitted samples in the lth transmission, $q_{0(l)}$ symbols are $-\sqrt{\varepsilon_b}$ and $g_{0(l)}$ of the $q_{0(l)}$ symbols are incorrectly decoded as $+\sqrt{\varepsilon_b}$. Similarly, assume that $k_{0(l)}$ symbols out of the $K - q_{0(l)}$ transmitted $+\sqrt{\varepsilon_b}$ are incorrectly decoded as $-\sqrt{\varepsilon_b}$. Thus, when PU is absent, the conditional overall error rate for all samples can be derived assuming independent samples. Similarly, when the licensed user is present, the error rate of each sample can be derived as $E_{H_1} = Q\left(\sqrt{\frac{2\gamma_s}{1+\gamma_p}}\right)$, where γ_p is the licensed user SNR. Assume that out of the K transmitted samples, $q_{1(l)}$ of them are $-\sqrt{\varepsilon_b}$ and $g_{1(l)}$ of the $q_{1(l)}$ samples are incorrectly decoded, and that $k_{1(l)}$ out of the $K - q_{1(l)}$ $+\sqrt{\varepsilon_b}$ are incorrectly decoded. Thus, when PU is present, one can also derive the conditional overall error rate for all samples.

After the cognitive radio signal is decoded and deducted from the received samples, they will be used for spectrum sensing. Using the previous assumptions, spectrum sensing becomes a binary hypothesis testing problem as

$$
\begin{aligned}
H_0 : D_l = {} & \sum_{k=1}^{g_{0(l)}} (-2\sqrt{\varepsilon_b} + N[k+lK])^2 + \sum_{k=g_{0(l)}+1}^{k_{0(l)}+g_{0(l)}} (2\sqrt{\varepsilon_b} + N[k+lK])^2 \\
& + \sum_{k=g_{0(l)}+k_{0(l)}+1}^{K} N[k+lK]^2 \\
H_1 : D_l = {} & \sum_{k=1}^{g_{1(l)}} (A[k+lK] - 2\sqrt{\varepsilon_b} + N[k+lK])^2 \\
& + \sum_{k=g_{1(l)}+1}^{k_{1(l)}+g_{1(l)}} (A[k+lK] + 2\sqrt{\varepsilon_b} + N[k+lK])^2 \\
& + \sum_{k=g_{1(l)}+k_{1(l)}+1}^{K} (A[k+lK] + N[k+lK])^2
\end{aligned} \tag{5}
$$

Using the Gaussian approximation based on the central limit theorem, the mean and variance of the decision variable in (5) for H_0 and H_1 can be derived as

$$
\begin{cases}
m_{0(l)} = K\left(1 + \frac{4g_{0(l)}}{K} + \frac{4k_{0(l)}\gamma_s}{K}\right) \\
\sigma_{0(l)}^2 = 2K + K\left(\frac{16g_{0(l)}}{K} + \frac{16k_{0(l)}\gamma_s}{K}\right)
\end{cases} \tag{6}
$$

and

$$\begin{cases} m_{1(l)} = K(1+\gamma_p + \frac{4g_{1(l)}\gamma_s}{K} + \frac{4k_{1(l)}\gamma_s}{K}) \\ \sigma^2_{1(l)} = 2K + K(4\gamma_p + \frac{16g_{1(l)}\gamma_s}{K} + \frac{16k_{1(l)}\gamma_s}{K}) \end{cases} \quad (7)$$

respectively. The conditional probability of detection is $P_{d(l)}(T, q_{1(l)}, g_{1(l)}, k_{1(l)}) = \frac{1}{2}erfc(\frac{T_3 - m_{1(l)}}{\sqrt{2\sigma^2_{1(l)}}})$, where T_3 is the detection threshold. The conditional probability of false alarm can be derived as $P_{f(l)}(T, q_{0(l)}, g_{0(l)}, k_{0(l)}) = \frac{1}{2}erfc\left(\frac{T_3 - m_{0(l)}}{\sqrt{2\sigma^2_{0(l)}}}\right)$.

After spectrum sensing, the following frame will be used for data transmission if no licensed user is detected. The probability when the lth frame is idle and the probability that the lth frame is busy but is mis-detected can be derived. Using all these derivations, the total achievable throughput of the L frames can finally be derived as $R = \sum_{l=1}^{L}(P_0(l)C_0 + P_1(l)C_1)$, where $C_0 = \ln(1+\gamma_s)$, $C_1 = \ln(1+\gamma_s/(1+\gamma_p))$ and other symbols can be explained in [9]. Next, some sample numerical examples will be shown to compare decision-feedback sensing with coherent sensing, as shown in Figs. 7, 8, 9 and 10. More detailed discussions could be found in [9].

Figure 7 shows the throughput for each frame. One sees that the throughput decreases when the frame index increases for decision-feedback sensing, as expected, as the probability that the lth frame can be used for data transmission depends on the sensing results of the previous $l-1$ transmissions. Due to mis-detection or false alarm, this probability decreases when l increases and therefore, the average throughput becomes smaller. On the other hand, the throughput of coherent sensing remains the same, as it does not accumulate any decoding errors. Also, the sensing accuracy decreases when the errors in decisions occur. Thus, the decrease of the throughput in the case when there is error is faster than that without error.

Figure 8 shows the total throughput of the L transmissions. Although they all increase with the value of L, the total throughput in coherent sensing increases lin-

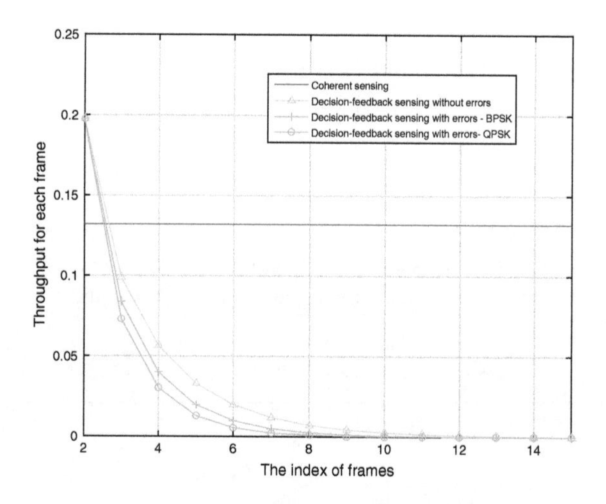

Fig. 7 Comparison of the individual throughputs of decision-feedback sensing and coherent sensing when $\gamma_p = -5$ dB and $\gamma_s = 0$ dB

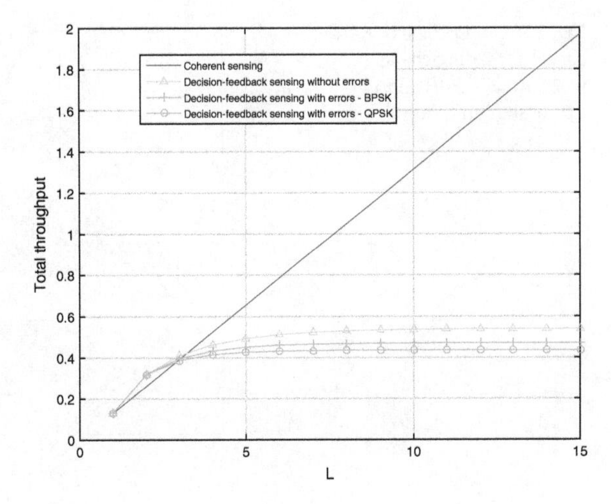

Fig. 8 Comparison of the total throughputs of decision-feedback sensing and coherent sensing when $\gamma_p = -5$ dB and $\gamma_s = 0$ dB

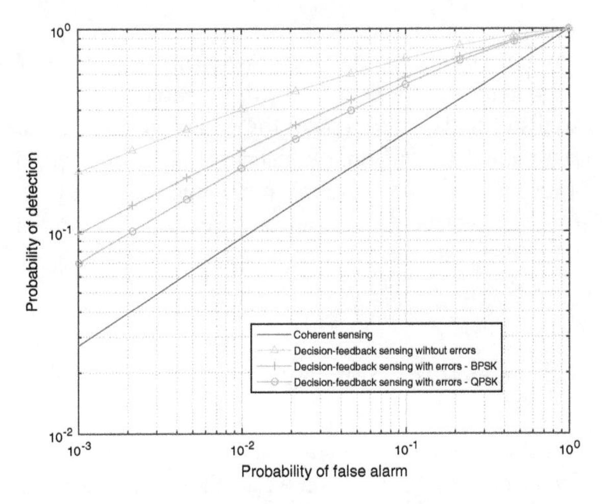

Fig. 9 Comparison of ROC curves for static channels when $\gamma_p = -5$ dB and $\gamma_s = 3$ dB for the Lth frame

early, while that in decision-feedback sensing approaches an upper limit. A threshold below which decision-feedback sensing has a larger throughput than coherent sensing can also be seen. The existence of the threshold is due to the dependence of the sensing performance on the accumulation of the decision errors over frames.

Figures 9 and 10 show the receiver operating characteristics (ROC) curves. One sees from these figures that decision-feedback sensing has a better sensing performance than the coherent sensing. This is due to the fact that coherent sensing only uses samples in the dedicated sensing period, while decision-feedback sensing can use all samples in the cognitive radio operations. For the same predetermined probability of false alarm, one sees that decision-feedback sensing has higher probability of detection. Also, the case without errors serves as a performance upper bound. Moreover, the sensing accuracy decreases when QPSK is used, as it incurs a larger probability of error which leads to more accumulated interferences in the samples.

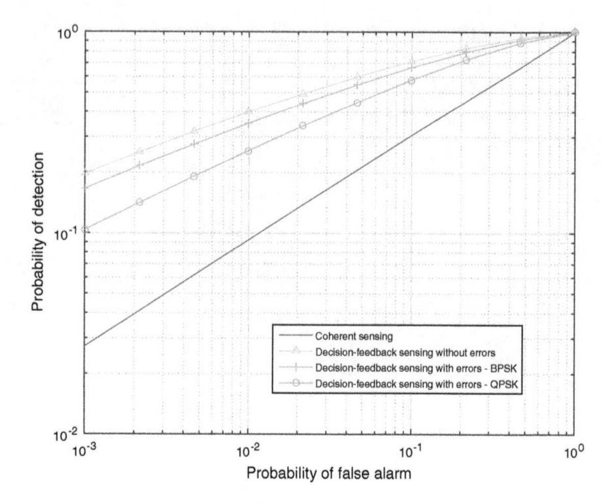

Fig. 10 Comparison of ROC curves for fading channels when $\gamma_p = -5$ dB and $\gamma_s = 3$ dB for the Lth frame

The above decision-feedback sensing has aimed to "clean" the received samples using data decisions. It is worth mentioning that there are other schemes that use statistics of the interferences to "clean" the received samples based on specific characteristics of orthogonal frequency division multiplexing (OFDM) signals [10–12]. These characteristics are inherent in the cognitive radio signals and thus can be used to lower interference power or cancel interference but they are only applicable to OFDM signals.

1.3 Censored Sensing

The previous two subsections have investigated non-coherent sensing and decision-feedback sensing as alternatives to the traditional coherent sensing to save the energy consumed by the dedicated sensing period. Another way of saving energy is to reduce the number of cognitive radios involved in spectrum sensing. In existing sensing schemes, all I cognitive radios will send their samples in (1) or decisions based on these samples to a fusion center. Thus, if the transmission of the K samples or the decision at each cognitive radio consumes an energy of E_C, the total energy consumed by sending them for sensing will be $I * E_C$. The idea of censored sensing is to do a quality check on the received samples $r_i[k]$ and only use those samples that meet some criterion and therefore, saves energy by not sending all available samples or decisions to the fusion center.

Censored detection have been used in ad hoc networks before [13–16]. In particular, [13] analyzed the censored sensing using the "OR" rule for both perfect and imperfect control channels, [14] analyzed the censored sensing using the "no-send" information, [15, 16] proposed censoring for cyclostationarity detection. These works derived either the analytical expression for the probability of detection, or the probability of false opportunity but not both and only considered an upper limit. Although the lower limit is approximately zero for some applications, it is non-zero in general and therefore, it is necessary to derive analytical expressions for both probability of detection and probability of false opportunity by taking into account the upper limit as well as the lower limit.

In this subsection, we derive analytical expressions for the probability of detection and the probability of false opportunity for censored sensing based on the Neyman-Pearson rule. Unlike the existing sensing schemes that use local decisions from all the cognitive radios for an overall decision, the censored sensing compares the samples with two pre-determined but optimized limits and only forward local decisions where the samples are either smaller than the lower limit or larger than the upper limit. In this case, both non-zero upper limit and non-zero lower limit are examined. We consider two typical decision fusion rules "OR" and "AND" where each cognitive radio sends a binary local decision to the fusion center. The study indicates that the censored sensing outperforms the existing sensing schemes when the optimum upper and lower limits are used to censor the samples before using them to make the local decisions. The performance gain depends on various parameters. In addition, the number of decisions forwarded to the fusion center is reduced and therefore, saves energy.

Consider the received samples in (2), where we assume that sensing is performed within a dedicated sensing period to simplify the analysis of censoring. Again, consider binary phase shift keying such that only the real parts of the signals in (2) are used for spectrum sensing, giving

$$\begin{aligned} H_0 &: R_i[k] = N_i[k] \\ H_1 &: R_i[k] = A_i[k] + N_i[k]. \end{aligned} \tag{8}$$

where $A_i[k]$ and $N_i[k]$ are Gaussian with means zero and variances σ_a^2 and σ_n^2, respectively. Then, the ith cognitive radio makes a local decision by using the traditional method of energy detection by comparing decision variable D_i with a threshold of T_{coh}. It can be easily derived that D_i follows a Gamma distribution with shape parameter $\frac{K}{2}$ and scale parameter $\frac{2}{K}$ under H_0, and it follows a Gamma distribution with shape parameter $\frac{K}{2}$ and scale parameter $\frac{2}{K}(1+\gamma_p)$ under H_1, where $\gamma_p = \frac{\sigma_a^2}{\sigma_n^2}$ is the licensed user signal SNR. Thus, one has $P_d^{(i)} = Gamcdf(T_{coh}, \frac{K}{2}, \frac{2}{K})$ and $P_{f.o.}^{(i)} = Gamcdf(T_{coh}, \frac{K}{2}, \frac{2}{K}(1+\gamma_p))$, where the Gamma cumulative distribution function (CDF) is $Gamcdf(x,k,\theta) = \int_0^x \frac{t^{k-1}e^{-t/\theta}}{\Gamma(k)\theta^k}dt$, and $\Gamma(\cdot)$ is the complete Gamma function.

In the decision fusion "AND" rule, the overall decision is H_0 only when all the local decisions from all the cognitive radios are H_0. Using these, T_{coh} under the Neyman-Pearson criterion can be derived as $T_{coh} = T_{AND} = Gaminv(\beta^{1/I}, \frac{K}{2}, \frac{2}{K}(1+\gamma_p))$, where $P_{f.o.}^{(i)} = \beta$ and $Gaminv$ is the inverse of $Gamcdf$. The overall probability of detection in this case using the decision fusion "AND" rule is $P_d = [Gamcdf(Gaminv(\beta^{1/I}, \frac{K}{2}, \frac{2}{K}(1+\gamma_p), \frac{K}{2}, \frac{2}{K})]^I$.

Similarly, using the decision fusion "OR" rule, the detection threshold under the Neyman-Pearson criterion is $T_{coh} = T_{OR} = Gaminv(1-(1-\beta)^{1/I}, \frac{K}{2}, \frac{2}{K}(1+\gamma_p))$ and the overall probability of detection is $P_d = 1 - [1 - Gamcdf$ $(Gaminv(1-(1-\beta)^{1/I}, \frac{K}{2}, \frac{2}{K}(1+\gamma_p)), \frac{K}{2}, \frac{2}{K})]^I$, as the overall decision in the "OR" rule is H_1 only when all the individual decisions are H_1.

The above traditional method uses local decisions from all cognitive radios. This energy consumption can be reduced by using only a few of them. Thus, one has

$$
\begin{aligned}
D_i &= \frac{1}{K\sigma_n^2}\sum_{k=1}^{K}(R_i[k])^2 < L_1 \Rightarrow H_0 \\
D_i &= \frac{1}{K\sigma_n^2}\sum_{k=1}^{K}(R_i[k])^2 > L_2 \Rightarrow H_1.
\end{aligned} \tag{9}
$$

In this case, a binary local decision will be sent from the cognitive radio to the fusion center only if the samples satisfy $D_i < L_1$ or $D_i > L_2$. Otherwise, the cognitive radio is not involved in spectrum sensing to save energy.

If the "AND" rule is used, the overall decision is H_0 only when all the individual decisions from the selected users are H_0. Thus, one has

$$
\begin{aligned}
P_{f.o.} = Pr\{H_0|H_1\} &= \left[Gamcdf(L_2, \frac{K}{2}, \frac{2}{K}(1+\gamma_p))\right]^I \\
&- \left[Gamcdf(L_2, \frac{K}{2}, \frac{2}{K}(1+\gamma_p)) - Gamcdf(L_1, \frac{K}{2}, \frac{2}{K}(1+\gamma_p))\right]^I \\
P_d = Pr\{H_0|H_0\} &= \left[Gamcdf(L_2, \frac{K}{2}, \frac{2}{K})\right]^I \\
&- \left[Gamcdf(L_2, \frac{K}{2}, \frac{2}{K}) - Gamcdf(L_1, \frac{K}{2}, \frac{2}{K})\right]^I
\end{aligned} \tag{10}
$$

where the second terms in these equations represent the probability that all values of D_i fall between L_1 and L_2 such that no decision will be sent for fusion. Solving the equation for L_1 and using the solution in the expression of P_d, ,P_d can be maximized with respect to L_2 with $L_2 > T_{AND}$.

If the "OR" rule is used, the overall decision is H_1 only when all the individual decisions from the selected users are H_1. In this case, one has

$$
\begin{aligned}
P_{f.o.} &= 1 - \left[1 - Gamcdf(L_1, \frac{K}{2}, \frac{2}{K}(1+\gamma_p))\right]^I \\
&+ \left[Gamcdf(L_2, \frac{K}{2}, \frac{2}{K}(1+\gamma_p)) - Gamcdf(L_1, \frac{K}{2}, \frac{2}{K}(1+\gamma_p))\right]^I
\end{aligned}
$$

$$P_d = 1 - \left[1 - Gamcdf(L_1, \frac{K}{2}, \frac{2}{K})\right]^I + \left[Gamcdf(L_2, \frac{K}{2}, \frac{2}{K}) - Gamcdf(L_1, \frac{K}{2}, \frac{2}{K})\right]^I. \quad (11)$$

Again, the second terms represent the probability that all D_i fall between L_1 and L_2 such that all samples are excluded from spectrum sensing. By solving the equation for L_2 and using L_2 in P_d, it becomes a function of L_1 which can be maximized with respect to L_1 and L_1 satisfies $L_1 \leq T_{OR}$.

Several observations can be made from the above equations for the decision fusion rules. First, if $L_1 = L_2$, one can obtain $L_2 = T_{AND}$ and $L_1 = T_{OR}$ from the first subequations in (10) and (11), respectively. Using $L_2 = T_{AND}$ and $L_1 = T_{OR}$ again in the second subequations in (10) and (11), respectively, one obtains the results for the conventional spectrum sensing using the "AND" rule and the "OR" rule given before. Thus, the censored sensing is a generalization of the conventional sensing without censoring using the decision fusion rules. Second, let $L_1 = 0$ in (11), one has the probability of detection for the conventional sensing using the "AND" rule. Let $L_1 = L_2$ in (11), one has the probability of detection for the conventional sensing using the "OR" rule. Thus, by choosing L_1 between 0 and L_2 and using the "OR" rule, one can achieve sensing performances between that of the conventional "OR" rule and that of the conventional "AND" rule. Similarly, let $L_2 = \infty$ in (10), one has the probability of detection for the conventional "OR" rule. If one lets $L_1 = L_2$ in (10), one has the probability of detection for the conventional "AND" rule. Again, by choosing L_2 between L_1 and ∞ and using the "AND" rule, sensing performances between that of the conventional "OR" rule and that of the conventional "AND" rule can be developed. This adds flexibility to the cognitive radio design. Finally, the average number of individual decisions sent to the fusion center equals $\bar{k}$ as given by $I[1 - Gamcdf(L_2, \frac{K}{2}, \frac{2}{K}) + Gamcdf(L_1, \frac{K}{2}, \frac{2}{K})]$ in H_0, and $I[1 - Gamcdf(L_2, \frac{K}{2}, \frac{2}{K}(1+\gamma_p)) + Gamcdf(L_1, \frac{K}{2}, \frac{2}{K}(1+\gamma_p))]$ in H_1, while in the conventional sensing, the number of transmitted decisions is always I. Thus, the censored sensing saves energy by sending less decisions, especially when the difference between L_1 and L_2 is large.

Next, some sample numerical results using the decision fusion rules are shown. For the results using the data fusion rule, one is referred to [17]. In the examples, we set $K = 1$ and the probability of false opportunity $\beta = 0.01$. The value of L_2 is tested from T_{AND} to $20\gamma_p$ with a step size of 0.001 to find the optimum P_d in (10), and the value of L_1 is tested from 0 to T_{OR} with a step size of 10^{-10} to find the optimum P_d in (11).

Figure 11 compares the conventional sensing and the censored sensing using the "AND" rule for different values of γ_p. Their sensing accuracies are graphically indistinguishable. However, the censored sensing reduces the number of local decisions. This is reflected by Fig. 12, where the probability of detection per user is compared. The performance gain is considerable in this case. Thus, censored sensing improves energy efficiency.

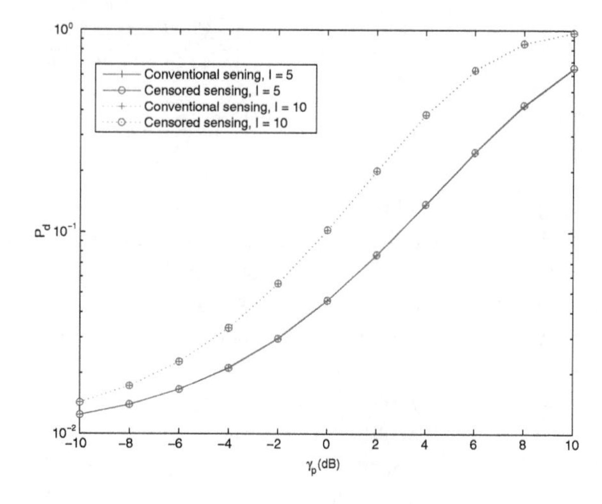

Fig. 11 P_d versus γ_p for censored sensing using the optimum limits and the conventional sensing for the "AND" rule

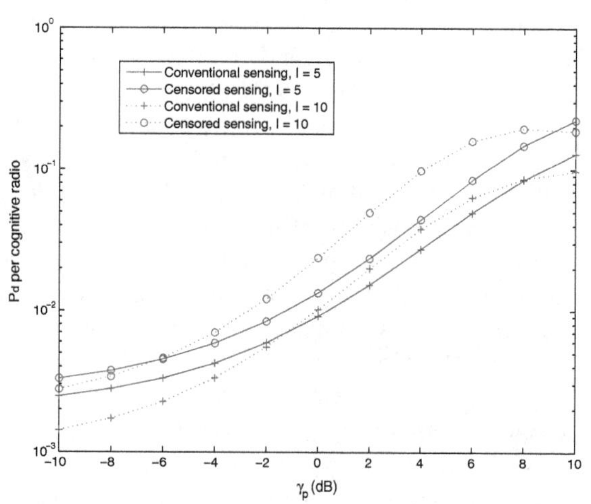

Fig. 12 P_d per user versus γ_p for censored sensing using the optimum limits and the conventional sensing for the "AND" rule

Figure 13 compares the conventional sensing and the censored sensing using the "OR" rule. In this case, the censored sensing has a large performance gain over the conventional sensing. The performance difference between $I = 10$ and $I = 5$ for the conventional sensing is too small compared with the performance gain of the censored sensing over the conventional sensing. Thus, they are graphically indistinguishable. From Fig. 14, the performance gains are even larger when the probability of detection per user is compared.

Finally, Figs. 15 and 16 show the found optimum values of L_1 and L_2 used to calculate the probability of detection in Figs. 11, 12, 13 and 14. The optimum value of L_2 increases with γ_p and I, while the optimum value of L_1 is very close to 0 but is not 0 in the cases considered.

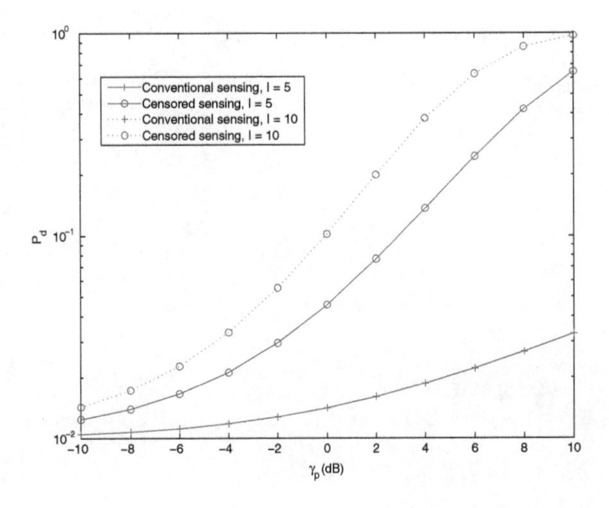

Fig. 13 P_d versus γ_p for censored sensing using the optimum limits and the conventional sensing for the "OR" rule

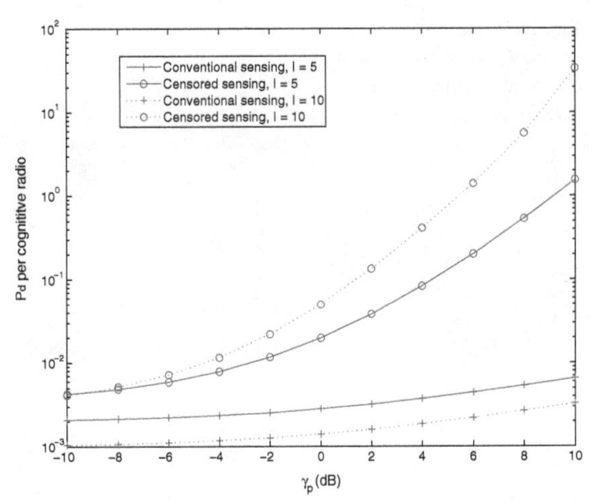

Fig. 14 P_d per user versus γ_p for censored sensing using the optimum limits and the conventional sensing for the "OR" rule

2 Cognitive Radio Energy Optimization

In this section, we investigate energy optimization problems for cognitive radios by taking both physical layer sensing and upper layer control into account. The optimization problem is explained as follows. On the one hand, the number of cognitive radios determines the sensing accuracy in the physical layer. As shown in [18] and many other works, the probability of detection for sensing increases when the number of cognitive radios increases. On the other hand, the number of cognitive radios also affects the throughput in the upper layer. For example, in IEEE 802.11 networks, its saturated throughput in distributed coordination function mode decreases as the number of cognitive radios increases [19], as a result of more overheads and packet collisions. Thus, the number of cognitive radios is a key design parameter for energy

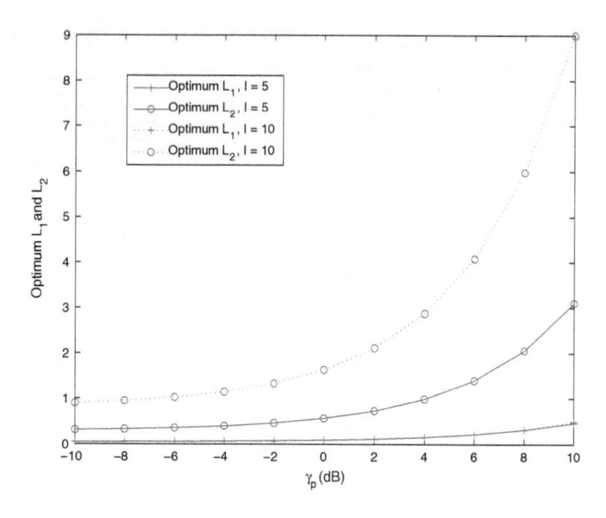

Fig. 15 Optimum values of L_1 and L_2 versus γ_p for censored sensing for the "AND" rule

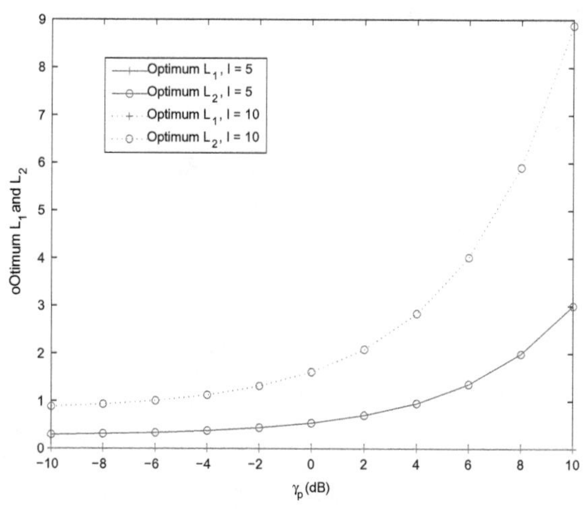

Fig. 16 Optimum values of L_1 and L_2 versus γ_p for censored sensing for the "OR" rule

efficiency and network efficiency. In [20], a linear combination of the detection probability and the resource usage efficiency was optimized with respect to the number of cognitive radios. The usage efficiency is a general metric of the physical layer resources, and it does not give any information on the network performance. Thus, in this section, we will try to find an optimum number of cognitive radios that can balance the needs for physical layer sensing accuracy as well as upper layer saturated throughput.

Assume that the cognitive radios communicate with their base station using the IEEE 802.11 WLAN protocol. This protocol has different coordination functions but if the distributed coordination function is used, the saturation throughput in the upper layer, defined as the ratio of the average payload information in a slot time to the average length of the slot time, has been derived as [19, Eq. (13)]

$J = \frac{Q_s Q_{tr} E\{Q\}}{(1-Q_{tr})\alpha + Q_{tr} Q_s T_s + Q_{tr}(1-Q_s) T_c}$ for I cognitive radios, where Q_s is the probability of a successful transmission given by $Q_s = \frac{Ip(1-p)^{I-1}}{1-(1-p)^I}$, Q_{tr} is the probability of transmission in the interested slot time given by $Q_{tr} = 1 - (1-p)^I$, p is the probability that each cognitive radio activates in the slot time given by $p = \frac{2(1-2c)}{(1-2c)(M+1)+cM(1-(2c)^m)}$, c is the conditional collision probability given by $c = 1 - (1-p)^{I-1}$, $E\{Q\}$ is the average size of the packet payload, α is the duration of the empty slot time, T_s is the average time of a successful transmission, T_c is the average time of a busy channel, M is the size of the contention window used in the protocol, m is the maximum backoff stage and I is the number of cognitive radios to be optimized. Essentially, this throughput is a function of the number of cognitive radios and it determines the transmissions of binary local decisions or samples from cognitive radios to the fusion center.

This throughput applies to conventional networks. However, for cognitive radio networks, the transmitted payload information could be useless or even harmful if a wrong sensing decision is made in spectrum sensing. For example, if the cognitive radios consider the licensed user to be absent while it is actually present, the transmission of payload from cognitive radios to fusion center or base stations will be considered as interferences to the licensed user. Thus, we have to take the physical layer sensing into account to define an effective throughput as

$$J_e = P_d \cdot J(d) + P_e \cdot J(e) \tag{12}$$

where P_d is the probability of detection, $P_e = 1 - P_d$ is the probability of missed-detection, $J(d)$ is the throughput when a correct decision is made, and $J(e)$ is the throughput when a wrong decision is made. In the IEEE 802.22 draft standard, $P_d = 0.9$ and $P_e = 0.1$ are used [21]. Thus, $J_e = 0.9J(d) + 0.1J(e)$ in this case. However, in other applications, P_e may be much smaller than P_d. In these applications, the second term in (12) may be ignored in the optimization. Next, the probability of detection needs to be obtained.

Assuming the Neyman-Pearson detection rule and following similar methods to those in [20], the probability of detection in shadowing channels can be derived as $P_d = P(H_0)(1-P_f) + P(H_1)Q\left(-\sqrt{\gamma_p \Delta} + Q^{-1}(P_f)\right)$, where $P(H_0)$ is the probability with a licensed user, $P(H_1)$ is the probability without licensed user, P_f is the probability of false alarm, $Q(x) = \frac{1}{\sqrt{2\pi}} \int_x^\infty e^{-t^2/2} dt$ is the Gaussian Q-function, $Q^{-1}(\cdot)$ is its inverse, γ_p is the licensed user signal-to-noise ratio, and $\Delta = \mathbf{1}^T \times \mathbf{V}^{-1} \times \mathbf{1}$, $\mathbf{1} = [11 \cdots 1]^T$, $[\cdot]^T$ represents the transpose operation, and $\mathbf{V}$ is the normalized covariance matrix of the noise samples. Also, the probability of detection in Rayleigh fading channels with unknown channel gains has been derived in [20], while the average probability of detection in Rayleigh fading channels with known channel gains is $P_d = P(H_0)(1-P_f) + P(H_1) \int Q\left(-\sqrt{\mathbf{h}^H \Sigma^{-1} \mathbf{h}} + Q^{-1}(P_f)\right) p(\mathbf{h}) d\mathbf{h}$.

The above results assume data fusion where samples are sent directly from cognitive radios to the fusion center. One can also use decision fusion where local decisions are sent to simplify the sensing structure, similar to those in censored sensing.

In shadowing channels, one has $P_d = P(H_0)(1 - P_f) + P(H_1)Q^I\left(-\sqrt{\gamma_p} + Q^{-1}(P_f^{1/I})\right)$ for the "AND" rule and $P_d = P(H_0)(1 - P_f) + P(H_1)[1 - (1 - Q(-\sqrt{\gamma_p} + Q^{-1}(1 - (1 - P_f)^{1/I})))^I]$ for the "OR" rule. Similarly, with unknown Rayleigh gains, one has $P_d = P(H_0)(1 - P_f) + P(H_1)Gamcdf^I(Gaminv(P_f^{1/I}, 1, 2 + 2\gamma_p), 1, 2)$ and $P_d = P(H_0)(1 - P_f) + P(H_1)[1 - (1 - Gamcdf(Gaminv(1 - (1 - P_f)^{1/I}, 1, 2 + 2\gamma_p), 1, 2))^I]$ for the "AND" and "OR" rule, respectively, and the average probability of detection in Rayleigh fading channels with known channel gains can be derived as by averaging them over γ_p. Using these equations, the effective throughput of the cognitive radio network in different channel conditions using different decision rules can be examined.

Next, some numerical examples are shown for the effective throughput. In these examples, assume that $J(d) = J$ and $J(e) = 0$. The (i, j)th element of the covariance matrix is given by $\mathbf{V}(i, j) = e^{-\frac{aD}{I-1}|i-j|}$, where $i, j = 1, 2, \ldots, I$, $a = 0.1$ and D denotes the length of the range over which cognitive radios are distributed [22]. Thus, the channel samples are either correlated with a finite value of D or independent with an infinite value of D. For convenience, we use the same system parameters as those given in [19, Table II]. Also, let $P(H_0) = 0.3$ and $P(H_1) = 0.7$.

Figure 17 shows J_e versus I in shadowing channels using data fusion. Several observations can be made. First, the effective throughput first increases quickly when I increases abut then decreases slowly when I further increases. There exists an optimum value of I in the cases considered. Second, comparing $D = 100$ m with $D = \infty$, one sees that J_e for independent shadowing is larger than that for correlated shadowing. Figure 18 shows the corresponding optimum values of I. It monotonically decreases as γ_p increases for data fusion, while it first increases then decreases for the decision fusion rules.

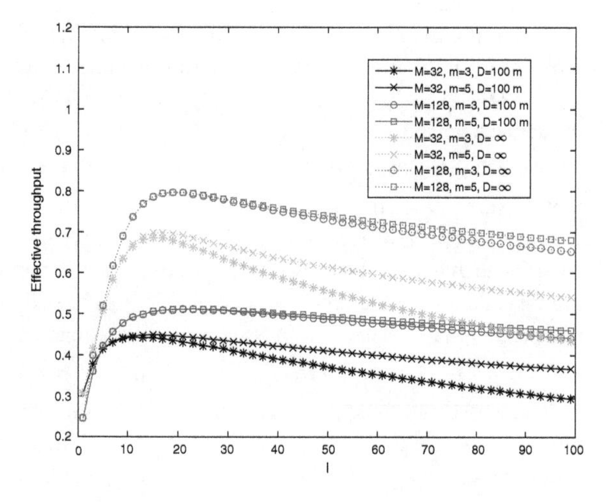

Fig. 17 J_e versus I in shadowing channels when data fusion is used, $\gamma_p = 0\ dB$ and $P_f = 0.01$

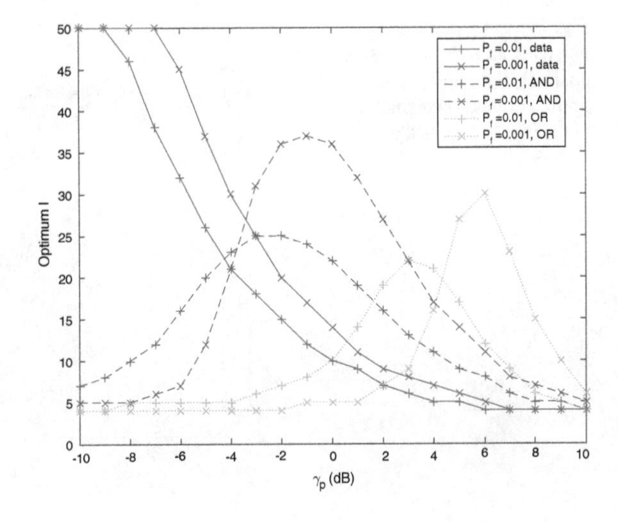

Fig. 18 The optimum I versus γ_p in shadowing channels with independent samples for different values of P_f and different decision rules when $M = 128$ and $m = 3$

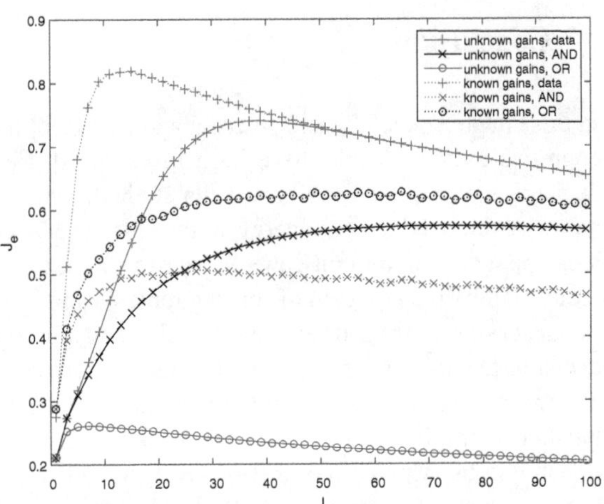

Fig. 19 J_e versus I in Rayleigh fading channels with independent known or unknown channel gains for different decision rules when $\gamma_p = 0\ dB$, $P_f = 0.01$, $M = 128$ and $m = 3$

Figure 19 shows J_e versus I in Rayleigh fading channel. Again, J_e first increases then decreases as I increases, indicating that there exists an optimum number of I. The data fusion rule has the highest effective throughput. Also, when the channel gains are unknown, the "OR" rule has the lowest effective throughput, while when the channel gains are known, the "AND" rule has the lowest effective throughput. Figure 20 shows the optimum value of I versus γ_p. From this figure, the optimum value decreases monotonically for known channel gains, while it increases then decreases for unknown channel gains for $P_f = 0.01$ and $P_f = 0.001$. There is an upper limit of 4 for the optimum I.

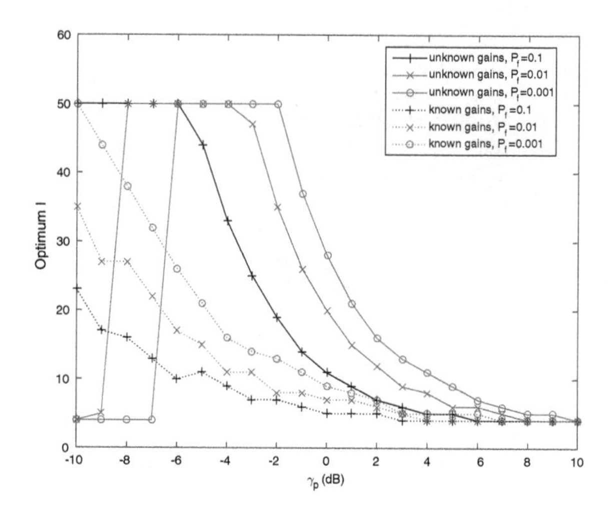

Fig. 20 The optimum I versus γ_p in Rayleigh fading channels with independent known or unknown channel gains when $M = 128$ and $m = 3$

3 Conclusions

In this chapter, several energy-saving and energy-optimization techniques in ad hoc cognitive radio networks have been investigated. First, non-coherent sensing and decision-feedback sensing have been studied, where the dedicated sensing period has been removed to save energy, at the cost of reduced sensing accuracy. However, it can be seen that, for non-coherent sensing, although the sensing accuracy has been reduced, the diversity gain of sensing remains and a similar accuracy to the conventional sensing can still be achieved by choosing appropriate sensing parameters. For decision-feedback sensing, its performance is actually better than the conventional sensing initially and only starts to degrade when more decision errors are accumulated. This implies that decision-feedback sensing is more suitable for short bursty transmissions, which offers energy-saving as well as performance-improvement. For longer periods of transmissions, the disadvantages of decision-feedback sensing may outweigh the benefits of decision-feedback sensing. Then, censored sensing has been studied. From this study, it can be seen that performing censoring before sensing offers performance gains in terms of probability of detection as well as reduces energy consumption. The reason is that it excludes very noisy samples from the sensing operation and therefore, offers higher accuracy. After this, energy optimization that takes both physical layer sensing and upper layer throughput into account has been examined. The examination has shown that there does exist an optimum choice of the number of cognitive radios involved in spectrum sensing in terms of energy efficiency. This number depends on a few important system parameters. From these studies, energy saving and optimization in cognitive radio networks are challenging but can bring a lot of benefits, and in some cases both performance and energy benefits. Thus, it is important to design proper cognitive radio techniques to achieve these benefits. In the future, works can focus on the dynamic topology of ad hoc networks as well as the use of these energy saving and energy optimization techniques in emerging networks such as wireless relaying.

Acknowledgments The author would like to thank Dr. Liang Tang for providing some of the materials used in this chapter.

References

1. Haykin, S.: Cognitive radio: brain-empowered wireless communications. IEEE J. Sel. Areas Commun. **23**, 201–220 (2005)
2. Cordeiro, C., Challapali, K., Birru, D., Shankar, S.: IEEE 802.22: an introduction to the first wireless standard based on cognitive radios. J. Commun. **1**, 38–47 (2006)
3. Chen, Y., Tang, L., Long, M.: Analysis of collaborative spectrum sensing without dedicated sensing period. IET Commun. **7**, 1617–1627 (2013)
4. Stüber, G.L.: Principles of Mobile Communication, 2nd edn. Kluwer Academic, Norwell (2001)
5. Abramowitz, M., Stegun, I.A.: Handbook of Mathematical Functions, with Formulas, Graphs, and Mathematical Tables. Dover, New York (1972)
6. Stotas S., Nallanathan A.: Overcoming the sensing-throughput tradeoff in cognitive radio networks. In: IIEEE International Conference on Communications (ICC'10). Cape Town, South Africa (2010)
7. Ma, J., Zhou, X., Li, G.Y.: Probability-based periodic spectrum sensing during secondary communication. IEEE Trans. Commun. **58**, 1291–1301 (2010)
8. Ghasemi, A., Sousa, E.S.: Optimization of spectrum sensing for opportunistic spectrum access in cognitive radio networks. In: IEEE 4th Consumer Communications and Networking Conference (CCNC 2007), pp. 1022–1026. Las Vegas. USA (2007)
9. Tang, L., Chen, Y., Nallanathan, A., Hines, E.L.: Performance evaluation of spectrum sensing using recovered secondary frames with decoding errors. IEEE Trans. Wirel. Commun. **11**, 2934–2945 (2012)
10. Jeong, S.S., Jeong, D.G., Jeon, W.S.: Nonquiet primary user detection for OFDMA-based cognitive radio systems. IEEE Trans. Wirel. Commun. **8**, 5112–5123 (2009)
11. Chen, D., Li, J., Ma, J.: In-band sensing without quiet period in cognitive radio. In: Proceedings of the IEEE Wireless Communications and Networking Conference (WCNC '08), pp. 723–728. Las Vegas, Nev, USA (2008)
12. Jeong D.G., Jeong S.S., Jeon W.S.: Channel sensing without quiet period for cognitive radio systems: a pilot cancellation approach. EURASIP Journal on Wireless Communications and Networking, (2011)
13. Sun, C., Zhang, W., Letaief, K.B.: Cooperative spectrum sensing for cognitive radios under bandwidth constraints. In: Proceedings of the WCNC, pp. 1–5. Hong Kong. China (2007)
14. Wang, W., Zou, W., Zhou, Z., Zhang, H., Ye, Y.: Decision fusion of cooperative spectrum sensing for cognitive radio under bandwidth constraints. In: Proceedings of the ICCIT'08, pp. 733–736 (2008)
15. Lundén, J., Koivunen, V., Juttunen, A., Poor, H.V.: Censoring for collaborative spectrum sensing in cognitive radios. In: Proceedings of the Asilomar Conference on Signals, Systems and Computers, pp. 772–776 (2007)
16. Lundén, J., Koivunen, V., Huttunen, A., Poor, H.V.: Collaborative cyclostationary spectrum sensing for cognitive radio systems. IEEE Trans. Signal Process. **57**, 4182–4195 (2009)
17. Chen, Y.: Analytical performance of collaborative spectrum sensing using censored energy detection. IEEE Trans. Wirel. Commun. **9**, 3856–3865 (2012)

18. Visotsky, E., Kuffner, S., Peterson, R.: On collaborative detection of TV transmissions in support of dynamic spectrum sharing. In: Proceedings of the IEEE DySPAN 2005, pp. 338–345. Baltimore, USA (2005)
19. Bianchi G.: Performance analysis of the IEEE 802.11 distributed coordination function. IEEE J. Select. Areas Commun. **18**, 535–547 (2000)
20. Chen, Y.: Optimum number of secondary users in collaborative spectrum sensing considering resources usage efficiency. IEEE Commun. Lett. **12**, 877–879 (2008)
21. Shellhammer S.: Spectrum sensing in IEEE 802.22. In: Proceedings First Workshop on Cognitive Information Processing (CIP 2008), Santorini, Greece (2008)
22. Gudmundson, M.: A correlation model for shadow fading in mobile radio. Electron. Lett. **27**, 2146–2147 (1991)

Part V
Energy Management in Emerging Wireless Networks

Visible Light Communications for Energy Efficient Heterogeneous Wireless Networks

Mohamed Kashef, Muhammad Ismail, Mohamed Abdallah, Khalid A. Qaraqe and Erchin Serpedin

Abstract The necessity of having energy efficient wireless communication networks emerges because of the related environmental and economical benefits. Achieving energy efficient wireless communications using visible light communications (VLC) has been recently studied due to its transmission properties. Moreover, the integration of VLC and radio frequency (RF)-based wireless networks has shown improved data rate and reliability for the mobile users. In this chapter, we investigate the energy efficiency of the integration of VLC and RF wireless networks. We formulate and solve the energy efficiency maximization problem by allocating the power and bandwidth of a heterogenous RF/VLC wireless network. We study the impact of various system parameters on the network performance. Numerical results are presented to demonstrate the performance gains of the hybrid system and to quantify the impact of the system specifications on the achieved energy efficiency. Moreover, several challenging issues due to the RF/VLC integration are addressed.

M. Kashef (✉) · M. Ismail · M. Abdallah · K.A. Qaraqe
Department of Electrical and Computer Engineering, Texas A & M University at Qatar,
P.O. Box No. 23874 Education City, Doha, Qatar
e-mail: mohamed.kashef@qatar.tamu.edu

M. Ismail
e-mail: m.ismail@qatar.tamu.edu

M. Abdallah
Department of Electronics and Communications Engineering, Cairo University,
Giza 12613, Egypt
e-mail: mohamed.abdallah@qatar.tamu.edu

K.A. Qaraqe
e-mail: khalid.qaraqe@qatar.tamu.edu

E. Serpedin
Electrical and Computer Engineering, Texas A & M University,
College Station, USA
e-mail: serpedin@ece.tamu.edu

© Springer International Publishing Switzerland 2016

M.Z. Shakir et al. (eds.), *Energy Management in Wireless Cellular and Ad-hoc Networks*, Studies in Systems, Decision and Control 50,
DOI 10.1007/978-3-319-27568-0_13

1 Introduction

Achieving energy efficient communications is motivated by the increasing energy consumption of wireless communication networks caused by the high demand for wireless communication services. The annual energy consumption of a mobile service operator is around 50–100 GWh [1]. From an environmental viewpoint, the CO_2 emissions of the telecommunications industry represent 2 % of the total CO_2 emissions worldwide and are expected to reach 4 % by 2020 [2].

Advances in the field of lighting and illumination, especially in solid state lighting, allow light emitting diodes (LEDs) to dominate the future lighting market. LEDs have been introduced to be used in illumination due to their superior properties over the existing light sources. Huge energy saving is achieved by exploiting LEDs as it consumes five times less power than the fluorescent light sources and twenty times less than the conventional light sources [3, 4]. The advantages of LEDs over the existing lighting technologies also include their long life time, their improved color rendering capability, and their environmental benefits [5]. The share of LEDs in the illumination market is expected to evolve dramatically during the current decade [6].

The capability of the LEDs light intensity to be modulated at high frequencies has motivated exploiting LEDs for wireless data transfer through what is referred to as visible light communications (VLC) [7]. The use of LEDs in communications does not affect their main functionality of illumination by guaranteeing that modulation does not generate flickering. The flicker threshold of human eye is typically less than 3 kHz such that flickering is avoided by modulating the light with frequencies which are greater than this flicker threshold [8]. Moreover, LEDs already exist in numerous electronic devices and hence exploiting them for communication purposes can be realized using the existing devices. The receiver for VLC signals is a photo sensitive detector that demodulates the light signal into an electrical signal. Thus, data transfer is performed using intensity modulation and direct detection (IM/DD) [9], which can be practically obtained by a number of pulsed modulation schemes [10].

The increased demand for wireless data services with the limited radio frequency (RF) spectrum makes it essential to investigate alternative wireless technologies. VLC is introduced as a wireless technology to augment the existing RF technologies due to its superiority over RF technology in certain aspects [11]. The visible light spectrum is huge, unlicensed, and currently unused that allows data transfer with high rates over this spectrum. The signal isolation of VLC systems because of its property of not penetrating through walls allows easy separation of communication cells and enhance communication security. VLC systems are built using simple and cheap components and have no electro-magnetic interference (EMI) to the existing wireless devices. The major limitations for VLC systems are performance degradation in absence of line of sight (LOS), and the noise increase due to the ambient light sources. Hence, VLC systems are more suitable to augment the existing RF networks than to completely replace them.

The applications of VLC systems are more concentrated in indoor environments where lights are switched on including industrial areas, medical environments, airports, and shopping centers. High data rate transmission in crowded areas is a major VLC application that has been proven by real-time systems. The first high speed VLC demonstration was of the European project OMEGA at February 2011 [12]. There are also some standardization efforts that have been worked on to allow the availability of VLC-enabled devices. Currently, Infrared Data Association (IrDA) group and an IEEE group work on the standardization process that the first standard, IEEE 802.15.7, was published September 2011. Other attractive environments for VLC applications are EMI-sensitive environments like aircraft cabins and hospitals. A different approach for exploiting VLC is navigation and localization in indoor environments as shopping centers where the LEDs light sources can be connected to users VLC (smart phone) for guiding and advertising about various stores or products based on users location. The market for indoor localization is expected shortly to reach $5 billion [13]. Various additional potential applications of VLC techniques have been proposed including broadband indoor communication, and military applications that demand anti-jamming. In [14], the use of LEDs in communications is reviewed where comparisons to other communication techniques are presented. The VLC has shown many advantages including their cheap transmitters and receivers, low power consumption, and good safety features. A more detailed comparison of the performance of VLC systems against RF communication systems indicates better area spectral efficiency is achieved for VLC systems [15]. Additionally, VLC has been introduced as an energy efficient wireless technology that exploits the illumination energy, which is already consumed for lighting, in data transmission with high achievable data rates. In [3], the improvement in energy efficiency of VLC data transmission is demonstrated as a result of using white LEDs for both illumination and communication.

Furthermore, wireless networks with different technologies can work together to enhance the overall system performance using heterogeneous networks principles. Such enhancements are achieved because of the diversity in fading channels, propagation losses, and the available resources at different networks. However, such gains can be only achieved by tackling the major challenge of developing resource allocation algorithms that assign the power and bandwidth among the heterogeneous networks to achieve different service requirements. Particularly, the ongoing work on fifth generation (5G) wireless access aims to achieve higher system data rates, network capacity, and reliability of communications [16]. Moreover, it aims to achieve lower latency, and energy consumption. In [16, 17], the suitability of VLC technology to serve indoor communication in future 5G networks was demonstrated. The main factors that help this idea are the availability of the huge visible light spectrum bandwidth, the non-exitance of the interference towards the existing RF transmissions, the spatial reuse capability the visible light characteristics, and the low cost for data transmission.

VLC and RF communication systems can work together to take advantage of the benefits yielded by both systems to enhance the communication energy efficiency while maintaining good reliability. In this chapter, we investigate the problem of resource allocation in a heterogeneous RF/VLC communication system from an energy efficiency viewpoint. We study the energy efficiency of an indoor heterogeneous network composed of a single RF access point (AP) and a single VLC AP transmitting to a number of MTs located in the coverage region of both APs. The VLC system employs its illumination power for data transmission while consuming additional power for data processing. On the other hand, the RF communication system consumes both data processing and transmission powers. We formulate and analyze the problem of maximizing the heterogeneous network energy efficiency constrained by the required data rates for the MTs and the maximum allowable transmission powers for the APs. MTs are equipped with multi-homing capability and can receive data from both VLC and RF communication systems. We compare the performance of the heterogeneous network consisting of VLC and RF communication systems to the benchmarks represented by an RF only network and a heterogeneous network composed of two RF communication systems. We compare the energy efficiency of these systems to quantify the impact of using mixed RF/VLC systems on the network energy efficiency.

2 Related Work

2.1 Energy Efficiency in VLC Networks

A major advantage of employing VLC is achieving high data-rates while the power consumption of the system is low [3]. Hence, optimizing VLC systems transmission parameters is commonly used in literature to enhance some performance measure such as the achievable rate, and the spectral efficiency because the consumed power is low due to the VLC system nature. The optimization of the energy efficiency of VLC systems by controlling the data transmission parameters has been considered in few papers. In [18], the transmission parameters of a pulse position modulation (PPM) scheme are controlled to minimize the total power consumption of a VLC system while achieving a certain communication and brightness requirements. These transmission parameters are the DC bias and amplitude range of the data sent using the PPM scheme. The transmission parameters control has reduced the consumed power significantly compared to the conventional fixed PPM modulation depths.

In [19], a solar cell is used to receive low-frequency VLC signals to implement a green VLC system. The proposed system is considered as an environment friendly as it exploits the power resulting from the solar cell and does not need an extra photo-detector for data reception. Moreover in [20], an intelligent lighting system that exploits VLC was introduced. In this system, location information of the lighting sources and a number of light sensors are transferred using VLC to achieve certain illumination requirements while minimizing the consumed electricity. It was shown that the proposed VLC-based strategy can achieve higher energy savings and faster illumination response to the environmental changes.

2.2 Energy Efficient Heterogeneous Networks

Employing multiple radio interfaces in communication systems has proven to enhance energy efficiency in RF communication systems. In [21], the cooperation between mobile terminals (MTs) is exploited by allowing the MTs to transmit their data efficiently to base stations using space-time coding over multiple radio interfaces. Also, MTs are used to relay source data using multiple radio interfaces in [22]. Moreover in [23], the authors have discussed the enhancement in the energy efficiency of MTs equipped with multi-homing capabilities where the MTs are allowed to aggregate the available resources from different networks in a downlink communication scenario.

Cooperation in VLC networks between multiple VLC nodes has not been investigated comprehensively yet. Cooperation in VLC systems has been considered in [24] where the total rate of a VLC network contains multiple interfering transmitters is improved by employing cooperative transmission power control. Additionally, few papers have discussed the complementary use of VLC and RF communication systems to achieve throughput and reliability gains. In [25, 26], the authors have discussed the potential benefits of the RF/VLC combination and the optimal handover techniques and have shown that lower data transfer delay and higher data rates can be achieved because of the nodes ability to switch their access between the VLC and RF networks. The hybrid simultaneous use of VLC and RF systems has been discussed in [27, 28] where the authors investigate the feasibility and potential benefits of RF/VLC hybrid systems in enhancing the throughput and increasing the coverage.

2.3 Employing VLC in 5G Networks

Various VLC benefits include the low energy consumption and the low interference are gained beside the currently reported high data rate of 3.5 Gb/s [29]. Due to the requirements of 5G networks to have ubiquitous connectivity and hence achieving high data rates for very dense networks, they need to be energy efficient to reduce the total cost per transmitted bit [30]. A promising proposal to achieve this goal is employing small communication cells that exploit low-cost base stations and have short-distance communication links [31]. VLC technology represents an excellent candidate to satisfy the requirements of having small cells with low energy cost per transmitted bit [32].

3 System Model

We consider an indoor downlink scenario in which M MTs equipped with VLC receivers and RF receivers are communicating with a single RF AP and a single VLC AP as shown in Fig. 1. Examples of RF wireless communication net-

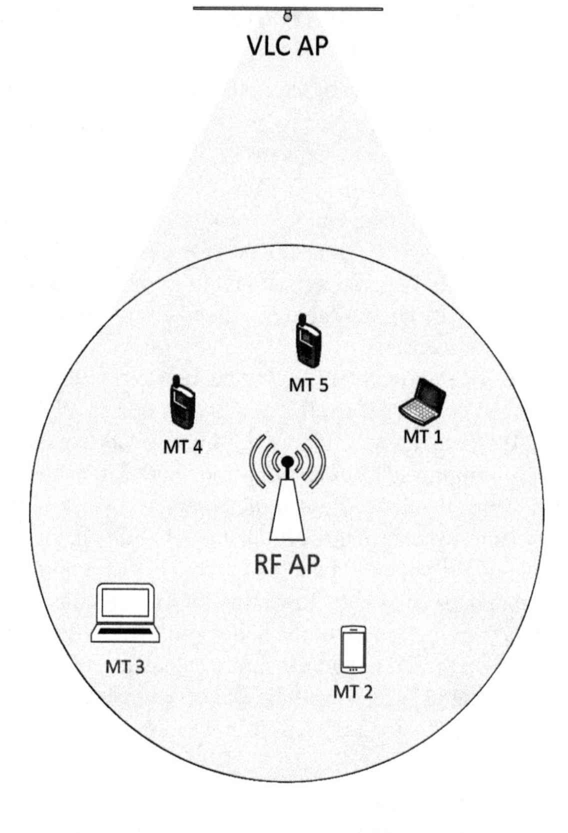

Fig. 1 Illustration of the APs coverage areas

works include cellular networks (e.g., femto-cells). The set of MTs is denoted by $\mathcal{M} = \{1, 2, \ldots, M\}$. All MTs are in the coverage areas of both APs. Each MT has multi-homing capabilities that allow simultaneous association with both networks and enable the MTs to aggregate the available resources from both networks to provide services with high performance requirements such as improving network capacity [23].

3.1 VLC System Model

The single-user VLC system transmitter and receiver models are shown in Fig. 2 as well as the signal flow through the system to calculate the electrical signal to noise power ratio at the receiver which controls the performance of the VLC system.

At the transmitter side, the driving current of the LED expressed in Amperes (Amp) is denoted by x and is proportional to the modulated data of the source. The driving power is denoted by P_{dr} and is calculated as the average of the squared driving current $\overline{x^2}$. In a multi-user system, the mth MT is allocated a proportion of

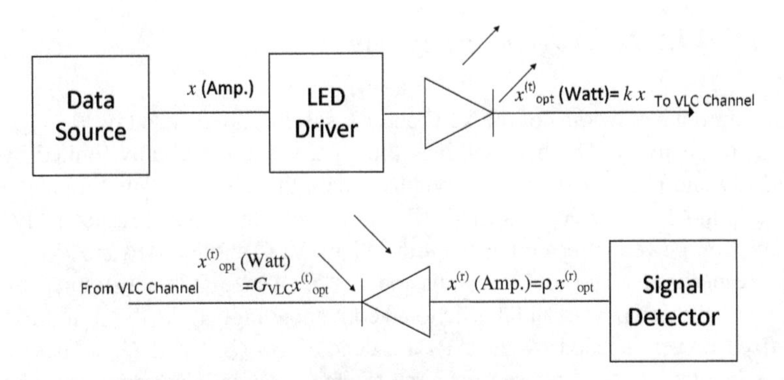

Fig. 2 VLC single-user transmitter and receiver models

P_{dr} that is denoted by $P_{\mathrm{VLC},m}$. These proportions of the driving power $P_{\mathrm{VLC},m}$ are to be controlled while optimizing the system performance. The output of the LED is an optical intensity signal which is proportional to the driving current, when operating in the LED dynamic range, and is denoted by $x_{\mathrm{opt}}^{(\mathrm{t})} = kx$ where k (Watt/Amp) is the proportionality factor of the electrical to optical conversion and is determined by the LED characteristics. The average optical transmitted power is denoted by $P_{\mathrm{opt}}^{(\mathrm{t})}$ and is calculated as follows

$$P_{\mathrm{opt}}^{(\mathrm{t})} = \overline{x_{\mathrm{opt}}^{(\mathrm{t})}}, \tag{1}$$

where $\overline{(.)}$ is the average of the signal. This optical transmitted power determines the illumination level of the LED.

At the receiver of the mth MT, the received optical signal $x_{\mathrm{opt},m}^{(r)}$ represents the light intensity at the photo-sensitive detector, and it is expressed as follows

$$x_{\mathrm{opt},m}^{(\mathrm{r})} = G_{\mathrm{VLC},m} x_{\mathrm{opt},m}^{(\mathrm{t})}, \tag{2}$$

where $G_{\mathrm{VLC},m}$ is the VLC channel power gain between the VLC AP to the mth MT. The photo-detector has a responsivity ρ (Amp/Watt) and converts the received optical signal into an electrical current which is proportional to the optical intensity. The received electrical signal, which is denoted by $x_m^{(\mathrm{r})}$, is calculated as follows

$$x_m^{(\mathrm{r})} = \rho x_{\mathrm{opt},m}^{(\mathrm{r})}. \tag{3}$$

The average electrical power of the received signal is calculated as follows

$$P_{\mathrm{elec},m} = \overline{\left(x_m^{(\mathrm{r})}\right)^2},$$

$$P_{\mathrm{elec},m} = \left(k\rho G_{\mathrm{VLC},m}\right)^2 P_{\mathrm{VLC},m}. \tag{4}$$

3.2 RF/VLC Heterogeneous System

The maximum bandwidths of the VLC and RF systems are denoted by $B_{\text{VLC,max}}$ and $B_{\text{RF,max}}$, respectively. The bandwidth of the VLC system is usually limited by the used LED and photo-detector bandwidths. The mth user is assigned a bandwidth of $B_{\text{VLC},m}$ and $B_{\text{RF},m}$ by the VLC and RF communication systems, respectively. The transmission powers allocated to the mth MT by VLC and RF APs are $P_{\text{VLC},m}$ and $P_{\text{RF},m}$, respectively. The total transmission power of an AP is constrained by the maximum allowed power, and it is denoted correspondingly by $P_{\text{VLC,max}}$ and $P_{\text{RF,max}}$. The fixed powers needed by the APs are denoted by Q_{VLC} and Q_{RF}. These fixed powers include the powers used for circuits' operation and data processing before transmission. The value of Q_{VLC} includes also any required power to compensate for the losses in the LED efficiency due to data transmission.

The data rates achieved by the mth MT are denoted by $R_{\text{VLC},m}$ and $R_{\text{RF},m}$, respectively, where the sum of the data rates of the mth MT is constrained to be larger than or equal to a required minimum data rate $R_{\min,m}$.

The power gains of the channels between the mth MT and the RF and VLC APs are denoted by $G_{\text{RF},m}$ and $G_{\text{VLC},m}$, respectively. The channel power gain for the RF communication system captures both the channel fading and path loss. For the VLC system, the channel power gain captures LOS path loss for the optical wireless signal. The distances between the APs to the mth MT are denoted by $d_{\text{RF},m}$ and $d_{\text{VLC},m}$. The thermal noise power spectral density affecting the RF receivers is denoted by $N_{0,\text{RF}}$, and is given by $N_{0,\text{RF}} = k_{\text{B}}T$, where k_{B} stands for Boltzmann's constant and T denotes the ambient temperature. The noise power spectral density affecting the VLC receivers is dominated by the light shot noise and is denoted by $N_{0,\text{VLC}}$. It is shown in [33], Chap. 2, that the noise affecting VLC systems can be well approximated by Gaussian noise independent of the received signal. The characteristics of the channel models can be found in [15].

The RF communication path loss, denoted by PL, typically takes the form

$$\text{PL[dB]} = A\log_{10}(d_{\text{RF},m}) + B + C\log_{10}\left(\frac{f_c}{5}\right) + X, \tag{5}$$

where f_c is the carrier frequency in GHz, A, B and C are constants depending on the propagation model, and X stands for an environment specific term. For the LOS scenario, $A = 18.7$, $B = 46.8$ and $C = 20$. For the non-LOS (NLOS) scenario, $A = 36.8$, $B = 43.8$, $C = 20$ and $X = 5(n_w - 1)$ in case of light walls or $X = 12(n_w - 1)$ in case of heavy walls, where n_w denotes the number of walls between the AP and the MT [15]. The channel power gain is defined as

$$G_{\text{RF},m} = 10^{-\text{PL[dB]}/10}. \tag{6}$$

We denote the channel power gain in the presence of LOS by $G_{\text{RF},m}^{\text{LOS}}$ and for the NLOS scenario by $G_{\text{RF},m}^{\text{NLOS}}$. For VLC systems, the channel power gain is given in [15] by

$$G_{\mathrm{VLC},m} = \frac{(n+1)\cos^n(\phi_m) A_m \cos(\theta_m)}{2\pi d^2_{\mathrm{VLC},m}}, \tag{7}$$

where A_m stands for the physical area of the photodetector at the mth MT, ϕ_m denotes the angle of irradiance from the LED to the mth MT, n is the order of the Lambertian emission defined by the LED's semi-angle at half power Φ, which is $n = \ln(1/2)/\ln(\cos(\Phi))$, and θ_m represents the angle of incidence.

The LOS availability probabilities for RF and VLC systems are defined as the probability that there are no obstacles in the communication link between the MT and the corresponding AP, and are denoted by ρ_{RF} and ρ_{VLC}, respectively. In the case of RF transmissions, the channel path loss exponent increases with the LOS absence as discussed earlier. For the case of VLC, the signal is degraded significantly in the absence of LOS that may result in unsuccessful data transmissions. In this work, we assume that the NLOS VLC transmissions are unsuccessful that we focus only on the system performance with LOS transmissions.

4 Energy Efficiency Maximization Problem

In this section, we formulate the energy efficiency maximization problem, where energy efficiency is defined as the total achieved data rate per unit power consumption.

The average received electrical signal to noise power ratios for the mth MT corresponding to the RF and VLC systems are denoted by $\gamma_{\mathrm{RF},m}$ and $\gamma_{\mathrm{VLC},m}$, respectively, and are expressed as [15]

$$\gamma_{\mathrm{RF},m} = \frac{P_{\mathrm{RF},m} G_{\mathrm{RF},m}}{B_{\mathrm{RF},m} N_{0,\mathrm{RF}}}, \tag{8}$$

$$\gamma_{\mathrm{VLC},m} = \frac{P_{\mathrm{VLC},m} \left(k\rho G_{\mathrm{VLC},m}\right)^2}{B_{\mathrm{VLC},m} N_{0,\mathrm{VLC}}}, \tag{9}$$

where the value of $G_{\mathrm{RF},m}$ can be substituted by $G^{\mathrm{LOS}}_{\mathrm{RF},m}$ or $G^{\mathrm{NLOS}}_{\mathrm{RF},m}$ to obtain the corresponding $\gamma^{\mathrm{LOS}}_{\mathrm{RF},m}$ and $\gamma^{\mathrm{NLOS}}_{\mathrm{RF},m}$ for the LOS and NLOS channels. Also, the value of $\gamma_{\mathrm{VLC},m}$ is calculated by dividing the received electrical power in (4) by the noise power.

The achieved data rates by the mth MT exhibited by different networks are denoted by $R_{\mathrm{VLC},m}$ and $R_{\mathrm{RF},m}$, respectively. Using the multi-homing capability, the sum of the achievable data rates of the mth MT via the VLC and RF APs should not be less than the required data rate $R_{\min,m}$. The expected values of the achievable data rates from each AP are calculated over the probability mass function of LOS availability and are calculated as follows

$$R_{\mathrm{RF},m} = B_{\mathrm{RF},m}\left(\rho_{\mathrm{RF}} \log_2\left(1+\gamma^{\mathrm{LOS}}_{\mathrm{RF},m}\right) + (1-\rho_{\mathrm{RF}})\log_2\left(1+\gamma^{\mathrm{NLOS}}_{\mathrm{RF},m}\right)\right), \tag{10}$$

$$R_{\mathrm{VLC},m} = B_{\mathrm{VLC},m}\rho_{\mathrm{VLC}} \log_2\left(1+\gamma_{\mathrm{VLC},m}\right). \tag{11}$$

The total achieved data rate in the heterogeneous RF/VLC network is denoted by R_{T} and its value is calculated as follows

$$R_{\mathrm{T}} = \sum_{m \in \mathcal{M}} R_{\mathrm{VLC},m} + \sum_{m \in \mathcal{M}} R_{\mathrm{RF},m}. \tag{12}$$

The total consumed power for communication is denoted by P_{T} and its value is calculated as follows

$$P_{\mathrm{T}} = Q_{\mathrm{VLC}} + Q_{\mathrm{RF}} + \sum_{m \in \mathcal{M}} P_{\mathrm{RF},m}. \tag{13}$$

where the first term in (13) represents the consumed power for the VLC AP and is calculated using the fact that the transmission power is the optical power used for illumination by design, and hence only the fixed power consumption Q_{VLC} is accounted for as a power cost. The second and third terms represents the RF power consumption which accounts for both the processing and transmission powers.

We study the resource allocation problem to maximize the energy efficiency of the heterogeneous network over the assigned transmission powers and bandwidths to the MTs by the APs. The hybrid RF/VLC network energy efficiency is denoted by $\eta = R_{\mathrm{T}}/P_{\mathrm{T}}$. The user total achieved data rates, i.e., $R_{\mathrm{VLC},m} + R_{\mathrm{RF},m}$ are constrained by the minimum required data rates for the MTs. For each AP, the total transmission power consumption, i.e., $\sum_{m \in \mathcal{M}} P_{\mathrm{VLC},m}$ and $\sum_{m \in \mathcal{M}} P_{\mathrm{RF},m}$, is constrained by the maximum allowable transmission power. The total allocated bandwidth of each AP is constrained by the maximum allowable bandwidth as well. Given these constraints, the problem is formulated as follows

$$\begin{aligned}
&\max_{P_{\mathrm{VLC},m}, P_{\mathrm{RF},m}, B_{\mathrm{VLC},m}, B_{\mathrm{RF},m}} \quad \eta \\
\text{s.t.} \quad & R_{\mathrm{VLC},m} + R_{\mathrm{RF},m} \geq R_{\mathrm{min},m}, \qquad \forall m \in \mathcal{M}, \\
& \sum_{m \in \mathcal{M}} P_{\mathrm{VLC},m} \leq P_{\mathrm{VLC,max}}, \\
& \sum_{m \in \mathcal{M}} P_{\mathrm{RF},m} \leq P_{\mathrm{RF,max}}, \\
& \sum_{m \in \mathcal{M}} B_{\mathrm{VLC},m} \leq B_{\mathrm{VLC,max}}, \\
& \sum_{m \in \mathcal{M}} B_{\mathrm{RF},m} \leq B_{\mathrm{RF,max}}, \\
& P_{\mathrm{VLC},m}, P_{\mathrm{RF},m}, B_{\mathrm{VLC},m}, B_{\mathrm{RF},m} \geq 0, \qquad \forall m \in \mathcal{M}.
\end{aligned} \tag{14}$$

The problem (14) is referred to as concave-convex fractional program [34] since the numerator of the objective function is concave with respect to the decision variables and the denominator is affine. The concavity of the numerator of the objective function can be proven by calculating the Hessian matrix of R_{T} with respect to the optimization variables. We found that both the diagonal elements of the Hessian matrix are negative and the principal minors are 0.

5 Numerical Results

In the following, we assess the performance of the proposed resource allocation strategy. We compare the energy efficiency of the proposed mixed RF/VLC heterogeneous networks to two benchmark systems. We refer to the proposed system in the following results by 'RF-VLC'. We compare it to a system consisting of a single RF wireless network, which is denoted by 'RF-Only', and hence no multi-homing is performed, and we also compare it to a system comprising two RF APs over different frequency bands and which will be denoted by 'RF-RF', and hence multi-homing is achieved only over RF links. In the system with two RF APs, one of the RF systems is assigned a bandwidth equal to that of the VLC system to ensure a fair comparison.

In the following results, we set $R_{\text{min},m} = 2$ Mbps, $P_{\text{RF,max}} = 1$ W, $Q_{\text{RF,max}} = 6.7$ W, $Q_{\text{VLC,max}} = 4$ W, $N_{0,\text{RF}} = 3.89\text{x}10^{-21}$ W/Hz, $N_{0,\text{VLC}} = 10^{-21}$ W/Hz, $B_{\text{RF}} = 10$ MHz, $B_{\text{VLC}} = 20$ MHz, $k = 10$ W/A, $\rho = 0.8$ A/W, and $M = 4$. The VLC AP is located such that the MTs are uniformly-distributed and randomly located such that the distance to the AP is in the range between 1.5 and 2 m. The RF AP is located such that the MTs are randomly and uniformly-distributed located such the distance to the AP is in the range between 1 and 1.5 m. Note that the RF AP practically has a larger coverage region which is covered by multiple VLC APs. In this work, we consider the performance of the MTs which are in the common coverage region of a single VLC and a single RF APs. The VLC system maximum power is the product of the number of LEDs used at the VLC source by the maximum power driving each LED, and we set the number of LEDs to 38 with the maximum power to drive a LED to 300 mW. The value of the maximum driving power is set to generate around 900 lumens from the VLC source which is practically a suitable value for lighting. The values of the VLC and RF systems are obtained from [15] and [35], respectively.

In Fig. 3, we show the energy efficiency of the different systems against the number of MTs. The performance of the RF-VLC system is significantly better than the performance of the RF-only system because of the multi-homing capability of the MTs and the energy efficient nature of the VLC systems. Using multi-homing, MTs can have links with better channel conditions with at least one AP, leading to high achieved data rates and low power consumption. Also, the performance of the RF-VLC system is better than that of the RF-RF system because of the low cost of VLC AP power consumption compared to the RF communication systems. The power consumption in RF systems is the sum of the fixed and transmission powers, while in VLC systems, the power is due only to the fixed power component since no power is dedicated for transmission as its transmission power is already used for illumination.

In Fig. 4, we show the energy efficiency performance versus the fixed power consumption of the VLC AP to investigate the impact of any increased power consumption in the VLC network. The energy efficiency of RF-VLC system is equal to

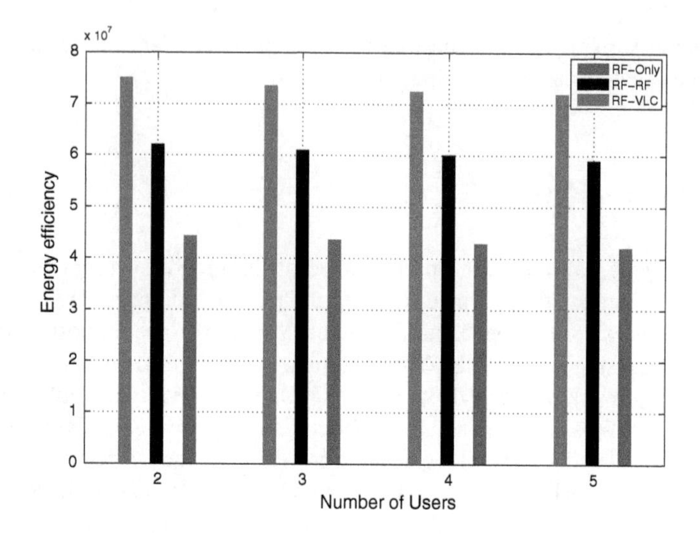

Fig. 3 Energy efficiency against the number of MTs

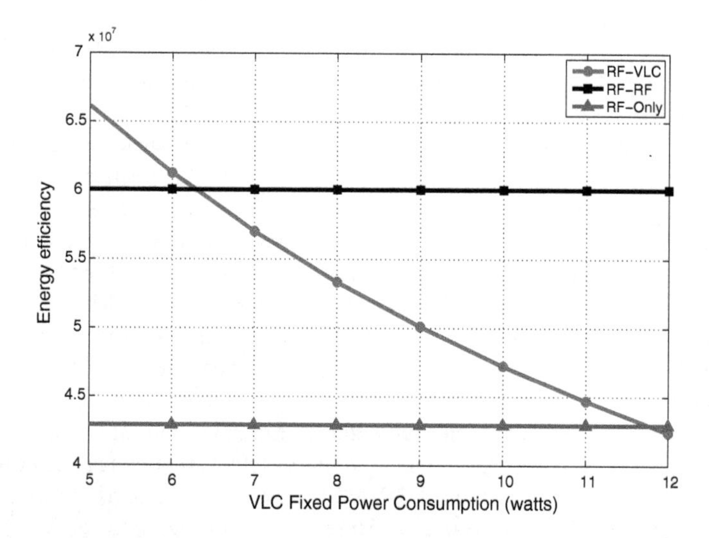

Fig. 4 Energy efficiency against the fixed power of the VLC system

that of the RF-RF system when the fixed power is 6 W which is nearly equal to the fixed power of an RF AP. As a result, the integration of a VLC system in a heterogeneous networking with RF communication will not be beneficial if the VLC AP fixed power is high compared to an equivalent RF AP.

We study in Fig. 5 the effect of the number of LEDs on the energy efficiency of the RF-VLC system. Increasing the number of LEDs allows higher transmission power for the VLC system which motivates the MTs to obtain most of their required data service from the VLC AP, reducing the transmission power consumption of RF

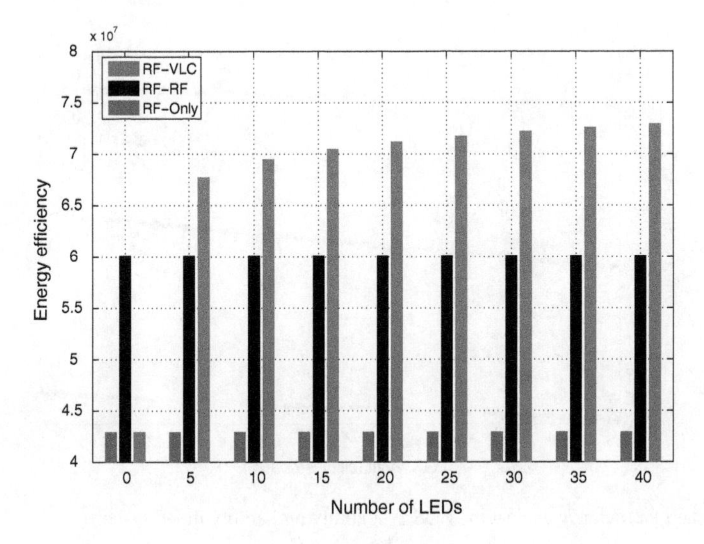

Fig. 5 Energy efficiency against the number of LEDs used by the VLC system

AP, and hence improving the overall energy efficiency. Also, the figure shows that introducing the VLC network even with a small number of LEDs (only 5 LEDs) enhances the energy efficiency significantly.

In Fig. 6, we consider the case in which the LOS availability probabilities for RF and VLC systems are equal. We show the energy efficiency versus the LOS availability probability. The performance for the RF-VLC system is better than the

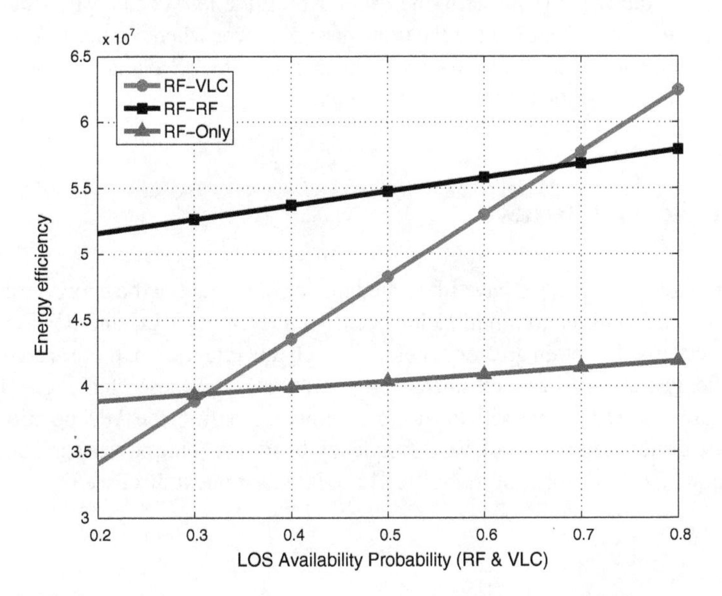

Fig. 6 Energy efficiency against the LOS availability probability in VLC and RF systems

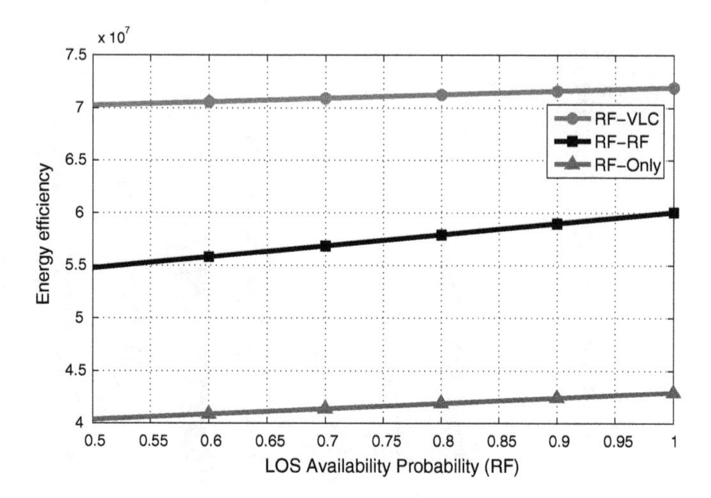

Fig. 7 Energy efficiency against the LOS availability probability in RF systems

benchmarks when the probability of LOS availability in the VLC system is higher than 0.7 because of the good energy efficiency properties of the proposed RF-VLC system. Also, the slope of the curve of the RF-VLC energy efficiency is higher than those of the benchmark systems because of the significance of the LOS availability to the VLC system compared with the RF systems.

Finally in Fig. 7, we discuss the effect of LOS availability probability in the RF system on the energy efficiency of the RF-VLC heterogeneous system when $\rho_{\text{VLC}} = 1$. In the RF-VLC system, the users exploit the less costly VLC energy for data transmission and exploit the RF transmission power when required. As a result, the enhancement of the performance with the increase of LOS availability probability presents a smaller slope than the RF benchmarks.

6 Challenging Issues

In the previous section, we have illustrated the improvement in the energy efficiency of an indoor downlink communication scenario integrating VLC and RF communication systems. To obtain greater benefits through this integration, there are a number of challenges that still remain unanswered. To overcome these challenges, further studies are required to deal with the users' spatial distribution, the APs' position planning, the unlicensed nature of VLC frequency band, the joint power and subcarrier allocation, and the impact of mobility. These issues are next discussed.

6.1 MT Spatial Distribution

High data rates in optical communication systems can be achieved when there is an unobstructed LOS between a transmitter and a receiver. The rate is reduced significantly in the absence of LOS. The obstruction of LOS can be due to the user lying outside the FOV of the VLC AP or due to existing objects such as furniture and walls. Therefore, the drop of the data rate can occur abruptly as soon as the LOS is obstructed, which is not the case for the RF networks where the signal quality typically degrades in a predictable continuous pattern. This specific feature of the VLC networks makes the performance sensitive to the probability of the LOS availability, which depends on the specific MT spatial distribution.

In particular, due to the abrupt change in the VLC signal quality, the probability of the LOS availability can vary significantly with any slight change in the location of the users. Thus, the exact knowledge of the spatial distribution of the MTs is a crucial parameter in assessing the performance of VLC systems. Using the same justification, the resource allocation problem in heterogenous networks integrating RF and VLC systems is highly affected by the spatial distribution of the MTs. Hence, allocating network resources in the RF and VLC networks assuming knowledge of the MTs spatial distribution can achieve better energy efficiency by, for example, usage of lower energy transmissions for the regions with a high density of MTs. As a result, examining the MTs' spatial distribution and the techniques to employ it in resource allocation problems plays an important role in the system design stage. Moreover in case that there are uncertainties about users' locations, there is a need to design a resource allocation strategy that is robust to such uncertainties. An example depicting the effect of user location on the performance of the cooperative VLC networks is discussed in [24], where it was shown that a small error in the user location information can degrade significantly the data rate.

Furthermore, VLC systems are characterized by non-uniform coverage areas [15]. Dark and light spots are identified within the VLC coverage region, based on the locations of the light sources and the illumination requirements. The achieved data rate at these dark spots is relatively low. By examining the spatial distribution of MTs and studying the expected required rates of the users located in the dark spots, more RF transmission power can be allocated to MTs in the dark spots while allocating more VLC transmission power to the light spots. An improved VLC coverage, and hence better reliability, can be obtained depending on the LED characteristics. Specifically, the field of view (FOV) of the used LEDs determines the coverage area of the light sources. Hence, increasing the FOV allows more coverage but introduces also more interference in the VLC network. Such interference can jeopardize the achieved energy efficiency. Thus, further investigation is required on the impact of users' spatial distribution on the resulting energy efficiency. Also, the resource allocation problem should consider the joint effects of VLC non-uniform coverage and the MTs spatial distribution.

6.2 APs Position Planning

As discussed, the MTs' spatial distribution affects the network planning for the VLC and RF systems. One of the major design issues is the APs' placement planning. The selection of the positions of the VLC and RF APs in both systems can considerably affect the system performance. While the RF AP placement is considered mainly by the communication requirements, the VLC AP is constrained by both the illumination and communication requirements [36].

In addition, the transmission powers in both VLC and RF communication systems are highly affected by the path loss, and hence finding the optimal placement of the APs can highly benefit the achieved energy efficiency. Further research is needed for the joint placement optimization of RF and VLC APs to improve the energy efficiency while maintaining the communication and illumination requirements.

6.3 Unlicensed VLC Spectrum

Although it is beneficial that the visible light spectrum provide access to several hundred Tera-Hertz of unlicensed spectrum, communication over the light spectrum can be exposed to different types of interference due to using any light source. The visible light spectrum is wide enough to allow high data rates and in indoor scenarios light does not penetrate walls and hence any portion of the visible light spectrum can be exploited inside some closed indoor spaces. Being unlicensed allows exploiting the spectrum for various services using different communication schemes. On the other hand, being unlicensed does not allow reserving certain portions of the spectrum for communication purposes only and hence any light source or light reflection is considered as an interference source in VLC networks. This problem can be addressed by employing coding approaches in VLC systems. However, this solution affects the energy consumption in VLC systems due to the extra processing requirements. By integrating both VLC and RF systems, interference mitigating solutions that exploit data transmissions over the two used spectra can improve energy efficiency, and it requires a further investigation.

6.4 Joint Power and Subcarrier Allocation

Orthogonal frequency division multiple access (OFDMA) is used in both VLC and RF communication systems to enable high data rates by mitigating the inter-symbol interference (ISI) effects. Consequently, OFDMA is a good signaling scheme for energy efficient VLC and RF communication networks. In this context, subcarrier and power allocation should be jointly optimized between the VLC and RF systems to enhance the system performance. The problem of jointly allocating the OFDMA subcarriers and power levels for different users has been discussed before in literature. The existing works mainly focus on homogeneous network scenarios. Further extensions are required for an integrated VLC and RF heterogeneous wireless system. In such a scenario, the main challenges are the required coordination between the VLC and RF systems, and the different channel characteristics in both systems. Furthermore, computational complexity should be addressed as the joint resource allocation is an NP-hard mixed integer programming problem [37].

6.5 Mobility of Users

VLC systems present channels with slow dynamics but their quality significantly depends on LOS availability between the transmitter and the receiver. Thus, the reliability of data transmission is affected heavily by the user mobility in case of indoor scenarios with many obstacles facing the light signals. The MTs mobility can cause frequent service dropping. On the other hand, RF channels are usually highly dynamic but do not require the LOS availability between the transmitter and the receiver. Consequently, one major challenge facing the resource allocation problem for designing an energy efficient heterogenous VLC and RF communication network is to deal with the user mobility impact such that there is no abrupt dropping in the users' perceived quality.

7 Conclusions and Future Research

The research in the area of energy efficient communications has been motivated by environmental and financial considerations. Integrating RF and VLC APs in heterogeneous wireless networking environment has shown promising improvements in the achieved energy efficiency. The multi-homing capability of the MTs in a heterogeneous network with VLC and RF APs allows users to benefit from the huge unlicensed bandwidth of the visible light spectrum and the low cost of the transmission power. It also allows for an improved VLC system reliability, as RF communications are employed in the absence of VLC LOS. Future research should deal with many unresolved challenging issues. These issues include studying the spatial distribution of the MTs and its impact on the different network parameters selection, and investigating the problem of joint power and subcarrier allocation.

Acknowledgments This publication was made possible by the NPRP award [NPRP 5 -980-2-411] from the Qatar National Research Fund (a member of The Qatar Foundation). The statements made herein are solely the responsibility of the author[s].

References

1. Mclaughlin, S., Grant, P.M., Thompson, J.S., Haas, H.: Techniques for improving cellular radio base station energy efficiency. IEEE Wirel. Commun. **18**(5), 10–17 (2011)
2. Ismail, M., Zhuang, W.: Network cooperation for energy saving in green radio communications. IEEE Wirel. Commun. **18**(5), 76–81 (2011)
3. Kavehrad, M.: Sustainable energy-efficient wireless applications using light. IEEE Commun. Mag. **48**, 66–73 (2010)
4. U.S. Dept. of Energy: Solid-state lighting portfolio energy savings potential of solid-state lighting in general illumination applications (2008). http://www1.eere.energy.gov/buildings/ssl/
5. Solid-state lighting research and development: multi year program plan (2013). http://apps1.eere.energy.gov/buildings/publications/pdfs/ssl/ssl_mypp2013_web.pdf
6. Lighting the way: perspectives on the global lighting market, 2nd edn. McKinsey & Company (2012). http://www.mckinsey.com
7. Tanaka, Y., Haruyama, S., Nakagawa, M.: Wireless optical transmissions with white colored LED for wireless home links. In: IEEE International Symposium on Personal. Indoor and Mobile Radio Communications (PIMRC), vol. 2, pp. 1325–1329 (2000)
8. Roufs, J.A.J., Blommaizt, F.J.J.: Temporal impulse and step responses of the human eye obtained psychophysically by means of a drift-correction perturbation technique. Vis. Res. **21**(8), 1203–1221 (1981)
9. Grubor, J., Gaete Jamett, O.C., Walewski, J.W., Randel, S., Langer, K.D.: High-speed wireless indoor communication via visible light. ITG Fachbericht **198**, 203–208 (2007)
10. Proakis, J.G.: Digital communication, 4th edn. McGraw-Hill, New York (2000)
11. Jovicic, A., Junyi, L., Richardson, T.: Visible light communication: opportunities, challenges and the path to market. IEEE Commun. Mag. **51**(12), 26–32 (2013)
12. The home Gigabit access prjoct (OMEGA). www.ict-omega.eu
13. Connolly, P., Bonte, D.: Indoor location in retail: where is the money?. ABI Research Report Mar. (2013). http://www.abiresearch.com/research/product/1013925-indoor-location in-retail-where-is-themon/
14. George, J.J., Mustafa, M.H., Osman, N.M., Ahmed, N.H., Hamed, D.M.: A survey on visible light communication. Int. J. Eng. Comput. Sci. **3**(2), 3805–3808 (2014)
15. Stefan, I., Burchardt, H., Haas, H.: Area spectral efficiency performance comparison between VLC and RF femto cell networks. In: 2013 IEEE International Conference on Communications (ICC), pp. 3825–3829 (2013)
16. Wang, C.-X., Haider, F., Gao, X., You, X.-H., Yang, Y., Yuan, D., Aggoune, H., Haas, H., Fletcher, S., Hepsaydir, E.: Cellular architecture and key technologies for 5G wireless communication networks. IEEE Commun. Mag. **52**(2), 122–130 (2014)
17. Wu, S., Wang, H., Youn, C.-H.: Visible light communications for 5G wireless networking systems: from fixed to mobile communications. IEEE Netw. **28**(6), 41–45 (2014)
18. Din, I., Kim, H.: Energy-efficient brightness control and data transmission for visible light communication. IEEE Photonics Technol. Lett. **26**(8), 781–784 (2014)
19. Kim, S., Won, J.: Simultaneous reception of visible light communication and optical energy using a solar cell receiver. In: 2013 International Conference on ICT Convergence (ICTC), pp. 896–897 (2013)
20. Miki, M., Asayama, E., Hiroyasu, T.: Intelligent lighting system using visible-light communication technology. In: 2006 IEEE Conference on Cybernetics and Intelligent Systems, pp. 1–6 (2006)
21. Zou, Y., Zhu, J., Zhang, R.: Exploiting network cooperation in green wireless communication. IEEE Trans. Commun. **61**(3), 999–1010 (2013)
22. Lim, G., Cimini, L.G.: Energy-efficient cooperative relaying in heterogeneous radio access networks. IEEE Wirel. Commun. Lett. **1**(5), 476–479 (2012)
23. Ismail, M., Zhuang, W.: Green radio communications in a heterogeneous wireless medium. IEEE Wirel. Commun. **21**(3), 128–135 (2014)

24. Kashef, M., Abdallah, M., Qaraqe, K., Haas, H., Uysal, M.: On the Benefits of Cooperation via Power Control in OFDM-Based Visible Light Communication Systems. In: IEEE 25th International Symposium on Personal. Indoor and Mobile Radio Communications—(PIMRC), Washington DC (2014)
25. Hou, J., OBrien, D.: Vertical handover-decision-making algorithm using fuzzy logic for the integrated radio-and OW system. IEEE Trans. Wirel. Commun. **5**(1), pp.176–185 (2006)
26. Vegni, A., Little, T.: Handover in VLC systems with cooperating mobile devices. In: International Conference in Computing, Networking and Communications (ICNC), Hawaii (2012)
27. Chowdhury, H., Ashraf, I., Katz, M.: Energy-efficient connectivity in hybrid radio-optical wireless systems. In: 10th International Symposium on Wireless Communication Systems (ISWCS), pp. 1–4. Ilmenau (2012)
28. Chowdhury, H., Katz, M.: Cooperative multihop connectivity performance in visible light communications. In: Wireless Days 2013, pp. 1–5. Valencia (2013)
29. Tsonev, D., Videv, S., Haas, H.: Light fidelity (LI-FI): towards all-optical networking. SPIE OPTO, Int. Soc. Opt. Photonics, 702–900 (2013)
30. Cavalcante, R.L.G., Stanczak, S., Schubert, M., Eisenblaetter, A., Tuerke, U.: Toward energy-efficient 5G wireless communications technologies: tools for decoupling the scaling of networks from the growth of operating power. IEEE Signal Process. Mag. **31**(6), 24–34 (2014)
31. Osseiran, A., Boccardi, F., Braun, V., Kusume, K., Marsch, P., Maternia, M., Queseth, O., Schellmann, M., Schotten, H., Taoka, H., Tullberg, H., Uusitalo, M.A., Timus, B., Fallgren, M.: Scenarios for the 5G mobile and wireless communications: the vision of the METIS project. IEEE Commun. Mag. **52**(5), 26–35 (2014)
32. Cheng, C., Tsonev, D., Haas, H.: Joint transmission in indoor visible light communication downlink cellular networks. In: 2013 IEEE Globecom Workshops (GC Wkshps), pp.1127–1132 (2013)
33. Komine, T.: Visible light wireless communications and its fundamental study, Ph.D. dissertation, Keio University (2005)
34. Frenk, J.B.G., Schaible, S.: Fractional programming. ERIM Rep. Ser. Res. Manage. (2004)
35. Ashraf, I., Boccardi, F., Ho, L.: Sleep mode techniques for small cell deployments. IEEE Commun. Mag. **49**(8), 72–79 (2011)
36. Stefan, I., Haas, H.: Analysis of optimal placement of LED arrays for visible light communication. 2013 IEEE 77th Vehicular Technology Conference (VTC Spring), pp. 1–5 (2013)
37. Kivanc, D., Guoqing, L., Hui, L.: Computationally efficient bandwidth allocation and power control for OFDMA. IEEE Trans. Wirel. Commun. **2**(6), 1150–1158 (2003)

A VANET Based Electric Vehicle Energy Management Information System

Muhammad Awais Javed, Jamil Yusuf Khan and Duy Trong Ngo

Abstract Electric vehicles (EVs) are an integral part of the future transportation systems due to enhanced fuel and energy conversion efficiency. The success of electric vehicle technology requires an efficient charging management system that ensures their timely fueling. To support such a service, vehicular ad hoc networks can be used to implement an information system for EV energy management. For this purpose, it is essential for the electric vehicles to reliably and timely exchange information with the information/control servers using infrastructure nodes (INs) deployed at different geographical locations. As the INs may be located farther away from the vehicles, robust multi-hop packet transmissions are required. In this chapter, we present the design considerations for an information system for EV energy management. We propose a vehicle-to-infrastructure and infrastructure-to-vehicle (V2I-I2V) information transmission system that efficiently delivers packets for EV energy management services by mitigating the broadcast storm and hidden node problems. Moreover, the proposed system provides a signaling mechanism to select the shortest path for the downlink I2V transmissions, a challenging task due to the mobile nature of vehicles. Simulation results show that the developed system offers a low delay and reduced number of packet transmissions for different vehicle densities and mobility conditions.

M.A. Javed (✉)
Comsats Intitute of Information Technology, Chak Shazad, Islamabad, Pakistan
e-mail: awais.javed@comsats.edu.pk

J.Y. Khan · D.T. Ngo
The University of Newcastle, Callaghan, Newcastle, Australia
e-mail: Jamil.Khan@newcastle.edu.au

D.T. Ngo
e-mail: Duy.Ngo@newcastle.edu.au

© Springer International Publishing Switzerland 2016

M.Z. Shakir et al. (eds.), *Energy Management in Wireless Cellular and Ad-hoc Networks*, Studies in Systems, Decision and Control 50,
DOI 10.1007/978-3-319-27568-0_14

1 Introduction

Electric Vehicles (EVs) are envisaged to dominate the transportation systems in the near future. Reduction of dependence on the fossil fuels, less emission of carbon dioxide and green house gases, higher efficiency of electric vehicle engines and the possibility to charge vehicles from renewable energy sources are some of the key benefits of electric vehicle systems [1, 2]. By adopting electric vehicles for traveling, the consumption of conventional energy resources can be reduced. This could increase the life time of other energy sources and help mitigate any possible future energy crisis. Vehicles are also a major source of green house gas emissions in the world. Using electric vehicles will be a big step towards a cleaner environment [3]. Studies have also shown that the electric vehicle engine is more efficient in terms of fuel efficiency [4]. Also, electric vehicles may be used in conjunction with many renewable sources, hence solving the issue of dependency on one fuel source [5].

One of the key issues related to EV deployment is how to design its charging/energy management systems. Since an EV could take longer time to recharge compared to the refueling of a conventional vehicle, a city wide or an area based charging infrastructure management system is required. It is therefore necessary to develop an information system that can interact with the mobile EV fleet and the spatially distributed charging stations with different occupancy levels. For example, in future, we may see charging station providers supply energy using multiple sources where price and availability of energy could be time and location dependent. EVs could also be used as an energy storage device when peak/average generation capacity exceeds the consumption load in certain areas. The primary challenge in enabling an electric vehicle information system (EVIS) is to efficiently and timely provide the vehicles with the desired service information such as the nearest charging station location, a lower cost station, quicker charging availability, etc.

To enable these energy management services, it is necessary to develop an electric vehicle information system (EVIS) that can interact with the mobile EV fleet and the spatially distributed information/control servers. Such a system should be able to support the data exchange from vehicles to infrastructure (V2I) on the uplink channel and from infrastructure to vehicles (I2V) on the downlink channel. In a VANET, infrastructure nodes (INs) in the form of road side units are deployed at strategic locations as shown in Fig. 1. Using an IN as a relay, a vehicle who intends to query an online energy management service sends request/query messages to the information server. Since the IN may be located outside the transmission range of a vehicle, multi-hop wireless transmissions are required for the information from the vehicle to reach the IN. Using the backbone network, the IN then relays request messages from the vehicle to the central information server (CIS). The CIS processes the request and sends a response message back to the vehicle via the IN.

The dissemination of service messages in an EVIS faces two technical challenges. First, the multi-hop transmissions in a VANET may suffer from the broadcast storm and the hidden node collisions [6, 7]. In this situation, a large number of packet losses may happen. Second, the vehicle may have moved closer to another IN during the time

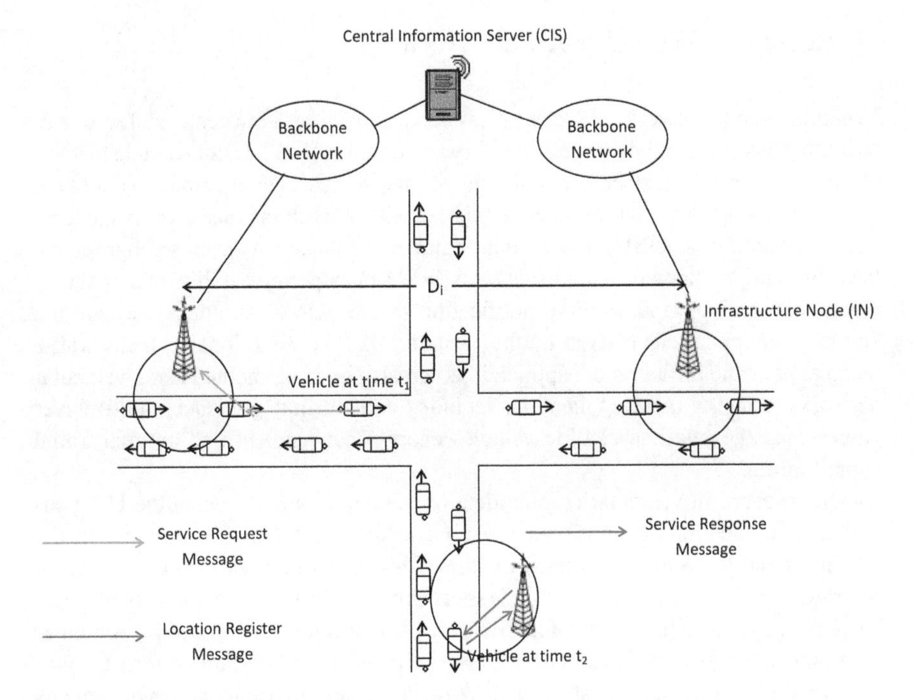

Fig. 1 A vehicle information transmission system

it waits for the response message from the CIS. Keeping track of the vehicle mobility on the downlink is thus required in order to select the shortest communication path.

In this chapter, we use simulations to analyze the performance of a proposed EVIS for energy management services in an urban road traffic topology. Here we propose to use the TSM protocol [6] to efficiently transmit the service messages on both uplink and downlink. We also propose a location register signaling mechanism between vehicles and INs. The purpose is to keep track of vehicle mobility and to select the shortest communication path for the downlink response messages. Using an OPNET-based simulation model, we evaluate the delay and the required number of transmissions for both downlink (request) and uplink (response) messages. Our simulation results confirm a significant improvement in the downlink delay with the proposed location register signaling mechanism.

The rest of this chapter is organized as follows: Sect. 2 presents a review of the current standards in vehicular ad hoc networks. Section 3 briefly explains various techniques for information dissemination in a VANET. Section 4 explains the communication architecture for the EVIS and the proposed packet transmission technique for the downlink and uplink messages. Section 5 evaluates the performance of the proposed technique using an OPNET simulation model in an urban road traffic. Finally, Sect. 6 concludes the chapter.

2 Review of Vehicular Ad hoc Network

Vehicular ad hoc networks utilize wireless communications between vehicles to realize numerous applications for traffic safety and comfort [8, 9]. Using vehicle to vehicle and vehicle to infrastructure communications, vehicles get information about the neighborhood traffic. Each vehicle is equipped with a wireless transceiver in the form of an on-board unit (OBU). Infrastructure nodes are placed at various geographical locations and are known as road side units (RSU). Cooperative collision avoidance, emergency braking and warning notification in the case of an emergency are the key safety applications offered by this system [10, 11]. Additionally, many traffic management and infotainment applications can also be implemented using vehicular networks. Some of these applications include efficient route guidance, multi-player games over the Internet, electric vehicles energy management and optimal traffic signal timing.

The research in vehicular communication gained popularity when the U.S Federal Communications Commission (FCC) allocated 75 MHz of bandwidth in the 5.9 GHz band for vehicular communication. This bandwidth was named Dedicated Short Range Communications (DSRC) spectrum [12]. Due to the popularity of Wireless LAN, an amended version of IEEE 802.11 was approved to be used for vehicular communication [13]. This launched a series of projects and research efforts to standardize vehicular communication. The Wireless Access in Vehicular Environments (WAVE) and the European Telecommunications Standards Institute (ETSI) laid down a set of proposals for vehicular networks [14, 15]. These standards define the application and communication requirements, protocol architecture and services provided by the vehicular communications.

The WAVE and ETSI standards are shown in Figs. 2 and 3 respectively. Both the standards allocate spectrum for vehicular applications in the 5.9 GHz range. While the 75 MHz allocated in the WAVE standard is divided into 7 channels, ETSI divides the 50 MHz spectrum into 5 channels. One of the channels in both standards is reserved for safety applications whereas the remaining channels are used for service applications.

In the WAVE standard, the MAC and PHY layer functionalities are provided by the IEEE 802.11p standard. The physical layer is similar to that of the IEEE 802.11 standard with the channel bandwidth reduced by half and hence, all timing parameters are doubled [16]. This is done to provide better protection against increased fading associated with the vehicular environment. The MAC layer is based on the IEEE 802.11e standard that uses CSMA/CA protocol to coordinate multiple access.

Since the WAVE standard proposes the use of multiple channels (control and service), the IEEE 1609.4 provides channel coordination and channel routing mechanism [17]. The Logic Link Layer (LLC) forwards the packet from the transport/network layer to the appropriate channel. At the transport/network layer, the TCP/UDP is used along with IP for non-safety messages. On the other hand, safety messages employ the WAVE short message protocol (WSMP) to transmit priority

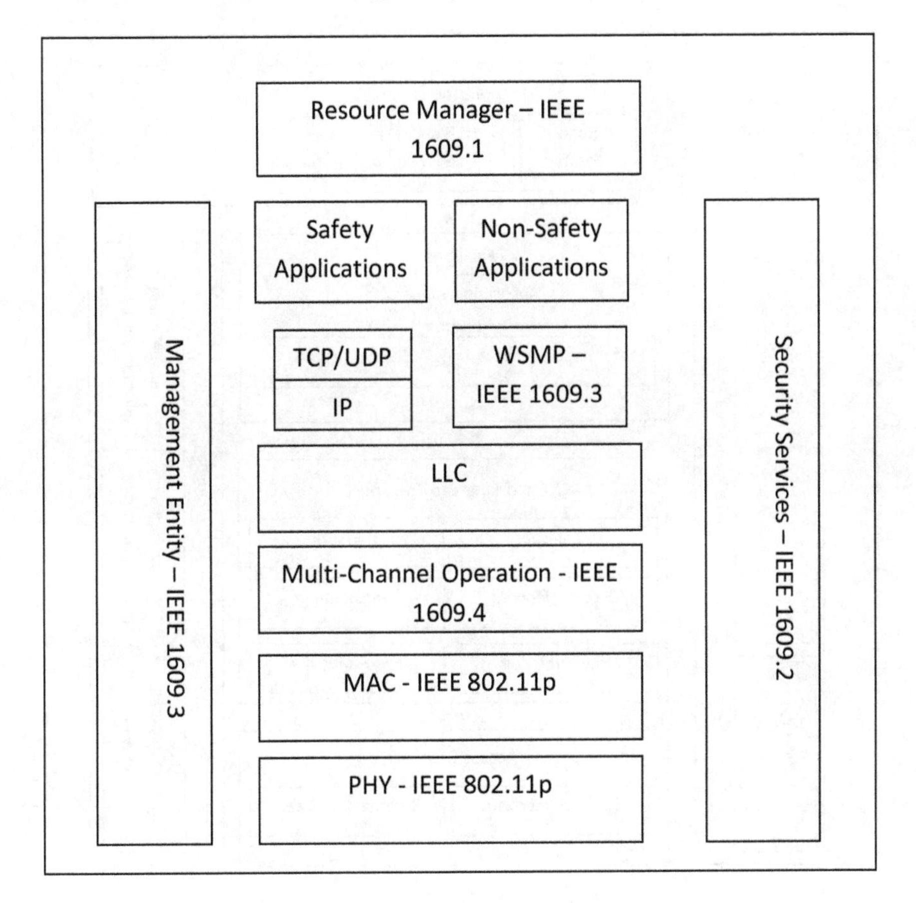

Fig. 2 WAVE standard

messages with a low overhead. The resource manager, security services and management services are defined in IEEE 1609.1, IEEE 1609.2 and IEEE 1609.3 standards respectively [18–20].

In the ETSI standard, the MAC and PHY layer incorporates different ad hoc and infrastructure based wireless technologies [21, 22]. For example, it allows the use of IEEE 802.11p, Wi-Fi, 3G and LTE. Such an integrated communication solution enables a large number of vehicular applications. Another difference in the physical layer of the ETSI from the WAVE standard is the use of different power spectral density values in accordance with the European regulations. At the Network and Transport layers, various protocols could be used such as TCP/UDP, IP and geographical routing based on the application requirements. The Facilities layer manages various tasks that are needed by several ITS applications. Finally, the ETSI standard also provides mechanism for management and security.

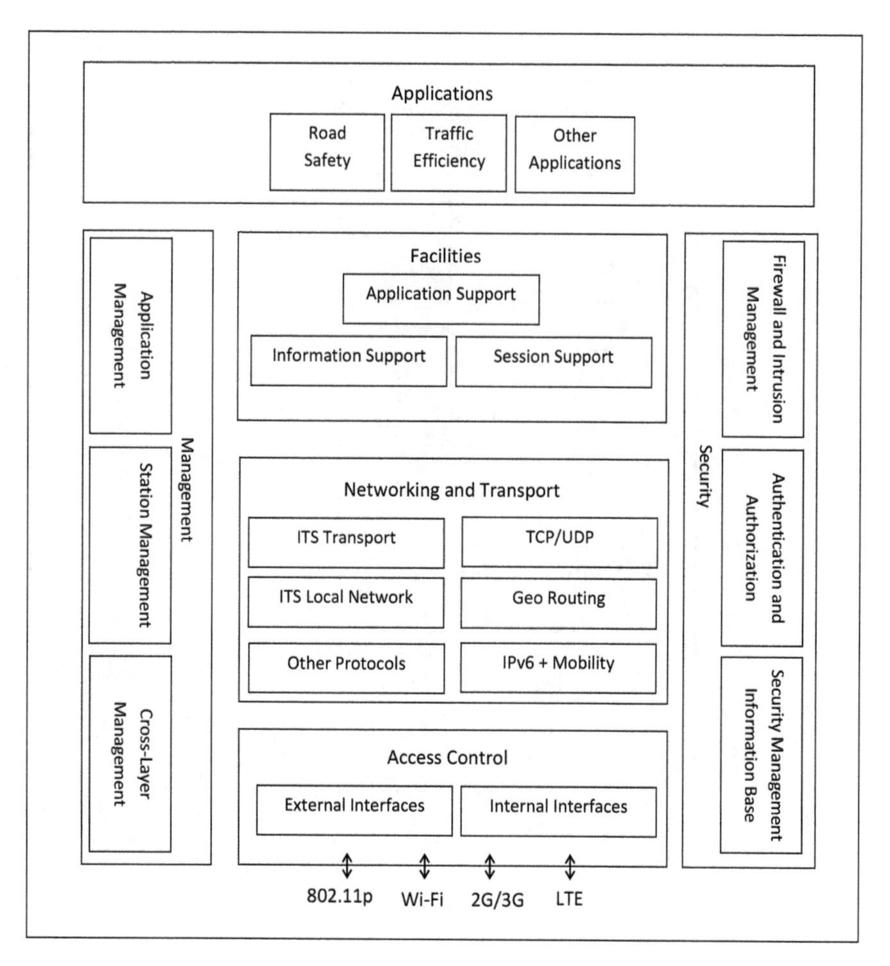

Fig. 3 ETSI standard

The WAVE standard defines a single type of safety message known as the basic safety message (BSM). This is a periodic message generated at a frequency of 1–10 Hz and used for cooperative awareness [23]. On the other hand, the ETSI standard uses two types of safety messages known as the cooperative awareness message (CAM) and the Decentralized Environmental Notification Message (DENM) [15]. While CAM is a periodic safety message similar to the BSM, DENM is an event-driven message generated in case of an emergency. The latter is transmitted to certain vehicles in a geographical area that could require the emergency information. The CAM or BSM are single-hop messages whereas DENM could use multiple hops to inform the relevant vehicles about the emergency.

3 Protocols for Inter-vehicle Communications

Applications supported by VANETs rely on the efficient dissemination of multiple traffic types to support various applications. For applications such as cooperative awareness, periodic single-hop safety messages are transmitted on the control channel. Other applications including warning notification and electric vehicle information system requires multi-hop transmissions on both control and service channels. The protocols for inter-vehicle communications in literature can be grouped into two broad categories, i.e., single-hop and multi-hop as shown in Fig. 4.

3.1 Single-Hop Protocols

Single-hop protocols efficiently transmit periodic safety message that are needed for cooperative awareness. Due to the stringent packet transmission requirements of safety messages, collisions due to hidden nodes severely degrade the packet reception rate [7]. To overcome this problem, single-hop protocols use different medium access techniques such as Carrier Sense Multiple Access with Collision Avoidance (CSMA/CA), Time Division Multiple Access (TDMA), Space Division Multiple Access (SDMA), Busy Tone Multiple Access (BTMA) and Code Division Multiple Access (CDMA). The strengths and weaknesses of each of these techniques are presented in Table 1.

The CSMA/CA protocol is the default MAC scheme in the IEEE 802.11p standard. It is popular due to its simplicity and multi-service capabilities. This technique has been shown to maintain a high packet success rate in dense traffic. However, it does not provide guaranteed channel access [15]. TDMA overcomes this problem of guaranteed channel access, hence reducing the idle time. However, distributed time slot allocation in a mobile ad hoc network is a challenging task. SDMA techniques

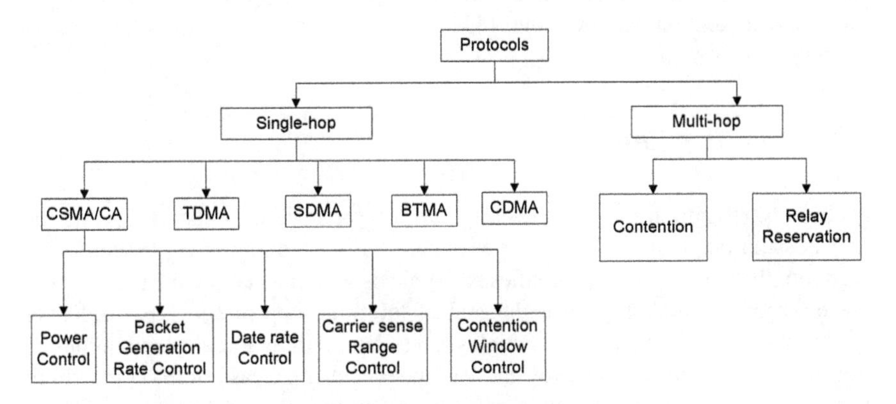

Fig. 4 Protocols for inter-vehicle communications

Table 1 Strengths and weaknesses of medium access techniques for single-hop communications in VANET

MAC Technique	Strengths	Weaknesses
CSMA/CA	Simplicity Distributed communications Multi-service capabilities Default IEEE 802.11p MAC High packet success rate in dense traffic	No guaranteed channel access
TDMA	Guaranteed channel access Reduced idle time	Distributed time slot allocation Reduced reception rate in dense networks
SDMA	Reduced hidden node collisions	Bandwidth underutilization
BTMA	Reduced hidden node collisions Prioritized messages	Separate channel for busy tone Expensive implementation
CDMA	Robust against jamming Robust against interference	Distributed code allocation Complex receiver

use geographical location to allocate time slots, however bandwidth underutilization is the main disadvantage [7]. BTMA uses busy tone mechanism to reduce the hidden node collisions, however its down side is the requirement of a separate busy tone channel. CDMA is a robust technique against jamming and interference, yet distributed code allocation and complex receive design makes it impractical to be used in a vehicular network.

Based on CSMA/CA, many techniques exist in literature that adapt the transmission parameters to improve the quality of service for different vehicular applications. These techniques include transmit power, packet generate rate, data rate, carrier sense range and contention window control [24–38].

3.2 *Multi-hop Protocols*

Multi-hop transmission techniques are used to propagate information beyond the transmission range in VANETs. For example, in a warning message dissemination scenario, the warning vehicle notifies the vehicles within a geographical area about the emergency situation. The main goal of the multi-hop protocols is to deliver the emergency information with a high success rate, minimum delay and minimum transmission overhead. However, hidden nodes and broadcast storm could result in a large number of packet collisions, causing an increased delay and number of transmissions.

To overcome the above problems, several multi-hop protocols have been proposed. They can be grouped into two categories, namely contention-based and relay reservation-based. The techniques proposed in this category use a contention mechanism to disseminate warning messages based on the distance between the receiver and the source node. Each vehicle that receives the warning message calculates a wait time or a probability of transmission which is inversely proportional to its distance from the source vehicle [39]. In this way, vehicles that are at a longer distance from the source vehicle are selected as the next relay vehicles and they transmit the warning message. All other vehicles that receive the duplicate warning message cancel their intention to rebroadcast. Distributed Vehicular broadcast (DV-CAST) [40], Contention Based Forwarding (CBF) [41] and Opportunistic Broadcast [42] are other popular protocols that use contention-based relay selection mechanism. In the second category of multi-hop protocols, the protocols use the relay reservation mechanism with the help of control packets to suppress the broadcast storm. Urban multi-hop (UMB) [43], Smart Broadcast (SB) [44] and Time-slotted Multi-hop (TSM) [6] are some protocols proposed in this category.

Related to electric vehicle information systems, [45] proposes a smart management system for electric vehicle recharge. The proposed scheme aims to optimize the use of distributed energy resources and control EV charging. Using a combination of IEEE 802.11p and LTE, simulation is carried for two different charging station assignments to evaluate the service delay. The work of [46] presents an algorithm for optimized electric vehicle charging route selection that improves the energy utilization and travel costs. Similarly, [2, 47] also proposed optimized charging route calculation algorithms. Most of the work in the literature is directed towards finding parameters for an optimal EV service. However, to the best of our knowledge, the performance of electric vehicle information system is not evaluated.

4 Electric Vehicle Information Systems

In this section, an Electric Vehicle Information System (EVIS) architecture to support energy management service is presented. As shown in Fig. 1, electric vehicles (EVs) equipped with a wireless transceiver (OBU) are traveling on a two-way road. Separated by a certain distance D_i, infrastructure nodes (INs) in the form of road side units are placed. INs are connected with a central information server (CIS) using data links of a communication technology such as a fixed broadband network or a high data rate wireless link. In this architecture, EVs and INs are connected via VANET communication links. It is assumed that all EVs are equipped with a DGPS unit that supplies the position information to INs. Also, INs are aware of the position of other INs located in the neighborhood area. The description of each node and its operation in the proposed EVIS is detailed as follows.

4.0.1 Electric Vehicles

While traveling on the road, Electric vehicles may require a number of energy management services such as location, the number of customers and the optimal choice of charging station, etc. To request a particular service, an EV contacts the CIS by sending an event-driven service request message using the nearest IN, as shown in Fig. 1. The purpose of the service request message is to ask the CIS to provide EV with the information about that service. The EV to CIS communication is supported by an IN using a multi-hop V2I link and a fixed cellular hop.

4.0.2 Infrastructure Nodes

Infrastructure nodes are in continuous communication with the CIS which updates it about different available EV services such as the location of different charging stations in the vicinity, the number of customers in each of the station and the special discounts offered. To update EVs with the information about different services, IN could periodically broadcast service information on the service channel (i.e., one of the many available SCHs available in VANETs).

Another function of the IN is to respond to a service request message by an EV. Upon receiving a service request, an IN forwards the request to the CIS and the corresponding reply is sent back to the EV via an IN. Due to vehicle mobility, an EV may move out of the coverage range of the IN when the reply is returned. Based on the reply message delay, the CIS may decide to return the reply to a number of neighboring INs who can simulcast the message.

4.0.3 Central Information Server

A Central information server is connected with all available EV service providers using a backbone network, receiving live updated information about different services. This information is periodically transmitted to each IN. For example, for an EV charging station application, the CIS finds the optimal charging station location when an EV charging request message is received through the IN. Based on desired selection criteria and an optimization algorithm, the CIS informs the IN of the selected station.

4.1 Service Messages in an EVIS

4.1.1 Request/Response Messages

Request/Response messages are exchanged between the vehicles and the CIS, using an IN as a relay node. On the uplink, vehicles transmit the service request messages to acquire information about a particular service from the CIS. After the CIS processes the request, it transmits the service response message back to the vehicles via an IN.

4.1.2 Location Register Messages

These messages are used to update the CIS about the position of the vehicle who initiates the request message. Depending on the vehicle speed and response time of the CIS, the vehicle may have moved a significant distance while waiting for the response message, as shown in Fig. 1. Consequently, the information of the closest IN is outdated at the time of response message transmission. Since the mobility of a vehicles depends on a number of factors such as road types and traffic conditions, it is difficult to accurately predict its position at the time when the response message is generated.

To overcome this problem, we propose that the vehicles who initiate a service request transmit location register messages. Once a vehicle moves a certain distance D_m, a single location register message is transmitted to the CIS using the nearest IN. When the response message is ready, the CIS uses the latest location register message by that vehicle to select the nearest IN. This proposal improves the response time because the shortest communication path on the downlink is utilized.

4.2 Segment Leader Selection

To improve the efficiency of multi-hop service messages, we propose to divide the urban road into fixed size road segments and to select a segment leader per segment as a potential relay node. By allowing only the segment leaders to forward the service messages, the broadcast storm is prevented and the number of transmissions is reduced. Using the periodic safety messages, we choose the vehicle with a longest remaining time in a given segment as the initial segment leader. Before the segment leader leaves the segment, it appoints the next segment leader—the vehicle with the current longest remaining time in that segment. A detailed description of the segment leader selection procedure and its implementation can be found in our previous work [6].

4.3 Multi-hop Time-Slot Reservation Mechanism

Since there are multiple vehicles transmitting different types of service messages on the service channel, hidden node collisions may be present and thus significantly degrade the quality of communication. To reduce such collisions, we propose to use the time slot reservation mechanism for the service messages. The proposed time slot structure is depicted in Fig. 5. Each time interval T is divided into two time slots T_s and T_r. While T_s is reserved for the request/response messages, T_r is used for the location register messages. Each of these two time slots is further divided into a

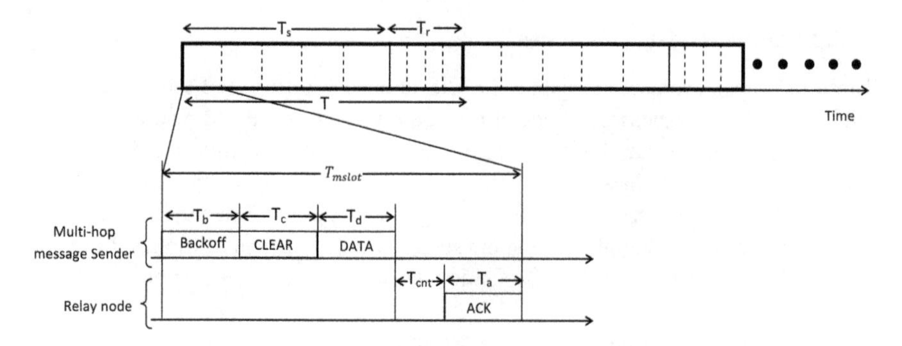

Fig. 5 Structure of a time slot in the proposed multi-hop broadcast protocol

number of multi-hop time slots whose structure is shown in Fig. 5. As the sizes of the request/response and location register messages are different, so are the sizes of the multi-hop time slots in T_s and T_r.

The proposed time-slot reservation mechanism involves following phases:

4.3.1 BACKOFF

At the start of a multi-hop time slot, each vehicle who intends to send a service message takes a random BACKOFF. Then, the vehicle with the smallest backoff value transmits a CLEAR message in the next phase to reserve the multi-hop time slot.

4.3.2 CLEAR

In this phase, a CLEAR packet is sent at a transmission range of R_t by the vehicle whose backoff timer expires first. The CLEAR packet reserves the remaining of the multi-hop time slot for the vehicle by letting other vehicles know of an upcoming service message transmission. As a result, other vehicles suspend their service messages until the start of the next multi-hop time slot.

4.3.3 DATA

After the CLEAR packet, the DATA (i.e., the service message) is sent at a half transmission range of $R_t/2$. The purpose is to notify the upcoming transmission to all hidden nodes located within two transmission hops from the service message sender. The DATA can be either a request/response message or a location register message. Depending on the DATA size, the size of multi-hop time slot can vary.

4.3.4 CONTENTION

The CONTENTION phase selects the segment leader farthest from the service message sender as the next relay node. This is implemented using distance-based wait time approach. All segment leaders who receive the service message initiate a wait time that is inversely proportional to the distance between the service message sender and receiver. The farthest segment leader has the shortest wait time and hence acts as a relay vehicle.

4.3.5 ACK

The relay vehicle selected from the last phase (i.e., the farthest segment leader) transmits an ACK packet. The purpose is to request other potential relay vehicles to abandon their intention of transmitting a service message.

5 Performance Evaluation of Electric Vehicle Information System

In this section, performance of the proposed time-slotted protocol for EVIS is evaluated for urban scenarios. We develop a network simulation model using OPNET Modeler 16.0. We consider a bi-directional urban road traffic scenario with a total size of 3.6 km × 2.4 km, and develop a model in MATLAB to simulate the traffic mobility. We separate X INs by a distance D_i within the road area. The number of transmitting vehicles that send a query message to CIS is varied between 1 and 20. The CIS response time is taken as 60 s.

The 40-byte location register message is sent by vehicles who initiate a service request message once they have moved a distance of 50 m. The background traffic is the data traffic required for other periodic services and it is sent by every vehicle to the nearest IN at the rate of 1 Hz. To represent a medium fading intensity, we assume Nakagami-m fading with $m = 3$. The practical parameters used in our simulations are listed in Table 2. Unless stated otherwise, we set vehicle density $\rho = 180$ vehicles/km^2 and allow the vehicle speeds to be in the range of 40–70 km/h. The proposed TSM protocol is compared with two other well known multi-hop techniques, namely, DV-CAST [40] and SB [44]. As mentioned in Sect. 3, DV-CAST uses contention based multi-hop technique where as the SB protocol belongs to the relay reservation category. These two protocols are selected to compare TSM protocol with a technique in each category of multi-hop protocols.

Figure 6 plots the average number of uplink (service request) transmissions as the number of transmitting vehicles (i.e., vehicles who send a request message to the CIS) varies. As seen from the figure, the required number of transmissions increases as more vehicles transmit. The average number of transmissions for the uplink messages are 20, 27, and 31 when there are five transmitting vehicles for the proposed TSM, SB and DV-CAST protocols, respectively. This figure increases to 104, 124 and 131, respectively, as the number of transmitting vehicles is 20.

In Fig. 7, we display the uplink delay for the three protocols under consideration. It is clear from the figure that the TSM protocol exhibits the lowest delay among the three schemes. Specifically, when the number of transmitting vehicles is 20, the uplink delay of TSM, SB and DV-CAST protocols is 63 ms, 74 ms and 76 ms, respectively. This improvement is due to the reduced number of potential relay vehicles by the TSM protocol with the help of the segment leader mechanism. In addition, the time slot reservation mechanism ensures an interference-free transmission of the service messages. As a result, the TSM protocol requires the smallest number of transmissions for the uplink messages (Fig. 8).

Table 2 Simulation parameters

Parameter		Value
Urban road	Road area	3.6 km × 2.4 km
	No. lanes	2 (one per direction)
Vehicle	Density	90–180 vehicles/km^2
	Speed	10–40 and 40–70 km/h
Service message	Size	500 bytes
	Data rate	6 Mbps
	Transmission range	300 m
	No. of transmitting vehicles	1–20
Background traffic	Size	500 bytes
	Data rate	6 Mbps
	Transmission range	300 m
	Generation freq.	1 Hz
Location register	Size	40 bytes
	Transmission range	300 m
message	Generation frequency	1 every 50 m
IN and CIS	No. of INs	3
	D_i	1km
	CIS response time	60s
parameters	IN transmission range	300 m
Multi-hop	T	100 ms
	T_s	95 ms
	T_r	5 ms
time slot	T_{mslot}	1 ms
Fading model		Nakagami-m ($m = 3$)
Reception Rx_{th}		−91 dBm
Background noise		−99 dBm

To evaluate the performance of the downlink messages, we compare our TSM protocol with and without the location register (LR) messages. The results are also compared with the SB technique. In Fig. 10, the average number of required transmissions are plotted. As seen, the location register mechanism significantly reduces the number of transmissions required. This is due to the updated exchange of location register messages between transmitting vehicles and the CIS, which selects the nearest IN to transmit the downlink response message. Particularly, the location register technique reduces the number of required transmission by 17 % compared with the scenario when the LR mechanism is not used. To provide an indication of the number of time slots required for the downlink messages by the multi-hop TSM protocol, Fig. 9 plots the cumulative distribution function (CDF) when there are 20 transmitting vehicles. Since the number of time slots are discrete values, the CDF plot is also discrete and shows the variability in the time slot allocation.

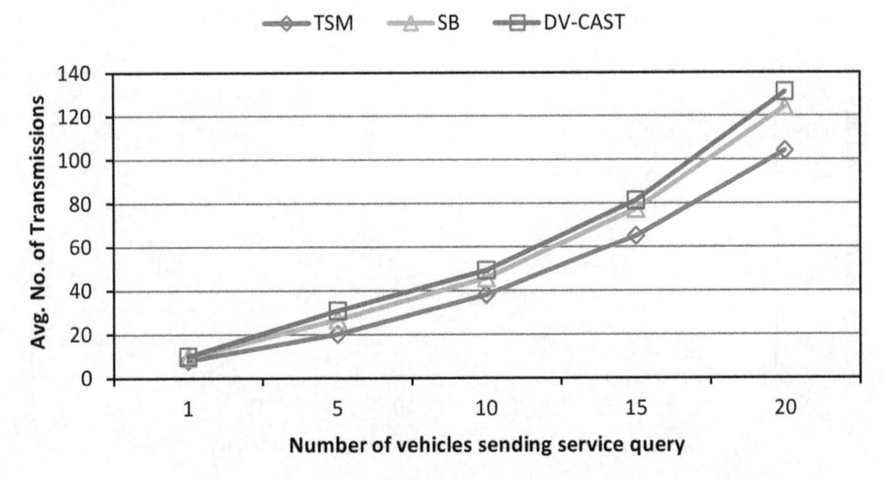

Fig. 6 Average number of required transmissions for uplink messages at $\rho = 180$ vehicles/km^2

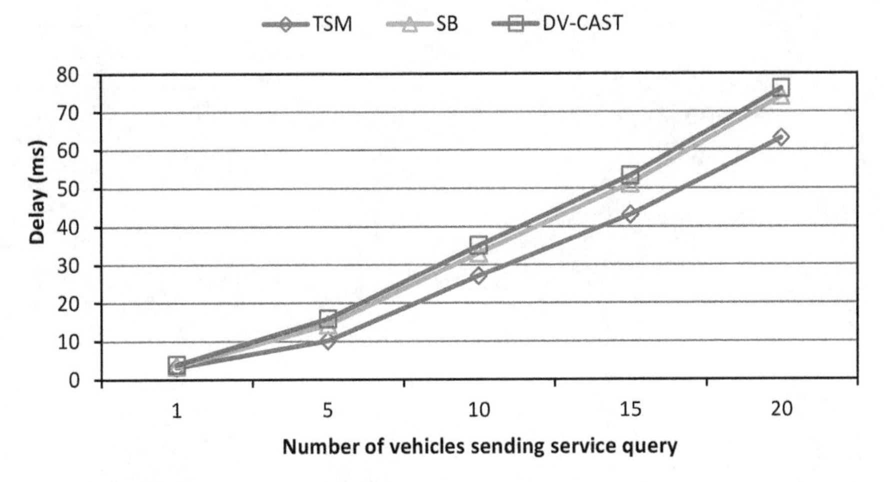

Fig. 7 Uplink delay for $\rho = 180$ vehicles/km^2

The downlink delay is displayed in Fig. 10. With location register mechanism included, the downlink delay is improved up to 16 ms when the number of transmitting vehicles are 20. This enhancement comes from the smaller number of required transmissions when the location register mechanism is used. The reason behind the improvement is that the CIS gets an updated information about the vehicle mobility and chooses the nearest IN when transmitting the downlink response message. As a result, the request/response messages are transmitted in less than 72 ms.

To evaluate the effect of road traffic on the performance of EVIS, we plot the downlink delay with the location register mechanism at different vehicle densities and speeds in Figs. 11 and 12. As the vehicle density increases, the downlink delay

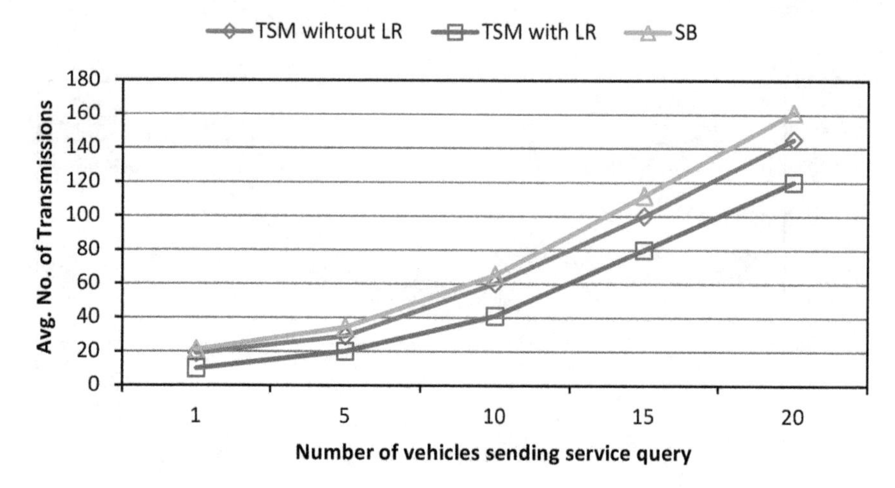

Fig. 8 Average number of required transmissions for downlink messages at $\rho = 180$ vehicles/km^2

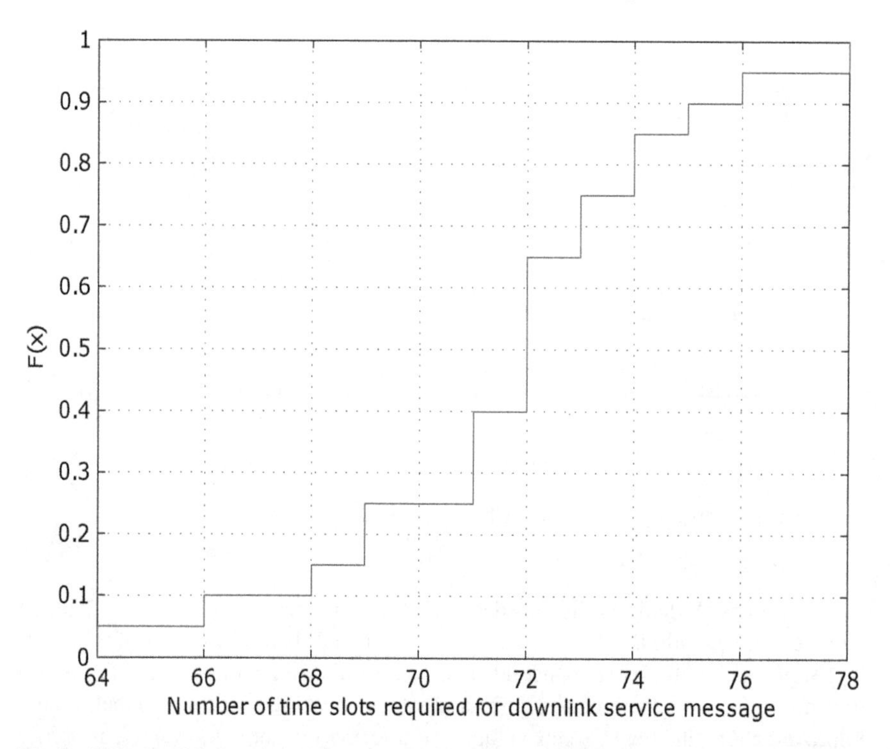

Fig. 9 CDF plot of the number of time slots required for downlink messages by the TSM protocol, assuming 20 transmitting vehicles and $\rho = 180$ vehicles/km^2

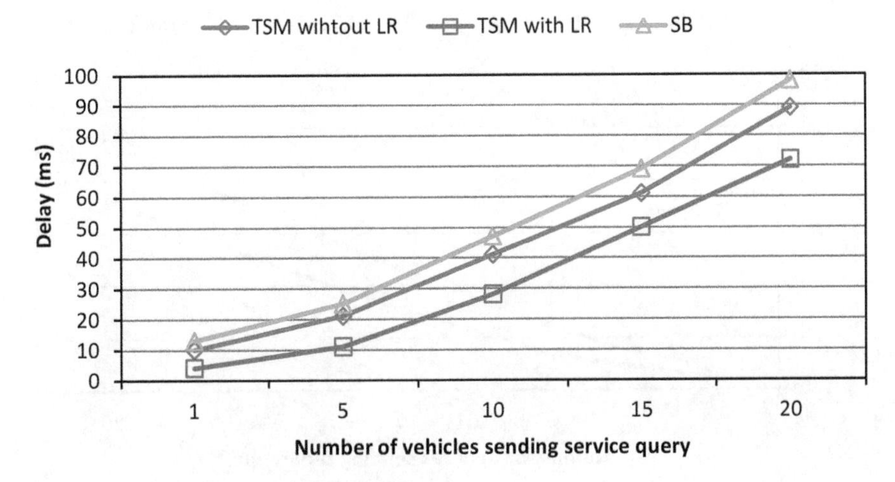

Fig. 10 Downlink delay for $\rho = 180$ vehicles/km^2

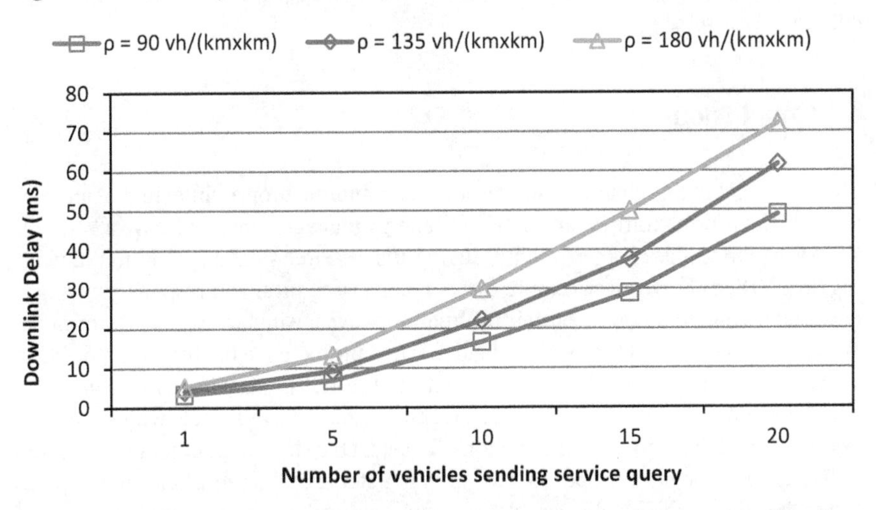

Fig. 11 Downlink delay performance of the proposed technique at different vehicle densities

also grows. This is due to a higher network load caused by the periodic background traffic generated by every vehicle. In particular, when the number of transmitting vehicles is 20, the downlink delay is 49, 61 and 72 ms at a vehicle density of 90, 135 and 180 vehicles/km^2, respectively.

Figure 12 plots the downlink delay at two different vehicle speed values. As can be seen, the downlink delay is improved as the vehicle speed increases. This is because at higher speeds, vehicles who transmit the query message move faster to another IN when the response message is ready. With 20 transmitting vehicles, the downlink delay at the vehicle speed of 10–40 km/h and 40–70 km/h is 99 and 72 ms, respectively.

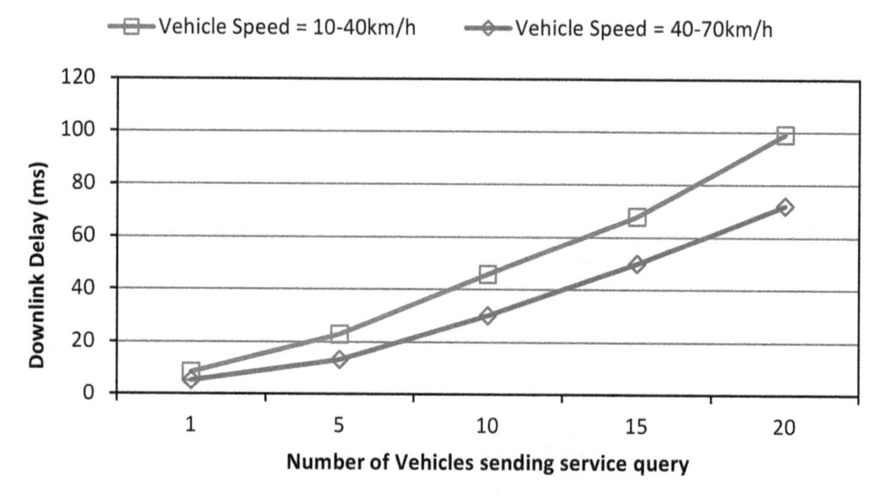

Fig. 12 Downlink delay performance of the proposed technique at $\rho = 180$ vehicles/km^2 and different vehicle speed

6 Conclusions

The main goal of this chapter is to present the communication architecture of an electric vehicle information system (EVIS) for energy management services in a VANET. We have started the chapter by introducing the research challenges in implementing an EVIS. Such services require the exchange of request and response messages between a vehicle and the central information server using an infrastructure node. We have then reviewed the current standards and protocols in the literature for inter-vehicle communications. To efficiently transmit the multi-hop service messages in an EVIS, we have presented a time-slotted technique that actively deals with the broadcast storm and the hidden node collisions. To select the shortest path for the downlink transmissions, we have proposed a location register signaling mechanism that helps select the nearest IN. Finally, simulation results show performance improvements of the proposed information transmission system in some key performance areas.

References

1. Su, H., Qiu, M., Wang, H.: Secure wireless communication system for smart grid with rechargeable electric vehicles. IEEE Commun. Mag. **50**(8), 62–68 (2012)
2. Wang, M., Liang, H., Deng, R., Zhang, R., Shen, X.S.: VANET based online charging strategy for electric vehicles. In: Proceedings of the IEEE Global Telecommunications Conference Workshops, Dec 2013, pp. 4804–4809
3. Uhrig, R.: Greenhouse gas emissions from gasoline, hybrid-electric, and hydrogen-fueled vehicles. In: Proceedings of the IEEE EIC Climate Change Technology, May 2006, pp. 1–6

4. Imai, S., Takeda, N., Horii, Y.: Total efficiency of a hybrid electric vehicle. In: Proceedings of the IEEE Power Conversion Conference, vol. 2, pp. 947–950, Aug 1997
5. Jin, C., Sheng, X., Ghosh, P.: Optimized electric vehicle charging with intermittent renewable energy sources. IEEE J. Sel. Top. Sign. Proces. **8**(6), 1063–1072 (2014)
6. Javed, M., Ngo, D., Khan, J.: A multi-hop broadcast protocol design for emergency warning notification in highway VANETs. EURASIP J. Wirel. Commun. Netw. (179) (2014)
7. Javed, M., Ngo, D., Khan, J.: Distributed spatial reuse distance control for basic safety messages in SDMA-based VANETs. Veh. Commun. **2**(1), 27–35 (2015)
8. Chen, R., Jin, W.-L., Regan, A.: Broadcasting safety information in vehicular networks: issues and approaches. IEEE Netw. **24**(1), 20–25 (2010)
9. Caveney, D.: Cooperative vehicular safety applications. IEEE Control Syst. **30**(4), 38–53 (2010)
10. Javed, M., Khan, J.: Performance analysis of a time headway based rate control algorithm for VANET safety applications. In: Proceedings of the IEEE International Conference on Signal Processing and Communication Systems, pp. 1–6, Dec 2013
11. Javed, M., Khan, J.: A cooperative safety zone approach to enhance the performance of VANET applications. In: Proceedings of the IEEE Vehicular Technology Conference, pp. 1–5, June 2013
12. Morgan, Y.: Notes on DSRC & WAVE standards suite: its architecture, design, and characteristics. IEEE Commun. Surv. Tutor. **12**(4), 504–518 (2010)
13. Kenney, J.: Dedicated short-range communications (DSRC) standards in the United States. Proc. IEEE **99**(7), 1162–1182 (2011)
14. Uzcategui, R., Acosta-Marum, G.: WAVE: a tutorial. IEEE Commun. Mag. **47**(5), 126–133 (2009)
15. Stanica, R., Chaput, E., Beylot, A.-L.: Properties of the MAC layer in safety vehicular ad hoc networks. IEEE Commun. Mag. **50**(5), 192–200 (2012)
16. The Institute of Electrical and Electronics Engineers: IEEE Std 802.11-2007 (Revision of IEEE Std 802.11-1999), IEEE Standard for information technology—Telecommunications and information exchange between systems—Local and metropolitan area networks—Specific requirements—Part 11: Wireless LAN medium access control (MAC) and physical layer (PHY) specifications, Tech. Rep., Dec 2007
17. The Institute of Electrical and Electronics Engineers: IEEE Std 1609.4-2011, IEEE standard for wireless access in vehicular environments (WAVE)—Multi-channel operation, Tech. Rep. (2011)
18. The Institute of Electrical and Electronics Engineers: IEEE Std 1609.1-2006, IEEE standard for wireless access in vehicular environments (WAVE)—Resource Manager, Tech. Rep. (2006)
19. The Institute of Electrical and Electronics Engineers: IEEE Std 1609.2-2012, IEEE standard for wireless access in vehicular environments (WAVE)—Security services for applications and management messages, Tech. Rep. (2012)
20. The Institute of Electrical and Electronics Engineers: IEEE Std 1609.3-2010, IEEE standard for wireless access in vehicular environments (WAVE)—Networking services, Tech. Rep. (2010)
21. The European Telecommunications Standards Institute: ETSI TS 102 637-2 v1.2.1—Intelligent transport systems (ITS)—Vehicular communications—Basic set of applications—Part 2: Specification of cooperative awareness basic service, Tech. Rep. (2011)
22. The European Telecommunications Standards Institute: ETSI ES 202 663 v1.1.0—Intelligent transport systems (ITS)—European profile standard for the physical and medium access control layer of intelligent transport systems operating in the 5 GHz frequency band, Tech. Rep. (2010)
23. Javed, M., Khan, J., Ngo, D.: Joint space-division multiple access and adaptive rate control for basic safety messages in VANETs. In: Proceedings of the IEEE Wireless Communication and Networking Conference, pp. 1–5 (2014)
24. Torrent-Moreno, M., Mittag, J., Santi, P., Hartenstein, H.: Vehicle-to-Vehicle communication: fair transmit power control for safety-critical information. IEEE Trans. Veh. Technol. **58**(7), 3684–3703 (2009)
25. Rawat, D., Popescu, D., Yan, G., Olariu, S.: Enhancing VANET performance by joint adaptation of transmission power and contention window size. IEEE Trans. Parallel Distrib. Syst. **22**(9), 1528–1535 (2011)

26. Artimy, M.: Local density estimation and dynamic transmission-range assignment in vehicular ad hoc networks. IEEE Trans. Intell. Transp. Syst. **8**(3), 400–412 (2007)
27. Sepulcre, M., Gozalvez, J., Harri, J., Hartenstein, H.: Contextual communications congestion control for cooperative vehicular networks. IEEE Trans. Wirel. Commun. **10**(2), 385–389 (2011)
28. Gozalvez, J., Sepulcre, M.: Opportunistic technique for efficient wireless vehicular communications. IEEE Veh. Technol. Mag. **2**(4), 33–39 (2007)
29. Li, J., Chigan, C.: Delay-aware transmission range control for VANETs. In: Proceedings of the IEEE Global Telecommunications Conference, pp. 1–6, Dec 2010
30. Huang, C.-L., Fallah, Y., Sengupta, R., Krishnan, H.: Adaptive intervehicle communication control for cooperative safety systems. IEEE Netw. **24**(1), 6–13 (2010)
31. Park, Y., Kim, H.: Application-level frequency control of periodic safety messages in the IEEE WAVE. IEEE Trans. Veh. Technol. **61**(4), 1854–1862 (2012)
32. Sahoo, J., Wu, E.-K., Sahu, P., Gerla, M.: Congestion-controlled-coordinator-based MAC for safety-critical message transmission in VANETs. IEEE Trans. Intell. Transp. Syst. **14**(3), 1423–1437 (2013)
33. Bharati, S., Zhuang, W.: CAH-MAC: cooperative ad hoc MAC for vehicular networks. IEEE J. Sel. Areas Commun. **31**(9), 470–479 (2013)
34. Omar, H., Zhuang, W., Li, L.: VeMAC: a TDMA-based MAC protocol for reliable broadcast in VANETs. IEEE Trans. Mob. Comput. **12**(9), 1724–1736 (2013)
35. Hafeez, K., Zhao, L., Liao, Z., Ma, B.: A new broadcast protocol for vehicular ad hoc networks safety applications. In: Proceedings of the IEEE Global Telecommunications Conference, pp. 1–5, Dec 2010
36. Hassan, M., Vu, H., Sakurai, T.: Performance analysis of the IEEE 802.11 MAC protocol for DSRC safety applications. IEEE Trans. Veh. Technol. **60**(8), 3882–3896 (2011)
37. Ma, X., Zhang, J., Yin, X., Trivedi, K.: Design and analysis of a robust broadcast scheme for VANET safety-related services. IEEE Trans. Veh. Technol. **61**(1), 46–61 (2012)
38. Blum, J.J., Eskandarian, A.: A reliable link-layer protocol for robust and scalable intervehicle communications. IEEE Trans. Intell. Transp. Syst. **8**(1), 4–13 (2007)
39. Wisitpongphan, N., Tonguz, O., Parikh, J., Mudalige, P., Bai, F., Sadekar, V.: Broadcast storm mitigation techniques in vehicular ad hoc networks. IEEE Wirel. Commun. **14**(6), 84–94 (2007)
40. Tonguz, O., Wisitpongphan, N., Bai, F.: DV-CAST: a distributed vehicular broadcast protocol for vehicular ad hoc networks. IEEE Wirel. Commun. **17**(2), 47–57 (2010)
41. Torrent-Moreno, M.: Inter-vehicle communications: Assessing information dissemination under safety constraints. In: Proceedings of the Conference on Wireless on Demand Network Systems and Services, pp. 59–64, Jan 2007
42. Li, M., Zeng, K., Lou, W.: Opportunistic broadcast of event-driven warning messages in vehicular Ad Hoc networks with lossy links. Comput. Netw. **55**(10), 2443–2464 (2011)
43. Korkmaz, G., Ekici, E., Özgüner, F., Özgüner, U.: Urban multi-hop broadcast protocol for inter-vehicle communication systems. In: Proceedings of the ACM International Workshop on Vehicular Adhoc Networks, pp. 76–85, Oct 2004
44. Fasolo, E., Zanella, A., Zorzi, M.: An effective broadcast scheme for alert message propagation in vehicular ad hoc networks. In: Proceedings of the IEEE International Conference on Communications, pp. 3960–3965, Jun 2006
45. Gharbaoui, M., Valcarenghi, L., Bruno, R., Martini, B., Conti, M., Castoldi, P.: An advanced smart management system for electric vehicle recharge. In: Proceedings of the IEEE International Electric Vehicle Conference, pp. 1–8, March 2012
46. Wang, M., Liang, H., Zhang, R., Deng, R., Shen, X.: Mobility-aware coordinated charging for electric. IEEE J. Sel. Areas Commun. (2014)
47. Hess, A., Malandrino, F., Reinhardt, M.B., Casetti, C., Hummel, K.A., Barceló-Ordinas, J.M.: Optimal deployment of charging stations for electric vehicular networks. In: Proceedings of the Workshop on Urban Networking, pp. 1–6 (2012)

RF Energy Harvesting Communications: Recent Advances and Research Issues

M. Majid Butt, Ioannis Krikidis, Amr Mohamed and Mohsen Guizani

Abstract Green radio communications has got a lot of attention in recent years due to its telling effects on telecom business and environment. On the other side, energy harvesting (EH) communication has emerged as a potential candidate to reduce the communication cost by tackling the problem in a contrasting fashion. While green communication techniques focus on minimization the use of radio resources, EH communication relies on environment friendly techniques to generate energy (from the renewable resources) and effective use of created energy conditioned on the fact that there is always energy available when required. Thus, the focus migrates from minimization of energy to optimal time domain distribution of energy and this causes a paradigm shift in radio resource allocation research. Instead of just focusing on average and maximum power constraint, the packet/energy arrival processes and packet/energy buffering interact in a challenging way to open new research opportunities. This chapter summarizes the major research work in the area of radio frequency (RF) energy harvesting resource allocation. First, we discuss the fundamental concepts related to energy harvesting communications. Then, we review the recent developments in this area and outline the major research challenges for the research community. We address the cooperation aspect of energy harvesting, which has emerged as an interesting area of research. Wireless powered communication networks allow energy and information transfer from the radio frequency waves and provide sustainable networks. Finally, we discuss a wireless powered relay network in detail and show the performance comparison of different relay selection techniques.

1 Introduction

Green communication has attracted a lot of attention due to rising electricity cost for network operation and its adverse effects on the environment because of CO_2 emissions. It is predicted that the overall Information and Communication (ICT)

An erratum of this chapter can be found under DOI 10.1007/978-3-319-27568-0_19

M.M. Butt (✉) · I. Krikidis · A. Mohamed · M. Guizani
Department of computer science and engineering, Qatar University, Doha, Qatar
e-mail: majid.butt@ieee.org

© Springer International Publishing Switzerland 2016

M.Z. Shakir et al. (eds.), *Energy Management in Wireless Cellular and Ad-hoc Networks*, Studies in Systems, Decision and Control 50,
DOI 10.1007/978-3-319-27568-0_15

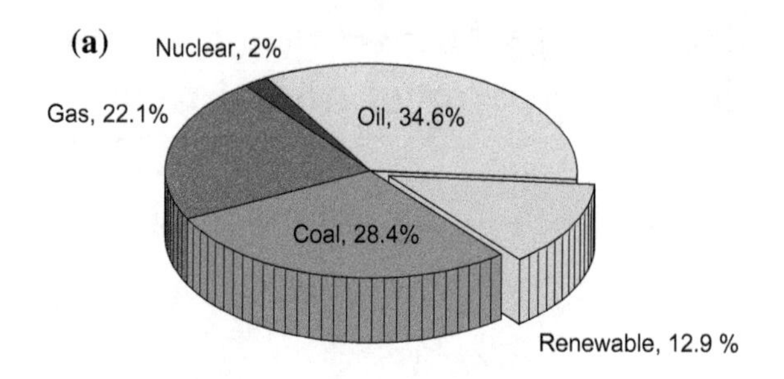

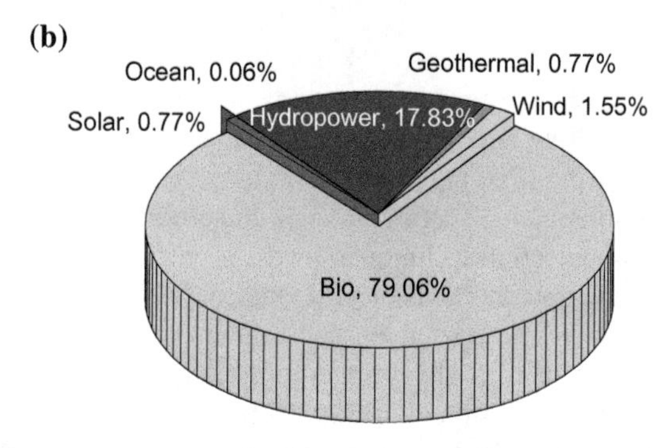

Fig. 1 Shares of energy sources in total global primary energy supply in 2008 [9]. **a** Energy contribution from the different sources. **b** Energy contribution for the renewable energy sources.

footprint will almost double or triple between 2007 and 2020 [7, 42]. In a mobile network, base stations alone are responsible for 80 percent of the network's power consumption [30].

The cost of operating a site in a remote area is extremely high due to difficulties in installation and maintenance. In some areas, the sites are operated by generators requiring a continuous supply of fuel and the cost of transportation of fuel is a significant part of the operational cost due to difficult terrain. Energy harvesting (EH) has emerged as a promising solution to reduce the cost of network operation and increase its sustainability. The research community has started to show some interest in deploying sites based on energy harvesting from natural sources like the sun and the wind. Figure 1 shows the breakdown of energy sources in 2008 in global primary energy supply [9]. Although, traditional means of energy still dominate yet renewable energy (RE) sources contribute about 13 % of the total energy supply. It is worthwhile to mention that the capital cost of EH solutions is greater than the traditional continuous power supply or generator counterpart, but the running cost compensates the cost in a few years of operation.

EH brings a few associated challenges for the network designers. The traditional power supply from the grid provides an uninterrupted energy supply for communication. In contrast, EH communication has to deal with a time dependent energy supply. For example, the sun light is available throughout the day but not at night. In this case, the harvesting period is constant for a long time. However, energy harvested by wind and other sources may not follow this long pattern of harvesting and harvesting periods may vary stochastically. In contrast to traditional communication system design for energy efficiency by maximizing the use of favorable channel

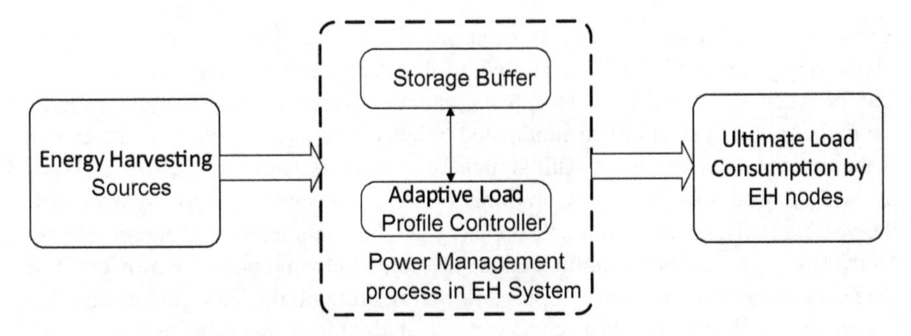

Fig. 2 The block diagram for a typical EH communication system

conditions, EH communication focuses on smart use of energy such that energy is always available for future use. This conversation of energy and adaptation of communication systems according to EH profiles such that energy expenditure is always less than the stored energy at any time is termed as neutrality constraint or causality condition in literature [3, 20]. Thus, energy profile (energy arrival process) impacts the underlying communication scheme enormously.

Figure 2 shows block diagram for a typical energy communication system. The system design depends on the energy harvesting profile (source), the data transmission requirements and storage buffer. The storage structure defines the storage management system for the harvested energy. For example, EH systems can be designed by considering ideal/non-ideal storage of energy (buffer). Similarly, the arrived energy can be modeled by using either online or offline assumptions. Offline schemes are of interest when the amount of harvested energy, channel conditions and the amount of incoming data for all transmission intervals are known in advance while online allocation schemes are only based on causal information regarding the channel conditions, harvested energy and the amount of data to be transmitted. Based on the fact that the EH profile and the channel conditions are random in nature and cannot be predicted in advance, online allocation schemes are more realistic in this case. However, offline schemes are important to provide performance upper bounds for the practical online schemes.

In the rest of this chapter, we introduce the fundamentals of radio resource allocation in EH communication. Then, we mainly focus on the energy harvesting communication where energy is harvested from radio frequency (RF) waves. We provide example of a system design where relay selection is performed for a single source, single destination system.

2 Modeling Energy Harvesting

EH profile model is one of the crucial and important factors in designing the resource allocation mechanisms for EH systems. The EH profile defines the harvested energy at an EH node as a time critical function. Depending on the types of different sources of energy which have significant environmental influence, one has to be very careful while designing the EH profile models in different scenarios. The energy sources can be categorized as fully deterministic and non-deterministic. The energy arrival from the sources can either be anticipated or it can be unpredictable as in the case of solar or wind energy sources. Although there are several natural sources for EH such as wind power, solar energy, vibrational energy, geothermal energy, hydroelectric power, bio fuels, natural gas, nuclear energy yet, the main renewable energy sources considered for wireless communications are solar and wind power. However, these EH sources need to be managed properly before utilizing the harvested energy. For example, practically there is no energy output at night from any solar cell. Therefore, solar energy which is harvested during day time can be stored in a storage buffer for later use when there is no available energy to harvest in the system. Thus, load profile controllers can play a vital role according to the availability of the harvested energy. To ensure that the neutrality constraint is met, not only we rely on the profile of the available energy, but we need to model and implement it in an efficient way as well.

A deterministic system model is the one where the model parameters such as energy arrival time and rate for EH profile are always considered to be determined and this type of model always produces unique outputs from a specific initial state. However, we cannot always consider the arrival process deterministic and have to model it as a stochastic or non-deterministic process. Traditionally, the approach of modeling EH profile treats stochastic process as a function of one or more deterministic attributes such as energy arrival time, amount of energy, etc., whose outputs are random variables or non-deterministic quantities and having certain probability distributions. Since EH nodes can replenish their supply of energy in an un-predictable way, they require more sophisticated energy arrival process modeling.

In the next section, we review the most common energy harvesting profile models employed in the literature.

2.1 Markovian Model

A Markov chain is a discrete-time process where the future behavior (state) only depends on the present (state) and not on the past. For example, at time epochs $n = 1, 2, 3, \ldots$ the Markov process changes from the current state i to the next state j in a system with transition probability p_{ij}. In a discrete-time Markov chain process, state transition mechanisms can be expressed by using a transition probability matrix. Many queuing models are in fact Markov processes. As the energy arrival process of an EH node can be characterized as a queuing model, Markov model is used extensively in the literature. For example, the energy state of an EH node is modeled as a discrete-time Markov chain in [45]. In this model, Yang et al. choose the Poisson process as the energy model and transfer it into a Bernoulli process for a slotted CSMA/CA system. Each state in the model, $e(t) \in \{0, 1, \ldots N-1\}$ denotes the amount of energy on a certain time slot. The authors assume that an EH node can transmit only when its energy exceeds the threshold energy and the energy to sense the channel status (busy or idle) is smaller than the recharging rate, so that the EH node can get positive net energy gain while it keeps listening and recharging.

2.2 *Storage Capacity Considerations*

The EH profile model without storage buffer is not suitable in the application of wireless networks. As the natural source of energy like sun or wind cannot be continuously supplied to an EH node, it is wise to store the finite energy in a finite storage buffer periodically. Most of the models mentioned in the literature consider a limited/infinite capacity storage buffer like a battery in their proposed designs. The main purpose of storing harvested energy is to minimize the delay in data transmission and to extend the energy sustainability of the wireless node.

We discuss example of an EH profile model which employs both Markovian model and energy storage [24]. Let us consider a single hop transmission over a replenishable sensor network. The stationary time-continuous Markov chain model is introduced to provide optimal transmission policy for wireless sensor nodes with different energy budgets to maximize the average reward. For an N-state Markov process, the reward is modeled as a physical quantity or economic unit relevant to the transition from one state to another. Figure 3 shows this Markov process of a sensor having its energy replenished from both replacement (Poisson) process and recharging (Poisson) process.

Another power management model is proposed in [14] with a two-stage algorithm. The authors assume known, but varying channel states and harvesting instants. However, the coherence time for the channel is considerably shorter than the constant harvesting period. Due to fairly long harvesting interval, the authors consider constant available harvested power for a transmission time slot in the first stage of the algorithm and apply power control as a function of channel gain to maximize the average data rate. In the second stage, *constant* power for the first phase is optimized by modeling energy neutrality operation as a function of the harvested power to maximize the average rate for the first stage.

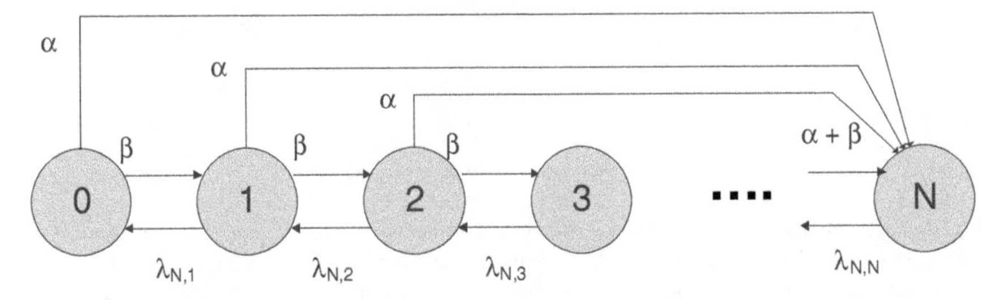

Fig. 3 Hybrid EH model for the energy state information based on Markov process [24] where α and β are old battery replacement rate and recharging rate, respectively

Another model in [33] defines the ambient energy supply by a two-state Markov chain where "GOOD" and "BAD" states correspond to an abundance and scarcity of ambient energy, respectively. The amount of energy at time $m + 1$ is defined as

$$E_{m+1} = min\{E_m - Q_m + B_m, c_{max}\}, \tag{1}$$

where B_m is the energy arrival process, Q_m is the action process and c_{max} is the battery capacity. $Q_m = 1$, if the current data packet is transmitted. $B_m \in \{0, 1\}$, models the randomness in the energy harvested in slot m, i.e., either one energy quantum is harvested, or no energy is harvested at all. In this model, the authors assume that each position in the storage buffer can hold one energy quantum and the transmission of one data packet requires the expense of one energy quantum.

Most of the recent work on communication systems with EH capability assumes that EH is the only source of energy. However, the concept of hybrid energy sources has also drawn interest. In general, a hybrid energy source is defined as a combination of a constant energy source, e.g., power grid, diesel generator, etc., and an EH source which harvests energy from the sun, wind, etc. Thus, allocating power for such a communication system is a challenging problem. For a point-to-point communication link, both offline and online power allocation schemes for an EH transmitter with a

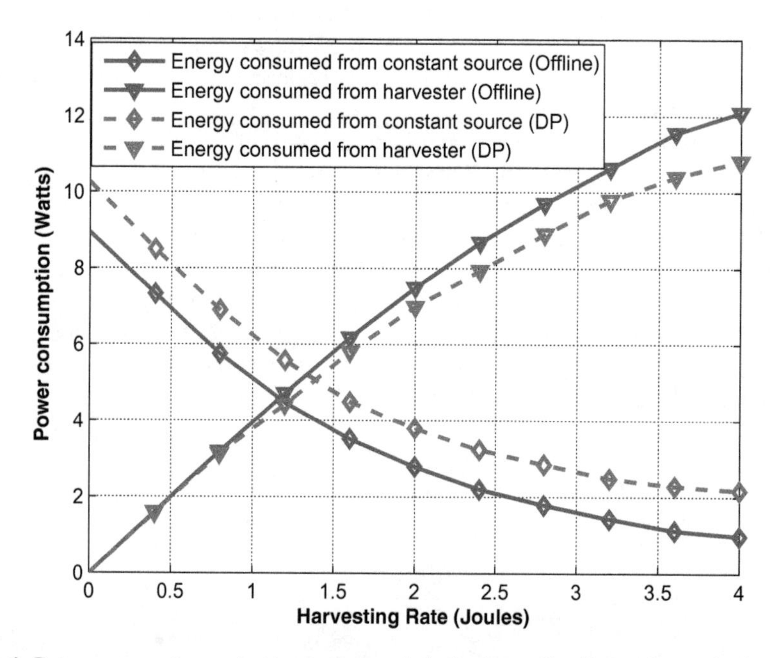

Fig. 4 Power consumption against harvesting rate in Joules (normalized by one harvesting interval) for the case where all the data packets have arrived before the transmission begins [2]. The results show that harvested energy is efficiently utilized for transmitting a given number of data packets over a finite number of transmission intervals by maximizing the power consumption rate from the harvester

hybrid energy supply have been proposed in [2]. The aim of this work is to minimize the amount of energy drawn from the constant energy source, such that the harvested energy is efficiently utilized for the transmission of data packets over a finite number of time slots. For the work in [2], Fig. 4 illustrates the performance gain quantitatively for the case where the data packets to be transmitted arrive in the system before the transmission begins and no packets arrive during the transmission. It is evident that the offline power allocation scheme performs better than the Dynamic Programming (DP) based online power allocation scheme. For the high harvesting rate, the offline scheme makes efficient use of the harvested energy whereas the online schemes may under-utilize the harvested energy and result in a lower consumption of the harvested energy.

3 Wireless Energy Transfer

Energy cooperation has emerged as an attractive research area in EH wireless systems where the EH network nodes share their energy resources to enhance the system energy efficiency. Like relay networks where user cooperation is introduced to enhance the system throughput by exploiting the broadcast nature of wireless communications, similar techniques are employed in EH systems by sharing the energy as well as information among the various nodes in the network. Thus, in addition to other natural harvesting sources like solar or wind, RF signals can also be used as a potential source of EH in wireless communication.

First, we survey the EH works where wireless energy transfer technique has been incorporated to prolong the network life time or develop an energy efficient system [1]. RF-based EH is quite suitable for low-power applications. RF signals that carry energy can be used as a carrier for transporting information at the same time [47]. This is termed as simultaneous wireless information and power transfer (SWIPT) in the literature. Zhang et al. study SWIPT by considering simplified scenarios with only one or two active user terminals in the network at any given time [47]. The authors investigate two practical receiver designs, termed as time switching and power splitting, for a three-node multiple-input multiple-output (MIMO) broadcast system where the EH and information decoding (ID) receivers harvest energy and decode information separately from the signal sent by a common transmitter. In the time switching protocol, the receiver spends some time for EH and the remaining for information processing while in the power splitting protocol, the receiver uses a part of the received power for EH and the remaining for information processing. Zhang et al. further focus on the power splitting scheme for a point-to-point single-antenna flat-fading channel in [26]. The authors propose a scheme called dynamic power splitting (DPS), where the receiver is capable of dynamically adjusting the power split ratio for ID and EH based on the known CSI at the receiver. They assume that the transmitter has a conventional constant power supply, whereas the receiver harvests energy from the received signal sent by the transmitter. For the single-input single-output case, the authors show that to achieve the optimal rate-energy trade-off, a fixed amount of the received signal power should be allocated to the information receiver and the rest of the power should be allocated to the energy receiver when the fading channel gain exceeds a given threshold [47]. For the single-input multiple-output case where the receiver is equipped with multiple antennas, they extend the result for DPS by considering a uniform power splitting scheme.

A more practical receiver design for SWIPT is studied in [48] for a point-to-point wireless link where three special cases namely time switching, static power splitting and on-off power splitting of DPS are investigated. The time switching scheme proposed in [47] is further investigated in [27] for a point-to-point single-antenna flat fading channel. Two energy efficient resource allocation algorithms for multi-carrier systems employing hybrid information and EH receivers are proposed in [35, 36]. This work has been extended in [37] by introducing an algorithm for power splitting receivers where the authors consider both continuous and discrete power splitting ratios. Data multiplexing of the users on different sub carriers is also incorporated in the algorithm design.

The resource allocation algorithm design for secure multiuser multiple-input single-output (MISO) systems with concurrent wireless information and power transfer is discussed in [38, 39], where the authors formulate an optimization problem with the objective of minimizing the total transmit power constrained by QoS guarantees in terms of minimum SINR at the desired receiver and minimum power transfer to the idle legitimate receivers. Secure communication in MISO SWIPT systems is also studied in [28] where a multi antenna transmitter sends information to one information and multiple energy receivers simultaneously. The secrecy rate for the information receiver is maximized for the constraints on minimum received energy by the energy receivers. As compared to a perfect CSI case in [28, 38, 39], the channel uncertainty case for MISO SWIPT is addressed in [8]. SWIPT in a multiuser OFDM system is also discussed in [49] where the users harvest energy and decode information using the same signals received from a fixed access point. In a recent work, SWIPT in a MISO multicast network is studied [18].

In the context of wireless energy transfer, *wireless powered communication networks* refer to the protocol of separate transmission of wireless energy and information. For example, energy from an access point is transferred to the terminals in downlink transmission and information scheduling is performed in uplink transmission as shown in Fig. 5. Ju et al. in [19] propose a protocol termed "harvest-then-transmit", where in the first phase, wireless energy is broadcasted by the hybrid access point to all the users in the downlink. Then in the second phase, the users send their independent information to the hybrid access point in the uplink using their individually harvested energy by time-division-multiple-access. The authors mainly focus on maximizing the uplink throughput of the wireless powered communication networks by optimally allocating the time for the downlink wireless energy transfer by the hybrid access point and the uplink wireless information transmissions by different users. Solution of this problem reveals an interesting new phenomenon in the wireless powered communication networks. Ju et al. termed it as "doubly near-far" phenomenon. It depicts that when a far user receives less amount of wireless energy from the hybrid access point than a nearer user in the downlink, then it has to transmit

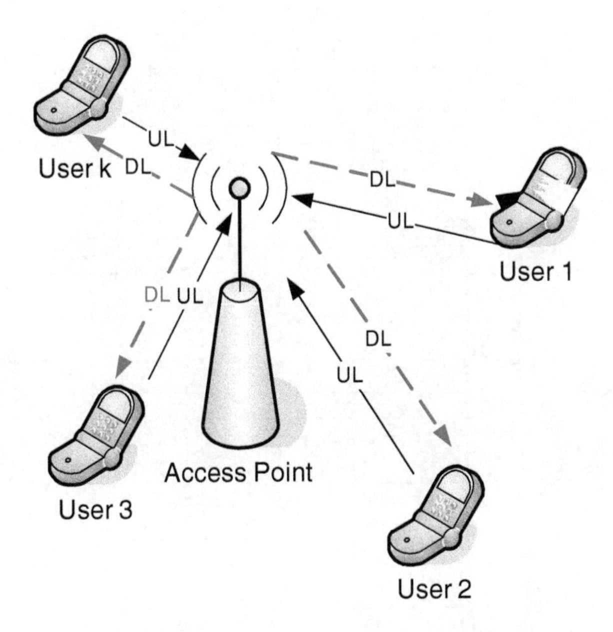

Fig. 5 Wireless power networks with energy transfer in downlink and information transfer in uplink

with more power in the uplink for achieving the same information rate. It occurs due to doubly distance-dependent signal attenuation in both the downlink and the uplink. The authors propose a new performance metric referred to as *common-throughput* to overcome the doubly near-far problem. This metric consists of an additional constraint which indicates that all users should be allocated an equal rate in their uplink wireless information transfers without considering the distance to the hybrid access point. This work has been extended to the case of multi-antenna systems in [25]. A similar work in [44] maximizes the transferred power to the users in downlink for a signal to noise and interference constraint for the multi-antenna settings. The rate-energy tradeoff for a SWIPT based multiuser wireless system is investigated in [23] for different collaboration schemes between the transmitters. In contrast to "harvest-then-transmit" protocol in [19], "harvest-use-store" architecture is proposed in [46] which helps to combat energy loss to storage device inefficiency.

EH from RF radiations is also investigated in [21] where a large scale network utilizing SWIPT is considered, in which the transmitters are connected to a power grid and the receivers employ the power splitting technique. The author studies the performance of the network by modeling the random location of nodes according to a homogeneous Poisson point process and derives the outage probability as well as the average harvested energy as a function of the power splitting ratio with the help of tools from stochastic geometry. Two protocols are examined, a non-cooperative and a cooperative, and it is shown that the cooperative protocol can significantly improve the performance of the system and achieve a better trade-off between the outage probability and the average energy transfer. Additionally, for the non-cooperative protocol, an optimization problem is formulated and solved which minimizes the transmitted power under the outage probability and EH constraints. Figure 6a, b show

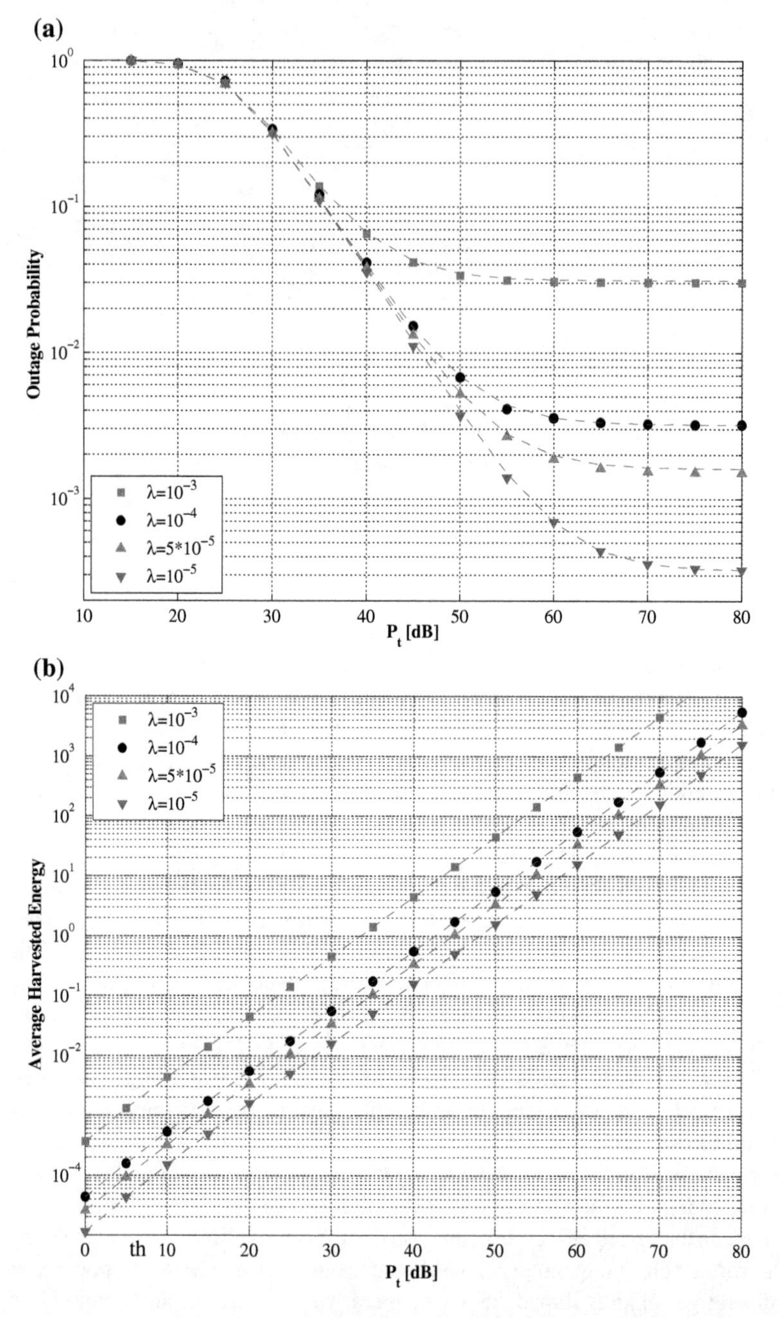

Fig. 6 Performance of the cooperative protocol versus transmission power P_t for different network densities λ. Analytical results are shown with *dashed lines* [21]. **a** Outage probability. **b** Mean harvested energy

the fundamental trade-off between information decoding and EH; the interference degrades the outage performance while it increases the average harvested energy.

A stochastic geometry approach for wireless energy transfer has also been studied in [15]. The authors model an uplink cellular network where the mobiles are recharged via microwave radiation by power stations. The base stations and power stations form independent homogeneous Poisson point processes and the mobile users are distributed uniformly in the corresponding cells with respect to the base stations. Using this model, the authors derive trade-offs between the network parameters. Specifically, under an outage constraint, it is shown that the minimum mobile transmission power increases super-linearly with the decreasing density of the base stations.

4 Wireless Energy Powered Relay Selection

Relay selection based on wireless energy transfer has received a lot of attention in the literature. A relay node performs SWIPT operation from the RF signals from the source and relays the information to the receiver. In [5], multiple source-destination pairs which communicate through an EH relay in a wireless cooperative network are considered. By assuming that the battery of the relay is large enough to accumulate a large amount of power for transmissions, the authors investigate how the power can be distributed efficiently among the receivers. Four scenarios are taken into account:

1. The relay transmits to the ith destination by only using the energy harvested from the ith source. This results in the outage performance decaying at a rate $\log(\text{SNR})/\text{SNR}$.
2. The relay distributes the accumulated power harvested from the sources evenly among the transmissions. This results in the outage probability decaying at a rate $1/\text{SNR}$.
3. The relay prioritizes the transmissions to receivers with a better channel (water filling principle). This scenario achieves optimal performance.
4. The relay allocates transmission power to each receiver according to a bid each receiver submits to the relay (auction based power allocation scheme). This scenario achieves a better trade-off between the system performance and the complexity.

RF-based EH relays are also investigated in [34]. An amplify-and-forward wireless cooperative network is considered, where the relay nodes harvest energy by employing one of the two relaying protocols: a time switching protocol and a power splitting protocol. A performance analysis for the throughput is derived for both delay-limited and delay-tolerant transmission modes and a comparison of the two protocols is provided with details of the effect of various system parameters. Wireless energy cooperation for the delay-limited communication system has also been studied in [40] with the objective of minimizing the loss probability due to violation of the packet delay deadline.

A three-node cooperative amplify-and-forward network is discussed in [22] where the relay node performs either harvesting or relaying operation according to a greedy switching policy based on the residual energy in the relay node. The energy transfer through RF signals between the source and the relay node is the key factor to establish the energy cooperation in this proposed system. The charging/discharging mode of the relay's battery is represented by a finite Markov chain and the performance is evaluated in terms of outage probability. In addition, the authors show that the greedy switching policy is an efficient solution by demonstrating that its performance is close to the performance of an optimal switching policy which incorporates a-priori knowledge of the channel coefficients for the whole transmission period.

A study for relay selection methods in an RF-based EH network with N relays is made in [32]. The relays are used for transmitting information to a designated receiver and for transmitting energy to a designated RF energy harvester. For the case $N > 2$, two relay selection methods are developed and analyzed in terms of the achievable trade-off between the outage probability and the average energy transfer as well as the trade-off between the ergodic capacity and the average energy transfer. Furthermore, for the case $N = 2$, an optimal relay selection method is developed which provides the maximum capacity and the minimum outage probability for a given energy transfer constraint as well as the maximum energy transfer for a given capacity or outage probability constraint.

In [11], for a simple multi-hop wireless communication system, an energy cooperation technique is discussed where the source and the relay nodes can harvest energy from natural sources. The authors assume that the source node can transfer a portion of its energy to the relay node. There is a separate wireless energy transfer unit installed in the source node which helps to send a portion of its energy to the relay so that the relay can forward more data. The relay here operates in full duplex mode. When the source transfers δ_i amount of energy to the relay through the wireless energy transfer unit, this amount of energy enters the energy queue at the next time slot. Thus, the queues of the relay are updated with one slot delay with respect to the queues of the source. Similar work in [10, 12] discuss a two-way communication channel where users can harvest energy from nature and the energy can be transferred in one-way from one of the users to the other by assuming that both users know the energy arrivals in advance.

Another cooperative transmission strategy for multiple source-destination pairs and one EH relay is proposed in [6]. In this strategy, the authors discuss the cooperative communication in two different phases. In phase one, each source sends its message to the relay and their power is shared under the total power constraint. In phase two, the relay first tries to decode the message and then carries out EH if there is any power left after decoding. Finally it delivers the correctly decoded message to the destination. In this model, the sources communicate with the relay via orthogonal channels, and no direct source-destination link is assumed. Similar work for direct and cooperative communication is also discussed in [17] where the performance of the transmission techniques is analyzed in terms of outage probability. Several transmission models are proposed for both direct and cooperative communication by considering the energy gathering and the energy salvage techniques during transmission. For a multi-access relay model, the transmit power allocation and energy transfer policies that jointly maximize the sum-rate are discussed in [43]. The authors consider all the nodes as EH nodes which are used for data transmission and they term this approach as bi-directional energy cooperation. Although Tutuncuoglu et al. derive the solution for the sum-rate maximization problem for multi-source model, they also consider special cases such as the single source model with a forward energy transfer capability for the same problem.

To demonstrate SWIPT based relay selection, we discuss a specific example in the next section, where relay selection is performed for a system with relays powered by RF signals from the source node.

4.1 System Model and Problem Formulation

We consider a Decode-and-Forward (DF) strategy based relaying communication system where a source node S communicates with a destination node D in the presence of N relays, represented by symbol L as shown in Fig. 7 [4]. The communication from the source to relay and relay to destination takes place in two orthogonal time slots where duration of each slot is denoted by T. We assume a fixed transmit power P_s at the source and a broadcast channel for the source-relay communication phase is considered.

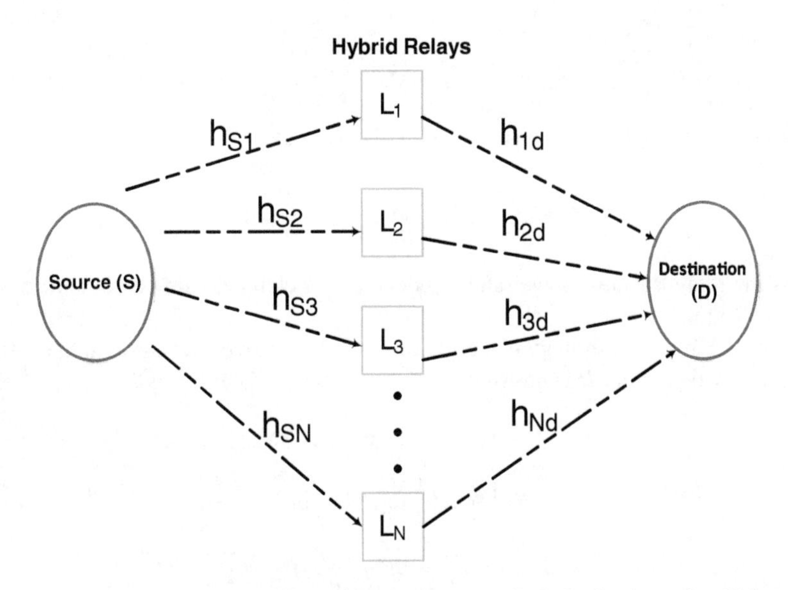

Fig. 7 System model for relay selection with a single source, single destination and multiple relays

The relay nodes are hybrid, i.e., they have the ability to harvest energy as well as retrieve the information from the source signal, but only one function can be performed in a given time slot t. The relay selected at time t to forward information to the destination is not available to harvest energy or forward information from the source at time slot $t+1$ due to assumption of orthogonal communication on the $S \rightarrow L$ and $L \rightarrow D$ links. However, the relays other than the selected one are free to receive data/energy, thereby mimicking a full-duplex relaying system [16]. The harvested energy is stored in a battery of an infinite capacity and the energy stored in the battery is assumed to increase and decrease linearly.

We assume independently and identically (iid) distributed fading channels on the $S \rightarrow L$ and $L \rightarrow D$ links which follow block fading model. The received signal $y_i(t)$ at the relay node L_i is expressed as:

$$y_i(t) = \frac{1}{\sqrt{d_i^2}} \sqrt{P_s} h_{si} x(t) + n(t) \tag{2}$$

where $x(t)$ and d_i denote the normalized information signal from the source and the distance between the transmitter and relay i, respectively. d_i is assumed to be unity throughout this work without loss of generality. $n(t) \sim Z(0, \sigma^2)$ is the Gaussian noise with zero mean and variance σ^2 while the channel gain coefficient for $S \rightarrow L_i$ link is represented by h_{si}.

The rate $R_{si}(t)$ provided on the $S \rightarrow L_i$ link in a time slot t is given by

$$R_{si} = \frac{1}{2} \log_2 \left(1 + |h_{si}|^2 \frac{P_s}{\sigma^2}\right). \tag{3}$$

Similarly, the rate R_{id} on the $L_i \rightarrow D$ link is given by,

$$R_{id} = \frac{1}{2} \log_2 \left(1 + |h_{id}|^2 \frac{P_r}{\sigma^2}\right). \tag{4}$$

P_r is the relay transmit power and h_{id} denotes the channel gain coefficient for the $L_i \rightarrow D$ link.

For a DF relaying strategy in a grid powered cooperative relay system, the outage probability that a rate R is not supported by the system is given by

$$P_{\text{out}} = \Pr\left(\min(R_{si^*}, R_{i^*d}) < R\right) \tag{5}$$

$$= \Pr\Big\{ \min\Big(\frac{1}{2} \log_2(1 + |h_{si^*}|^2 \frac{P_s}{\sigma^2}), \frac{1}{2} \log_2(1 + |h_{i^*d}|^2 \frac{P_r}{\sigma^2})\Big) < R\Big\} \tag{6}$$

where L_{i^*} denotes the selected relay node. For an energy harvesting DF relay system, (5) is a lower bound on outage probability because additional outages occur due to power limitation of the energy harvesting relays.

The energy harvested by the ith relay node during a single time slot is given by [34, 48]

$$E_i^h = \eta P_s \, |h_{si}|^2 \, T \tag{7}$$

where $0 < \eta \leq 1$ is the energy conversion efficiency which depends on the receiver circuit hardware and antenna sensitivity.

4.2 *Problem Settings*

We express the relay selection problem with the goal to minimize the outage probability and formulate the outage probability minimization problem for a multiple relay network:

$$\min_{\pi} \; P_{\text{out}} \tag{8}$$

$$\text{s.t.} \quad \begin{cases} N = c_1, \quad c_1 \in \mathbb{N} \\ P_s = c_2, \\ R \geq 0 \\ R_{si} = \frac{1}{2}\log_2(1 + |h_{si}|^2 \frac{P_s}{\sigma^2}) \\ R_{id} = \frac{1}{2}\log_2(1 + |h_{id}|^2 \frac{P_r}{\sigma^2}) \\ E_i^h = \eta P_s \, |h_{si}|^2 \, T \end{cases} \tag{9}$$

where π is the relay selection scheme and c_1 and c_2 are the constants representing a fixed number of relays in the system and fixed source power, respectively. As the network is operated by the energy harvested from the source RF signals, there must be sufficient (or at least one) charged relay nodes in a given time slot to be able to forward the signal successfully in order to avoid the outage event. Thus, there is a tradeoff between the number of relay nodes in EH mode and the number of nodes available for information transfer for a given transmission. The larger the number of nodes in EH mode, the more inefficient is the use of $L \rightarrow D$ link for information transfer in the current time slot, but more energy is available for information transfer in future. This is the main reason that EH communication focuses on meeting the neutrality constraint in contrast to making the best use of available resources using opportunistic communications solely in the current time slot.

5 Relay Selection Schemes

We assume that the CSI is not available both at the source and the relay for the $S \rightarrow L$ link. Similarly, the CSI is not available at the destination node for the $L \rightarrow D$ link. In relation to availability of the CSI at the relay node on the $L \rightarrow D$ link, we consider two cases which govern the relay selection strategy:

- The CSI at relay is causal and not available.
- The CSI is known at relay node before transmission on the $L \rightarrow D$ link.

The proposed schemes act in a centralized manner as they are based on the shared information of the stored energy levels for the relays and the CSI on the $L \rightarrow D$ links (when available).

5.1 Single Relay Selection (SRS)

First, we assume that the CSI is not available at the relay for transmission to the destination and therefore, the selected relay transmits with a fixed power P_r. The forwarding relay is selected solely based on the stored energy at the relay nodes as no other information is available to be exploited. In this case, only a single relay node L_{i*} is selected out of N nodes to decode and forward the information. We take this case as a baseline and compare results with our scheme in the next section. Similar to [29], the node L_{i*} with the maximum stored energy from the N candidate nodes is selected such that:

$$i^* = \arg\max_i \left(E_i^{\text{store}}(t) - E_r\right)^+ \tag{10}$$

where E_r is the energy spent due to transmission with fixed power P_r and $E_i^{\text{store}}(t)$ denotes the stored energy for the relay L_i at time t. The notation x^+ represents $\max(x, 0)$. Note that the relay selection is performed before the signal reception from the source and therefore, all other relays can harvest energy from the received RF signal using harvesting circuit. If $E_i^{\text{store}}(t) < E_r, \forall i$, no node is selected and all N nodes harvest energy. For the case $R_{si*} < R$, node i^* is unable to decode information from the source and results in an outage without making a transmission on the $L \rightarrow D$ link.

All the nodes except L_{i*} harvest energy depending on the received signal strength from the source such that

$$E_j^{\text{store}}(t+1) = E_j^h(t) + E_j^{\text{store}}(t), \quad j \neq i^* . \tag{11}$$

As mentioned in Sect. 4.1, the selected node i^* is not a candidate for selection in time slot $t+1$ for the both proposed schemes and therefore, the energy update is only meaningful for time slot $t+2$. Note that $E_{i*}^{\text{store}}(t+1) = E_{i*}^{\text{store}}(t)$.

Thus, the corresponding stored energy after transmission for the node i^* is given by,

$$E_{i*}^{\text{store}}(t+2) = E_{i*}^{\text{store}}(t+1) - E_r . \tag{12}$$

If we increase N, we have more relays to choose i^* for data transfer to the destination. This results in decrease in outage.

5.2 Multiple Relay Selection (MRS)

In this case, we assume that the CSI is known at the relay node for transmission to the destination. However, signal from the source is received in time slot t and transmitted to the destination in time slot $t + 1$. On one side, it is important to select a relay for forwarding with relatively high stored energy. On the other side, it is important to exploit availability of CSI on the $L \rightarrow D$ link for optimal power allocation, but it is available only at time $t + 1$. Based on the available information in two successive time slots, we propose a 2-step relay selection policy where contrast to SRS, multiple relays are selected in first step.

In the first step, a subset Γ of M relays is selected out of N relays such that

$$\Gamma^{M\times 1} = \{i : E_i^{\text{store}} \geq \gamma_M\} \tag{13}$$

where γ_M defines the stored energy of the node with Mth largest stored energy. Equation (13) states that Γ contains elements with M largest stored energies out of N relays. As the fading distribution is iid and the CSI for the next time slot is not available at time t, the selection is based on the known stored battery condition for the relays. All the nodes $i \in \Gamma$ (attempt to) decode the information from the source and cannot harvest energy in time slot t while rest of the $N - M$ nodes harvest energy. We limit the cardinality of the set Γ to a fixed value $M \leq N$ where M is a system parameter to be optimized.

Then, a set Λ is selected out of M nodes that can retrieve the information from the signal on the $S \rightarrow L$ link such that

$$\Lambda = \{i : i \in \Gamma, R_{si} > R\} \tag{14}$$

As the CSI at relay nodes in Λ is available at the time of transmission in time slot $t + 1$, a single relay L_{i*} from the set Λ is selected such that

$$i^* = \arg\max_{i\in\Lambda} \left(E_i^{\text{store}}(t+1) - E_r^i(t+1)\right)^+ \tag{15}$$

where $E_r^i(t + 1)$ results from power allocation $P_r^i(t + 1)$ and given by,

$$P_r^i = \frac{(2^{2R} - 1)\sigma^2}{|h_{id}|^2} \tag{16}$$

If $E_i^{\text{store}}(t + 1) < E_r^i(t + 1), \forall i \in \Lambda$, no node is selected for transmission which results in outage, but avoids energy loss due to unsuccessful transmission from node i^*.

If $E_{i^*}^{\text{store}}(t + 1) \geq E_r^{i^*}(t + 1)$, the stored energy for node L_{i*} is updated such that

$$E_{i^*}^{\text{store}}(t+2) = E_{i*}^{\text{store}}(t+1) - E_r^{i^*}(t+1) \tag{17}$$

The rest of the nodes harvest and the stored energy depending on the received signal strength from the source is given by,

$$E_j^{\text{store}}(t+1) = \begin{cases} E_j^h(t) + E_j^{\text{store}}(t), & j \notin \Gamma \\ E_j^{\text{store}}(t), & j \in \Gamma, j \neq i^* . \end{cases} \tag{18}$$

We notice that parameter M controls the outage probability for a fixed N. There is a tradeoff associated with selection of M. Increasing M makes the relay selection in (15) more opportunistic due to large cardinality of set Λ and more freedom in choosing i^* in (15). However, note that all $i \in \Gamma$ do not harvest energy and their storage level remains the same. In this work, we assume no leakage factor but practically, there is a leakage in storage for every node in each time slot even if the node is not transmitting. Large M implies that less number of nodes are charging their batteries and therefore, the storage at system level keeps on decreasing and causes more outage (network failure). Therefore, there is an optimal $M \leq N$ for the proposed scheme which maximizes the performance.

Given that we have MRS policy $\pi(M, N)$ for relay selection, the parameter optimization problem is formulated by

$$M^*(R, \eta) = \arg \min_{\pi(M,N), 0<M\leq N} P_{\text{out}} \tag{19}$$

with the same constraints as in (9). The value of M^* depends on the number of relays in the system N and energy harvesting efficiency η.

5.3 Numerical Results

We assume independent Rayleigh fading channels with mean 1 for the channels on the both $S \rightarrow L$ and $L \rightarrow D$ links. 20000 iterations are performed to compute outage probability numerically for the simulation results. The relays are assumed to be equidistant from the source with d equals one. Signal to noise ratio (SNR) on the $S \rightarrow L$ link is 10 db with $\sigma^2 = 1$.

Figure 8 shows the outage probability for the SRS scheme when N is fixed. As the CSI is not available at the relay node, P_r is fixed to 10 dBW. As expected, the outage probability increases as R increases. The number of relays N and energy harvesting efficiency η are important factors to characterize the scheme. For a fixed value of N, a decrease in η results in decreased harvested energy for the relay nodes. When η is decreased initially, P_{out} remains the same as for $N = 5$ case with $\eta = 0.5$ and $\eta = 0.4$, which implies that at least a single node is always available with enough harvested energy. However, if η is too small, the probability that no relay has enough energy to make a successful transmission increases as evident for the case $N = 5$, $\eta = 0.2$ in Fig. 8. When η is very small, the outage is observed even for very small R as the selected relay must have enough energy to transmit with power $P_r = 10$

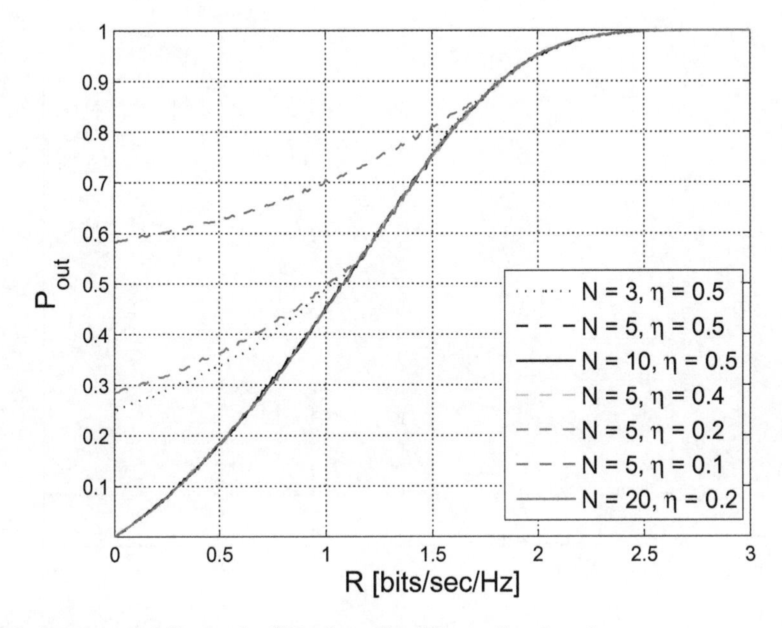

Fig. 8 Outage probability for the SRS scheme for different N and η values

dBW regardless of R. This region can be termed as power limited region where the outage performance is dominated by the harvested energy as compared to the large rate region where the channel distribution determines the outage behaviour and the power limitation effect almost vanishes.

The same effect is observed with small N where the effect of small η is even more pronounced as $N - 1$ relays harvest energy in a single time slot. Limiting N exaggerates the power limitation effect due to poor energy harvesting efficiency. As η in practically available systems is too low, it is important to have large N to reduce the effect of small η. For example, $N = 20$ in Fig. 8 improves outage performance considerably at small R as compared to $N = 5$ case when $\eta = 0.2$.

Figure 9 shows the outage performance for the MRS case. As the CSI is known at the relay node for transmission on the $L \rightarrow D$ link, P_r^i is determined by (16). For a fixed N, we plot the outage probability curves for different values of M and determine the optimal value M^* numerically. As discussed in Sect. 5.2, the outage probability for $M > M^*$ is not optimal due to sub-optimality in energy harvesting from the RF signals while $M < M^*$ results in too small group of candidate relays to exploit multiuser diversity significantly in time slot $t + 1$ on the $L \rightarrow D$ link. For the numerical example with $N = 10$, $M = 5$ provides the optimal outage performance.

Figure 10 compares SRS and MRS schemes for the same value of N. MRS outperforms SRS scheme even for $M = 1$ case thanks to power allocation according to available CSI at relay on the $L - D$ link. However, the performance improves considerably when $M = M^*$ for the MRS scheme.

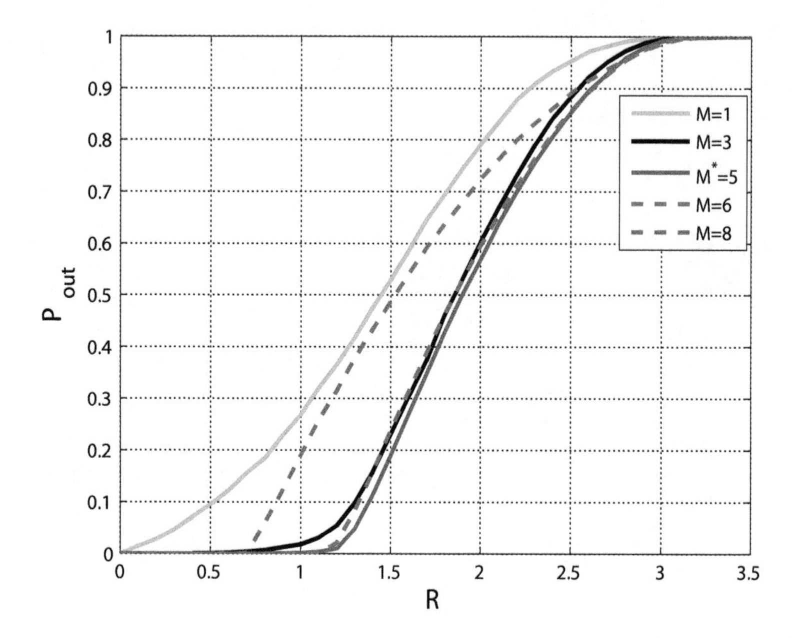

Fig. 9 Outage probability for MRS scheme for parameters $N = 10$, $\eta = 0.1$ and different M

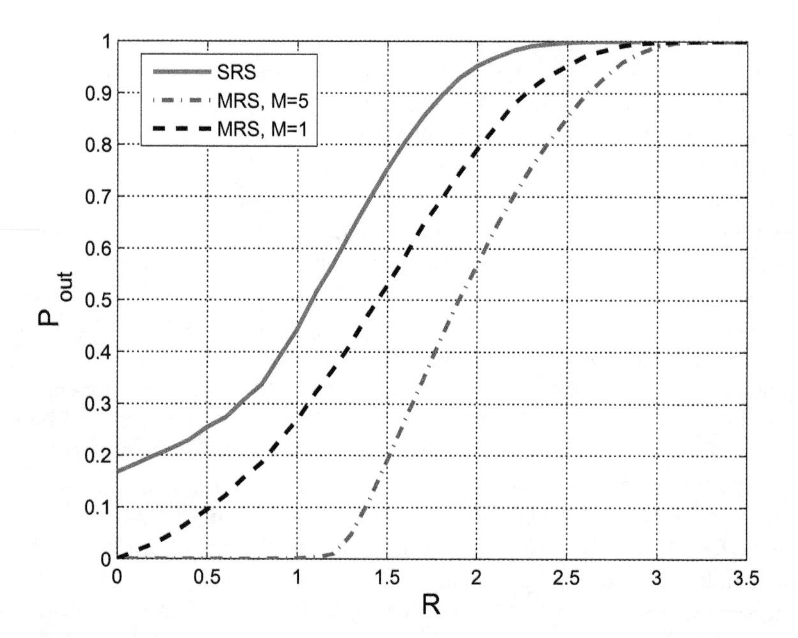

Fig. 10 Outage comparison of SRS and MRS schemes for a fixed $N = 10$ and $\eta = 0.1$

6 Radio Resource Allocation: Challenges and Potential

A key aspect that affects the link performance of EH wireless nodes is the energy profile, which models the availability of energy with time. In other words, we can define the energy profile as a statistical analysis of available energy during a specific course of time. During the analysis of EH profile, the following common assumptions have been made:

- Infinite energy and data storage to store unlimited energy/data.
- Ideal storage buffer without any leakage due to inefficiency in storage.
- The profile of the harvested energy is known in advance (offline).

These assumptions influence the performance of an EH profile greatly. For instance, ideal and infinite energy buffer should not be the appropriate selection for practical implementation of such systems. Though this issue has been addressed in [2, 22], it would be interesting to investigate the performance without making idealistic assumptions for more complex EH wireless communication applications.

Energy profile and its parameters (transmit energy per measurement to the energy profile, battery capacity of the EH node, etc.), in general, depend on the characteristics of the device physics. One of the important factors for allocating resources is to tune these parameters accurately. Various analytical models have been considered to tune these parameters for the EH profile. For example, the Markov model is one of the most common techniques used in the literature [13, 31, 41]. For possible future research scope, we need to think of modifying the proposed models by incorporating the multiple levels of harvested energy for a given time interval.

Recent work on EH has produced some useful analysis techniques and novel algorithms for managing resources with uncertainty about their availability. For example, EH cognitive radio and cognitive relays bring interesting challenges in terms of analysis. The interaction between energy queues and data queues makes the analysis of such systems very involved. Mechanisms such as, dominant system approach and moment generating functions are promising tools to tackle these type of problems. However, such mechanisms only promise to provide upper/lower bounds on the estimated delays and require further investigation.

EH cognitive radio has the potential benefit of an uninterrupted life without requiring external power cables or periodic battery replacements. There are open issues regarding optimal spectrum sensing policies under energy causality and collision constraints that need to be investigated deeply. For example, the issues related to bursty data traffic, finite battery capacity, inefficiency in storing energy in the battery, and temporal correlation of energy sources for designing the spectrum access policy need to be investigated carefully.

An interesting topic for future work is the employment of base station cooperative systems to exchange energy. In a cooperative system, the base stations exchange information via wireline backhaul links to coordinate their transmissions and thus, reduce interference. In an EH setting, the base stations could be self-powered using conventional energy sources and would also exchange energy through power lines to achieve sustainability of the network. Moreover, they could utilize SWIPT and transfer energy to a receiver by coordinating through their cooperative system and thus, allocating their resources efficiently. An additional approach would be to consider full duplex mode and study the effects on the battery charging and resource allocation.

In the domain of wireless energy transfer, one of the key challenges is to think about simultaneous information collection and power transfer. In the literature, we find that this technique can be implemented by exploiting the broadcast nature of wireless channels. In that case, proper receiver architecture designs for SWIPT can play a vital role. Moreover, the development of new signal processing techniques for SWIPT such as simultaneous information and energy transfer in the spatial domain are very important.

Another main challenge in energy cooperation is the low wireless energy transfer efficiency due to path-loss effects. A number of different directions have been proposed to address this issue. Most of the previous discussion in this chapter considers point-to-point and two-hop communication systems which are not practical since the path-loss effects increase with distance. Therefore, a reasonable approach to this challenge is the study of multi-hop communication systems which utilize SWIPT. Furthermore, massive MIMO systems have also been proposed to improve the wireless energy transfer efficiency by exploiting large array gain due to the large number of antennas deployed at the transmitter. Also, new network architectures such as cellular networks where mobiles and sensors are charged by dedicated power stations are desirable.

In wireless networks, the presence of signal interference is inevitable. However, the exploitation of this interference is beneficial in EH systems. The development of new techniques to exploit the interference are essential. For example, it has been shown that the interference alignment technique can be applied to use the interference as energy rather than discarding it. Also, the use of full duplex radio in all the aforementioned directions is desirable as it can essentially double the energy transfer. A well known disadvantage of the full duplex mode is the loop interference caused by the output and input antennas but as mentioned above, the exploitation of this extra interference will benefit an EH system even more.

7 Conclusions

Energy harvesting is a promising research area that addresses the issue of limited network lifetime for energy-constrained wireless networks. The recent prominence on green communications also indicates a strong motivation and desire for developing energy harvesting based communication systems. However, considerable research effort still needs to be given to investigate various problems in order to make energy harvesting based wireless communication more practically feasible. Specifically, the techniques of allocating resources for energy harvesting devices largely differ from the conventionally powered devices. The fundamental rule that manages the operation of an energy harvesting network is the energy neutrality constraint. The main problem of being too conventional in allocating the harvested energy to the energy harvesting nodes simply waste the harvested energy. On the other hand, applying a too aggressive approach in allocating the resources leads to energy harvesting nodes unnecessarily starve of the energy for a longer period of time. After reviewing the fundamental building blocks of an EH system, we mainly focus on wireless energy transfer based communication in this chapter. First, we briefly review (without being exhaustive) the recent literature in this area and then, present example of a wireless energy based relay selection problem. Finally, we outline the main result challenges in energy harvesting wireless communication and introduce some open research issues.

Acknowledgments This publication was made possible by NPRP 5-782-2-322 and NPRP 4-1034-2-385 from the Qatar National Research Fund (a member of The Qatar Foundation). The statements made herein are solely the responsibility of the authors.

References

1. Ahmed, I., Butt, M.M., Psomas, C., Mohamed, A., Krikidis, I., Guizani, M.: Survey on energy harvesting wireless communications: Challenges and opportunities for radio resource allocation. Elsevier Comput. Netw. (2015). doi:10.1016/j.comnet.2015.06.009
2. Ahmed, I., Ikhlef, A., Ng, D., Schober, R.: Power allocation for an energy harvesting transmitter with hybrid energy sources. IEEE Trans. Wirel. Commun. **12**(12), 6255–6267 (2013). doi:10.1109/TWC.2013.111013.130215
3. Antepli, M., Uysal-Biyikoglu, E., Erkal, H.: Optimal packet scheduling on an energy harvesting broadcast link. IEEE J. Sel. Areas Commun. **29**(8), 1721–1731 (2011). doi:10.1109/JSAC.2011.110920
4. Butt, M.M., Nasir, A., Mohamed, A., Guizani, M.: Trading wireless information and energy transfer: relay selection schemes to minimize the outage probability. In: IEEE Global Conference on Signal and Information Processing (GlobalSIP). Atlanta, Georgia (2015)
5. Ding, Z., Perlaza, S.M., Esnaola, I., Poor, H.V.: Power allocation strategies in energy harvesting wireless cooperative networks. IEEE Trans. Wirel. Commun. **13**(2), 846–860 (2014). doi:10.1109/TWC.2013.010213.130484
6. Ding, Z., Poor, H.: Cooperative energy harvesting networks with spatially random users. IEEE Signal Process. Lett. **20**(12), 1211–1214 (2013). doi:10.1109/LSP.2013.2284800
7. Fehske, A., Fettweis, G., Malmodin, J., Biczók, G.: The global footprint of mobile communications: The ecological and economic perspective. IEEE Commun. Mag. **49**(8), 55–62 (2011)
8. Feng, R., Li, Q., Zhang, Q., Qin, J.: Robust secure transmission in MISO simultaneous wireless information and power transfer system. IEEE Trans. Veh. Technol. **PP**(99), 1-1 (2014). doi:10.1109/TVT.2014.2322076
9. Group, C.I.R.: Renewable energy sources and climate change mitigation. http://srren.ipcc-wg3.de/ipcc-srren-generic-presentation-1 (2012)
10. Gurakan, B., Ozel, O., Yang, J., Ulukus, S.: Two-way and multiple-access energy harvesting systems with energy cooperation. In: Forty Sixth Asilomar Conference on Signals, Systems and Computers (ASILOMAR). Pacific Grove, CA (2012). doi:10.1109/ACSSC.2012.6488958
11. Gurakan, B., Ozel, O., Yang, J., Ulukus, S.: Energy cooperation in energy harvesting communications. IEEE Trans. Commun. **61**(12), 4884–4898 (2013). doi:10.1109/TCOMM.2013.110113.130184

12. Gurakan, B., Ozel, O., Yang, J., Ulukus, S.: Energy cooperation in energy harvesting two-way communications. In: IEEE International Conference on Communications (ICC). Budapest (2013)
13. Ho, C.K., Zhang, R.: Optimal energy allocation for wireless communications with energy harvesting constraints. IEEE Trans. Signal Process. **60**(9), 4808–4818 (2012)
14. Hossain, E., Bhargava, V.K., Fettweis, G.P.: Green Radio Commun. Netw., 1st edn. Cambridge University Press, Cambridge (2012)
15. Huang, K., Lau, V.: Enabling wireless power transfer in cellular networks: Architecture, modeling and deployment. IEEE Trans. Wirel. Commun. **13**(2), 902–912 (2014). doi:10.1109/TWC.2013.122313.130727
16. Ikhlef, A., Kim, J., Schober, R.: Mimicking full-duplex relaying using half-duplex relays with buffers. IEEE Trans. Veh. Technol. **61**(7), 3025–3037 (2012)
17. Ishibashi, K., Ochiai, H., Tarokh, V.: Energy harvesting cooperative communications. In: IEEE 23rd International Symposium on Personal Indoor and Mobile Radio Communications (PIMRC). Sydney, Australia (2012). doi:10.1109/PIMRC.2012.6362646
18. Ju, H., Zhang, R.: A novel mode switching scheme utilizing random beamforming for opportunistic energy harvesting. IEEE Trans. Wirel. Commun. **13**(4), 2150–2162 (2014). doi:10.1109/TWC.2014.030314.131101
19. Ju, H., Zhang, R.: Throughput maximization in wireless powered communication networks. IEEE Trans. Wirel. Commun. **13**(1), 418–428 (2014). doi:10.1109/TWC.2013.112513.130760
20. Kansal, A., Hsu, J., Zahedi, S., Srivastava, M.B.: Power management in energy harvesting sensor networks. ACM Trans. Embed. Comput. Syst. **6**(4) (2007)
21. Krikidis, I.: Simultaneous information and energy transfer in large-scale networks with/without relaying. IEEE Trans. Commun. **62**(3), 900–912 (2014). doi:10.1109/TCOMM.2014.020914.130825
22. Krikidis, I., Timotheou, S., Sasaki, S.: RF energy transfer for cooperative networks: data relaying or energy harvesting? IEEE Commun. Lett. **16**(11), 1772–1775 (2012). doi:10.1109/LCOMM.2012.091712.121395
23. Lee, S., Liu, L., Zhang, R.: Collaborative wireless energy and information transfer in interference channel. IEEE Trans. Wirel. Commun. **14**(1), 545–557 (2015). doi:10.1109/TWC.2014.2354335
24. Lei, J., Yates, R., Greenstein, L.: A generic model for optimizing single-hop transmission policy of replenishable sensors. IEEE Trans. Wirel. Commun. **8**(2), 547–551 (2009). doi:10.1109/TWC.2009.070905
25. Liu, L., Zhang, R., Chua, K.: Multi-antenna wireless powered communication with energy beamforming. IEEE Trans. Commun. **62**(12), 4349–4361 (2014). doi:10.1109/TCOMM.2014.2370035
26. Liu, L., Zhang, R., Chua, K.C.: Wireless information and power transfer: a dynamic power splitting approach. IEEE Trans. Commun. **61**(9), 3990–4001 (2013). doi:10.1109/TCOMM.2013.071813.130105
27. Liu, L., Zhang, R., Chua, K.C.: Wireless information transfer with opportunistic energy harvesting. IEEE Trans. Wirel. Commun. **12**(1), 288–300 (2013). doi:10.1109/TWC.2012.113012.120500
28. Liu, L., Zhang, R., Chua, K.C.: Secrecy wireless information and power transfer with MISO beamforming. IEEE Trans. Signal Process. **62**(7), 1850–1863 (2014). doi:10.1109/TSP.2014.2303422
29. Luo, Y., Zhang, J., Letaief, K.B.: Relay selection for energy harvesting cooperative communication systems. In: IEEE Global communications conference (Globecom). Atlanta, GA, USA (2013)
30. Mancuso, V., Alouf, S.: Reducing costs and pollution in cellular networks. IEEE Commun. Mag. **49**(8), 63–71 (2011)
31. Medepally, B., Mehta, N., Murthy, C.: Implications of energy profile and storage on energy harvesting sensor link performance. In: IEEE Global Telecommunications Conference (GLOBECOM). Honolulu, HI (2009). doi:10.1109/GLOCOM.2009.5425655

32. Michalopoulos, D.S., Suraweera, H.A., Schober, R.: Relay selection for simultaneous information transmission and wireless energy transfer: a tradeoff perspective. In IEEE journal on Selected Areas in Communications. **33**(8), 1578–1594 (2015)
33. Michelusi, N., Stamatiou, K., Zorzi, M.: Transmission policies for energy harvesting sensors with time-correlated energy supply. IEEE Trans. Commun. **61**(7), 2988–3001 (2013). doi:10.1109/TCOMM.2013.052013.120565
34. Nasir, A.A., Zhou, X., Durrani, S., Kennedy, R.A.: Relaying protocols for wireless energy harvesting and information processing. IEEE Trans. Wirel. Commun. **12**, 3622–3636 (2013)
35. Ng, D., Lo, E., Schober, R.: Energy-efficient power allocation in ofdm systems with wireless information and power transfer. In: IEEE International Conference on Communications (ICC). Budapest (2013). doi:10.1109/ICC.2013.6655208
36. Ng, D., Lo, E., Schober, R.: Energy-efficient resource allocation in multiuser ofdm systems with wireless information and power transfer. In: IEEE Wireless Communications and Networking Conference (WCNC). Shanghai (2013). doi:10.1109/WCNC.2013.6555184
37. Ng, D., Lo, E., Schober, R.: Wireless information and power transfer: energy efficiency optimization in ofdma systems. IEEE Trans. Wirel. Commun. **12**(12), 6352–6370 (2013). doi:10.1109/TWC.2013.103113.130470
38. Ng, D., Lo, E., Schober, R.: Robust beamforming for secure communication in systems with wireless information and power transfer. IEEE Trans. Wirel. Commun. **PP**(99), 1-1 (2014). doi:10.1109/TWC.2014.2314654
39. Ng, D., Xiang, L., Schober, R.: Multi-objective beamforming for secure communication in systems with wireless information and power transfer. In: IEEE 24th International Symposium on Personal Indoor and Mobile Radio Communications (PIMRC). London, UK (2013).doi:10.1109/PIMRC.2013.6666095
40. Niyato, D., Wang, P.: Delay-limited communications of mobile node with wireless energy harvesting: performance analysis and optimization. IEEE Trans. Veh. Technol. **63**(4), 1870–1885 (2014). doi:10.1109/TVT.2013.2285922
41. Sharma, V., Mukherji, U., Joseph, V., Gupta, S.: Optimal energy management policies for energy harvesting sensor nodes. IEEE Trans. Wirel. Commun. **9**(4), 1326–1336 (2010). doi:10.1109/TWC.2010.04.080749
42. The Climate Group: SMART2020: Enabling the low carbon economy in the information age. http://www.smart2020.org/_assets/files/02_Smart2020Report.pdf (2008)
43. Tutuncuoglu, K., Yener, A.: Cooperative energy harvesting communications with relaying and energy sharing. In: IEEE Information Theory Workshop (ITW). Sevilla, Spain (2013). doi:10.1109/ITW.2013.6691280
44. Xu, J., Liu, L., Zhang, R.: Multiuser MISO beamforming for simultaneous wireless information and power transfer. IEEE Trans. Signal Process. **62**(18), 4798–4810 (2014). doi:10.1109/TSP.2014.2340817
45. Yang, G., Lin, G.Y., Wei, H.Y.: Markov chain performance model for IEEE 802.11 devices with energy harvesting source. In: IEEE Global Communications Conference (GLOBECOM). Anaheim, CA (2012). doi:10.1109/GLOCOM.2012.6503948
46. Yuan, F., Zhang, Q., Jin, S., Zhu, H.: Optimal harvest-use-store strategy for energy harvesting wireless systems. IEEE Trans. Wirel. Commun. **14**(2), 698–710 (2015). doi:10.1109/TWC.2014.2358215
47. Zhang, R., Ho, C.K.: MIMO broadcasting for simultaneous wireless information and power transfer. IEEE Trans. Wirel. Commun. **12**(5), 1989–2001 (2013). doi:10.1109/TWC.2013.031813.120224
48. Zhou, X., Zhang, R., Ho, C.K.: Wireless information and power transfer: architecture design and rate-energy tradeoff. IEEE trans. Commun. **61**(11), 4754–4767 (2013). doi:10.1109/TCOMM.2013.13.120855
49. Zhou, X., Zhang, R., Ho, C.K.: Wireless information and power transfer in multiuser OFDM systems. IEEE Trans. Wirel. Commun. **13**(4), 2282–2294 (2014). doi:10.1109/TWC.2014.030514.131479

Part VI
Energy Management Practices in Wireless Networks

EMrise: An Energy Management Platform for WSNs/WBANs

Nanhao Zhu and Athanasios V. Vasilakos

Abstract Due to reliance on batteries, energy consumption has always been of significant concern for sensor node networks. This work presents the design and implementation of a house-build experimental platform, named EMrise (**E**nergy **M**anagement System fo**r** W**i**reless **S**ensor N**e**tworks) for the energy management and exploration on wireless sensor networks. Consisting of three parts, the SystemC-based simulation environment of EMrise enables the HW/SW co-simulation for energy evaluation on heterogeneous sensor networks. The hardware platform of EMrise is further designed to facilitate the realistic energy consumption measurement and calibration as well as accurate energy exploration. In the meantime, a generic GA (genetic algorithm) based optimization framework of EMrise is also implemented to automatically, quickly and intelligently fine tune hundreds of possible solutions for the given task to find the best suitable energy-aware tradeoffs.

Keywords Wireless sensor networks · Energy consumption · Energy management/evaluation · Simulation · SystemC · Measurement · Optimization · Genetic algorithm

1 Introduction

With the great development in embedded systems and wireless communication technologies, wireless sensor networks (WSNs) have gained worldwide attention and have been developing rapidly in the past decade. Wireless sensor networks are large-

He received his Ph.D. from Ecole Centrale de Lyon, France. Now he is working as a senior researcher at CETC Group, GCI Science and Technology Co., Ltd. Canton, China.

N. Zhu (✉)
CETC Group, GCI Science and Technology Co., Ltd, No. 381, Xingangzhong Road, Canton 510310, Haizhu District, China
e-mail: nanhaozhu@foxmail.com

A.V. Vasilakos
Department of Computer Science, Electrical and Space Engineering, Lulea University of Technology, 97187 Skellefteå, Sweden
e-mail: vasilako@ath.forthnet.gr

© Springer International Publishing Switzerland 2016

M.Z. Shakir et al. (eds.), *Energy Management in Wireless Cellular and Ad-hoc Networks*, Studies in Systems, Decision and Control 50,
DOI 10.1007/978-3-319-27568-0_16

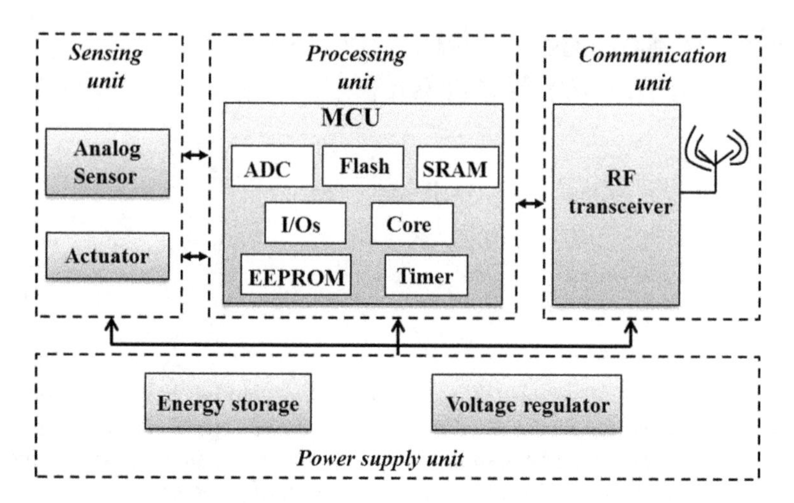

Fig. 1 Sensor node hardware architecture

scale networks with low-cost, small-size, low-power and limited processing sensor nodes deployed in various kinds of environments. As the basic part of WSNs, a typical sensor node comprises several components: microcontroller, transceiver, sensor and power supply (Fig. 1). By combining these different components into a miniaturized device, these sensor nodes become multi-functional and have been applied to a wide variety of application scenarios: medical and health care [1], environment and ecological monitoring [2], home and building automation [3], industrial monitoring [4] and battlefield [5]. Thus, such highly diverse scenarios impose application-specific requirements on WSN design and distinguish them from conventional networks.

Generally, WSNs are easy to achieve great scalability and a wide range of densities, due to the low cost and small dimensions of sensor nodes. In the meantime, specific sensor network protocols and algorithms with self-organizing capabilities are usually designed and applied to sensor nodes, so the whole sensor network is low-maintenance and tolerant against communication failure and topology changes which could be caused by node malfunction, node mobility or energy depletion of the nodes. Furthermore, sensor nodes can also be deployed in harsh environments with self-organized network to carry out given tasks in unmanned manners.

However, further and potential applications are limited due to inherent WSN disadvantages. For instance, the limited processing ability and low data rate of sensor nodes cannot guarantee high performance in some scenarios especially for real-time applications. Short communication range can cause energy waste and network inefficiency, since multi-hop communications are always required for data transport between source node and sink node. The severe energy constraints also lead to a worse observation: the improvement of data processing ability by using powerful processors are not expected, since the energy will be depleted very quickly and renders the network useless on the given task. Limited energy also means it is not possible to also maintain the operation of multi-hop communications for a long time. Meanwhile, sensor nodes are expected to be used heavily in many remote area applications, where the frequent changing or recharging of batteries is inconvenient.

Hence, an energy-efficient sensor network is necessary to carry out the work in such conditions. Taking energy consumption as the major consideration, while not undermining other network performances, is therefore desired as the priority strategy in the design of current wireless sensor networks.

2 Related Works

The emerging field of wireless sensor networks and their applications promises a higher quality of life in many aspects of our daily activities, but the strict energy constrains stated above significantly limit their functionality expansions and possible applications, which makes energy consumption one of the most critical concerns in WSNs. Therefore, energy consumption analysis and evaluation is a critical process in the design and implementation of the energy-efficient sensor network.

In recent years, great efforts which have been made on energy saving management for wireless sensor networks can be grouped into three categories: (1) simulation/emulation based approach for energy consumption calculation and analysis; (2) hardware based method for real-world energy measurement and calibration; (3) optimization based strategy for the exploration of energy-efficient behaviors and configurations.

First of all, as a useful tool for the energy analysis and management on wireless sensor networks, simulation/emulation based method is widely accepted and commonly used, it can provide a more realistic model for the evaluation of energy consumption rather than the oversimplified and idealized hypotheses that are assumed in the mathematics based theoretical models. Although relatively slowly, it offers the tradeoff between accuracy and efficiency. General simulation tools like NS-2 [6], OMNeT++ [7], Prowler [8], written respectively in C++, Java and MATLAB, provide satisfactory efficiency at the early design stages, since the models are usually at a high abstraction level to facilitate the testing and verification of algorithms, protocols and strategies. However, only a relative coarse energy consumption evaluation can thus be expected due to the lack of realistic low level models. Emulation tools like TOSSIN [9], ATEMU [10] and Avrora [11] can compensate this drawback on energy consumption evaluation but at the cost of efficiency, and they are basically limited to specific hardware platforms and operating systems (TOSSIN is used for sensor nodes with TinyOS [12] operating system, both ATEMU and Avrora are used for AVR microcontroller based node platforms).

From the aspect of hardware based method, an increasing number of research works have already focused on energy consumption in real-world sensor nodes. A detailed energy consumption analysis based on the MICAz mote [13] is presented in [14], several benchmarks are used for energy estimation, battery charge effect and battery lifetime are also considered. Authors in [15] present AEON (Accurate Prediction of Power Consumption) and evaluate detailed energy profiling on a MICA2 sensor node [16]. Elaborate energy consumption data on the TelosB mote [17] is provided based on the authors' measuring methodology and power consumption measuring system (PCMS) in [18]. Sensor Node Management Devices (SNMD) [19, 20] is presented with detailed performance parameter configuration for the energy data measurements.

In the view of optimization based energy management strategy, corresponding research works have been done from both hardware and software: (1) From the hardware perspective, better energy efficiency can be achieved by optimizing the power consumption of related hardware components [21–26]. (2) From the software perspective, the optimization method can be grouped into three categories: the development of new communication protocols (MAC and routing) for optimization, the adoption of energy-aware management methods for optimization, and the configuration/exploration of the optimal set of existing protocols for optimization [25, 27–29].

3 Motivation

Despite the tradeoff can be achieved between accuracy and efficiency with simulation/emulation based method for energy management and evaluation on wireless sensor networks as mentioned above, without hardware-based instruction level model for energy evaluation only unrealistic result can be acquired by simulation method. Meanwhile the inefficient and platform-specific based emulation method cannot be considered as a better solution. Since SystemC [30] is a system-level and hardware description language offering support for HW/SW co-simulation, concurrency, modeling at different abstraction levels as well as other flexible and diverse modeling advantages, it is therefore regarded as a good alternative to model and simulate WSNs for accurate, flexible and quick energy evaluation.

Although measurements from large-scale of sensor nodes are hard to acquire and sometimes these nodes are placed in harsh and inaccessible areas, the hardware based measurements from real world testbed nodes are still considered necessary and significant since they offer a more realistic environment to the real deployment and are able to calibrate and improve the accuracy of energy evaluation. Research works mentioned above can only provide simple energy measurements on specific node platforms and with costly measurement equipment (high-performance oscilloscope [15, 31], acquisition cards [18, 32]) on the sensor nodes to sample the varying low currents and voltages, which limit the scale of deployment for energy measuring beyond the lab environment. Therefore, the building of a cost-effective, multi-functional and measurement reliable hardware platform is currently required.

Compared with the hardware measurement method and software simulation method for optimization, in spite of the efficiency provided by software based simulation, the exhaustive and full simulation process on all the possible configurations of the protocol are not only time-consuming, but also unnecessary most of the time. Thus, an approach that can effectively and automatically select an appropriate protocol parameter setting out of hundreds or thousands of possible options is urgently needed. With the fast and intelligent search ability, Genetic Algorithms (GAs) [33] enables the fine tuning of parameter space of the protocol to find the most suitable configuration for energy saving and management.

4 The Design and Implementation of EMrise

Based on the limitations and suggestions of the three aspects mentioned above, our EMrise (**E**nergy **M**anagement System fo**r** W**i**reless **S**ensor N**e**tworks) is proposed in this chapter. Section 4.1 describes the design and implementation of EMrise_SS (EMrise_**S**imulation **S**ystem) which is a SystemC-based simulation environment at the system-level and transaction-level, and the energy performance of several different sensor motes under various case scenarios are also investigated by EMrise_SS in this section. Section 4.2 presents EMrise_MS (EMrise_**M**easurement **S**ystem), a cost-effective and flexible hardware based energy measurement platform for the calibration and management of realistic energy data. Section 4.3 focuses on the introduction and design of EMrise_OpS, a generic and genetic algorithm based optimization framework for the optimized energy management on WSNs. Finally, Sect. 5 concludes this book chapter and discusses the possible future works.

4.1 EMrise_SS (EMrise Simulation System)

EMrise_SS, also known as iWEEP_SW [34], is a system-level, transaction-level and energy-aware SystemC based generic sensor network simulation which implements several popular commercial off-the-shelf (COTS) low data rate based sensor node models including the Telos series (e.g., TelosB [17], Tmote Sky [35], Shimmer node [36]), the MICA series (e.g., MICA2 [16], MICAz [13]), and a house-built high data rate and ultra-low power testbed node iHop@Node [37] (PIC16F microcontroller with nRF24L01+ transceiver, maximum data rate up to 2 Mbps). Four MAC protocols are integrated and modeled in EMrise_SS which are unslotted and slotted CSMA/CA protocols in IEEE 802.15.4 [29], as well as Enhanced ShockBurst (ESB) and ShockBurst (SB) protocols [38] embedded in the high data rate iHop@Node.

4.1.1 The Framework of EMrise_SS

EMrise_SS is composed of SystemC defined components, ports, channels, interfaces and connections shown in Fig. 2. Components correspond to basic hardware modules such as microcontroller, transceiver, sensor and battery as well as various kinds of peripherals. Each component in EMrise_SS is modeled as an individual module of SystemC. In these hardware component models, the sensor is modeled

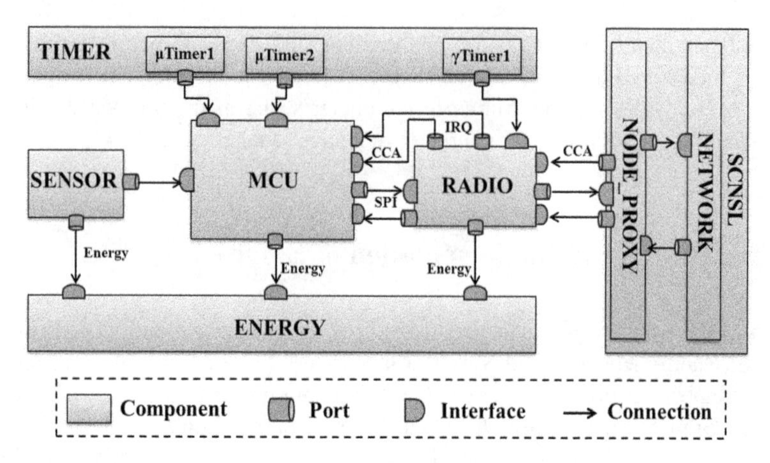

Fig. 2 EMrise_SS framework

to generate sensed data periodically from physical or environment conditions. The microcontroller (MCU) collects these analog data and converts them into digital signals by the built-in Analog to Digital Converter (ADC). The converted digital data are then sent to the transceiver according to different application scenarios and communication strategies. The transceiver (Radio) takes the responsibility of over-the-air packet transmission and reception, as well as clear channel assessment (CCA). The energy module functions as a power consumption monitor which can track and assess the energy consumption of other hardware components at different abstraction levels from bit-accurate/cycle-accurate to transaction-accurate based on designer's requirements. In the timer module, several sub-timer components can be defined for different purposes such as timer for sample interval, timer for each transmission interval, protocol related timer, etc. For some other peripherals such as the memory model and UART model inside the MCU, their modeling levels are optional, which depend on the detailed design requirements. In addition, the network module is also modeled as a component since it is inherited from the TLM-based SystemC Network Simulation Library (SCNSL) [39], which is used to establish network topology, implement packet transmission and handle network collisions.

Channels are responsible for the communications between different components, such as the internal bus of the microcontroller connecting CPU core and peripherals, the SPI (Serial Peripheral Interface) bus connecting microcontroller and transceiver. The channels' implementations are not mandatory and they are always abstracted to simplify the development process according to the designer's requirements. Each component and channel can have one or more interfaces to distinguish and specify a set of functions (or transactions), which are used to interact with the relevant components or channels. Like interfaces, multiple ports can be defined by components and channels, and they are utilized to specify the types of the interfaces, so each function-specific port can be connected to the related component or channel as long as the corresponding interface is implemented by that component or channel [40].

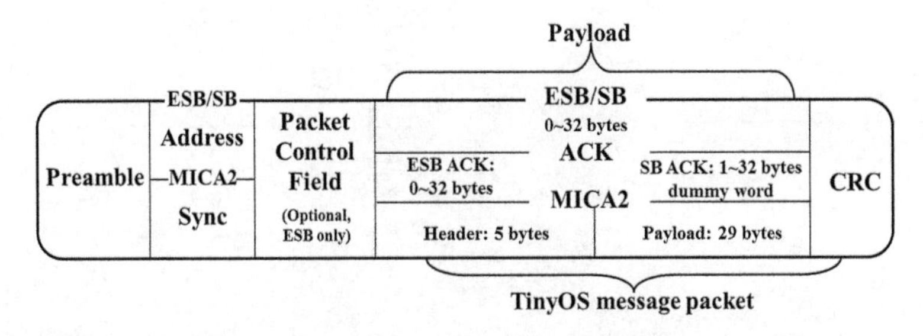

Fig. 3 Generic packet format

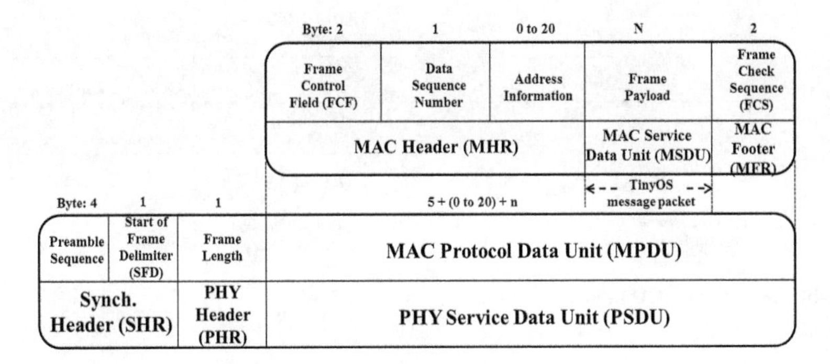

Fig. 4 IEEE 802.15.4 packet frame format

In addition, in EMrise_SS framework, the whole sensor network system maintains two data items for packet frames. One is a generic packet format for Enhanced ShockBurst (ESB)/ShockBurst (SB) protocol-based sensor node network and the other is for IEEE 802.15.4 based sensor network, which are presented in Figs. 3 and 4 respectively.

4.1.2 Microcontroller Modeling

Detailed model of the microcontroller is presented in Fig. 5. The microcontroller component defines the *sensordata_read_if* interface consisting of a single *SensorDataRead* transaction (or function) which allows the microcontroller to first read data from the target sensor module and then convert these analog data into digital data by the built-in Analog to Digital Converter (ADC). In some application scenarios, if an external ADC is needed for a more accurate and quicker conversion, the microcontroller defines another *extADC_read_if* interface. This is maintained by an *extADC_read* transaction (function) that emulates the process whereby the microcontroller reads converted ADC values through its digital I/O port pins or through SPI pins from an external ADC chip. The microcontroller component also links ADC's *extADC_config_if* interface by the defined output port on the microcontroller side, which is composed of a single *extADC_config* transaction to receive configuration

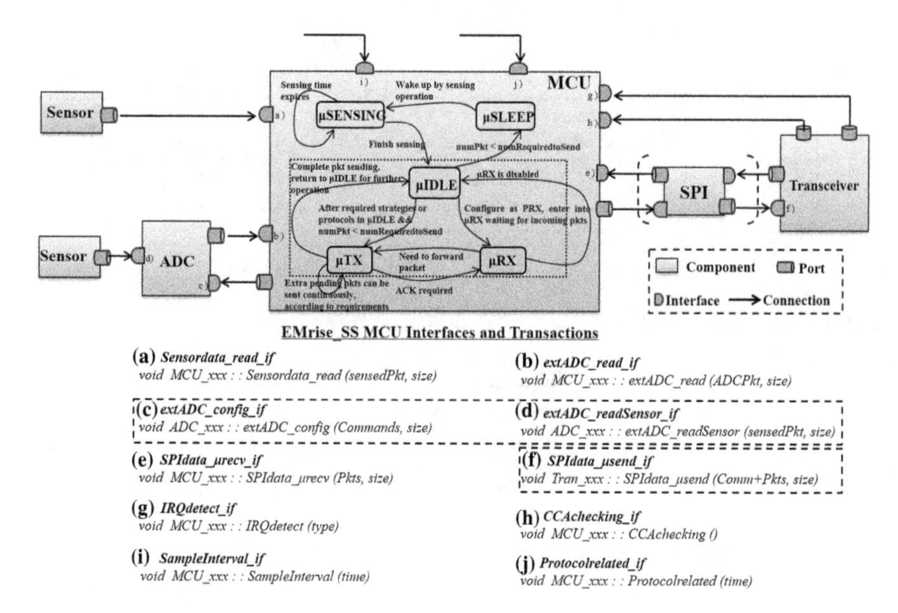

Fig. 5 EMrise_SS MCU model

information from microcontroller. Moreover, the *extADC_readSensor_if* interface and *extADC_readSensor* transaction defined on the external ADC are used to acquire sampled sensor data. Considering that cycle-by-cycle based energy consumption on the sensor node would be evaluated in later work, the above interfaces and transactions can be easily extended to incorporate detailed communication mechanism and device driver code implementation.

Another two transactions named *IRQdetect* and *CCAchecking*, which are defined by *IRQdetect_if* and *CCAchecking_if*, is responsible for interrupts detecting from transceiver and clear channel assessment processes respectively. Several timer transactions can be defined for varying functions such as the examples that are shown in Fig. 5. In addition, interactions between the microcontroller and transceiver are handled via SPI, considering that SPI communications between the microcontroller and transceiver are bidirectional. Hence, the SPI communication interface defines two transactions named *SPIdata_μrecv* and *SPIdata_μsend* on the microcontroller and transceiver components respectively. On the other hand, the operation of the microcontroller is performed with a finite state machine (FSM) in approximate-timed manner. A generic microcontroller model shown in Fig. 5 acts as the basic modeling for other specific microcontrollers.

Modeling of PIC16

Developed by Microchip, MiWi [41] as a wireless networking protocol stack is especially designed for PIC microcontroller families from PIC16 to dsPIC33. Currently, only the non-beacon mode of IEEE 802.15.4 is implemented as MiWi's MAC layer, and our PIC16F microcontroller is modeled following MiWi's unslotted algorithm as a FSM in previously mentioned μIDLE, μTX and μRX states, and divides these three states into many sub-states, which is shown in Fig. 6.

In Fig. 6, the microcontroller first performs a random backoff duration and checks the channel status. If the channel is detected to be busy, and the number of CCA attempts is larger than the protocol parameter *macMaxCSMABackoffs*, then a Channel Access Failure (CAF) is reported. If on the other hand the number of CCA attempts is smaller than *macMaxCSMABackoffs*, the microcontroller will go back for a new round of random backoff process. When the channel is indicated as free by CCA, the microcontroller will trigger over the air transmission to send a packet to the transceiver. After the transmission of an ACK required data packet, the protocol will make the microcontroller wait within a fixed time period (0.864 ms or 54 symbols) for the ACK frame confirmation. If ACK is received in time, this transmission is regarded as successful. Otherwise, the packet retransmission process will start. A Collision Failure (CF) occurs only after failure to receive ACK frame *macMaxFrameRetries* times, which might be caused by collisions of data packets or collisions between data packets and the ACK frame. In addition, as long as there are pending data packets in the MCU, they will be uploaded to the transceiver immediately, once the transmission process is over (no matter whether a success or a failure), which is to guarantee the timeliness and reliability of data transmission.

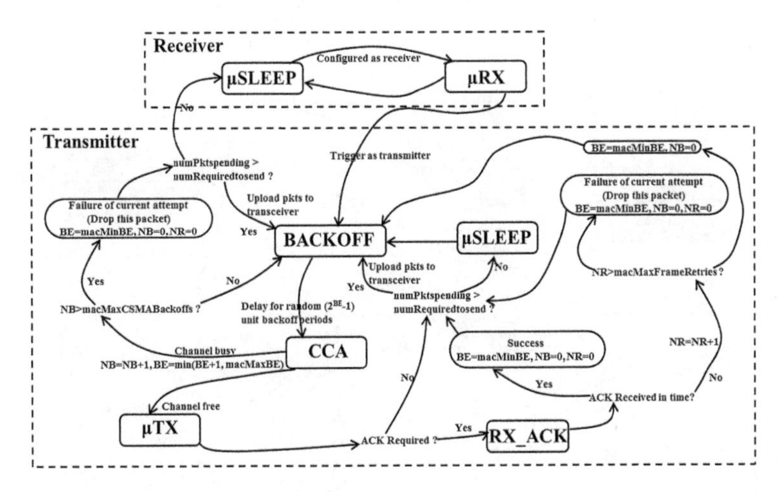

Fig. 6 Unslotted CSMA/CA algorithm model

Modeling MSP430

Compared with PIC16F, a powerful 16-bit MSP430 microcontroller is widely used in some current COTS (commercial-off-the-shelf) sensor motes such as the previous mentioned Telos, Tmote Sky and Shimmer nodes. A complete MAC protocol stack of IEEE 802.15.4 such as TKN15.4 [42] can be implemented in the TinyOS operating system on these popular sensor motes, which integrates both slotted and unslotted CSMA/CA algorithms (used in beacon-enable and non-beacon enable modes respectively). Hence, the MSP430 microcontroller component in EMrise_SS is modeled as a slotted and unslotted compatible FSM in the µIDLE, µTX and µRX states shown in Fig. 7.

Compared with the unslotted algorithm, the slotted CSMA/CA algorithm needs to be accurately timed, where a beacon packet broadcast by the coordinator at the beginning of the defined superframe [29] time period is used for synchronization between coordinator and sensor motes. After the reception of the beacon packet, sensor motes with enough packets will start a slotted-based CSMA/CA algorithm. This requires that the backoff period boundaries should be aligned with the superframe slot boundaries (simulation logs starting with '#####' in Fig. 19 show that the backoff period boundaries are accurately aligned with slot boundaries). Note that before each new backoff period there is a time evaluation process of checking whether remaining CSMA/CA can be undertaken before the end of the contention access period (CAP). If the number of backoff periods is less than the remaining number of backoff periods, the slotted algorithm will continue this backoff delay

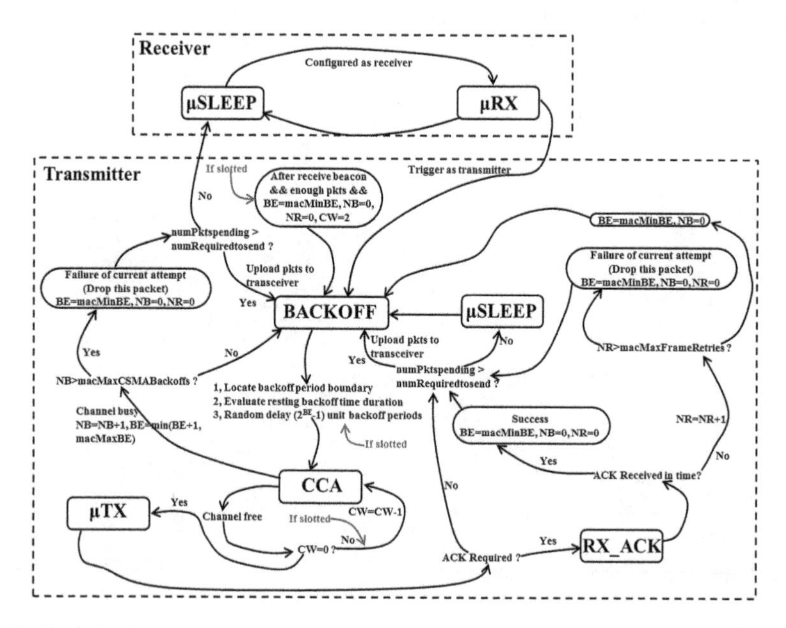

Fig. 7 Slotted and unslotted CSMA/CA model

under the condition that two CCA analyses, the frame transmission and acknowledgement can be completed before the end of the CAP. Otherwise, it must wait until the start of the next superframe CAP. If the number of backoff periods is greater than the number of remaining backoff periods, the slotted algorithm shall first pause the backoff countdown at the end of CAP and then resume it at the beginning of the CAP of next superframe. Except for the time evaluation mechanism, another difference between the slotted and unslotted algorithms is that two times of idle channel checks before packet transmission are required for the slotted algorithm.

4.1.3 Transceiver Modeling

The behavior of the RF transceiver is handled via two interfaces. As mentioned previously, the *SPIdata_μsend_if* interface is defined on the transceiver component and consists of a single transaction *SPIdata_μsend*, which manages events from the microcontroller such as receiving configuration commands and payload data. The *Netdata_rInput_if* interface defines the transaction *Netdata_rInput* that allows the transceiver to detect network related events such as the validation of incoming packets from the RF channel and the transaction *Netdata_channelstatus* can perform CCA operation. If the communication protocol is embedded in hardware, then some protocol-related timer transactions are also required such as ACK waiting transaction and auto retransmission delay (ESB mode). A typical transceiver is modeled shown in Fig. 8.

A specific feature of transceiver model in EMrise_SS is that its configuration is based on register settings like real hardware, which means that related registers are defined in each specific transceiver model. With the register configuration mechanism, the simulation of the network becomes flexible because different nodes could keep their own settings during the whole simulation process shown in Fig. 9. The final results could be acquired from each node to evaluate the performance of such a heterogeneous configured network scenario.

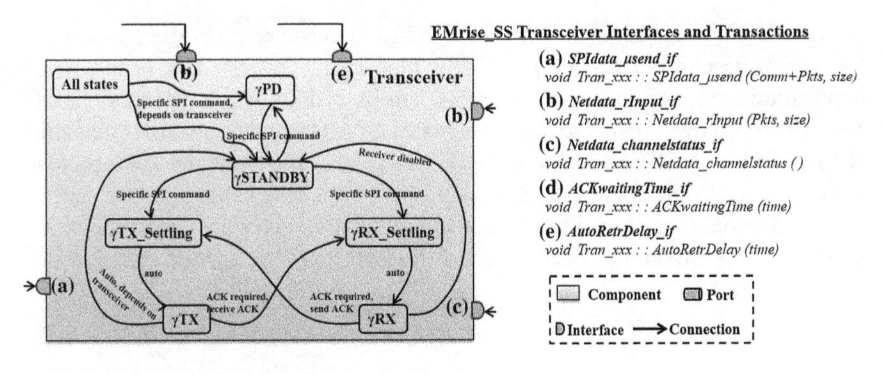

Fig. 8 Typical transceiver model

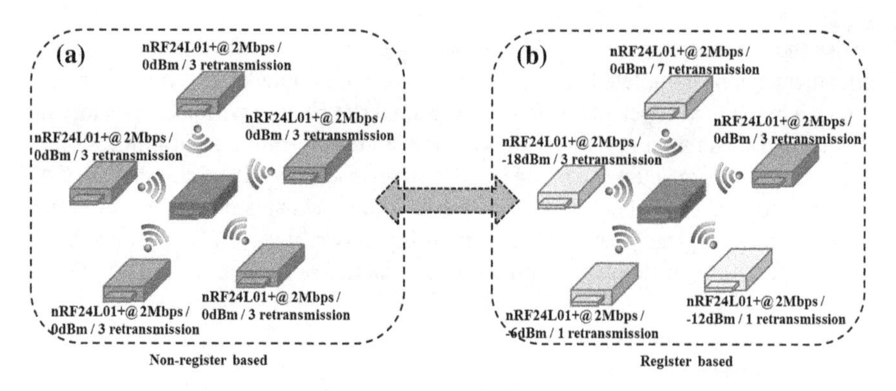

Fig. 9 Advantage of register configuration mechanism

Modeling of CC2420

The CC2420 is a 2.4 GHz IEEE 802.15.4 compliant RF transceiver chip with a maximum data rate of 250 Kbps. CC2420 is designed to support low power applications, it provides on-chip packet handling by means of automatically adding a preamble sequence, frame check sequence and frame delimiter to the data packet. It also provides data buffering, clear channel assessment, link quality indication and packet timing information, all of which reduce the load and power consumption on the host microcontroller. Since for CC2420 based sensor motes (e.g., Telos, Tmote Sky and Shimmer), MAC algorithms for channel access are programmed by software and embedded in the microcontroller, the CC2420 acts only as a TX/RX device for transmission and reception of the IEEE 804.15.4 format based data packet. Therefore, the model of CC2420 in EMrise_SS employs the generic model presented in Fig. 8.

Modeling of nRF24L Transceiver Series

Designed by Nordic Semiconductor, the nRF24 transceiver series [43] represent ultra-low-power RF chips that provide available solutions for 2.4 GHz ISM band wireless applications. As a typical type in nRF24 family, nRF24L01+ supports Enhanced ShockBurst (ESB) with a maximum data rate of 2 Mbps and backwards compatible with ShockBurst (SB). Designed both by Nordic, ESB is a baseband MAC protocol engine embedded in transceiver chips with some basic channel access mechanisms, while SB is only a simple TX/RX protocol without any channel access algorithm. ESB and SB protocol has been modeled in EMrise_SS via a finite state machine shown in Fig. 10.

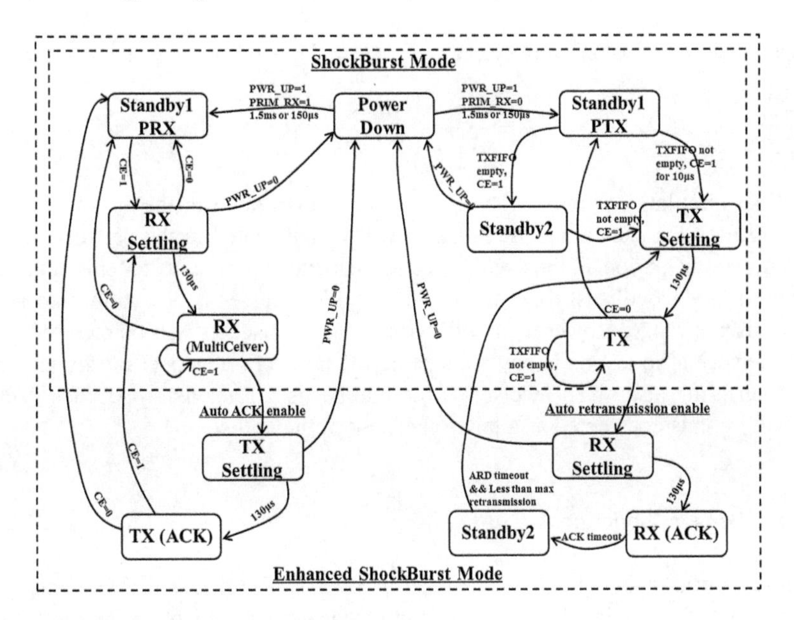

Fig. 10 Enhanced ShockBurst and ShockBurst model

4.1.4 Network Modeling

The network model is derived and modified from the SystemC Network Simulation Library (SCNSL). In network modeling, SystemC provides primitives such as concurrency mechanism as well as events that are used to simulate network transmission and reception behaviors. Before the start of simulation (in the elaboration process), *nodeProxy* class objects are instantiated with the same number of node instances in the SystemC entrance function *sc_main()*. The use of *nodeProxy* is to decouple each node implementation of the network simulation, and each *nodeProxy* instance uploads some significant information to the network object, including node identity, two-dimensional position information, TX output power and receiver sensitivity. Combined with the channel model, these parameters are used to reproduce the network scenario and perform network connection assessment. At present, two types of channel model are supported in EMrise_SS, which are free space propagation model [44] and human's on-body Line Of Sight channel model. The most fundamental free space model in EMrise_SS is presented as follows.

$$P_r = \frac{P_t \cdot G_t \cdot G_r \cdot \lambda^2}{(4\pi D)^2} \tag{4.1}$$

where P_r is the receiver power sensitivity measured in dBm, P_t represents TX output power at transmitter, D is the distance (m) between the transmitter and receiver in two dimensional space, G_r and G_t are antenna gains of the receiver and transmitter respectively (it is common to select $G_r = G_t = 1$), and λ denotes the wavelength, calculated as

$$\lambda = \frac{c}{f_c} = \frac{3 \times 10^8 \text{ (m/s)}}{2.4 \times 10^9 \text{ (Hz)}} = 0.125 \text{ (m)} \tag{4.2}$$

When simulation is initiated by *sc_start()* function, the over-the-air transmission time of the data packet and ACK frame will be calculated in the network model, and a waiting period is thus required for each transmission. After this time, the network model will distribute the packets to the receiver nodes within the radio transmission range. However, if collisions occur, no packets will be received. The collision scenario is simulated by checking the numbers of active transmitters that are interfering at a given receiver. If the number is greater than one, then packet transmission is considered as a failure because of the collision.

4.1.5 Sensor Modeling

A generic sensor model can be designed in 3 ways. The first is periodic data generation, which samples the signals of the physical environment by an application defined time interval (sampling period). The second is to read data from an input file where this file can be a designer-specific data input file at the early design stage, if the actual sensor has not yet been chosen. This input file can also be the actual measurements from experiments. The third way is to model the sensor in a specific function.

In this work, sensor component modeling is mainly focused on the generic model development for periodic data sensing which consistent with the high-level design philosophy of EMrise_SS.

4.1.6 Energy Modeling

As a significant part of EMrise_SS, the energy model plays an important role in energy consumption calculation and therefore it must be well designed and optimized to adjust accurate, realistic and flexible energy consumption evaluation. The characteristics of design method of the energy model is illustrated as follows.

Elaborate Lib and Register Based: An energy model is developed to incorporate an elaborate library with the current consumption of different operation modes of each hardware component. The working mechanism of this energy model is register-based. In the EMrise_SS energy model, the hardware registers of each component are mapped into the energy model as the mirrors, and the updates of these values are synchronous with the corresponding register prototypes in related hardware components, as shown Fig. 11. Developers and researchers can benefit greatly from this design method. Since the tracing of the correct energy value is always consistent with the change of component configuration during the whole simulation process, the energy model is therefore flexible enough to adjust scenarios where the hardware

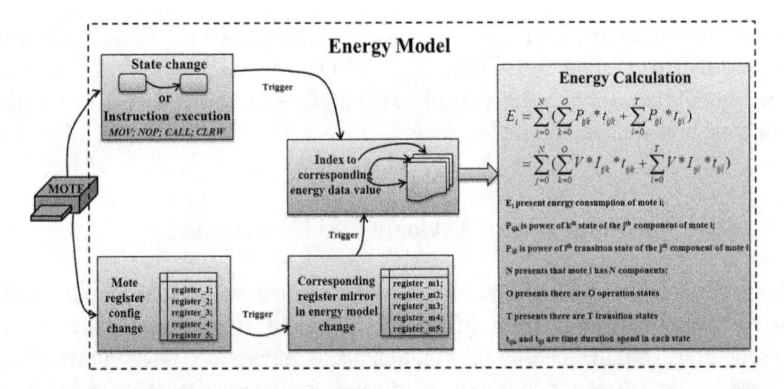

Fig. 11 Workflow of energy model

configurations need to be changed during the simulation. In that sense, the energy model is able to proceed without any interruptions to the whole simulation process, and no re-compilations are required for the new scenarios after configuration changes.

Energy Evaluation for Multi-abstraction levels: The energy model in EMrise_SS is able to trace the energy consumption at different abstraction levels. If the energy evaluation is focused on high-level, the change in operation state of each hardware component is linked to the energy model, and then the duration and current consumption of each state can be identified for the energy consumption calculation by applying the following basic formula.

$$E = P^*t = (V^*I)^*t \tag{4.3}$$

If the energy evaluation is at low-level, the energy model can track each executed instruction in different operation states, record their execution time and associate them with corresponding current loads in the energy library for the accurate energy estimation.

Calibrated Energy Support: The energy model incorporates calibrated energy values from real-world measurements for realistic and accurate energy evaluation. These measurements can be acquired from any COTS sensor motes and in-house testbed motes. However, accurate power consumption data are not easy to extract from product specifications, since a sensor mote is composed of various chips and some chips need additional peripherals to guarantee running operation. These peripherals on chip modules and sensor mote boards can cause extra energy consumption (nRF24L01+ module in Sect. 4.2.2). Thus, the calibrated energy model in EMrise_SS becomes necessary.

Transition State Energy Consumption: The energy model considers state transitions in each hardware component, because each transition introduces extra cost to the overall energy consumption that cannot be ignored. So the total energy consumption is calculated by adding the individual energy consumption of each operation state of every hardware component as well as the energy consumption of each transition state of every hardware component.

Multi Performance Metrics: The energy model supports various performance metrics and can be calculated in *mW, mJ, mAh*. On the other hand, if using a specific battery module measured in capacity, then the lifetime performance can be estimated in minutes, hours, days, months or years.

4.1.7 Case Studies for Energy Evaluation and Management

In this section, EMrise_SS is used for the energy evaluation and management on four mote type based networks (Telos, MICAz, MICA2 and iHop@Node) under different case scenarios. Detailed energy information and comparison data are presented and analyzed for the corresponding mote platform in each case scenario. The impact of sensing start time and payload collecting strategy on the energy performance are studied, while the energy consumption of ESB and SB can be refer to [34].

Sensing Start Time

Due to the spatially distributed characteristic of sensor node networks and cheap manufacturing of hardware components, it is difficult to guarantee the same start up time or clock period for each sensor node. Therefore, energy consumption with simultaneous sensing start time and random sensing start time (each node starts sensing first data at a random instant between the beginning of the simulation experiment and the first sample interval) are analyzed via a medical application scenario which belongs to the field of Wireless Body Area Networks (WBANs) [45].

A typical health monitoring based network for human body area [1, 46] (Fig. 12) is studied to explore the impact of sensing start time on the energy consumption of sensor network. Consisting of 3–10 nodes, the network is attached to the human body for different physiological signal monitoring purpose (10 Hz for temperature sensor, 50 Hz for glucose sensor, 100 Hz for ECG-electrocardiogram). Three types of sensor mote platforms applied in the simulation experiments are iHop250K (iHop@Node with 250 Kbps), Telos (250 Kbps) and MICAz (250 Kbps). For other parameter settings, a 2 byte payload is used to store 12 bits of sensed data. The default 0 dBm

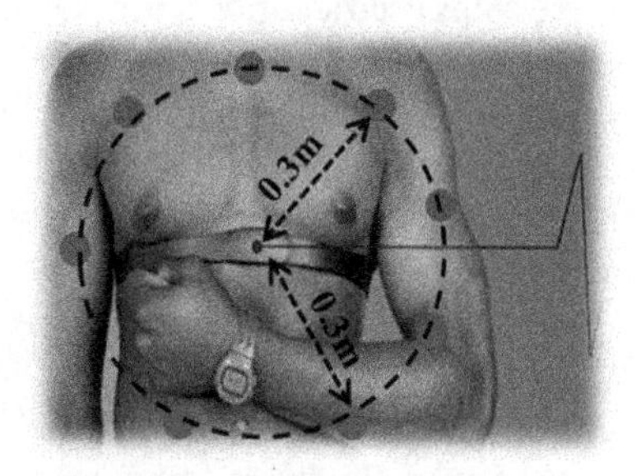

Fig. 12 Sensor node deployment on human body

output power is selected on the transceiver for over-the-air packet transmission. The packets format of Telos and MICAz are shown in Fig. 4. The packet format of ShockBurst (SB) shown in Fig. 3 is applied by iHop250K (1 byte preamble, 3 byte address as network id, 1 byte CRC, 4 bytes payload field including 2 bytes source node id and 2 bytes real payload data). Since only the unslotted CSMA/CA based MiWi communication protocol is supported by iHop@Node, both Telos and MICAz motes also need to apply the same unslotted CSMA/CA for the comparison. Four parameters of the unslotted algorithm are set as default values, which are 3, 5, 4 and 3 for *macMinBE*, *macMaxBE*, *macMaxCSMABackoffs* and *macMaxFrameRetries* respectively. In addition, each case runs for the same simulation time (4 seconds in this work) with different seeds to generate random backoff slot numbers, and the average value of 50 times of runs are employed.

Taking the results of Telos mote for analysis, Fig. 13 reveals that the case with random start time is more energy-efficient. Compared with the random start sensing case, simultaneous sensing consumes more energy due to protocol overhead. This is because all sensor nodes start data sensing at the same time under the simultaneous case, and before over-the air-transmission $2^{BE}-1$ (BE initialize as *macMinBE*) unit backoff periods are required. However, when *macMinBE* is initialized to a smaller value (e.g., default value 3), it is easy for different sensor nodes to get the same number of backoff slots (0–7). Thus, with the same sensing start point and the same/similar backoff period, the transmission time of many packets can overlap with each other, which causes serious packet collisions during a short period of time and therefore extra energy is consumed. On the contrary, the transmissions of the packets are spread out and channel competition is light with random sensing start time. Much energy is saved since it is more likely that packets can be successfully delivered with no collisions and therefore taking low protocol overhead.

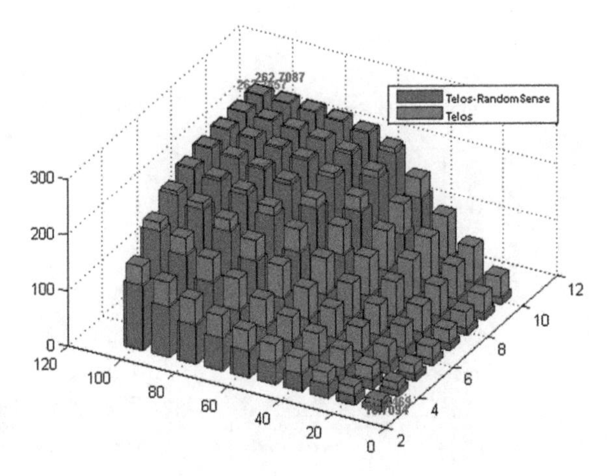

Fig. 13 Impact of random sensing start time on energy compariso

Packet Collecting Strategy

A typical lab-sized network (with a size of about 20 sensor nodes [47]) is under investigation with payload collecting strategy for energy evaluation. The comparison experiments are made on iHop250K, MICAz and Telos-based networks. In the first experiment, a packet with a 2 byte payload is transmitted every 10 ms, then a 4 byte payload packet every 20 ms, a 6 byte payload packet every 30 ms until the transmission of a 20 byte payload packet every 100 ms. Each experiment runs for the same time (10 s) and the average value of 50 runs is adopted. The results are presented in Fig. 14.

Since the increasing payload data can be attached to the same size of packet overhead and when this data collecting strategy is used, fewer data packets are transmitted when the same amount of payload data is required. Therefore, the energy wasted in packet overhead under frequent transmission conditions can be greatly saved, while energy cost is also reduced with fewer ACK frames. Fewer transmission also saves the energy wasted in active mode due to possible packet overflow preventing the sensor node entering into sleep mode. Besides, the experimental results show the packet loss probability drastically reduces from 96.3 % in the 2-byte, 90.1 % in 4-byte cases to 3.98 % in the 20-byte case for both MICAz and Telos (MICAz and Telos share the same protocol model). For iHop250K, this rate is from 93.9 % in the 2-byte case, 36.0 % in the 4-byte case, and is reduced to 1.97 % in the 20-byte case.

Although the payload collecting method can conserve energy, it has some drawbacks. When the channel is of low quality with a high bit error rate (BER)/packet error rate (PER), longer data packets are more likely to subsume errors and necessitate retransmissions. This will definitely cause extra energy consumption and longer packet latency.

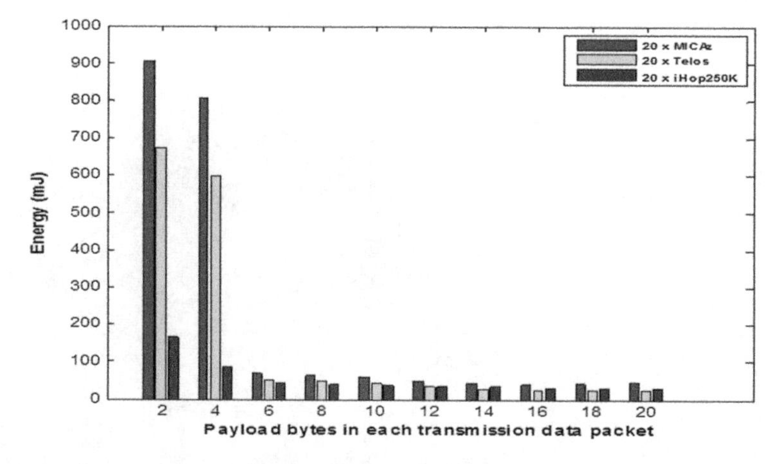

Fig. 14 Energy consumption under payload collecting strategy

4.2 *EMrise_MS (EMrise Measurement System)*

EMrise_MS, known as iWEEP_HW [34, 37], is the hardware based energy measurement and management platform that is made up of a multi-channel energy measurement device (MEMD) and energy data management software platform (EDMSP) to be able to measure the energy related data of iHop@Node, MICA mote, Telos mote and other similar sensor motes. For MEMD, it can provide simultaneous measurements on different components of a sensor node or simultaneous measurements on several different nodes by using its multiple channel advantage. For EDMSP, it offers an easy to use GUI to help users manage and save energy data from the sensor node.

4.2.1 IHop@Node Testbed Node

In order to explore energy performance of future sensor network which has high data rate and ultra-low power features, iHop@Node [34, 37] is designed and built by Microchip PIC16F88 [48] microcontroller with 0.93–1.30 mA current load in active mode and 0.3–0.5 μA in sleep mode, as well as high data rate (2 Mbps/1 Mbps) ultra low power transceiver Nordic nRF24L01+.

iHop@Node has a small-dimension of $4.1'' \times 2.4''$ and it also has multi-layer architecture design, which means that the microcontroller layer, transceiver layer, sensor layer and other components layer (e.g., LEDs, external memory) are all separate and can be fed either by the same power supply or by different supplies. When different power supplies are adopted, each hardware component is considered as an independent load under test and detailed energy information can be measured respectively and synchronously by using MEMD. Compared with the software based functional components decomposition approach [18], this means is at hardware level and provide much more reliable results.

4.2.2 MEMD

Without introducing side-effects to the sensor node hardware or software for energy measurements, and saving cost from using expensive measurement equipment such as oscilloscope and acquisition card [18]. A dedicated MEMD (**M**ultichannel **E**nergy **M**easurement **D**evice) is therefore designed and implemented.

Figure 15 is a simple architecture of one channel of MEMD. A shunt resistor (Rsense) of the known value is used between power supply and sensor node for energy consumption measurements, the voltage drop (Vsense) across the resistor will be amplified as the final measured energy usage data (Vout). Then, a 10-bit ADC chip [49] is employed to sample this Vout, and the sampled values can further be stored in the buffer of PIC18 [50]. The sampled data can be saved continuously in the whole buffer and then sent out in a buffered sampling mode. Alternatively, the data can also be sent out immediately after each sampling, in an unbuffered sampling

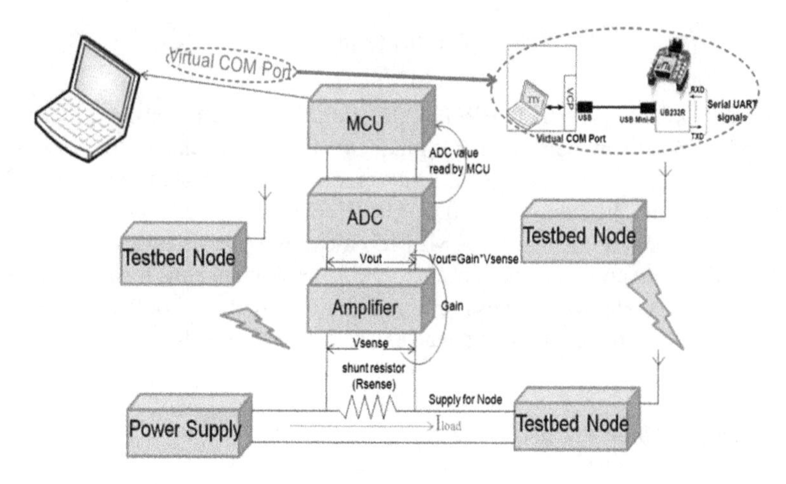

Fig. 15 Architecture of MEMD (one channel)

mode. To simplify the development, we use commercial virtual COM port module UR232R [51] for the communication between PIC18 and data terminal. Serial UART signals generated by the MCU is able to be sent over UB232R to data terminal for later evaluation and analysis. However, a large amount of raw measurements are likely to cause memory problem if the experiments run for a long time.

With the multi-layer architecture design of iHop@Node, MEMD can measure energy consumption on each hardware component separately and synchronously as shown in case 1 of Fig. 16a). While for case 2, instead of separate hardware components, MEMD is used to measure the energy consumption of four independent sensor motes. In Fig. 16b), the MEMD prototype is presented. Channel 1 of MEMD is responsible for the energy measurement of nRF24L01+ transceiver of

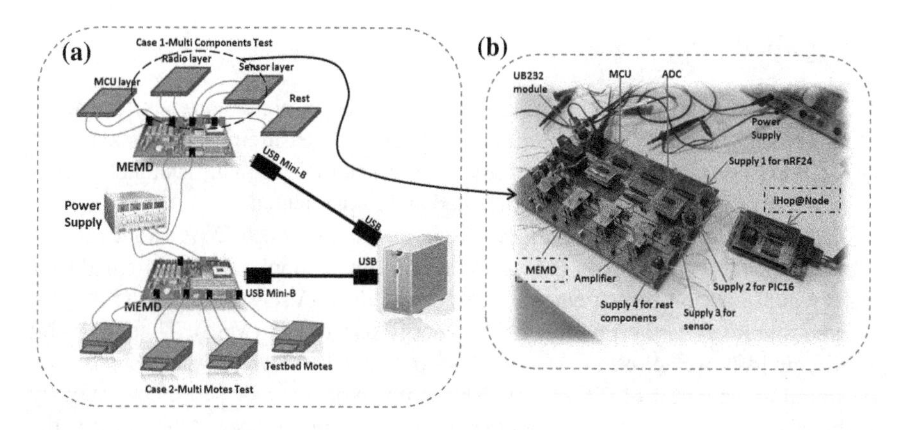

Fig. 16 MEMD prototype

Table 1 MEMD capabilities

MEMD capabilities
√ Four channels for sampling energy usage data
√ Synchronous energy usage data sampling
√ 10-bit resolution for ADC value
√ 1900 samples buffers (each channel)
√ Configurable gain (by tuning R1 and R2 in schematics)
√ 150 KHz buffered sampling rate
√ 5.5 KHz unbuffered sampling rate
√ Measurement range 0 ∼ 40 mA
√ Configurable data rate via virtual COM port interface (UB232R module)

iHop@Node, channel 2 is for PIC16 microcontroller energy measurements, channel 3 can be attached to the sensor chip for energy measurement, and channel 4 is used for other hardware components such as LEDs or external memory. All the channels run samplings simultaneously, and measurements are stored in respective microcontrollers via buffered sampling mode for high speed. These data will be sent and saved separately in data terminal according to the channel number for later evaluation. In addition, the cost of this MEMD is much lower than an oscilloscope, an acquisition card and recently proposed Sensor Node Management Device (SNMD) built by KIT [52]. Table 1 lists the detailed measurement capabilities of MEMD.

4.2.3 EDMSP

Compared with [53, 54], which provide SD or MMC card memory interfaces for energy measurement storage, EMrise_MS provides an alternative means for energy data saving and management.

Since huge storage space is needed for the collection of raw measurements, the on-board memory card based method cannot even support the experiment running for several minutes or hours. Therefore, saving measurements onto hard disk with TB capacity level would be a better choice. Thus, EDMSP is built to facilitate the raw measurement sampling process as well as the data sending to PC process for storage and future evaluation. A GUI-based Energy Data Management Software Platform (EDMSP) is designed and presented in Fig. 17.

The implementation of EDMSP is based on the well-known CSerialPort class [55] developed by MFC and Windows API. Many configurable parameters can be set via EDMSP GUI such as port number, baud rate, data bit and stop bit. Once the selected COM port receives data, the number of received data will be recorded and shown on the configuration table, in the meantime the data values will also be shown in the display window. Since EDMSP supports multi-type display functions, the received data can be presented in decimal value, hex value, voltage value, current

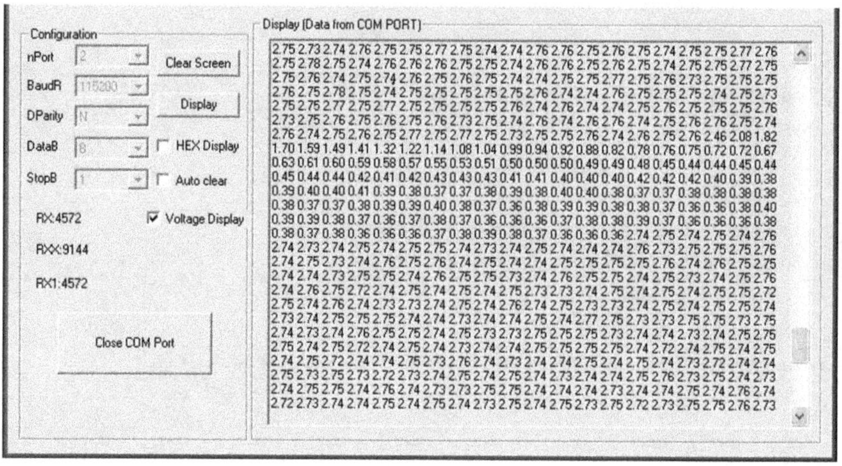

Fig. 17 EDMSP

value and even power consumption value when the corresponding display check-box is selected. Other specific functions can also be easily extended to EDMSP according to users' requirements. Further, all the displayed energy values can be printed out into a TXT file for the convenience of subsequent data processing and evaluation.

4.2.4 Measurements

In this part, MEMD and EDMSP are applied for the measurements and calibration of the iHop@Node testbed. The current consumption measurements of iHop@Node are presented in Fig. 18 on both PTX and PRX devices. Since the microcontroller and transceiver are typically the most power consuming parts, the measurement results are mainly focused on these two parts in order to simply the investigations.

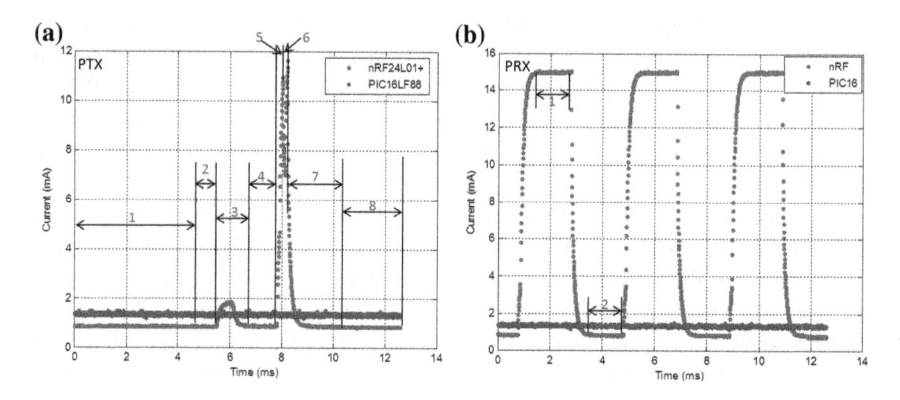

Fig. 18 PTX and PRX measurements

Figure 18a represents the process of PTX device under Enhanced ShockBurst mode, where the iHop@Node testbed is configured as 2 Mbps and 0 dBm output power. Each operation step in the figure is marked with a corresponding number for the clarity of illustration.

Interval 1: The PIC16 microcontroller is in active mode, while the nRF transceiver is in power down mode. The microcontroller demands about 1.4 mA current consumption, and the transceiver consumes about 0.8–0.9 mA current. During the whole experiment, the PIC16 microcontroller runs in active mode all the time.

Interval 2 and 4: During interval 2, the microcontroller sends commands via SPI to the transceiver to configure it from the power down mode to the standby mode. For interval 4, the microcontroller sends a 10 byte payload to the nRF transceiver's TXFIFO also by SPI. Both intervals show that SPI communication will not increase the current consumption of PIC16 microcontroller.

Interval 3: After reception of SPI commands from microcontroller, the crystal of nRF transceiver starts up and the transceiver enters into standby mode from power down mode. The maximum current consumption for this interval can reach about 1.83 mA.

Interval 5: The CE (Chip Enable) signal is set high for at least 10 μs, which triggers the transceiver first into the transition state TX_Settling. After 130 μs in this TX_Settling state, the transceiver will automatically enter into the TX mode for packet transmission. An average current consumption for this 130 μs settling state is given in the product specification. Based on the measurement results, there are actually two steps consuming different currents in this transition state. In the first 50 μs, the increase rate of current consumption is slower than the last 80 μs, which is probably because in the first 50 μs the transceiver only performs packet assembly, while for the next 80 μs many internal components of the transceiver are turned on for over the air packet transmission. Therefore, a much more rapid increase in current consumption can be detected during the second step.

Interval 6: With the ESB mode, the nRF transceiver will automatically enter into RX_ACK mode waiting for the acknowledgement packet. A transition state RX_Settling is therefore required to make the transceiver from PTX to PRX. Since the current consumption in RX_Settling is lower than that consumed in TX and RX states, so the consumed current during interval 6 will first reduce and then increase.

Interval 7 and 8: After the successful of reception ACK frame, the nRF transceiver automatically returns to standby mode. During interval 8, the microcontroller sends a command via SPI to configure the transceiver into power down mode.

Figure 18b represents the process of periodical listening of PRX device.

Interval 1: The transceiver is in RX mode for channel listening and is ready to receive possible incoming packets. The microcontroller works is in active mode.

Interval 2: CE is set low by the microcontroller, and the transceiver returns to standby mode.

Note that the nRF24L01+ transceiver is integrated into a module with an onboard 3.3 V regulator, RP-SMA 2.4GHz antenna and some peripheral circuits, so as shown in Table 2, the measured current values of nRF24L01+ are much greater than those listed in the product specification. This is especially true in power down and standby

Table 2 Measurements of iHop@Node (oscilloscope: Tek 1012B)

iHop@Node	Vout (scope) (V)	Vout (MEMD) (V)	Error of Vout	Current (scope) (mA)	Current (MEMD) (mA)	Current datasheet
PIC16 (sleep)	0.048	0.047	0.001 V; 2.08 %	0.13	0.12	0.5 μA
PIC16 (active)	0.36	0.35	0.01 V; 2.80 %	1.399	1.34	1.2 mA
PIC16 (SPI)	0.36	0.35	0.01 V; 2.80 %	1.399	1.34	N/A
PIC16 (UART)	0.36	0.35	0.01 V; 2.80 %	1.399	1.34	N/A
PIC (instruction)	0.36	0.35	0.01 V; 2.80 %	1.399	1.34	N/A
nRF (power down)	0.23	0.21	0.02 V; 8.70 %	0.89	0.82	0.9 μA
nRF (Standby)	0.24	0.22	0.02 V; 8.33 %	0.93	0.85	26 μA
nRF (Startup) max	0.48	0.47	0.01 V; 2.08 %	1.86	1.83	0.4 mA (ave)
nRF (TX_Settling) max	2.24	2.23	0.01 V; 0.45 %	8.70	8.66	8.0 mA (ave)
nRF (TX@0 dBm)	2.84	2.80	0.04 V; 1.41 %	11.03	10.88	11.3 mA
nRF (RX_Settling) max	2.76	2.75	0.01 V; 0.36 %	10.72	10.68	8.9 mA (ave)
nRF (RX@2 Mbps)	3.85	3.82	0.03 V; 0.78 %	14.96	14.84	13.5 mA

modes, because significant current loads are required by other hardware components on the module except for the nRF24L01+ chip itself. Table 2 also shows that SPI and UART communications in the microcontroller do not consume extra energy. This is also true for the instruction execution of microcontroller, where the instructions under investigation are AND, OR, XOR, ADD, DIV, MUL, WHILE and FOR loop [56]. Besides, the measured current consumption in the sleep mode of microcontroller is found to be significantly different (three orders of magnitude) from the values in the specification. This discrepancy could be related to some other scenarios: (1) The enable/disable of interrupt, SPI, ADC and Timer functions on the MCU chip (about 0.2 mA current consumption can be saved if these functions are disabled). (2) The enable/disable of watchdog function. (3) The configuration of I/O ports.

In addition, considering other popular sensor motes, such as Telos, Shimmer Node, Tmote (TelosB), MicaZ, Mica2 and N@L, the measurement range of the proposed MEMD and EDMSP is still available to use.

4.3 *EMrise_OpS (EMrise Optimization System)*

In this section, iMASKO [57], a generic GA-based optimization framework, is integrated into EMrise system and regarded as EMrise_OpS. Due to the global search and intelligent properties of genetic algorithms (GAs), EMrise_OpS is able to automatically and effectively fine tune hundreds of possible solutions to find the best tradeoff solution.

In order to facilitate the configuration of EMrise_OpS and make the process of simulation and optimization visualizable, a MATLAB based GUI has been designed to link both EMrise_SS and EMrise_OpS. The GUI is shown in Fig. 19, where all the corresponding parameters can be set easily via the interface. For EMrise_OpS, it is designed to be very generic of use. The fitness function in EMrise_OpS can be of multiple types as long as it provides parameter space inputs and performance metrics outputs. Thus, results from other well-known WSN simulators such as NS-2, OMNeT++ and Prowler can also be used under the evaluation of EMrise_OpS, even if the detailed implementation and knowledge of such simulations are unknown. In this

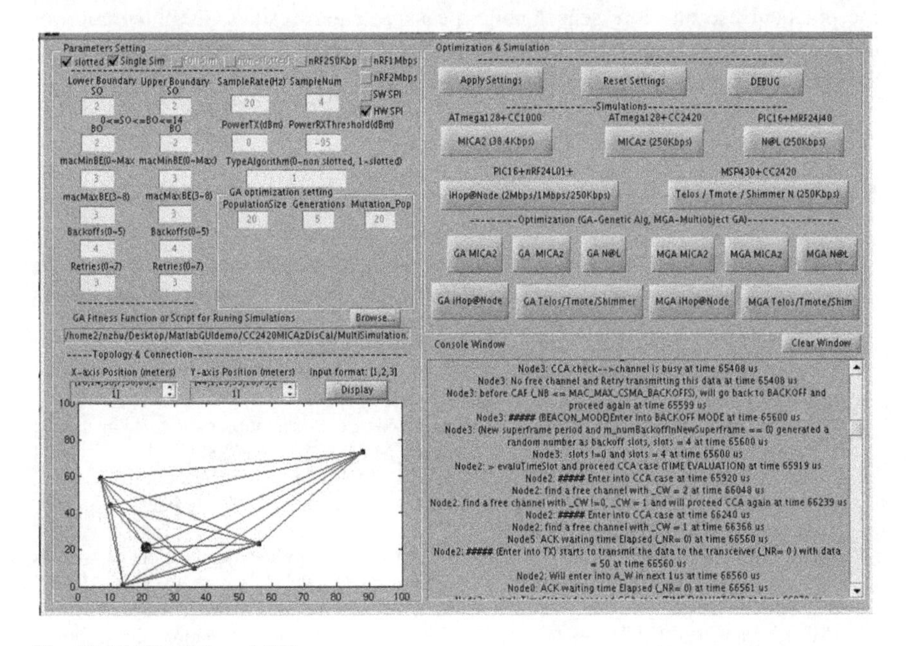

Fig. 19 MATLAB-based GUI

work, we use EMrise_SS simulation results as the fitness function in EMrise_OpS. A path loading for the selecting of different types of fitness function is therefore provided on the GUI (in the mid-left of GUI).

For the other information such as the description of GAs, the architecture of EMrise_OpS, multi-objective optimization and detailed case studies can be referred to [57].

5 Conclusion

In this book chapter, the design and implementation of a house-build platform EMrise is presented for the energy management and evaluation of WSNs. Benefiting from SystemC-based simulation, EMrise is able to support energy consumption evaluation and exploration of heterogeneous sensor network at system-level and transaction-level. With a cost-effective and flexible hardware measurement platform, simulation/emulation models can be calibrated and verified for the accurate energy prediction. In addition, a generic genetic algorithm-based optimization framework is also integrated in the EMrise for the fast, multi-objective and multi-scenario energy optimization.

For the future better work, more detailed simulation models and communication protocols models will be integrated into EMrise for more realistic energy evaluation. The re-design of hardware platform in a smaller size will also be necessary for the practical use on more sensor node measurements. Besides, the integration of other optimization algorithms will help the exploration of more efficient methods of achieving energy-aware and manageable WSNs.

References

1. Lorincz, K., Chen, B., Challen, G.W., Chowdhury, A.R., Patel, S., Bonato, P., Welsh, M.: Mercury: a wearable sensor network platform for high-fidelity motion analysis. In: Proceedings of the 7th ACM Conference on Embedded Networked Sensor Systems (SenSys'09), pp. 183–196 (2009)
2. Lloret, J., Bosch, I., Sendra, S., Serrano, A.: A wireless sensor network for vineyard monitoring that uses image processing. Sensors **11**(6), 6165–6196 (2011)
3. Yeh, L.-W., Wang, Y.-C., Tseng, Y.-C.: ipower: an energy conservation system for intelligent buildings by wireless sensor networks. Int. J. Sens. Netw. **5**(1), 1–10 (2009)
4. Mikhaylov, K., Tervonen, J., Heikkila, J., Kansakoski, J.: Wireless sensor networks in industrial environment: Real-life evaluation results. In: 2nd Baltic Congress on Future Internet Communications (BCFIC 2012), pp. 1–7 (2012)
5. Durisic, M.P., Tafa, Z., Dimic, G., Milutinovic, V.: A survey of military applications of wireless sensor networks. In: 2012 Mediterranean Conference on Embedded Computing (MECO 2012), pp. 196–199 (2012)
6. Fall, K., Varadhan, K.: The ns manual formerly ns notes and documentation (2011). http://www.isi.edu/nsnam/ns/doc/ns_doc.pdf
7. OMNeT++ Network Simulation Framework. http://www.omnetpp.org/

8. Simon, G., Volgyesi, P., Maroti, M., Ledeczi, A.: Simulation-based optimization of communication protocols for large-scale wireless sensor networks. In: Proceedings of 2003 IEEE Aerospace Conference, vol. 3, pp. 1339–1346 (2003)
9. Levis, P., Lee, N., Welsh, M., Culler, D.: Tossim: accurate and scalable simulation of entire tinyos applications. In: Proceedings of the 1st International Conference on Embedded Networked Sensor Systems (SenSys'03), pp. 126–137 (2003)
10. Polley, J., Blazakis, D., McGee, J., Rusk, D., Baras, J.S.: Atemu: a fine-grained sensor network simulator. In: 2004 First Annual IEEE Communications Society Conference on Sensor and Ad Hoc Communications and Networks (IEEE SECON'04), pp. 145–152 (2004)
11. Titzer, B.L., Lee, D.K., Palsberg, J.: Avrora: scalable sensor network simulation with precise timing. In: Processing of Fourth International Symposium on Information Sensor Networks (IPSN'05), pp. 477–482 (2005)
12. Levis, P., Madden, S., Polastre, J., Szewczyk, R., Whitehouse, K., Woo, A., Gay, D., Hill, J., Welsh, M., Brewer, E., Culler, D.: Tinyos: an operating system for sensor networks. In: Ambient Intelligence, pp. 115–148 (Springer, Berlin, 2005)
13. Crossbow Technology Inc. MicaZ datasheet. http://www.openautomation.net/uploadsproductos/micaz_datasheet.pdf
14. Barboni, L., Valle, M.: Experimental analysis of wireless sensor nodes current consumption. In: The Second International Conference on Sensor Technologies and Applications, SENSORCOMM 2008, pp. 401–406 (2008)
15. Landsiedel, O., Wehrle, K., Gotz, S.: Accurate prediction of power consumption in sensor networks. In: The Second IEEE Workshop on Embedded Networked Sensors, pp. 37–44 (2005)
16. Crossbow Technology Inc, "Mica2 datasheet". https://www.eol.ucar.edu/rtf/facilities/isa/internal/CrossBow/DataSheets/mica2.pdf
17. Crossbow Technology Inc, "Telosb datasheet". http://www.willow.co.uk/TelosB_Datasheet.pdf
18. Prayati, A., Antonopoulos, Ch., Stoyanova, T., Koulamas, C., Papadopoulos, G.: A modeling approach on the TelosB WSN platform power consumption. J. Syst. Softw. **83**(8), 1355–1363 (2010)
19. Philipp, H., Ny_enegger, B., Braun, T., Hergenroeder, A.: On the accuracy of software-based energy estimation techniques. Lecture Notes in Computer Science **6567**, 49–64 (2011)
20. Hergenroder, A., Wilke, J., Meier, D.: Distributed energy measurements in WSN testbeds with a sensor node management device (SNMD). In: International Conference on Architecture of Computing Systems (ARCS), pp. 1–7 (2010)
21. Texas Instruments Inc, "Msp430 datasheet" (2011). www.ti.com/lit/ds/symlink/msp430f1611.pdf
22. Mackensen, E., Lai, M., Wendt, T.M.: Bluetooth low energy (ble) based wireless sensors. IEEE Sens. 1–4 (2012)
23. Zhang, J., Orlik, P.V., Sahinoglu, Z., Molisch, A.F., Kinney, Patrick: Uwb systems for wireless sensor networks. Proc. IEEE **97**(2), 313–331 (2009)
24. Buratti, C., Conti, A., Dardari, D., Verdone, R.: An overview on wireless sensor networks technology and evolution. Sensors **9**(9), 6869–6896 (2009)
25. Zhu, N., O'connor, I.: 'Performance evaluations of unslotted CSMA/CA algorithm at high data rate WSNs acenario. In: The 9th IEEE International Wireless Communications and Mobile Computing Conference (IWCMC 2013), pp. 406–411 (2013)
26. Castagnetti, A., Pegatoquet, A., Belleudy, C., Auguin, M.: A framework for modeling and simulating energy harvesting wsn nodes with efficient power management policies. EURASIP J. Embed. Syst. **8**1 (2012)
27. Ye, W., Heidemann, J., Estrin, D.: Medium access control with coordinated adaptive sleeping for wireless sensor networks. IEEE/ACM Trans. Netw. **12**(3), 493–506 (2004)
28. van Dam, T., Langendoen, K.: An adaptive energy-efficient mac protocol for wireless sensor networks. In: Proceedings of the 1st International Conference on Embedded Networked Sensor Systems (SenSys'03), pp. 171–180 (2003)

29. IEEE Computer Society, Part 15.4: Wireless medium access control (mac) and physical layer (phy) specifications for low-rate wireless personal area networks (wpans). *IEEE Std 802.15.4-2006* (2006)
30. Accellera Systems Initiative, "About systemc". http://www.accellera.org/downloads/standards/systemc/about_systemc/
31. Barboni, L., Valle, M.: Experimental analysis of wireless sensor nodes current consumption. In: The Second International Conference on Sensor Technologies and Applications, SENSORCOMM 2008, pp. 401–406 (2008)
32. Agarwal, R., Martinez-Catala, R.V., Harte, S., Segard, C., O'Flynn, B.: Modeling power in multi-functionality sensor network applications. In: The Second International Conference on Sensor Technologies and Applications, SENSORCOMM 2008, pp. 507–512 (2008)
33. Sivanandam, S.N., Deepa, S.N.: Introduction to Genetic Algorithms, Chapter 2 Genetic Algorithms. Springer Publishing Company, Berlin (2007)
34. Zhu, N.: Simulation and Optimization of Energy Consumption on Wireless Sensor Networks. Ph.D. Thesis, EEA Department, Ecole Centrale de Lyon (2013)
35. Moteiv Corporation, "Tmote sky datasheet" (2006). http://www.eecs.harvard.edu/konrad/projects/shimmer/references/tmote-sky-datasheet.pdf
36. Shimmer Research, Shimmer—wireless sensor platform for wearable applications. http://www.shimmer-research.com/
37. Zhu, N., O'connor, I.: Energy measurements and evaluations on high data rate and ultra low power wsn node. In: IEEE International Conference on Networking, Sensing and Control (ICNSC'13), pp. 232–236 (2013)
38. Nordic Semiconductor Inc, nrf24l01+ product specification (2008). http://www.nordicsemi.com/eng/content/download/2726/34069/file/nRF24L01P_Product_Specification_1_0.pdf
39. Fummi, F., Quaglia, D., Stefanni, F.: A systemc-based framework for modeling and simulation of networked embedded systems. In: Forum on Specification, Verification and Design Languages (FDL'08), pp. 49–54 (2008)
40. Hiner, J., Shenoy, A., Lysecky, R., Lysecky, S., Ross, A.G.: Transaction-level modeling for sensor networks using systemc. In: 2010 IEEE International Conference on Sensor Networks, Ubiquitous, and Trustworthy Computing (SUTC'10), pp. 197–204 (2010)
41. Microchip Technology Inc, Microchip miwi wireless networking protocol stack (2010). Application Note http://ww1.microchip.com/downloads/en/AppNotes/AN1066MiWiAppNote.pdf
42. Hauer, J.-H.: Tkn15.4: an IEEE 802.15.4 mac implementation for tinyos2 (2009). TKN Technical Report TKN-08-003. http://www.tkn.tuberlin.de/fileadmin/fg112/Papers/TKN154.pdf
43. Nordic Semiconductor Inc, 2.4 ghz rf ultra low power 2.4 ghz rf ics/solutions. http://www.nordicsemi.com/eng/Products/2.4GHz-RF
44. Tom Henderson, Free space model. http://www.isi.edu/nsnam/ns/doc/node217.html
45. Latre, B., Braem, B., Moerman, I., Blondia, C., Demeester, Piet: A survey on wireless body area networks. Wirel. Netw. **17**(1), 1–18 (2011)
46. Jovanov, E., Milenkovic, A., Otto, C., de Groen, P.C.: A wireless body area network of intelligent motion sensors for computer assisted physical rehabilitation. J. Neuroeng. Rehab. **2**(6) (2005)
47. Ge, Y., Liang, L., Ni, W., Wai, A.A.P., Feng, G.: A measurement study and implication for architecture design in wireless body area networks. In: 2012 IEEE International Conference on Pervasive Computing and Communications Workshops (PERCOM'12), pp. 799–804 (2012)
48. Microchip Technology Inc, Pic16f87/88 data sheet (2005). http://ww1.microchip.com/downloads/en/devicedoc/30487c.pdf
49. Texas Instruments Inc, "Adc10664 datasheet" (2013). http://www.ti.com/lit/ds/symlink/adc10664.pdf
50. Microchip Technology Inc, PIC18F2525/2620/4525/4620 Datasheet. http://ww1.microchip.com/downloads/en/devicedoc/39626b.pdf
51. Future Technology Devices International Limited, Ft232r usb uart ic (2010). http://www.ftdichip.com/Support/Documents/DataSheets/ICs/DS_FT232R.pdf

52. Hergenroder, A., Wilke, J., Meier, D.: Distributed energy measurements in wsn testbeds with a sensor node management device (snmd). In: 23rd International Conference on Architecture of Computing Systems (ARCS'10), pp. 1–7 (2010)
53. Hergenroder, A., Horneber, J. Meier, D., Armbruster, P., Zitterbart, M.: Distributed energy measurements in wireless sensor networks. In: Proceedings of the 7th ACM Conference on Embedded Networked Sensor Systems (SenSys'09), pp. 299–300 (2009)
54. Selavo, L., Zhou, G., Stankovic, J.A.: Seemote: in-situ visualization and logging device for wireless sensor networks. In: 3rd International Conference on Broadband Communications, Networks and Systems (BROADNETS'06), pp. 1–9 (2006)
55. Spekreijse, R.: A communication class for serial port. http://www.codeguru.com/cpp/in/network/serialcommunications/article.php/c2483/Acommunication-class-for-serial-port.htm
56. Morton, G.: Msp430 competitive benchmarking. Application Report (2004). http://www.gaw.ru/pdf/TI/app/msp430/slaa205.pdf
57. Zhu, N., O'connor, I.: iMASKO: A genetic algorithm based optimization framework for wireless sensor networks. J. Sens. Actuator Netw. **2**(4), 675–699 (2013)

Energy Management in Mobile Networks Towards 5G

Dario Sabella, Damiano Rapone, Maurizio Fodrini, Cicek Cavdar, Magnus Olsson, Pal Frenger and Sibel Tombaz

Abstract The evolution of mobile networks from the introduction of the first generation systems until today, and the forecasts for the next decade [1], clearly indicate a growth of both the network itself in terms of installed equipment and carried traffic in terms of transmitted bits. The deployment of new generation systems upon existing ones unavoidably increases the energy consumption, even if new systems are more efficient than the older ones. More consumption means more costs, i.e., less margins for the operators, and greater carbon footprint from the entire Planet. On the other hand, for operators, it would not be possible to dismiss old generation systems in lieu of the new ones, due to the presence of legacy terminals in the network. For these reasons operators need to perform accurate assessment of the energy performance of 2G, 3G and 4G networks by looking in perspective at the evolution of the network in terms of traffic growth, change of paradigms/business models, introduction of next generation networks (i.e. 5G) and so on. This book chapter focuses on the energy efficiency aspects relevant for a sustainable evolution of mobile networks towards 5G from an operator perspective. The conducted analysis will cover both network deployment aspects, equipment evolution, introduction of energy efficiency features, cost analysis, network-level energy efficiency assessment and related standardization initiatives.

Some of the material included in the present chapter also appeared, in a preliminary form, in [2].

D. Sabella (✉) · D. Rapone · M. Fodrini
Telecom Italia, Roma, Italy
e-mail: dario.sabella@telecomitalia.it

C. Cavdar
KTH, Stockholm, Sweden

M. Olsson · P. Frenger · S. Tombaz
Ericsson, Stockholm, Sweden

© Springer International Publishing Switzerland 2016

M.Z. Shakir et al. (eds.), *Energy Management in Wireless Cellular and Ad-hoc Networks*, Studies in Systems, Decision and Control 50,
DOI 10.1007/978-3-319-27568-0_17

1 Acronyms

3GPP	3rd Generation Partnership Project
BCH	Broadcast Channel
BS	Base Station
BW	Bandwidth
CAGR	Compound Annual Growth Rate
CD	Component De-activation
CSRS	Cell-Specific Reference Signals
DMRS	DeModulation Reference Signal
DTX	Discontinuous Transmission
EE	Energy Efficiency
eNB	Evolved NodeB
GSM	Global System for Mobile Communications
HSPA	High Speed Packet Access
LTE	Long Term Evolution
M2M	Machine-to-Machine
MBSFN	Multicast-Broadcast Single-Frequency Network
MIMO	Multiple Input Multiple Output
OFDM	Orthogonal Frequency-Division Multiplexing
OPA	Operating-Point Adjustment
PDCCH	Physical Downlink Control Channel
PSS	Primary Synchronization CHannel
QoS	Quality of Service
RAN	Radio Access Network
RAT	Radio Access Technology
SSS	Secondary Synchronization CHannel
UMTS	Universal Mobile Telecommunications System

2 Introduction and Motivation

This chapter contains a summary of some possible actions that could be considered by the operator for the evolution of mobile networks from an Energy Efficiency and sustainability perspective.[1] To this end, mobile network's evolution towards 5G is analysed by taking into account the energy consumption of the Radio Access Network (RAN) when considering different load conditions (based on actual daily traffic profiles extracted from live network) and different years (according to current traffic forecasts). Thus, the study here reported provides an overview of different

[1]The present description is also extracted from the work performed in 2014 in the framework of Task A1406 "Energy Efficient experimental analysis for network operation" of the 5GrEEn project (EU project funded by EIT ICT Labs)".

"*what-if*" scenarios towards 2020 and beyond, as useful insights for operators aiming at analysing evolutionary steps of their mobile networks.

For the sake of clarity, in this chapter neither details regarding specifications of future 5G system, nor how new radio interfaces should perform are presented and discussed. On the contrary, the impact of network-wide actions on the overall energy consumption is analysed, regardless of the particular air interface to be defined for 5G systems. In this sense, in order to elaborate these "*what-if*" scenarios, it is not necessary to wait for the standardization of 5G new air interface. Instead, network-level energy efficiency assessments are essential to establish evolutionary trends and operators needs in terms of energy efficiency, that may lead to future design requirements for the upcoming 5G platforms.

In order to estimate the energy consumption of a future network, a model of the power consumption related to future Base Stations (BSs) is needed. To this end, following the methodology used in [3], results from the EARTH project [4] have been used, considering different power models for different types of BSs. Evolutionary power models have also been derived for each Radio Access Technology (RAT) through the years, providing an essential framework for network-level EE assessments. Moreover, as many different features which may be implemented within the mobile network might have different time scales), the evaluations have been decoupled as follows:

- at shorter time scale (i.e., milliseconds) **micro sleep** has been considered and assessed with a MATLAB-based system-level simulator;
- at higher time scale (i.e., from minutes to years), progressive **network renewal** with **traffic steering** features and **phase-off** policies has been evaluated by using a specific evaluation tool fed by actual traffic coming from the live network.

To be specific, for the latter set of features the aim was to assess energy efficiency performance at network level, also by taking into account the work performed by the Environmental Engineering Technical Committee of the European Telecommunications Standards Institute (ETSI TC EE) that has recently introduced a set of specifications on energy efficiency. These important standards are based on homogeneous clusters' evaluations, that may be easily extrapolated at country-level [5] to provide useful information to the operator on the possible evolutions of the mobile networks in the view of future 5G systems.

In next sections the main characteristics of the reference system considered for network-level energy efficiency assessment are described, in accordance with typical mobile operator network assumptions. Then a brief overview of the energy efficiency metrics considered for the evaluations is reported, together with elaborated BSs' power models, as essential analytical tools for the evolutionary energy efficiency assessments. Finally, some energy efficiency features selected as effective for the network-level energy efficiency improvement are described and the related cluster-level energy efficiency performance are showed and discussed.

3 Reference Scenario, Metrics and Power Models

In this section the main characteristics of the reference system considered for network-level energy efficiency assessment are described, in accordance with typical mobile operator network assumptions. Then a brief overview of the energy efficiency metrics considered for the evaluations is reported, together with elaborated BSs' power models, as essential analytical tools for the evolutionary energy efficiency assessments.

3.1 Description of Baseline Reference System

Energy consumption for the mobile operator is a practical problem involving not only new technologies and RATs but also legacy networks, hence an evaluation of 5G systems from a *green* perspective should necessarily go with a comprehensive assessment through the years (from today to 2020 and beyond) of the energy efficiency performance of the overall network. In addition to that, the operator will have to carefully evaluate the opportunity and convenience of new RATs' introduction in future systems to satisfy the increasing traffic demand.

Since multi-RAT environments should always be evaluated with realistic traffic assumptions and in accordance with methodologies currently considered as references for operators, the assumed baseline system is characterized by the following aspects:

- presence of 2G, 3G and 4G (in which potentially higher order MIMO systems and Carrier Aggregation are taken into account) and introduction of 5G (modeled in terms of capacity and energy performance);
- alignment with methodology currently under discussion (and finalization) in the framework of ETSI TC EE (EEPS subgroup) for the definition of the ES 203 228 specification,[2] in which clustered assessments based on measurements from the live network are considered;
- homogeneous layout within assessed clusters of network sites, as an easy way to model the network in contrast with exhaustive assessment methods (often too costly);
- performance evaluated by considering commonly accepted energy efficiency metrics, by exploiting the outcomes of EARTH project [4] and current standardization work in ETSI TC EE.

---

[2]The specification also defines metrics for RAN energy efficiency and methods for assessing (measuring) energy efficiency in live networks. The covered technologies are GSM, UMTS, LTE, but the methodology can be easily applied to new radio interfaces also.

3.2 Energy Efficiency Performance Metrics

Energy efficiency and power consumption are the most widely used power consumption metrics in the literature. Among the available, two main energy efficiency metrics can be considered:

bit/Joule: this is the most commonly used efficiency metric,[3] in particular for the evaluation of a single wireless link. Its use has naturally been extended for performance assessment of the whole wireless access network [6, 7]. Let Ψ denote the bit/Joule efficiency of the network, then it can be written as below:

$$\Psi = \frac{C_{net}}{P_{net}}$$

where, C_{net} is defined as the aggregate network capacity in [bit/s] and P_{net} is the total power consumption of the network in [W].

$\boldsymbol{W/km^2}$: another widely accepted energy efficiency metric is the area power consumption denoted by Ω [6, 7]. It relates the total power consumption of the network P_{net} to the size of the covered area A and is given by

$$\Omega = \frac{P_{net}}{A}$$

Note that the optimal energy efficiency is achieved when the bit/Joule metric is maximized or the power per unit area W/km^2 is minimized.

Figure 1 demonstrates the variation of both the bit/Joule and W/km^2 metrics with respect to the number of BSs when the capacity requirement is not considered. It can be observed that bit/Joule is monotonically increasing with network densification, while, on the contrary, the W/km^2 metric indicates that reduced transmit power cannot compensate the additional power consumption for idle state. Therefore, the W/km^2 metric increases with the number of BSs after reaching the optimum point. This suggests that maximizing the energy efficiency is not always equivalent to minimizing the energy consumption. That is the reason why the capacity requirement must be considered in order to prevent contradictory conclusions with different performance metrics. Otherwise the usage of bit/joule might be misleading, since adding more capacity into the network will always reflect an increase in energy efficiency.

Within this chapter both the described metrics (bit/Joule and W/km^2) can be applied to all simulations described in Sect. 5. In particular, performances evaluations reported in Sect. 5.2 (intra-sector traffic steering) are devoted to the minimization of area power consumption: in fact, while results provided by the tool are expressed in terms of bit/Joule, "*what-if*" scenarios analyzed in this section are compared in terms of power consumption of a single cluster of network sites, which is the same for every simulation

[3]Even if the [bit/J] is a well-known efficiency metric, common practice in network performance evaluations is to use in alternative a consumption metric expressed in [J/bit], as suggested by EARTH project in [6].

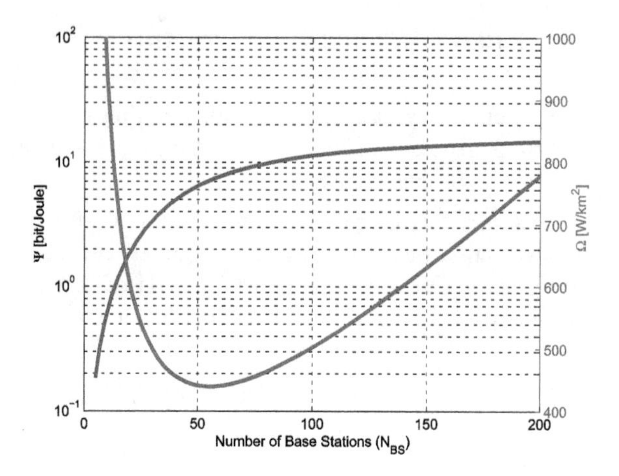

Fig. 1 Bit/Joule (*blue curve*), area power consumption (*red curve*) versus number of BSs [7]

3.3 Power Models

In order to estimate the energy consumption of a future network, a model of the power consumption related to future BSs is needed. To this end, following the methodology used in [3], results from the EARTH project [4] have been used. One of the major contributions of the EARTH project was the work performed to derive different power models for different types of BSs. Since the EARTH project derived BSs' power models for 4G (LTE) only, similar models for 2G (GSM) and 3G (UMTS Rel.'99 and HSPA) BSs have been derived by selecting the closest equivalent EARTH model in terms of system bandwidth (BW), output power and so on. This assumption seems to be reasonable due to the fact that BSs' hardware is becoming more and more multi-standard. Nowadays multi-RAT capable hardware is already available on the market and, when they will be deployed in large scale, it is expected that they will be able to provide additional and significant energy saving. However, in this study, these gains are not taken into account thus all BSs are assumed to be single-RAT.

3.3.1 EARTH Power Models

The EARTH project derived classes of power models corresponding to state-of-the art BSs manufactured in the years 2010 and 2012. These models are denoted as "EARTH class 1" and "EARTH class 2", respectively. In addition, the EARTH project derived a class of power models representing BSs' hardware manufactured during 2012 including hardware improvements that have been proposed by the project and these models are denoted as "EARTH class 3". The "EARTH improvements" introduced in the "EARTH Class 3" models are Operating-Point Adjustments (OPA) and Component De-activation (CD) (refer to [4, 8] for further details). Different power models for 1.4, 5, 10 and 20 MHz BWs and for Class 1, 2 and 3 are depicted in Fig. 2.

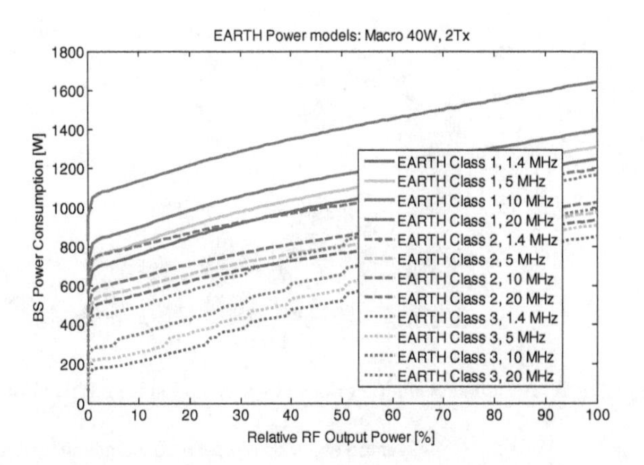

Fig. 2 Example of EARTH power models for different BWs (1.4, 5, 12 and 20 MHz) and classes (1, 2 and 3)

3.3.2 Power Consumption of Future Radio Base Stations

A follow-up of EARTH work on power models has been performed also by GreenTouch consortium, that developed an evolutionary power model for future BSs (for further details please see [9, 10]). Nevertheless, while this power model is more focused on detailed technological aspects of the radio components (especially in the view of assessing future LTE and 5G BSs), our goal is to consider different RATs (including legacy equipment) in accordance with the operator needs to evaluate the overall power consumption. For that reason the EARTH work has been considered as a starting point to derive also 2G and 3G power models toward 2020.

Here the power consumed by a typical BS installed during 2012 is assumed to be in line to the EARTH Class 1 of power models. Remember that the EARTH Class 1 represents a state-of-the art BS of 2010, which is assumed to represent a typical BS deployed two years later, i.e., in 2012. Since there is always a distribution of BSs installed in every year, how good the *typical* BS is represents the only interesting aspect, rather than how good the *best* one is. With a similar reasoning, the EARTH class 2 models are assumed to be used to represent a *typical* BS installed during 2014, even though it corresponds to a state-of-the art BS of 2012. The EARTH improvements that are included in the EARTH Class 3 models are assumed to be available from the year 2016. For the intermediate years we assume an 8 % improvement per year, which is in accordance with historical improvements also used in [11].

Combining this, power model parameters for a typical future 3G site are derived, as shown in Fig. 3. As a reference, in the leftmost bar, the consumption of one of the most common 3G BS in today's deployed networks is shown. It is an Ericsson RBS 3202 with 3 sectors and one 5 MHz carrier per sector (denoted as (1/1/1) in Fig. 3) which was a state-of-the-art BS approximately 10 years ago. Future 3G BS's power consumption is derived by considering the EARTH power model of a macro BS, 5 MHz BW, 1 transmit antenna, 20 W RF power per sector, 3 sectors per site and an average RF load of 20 %.

For 4G (LTE), a power model according to a macro BS, 10 MHz BW, 2 transmit antennas, 40 W RF power per sector, 3 sector/site and an average RF load of 20 % has

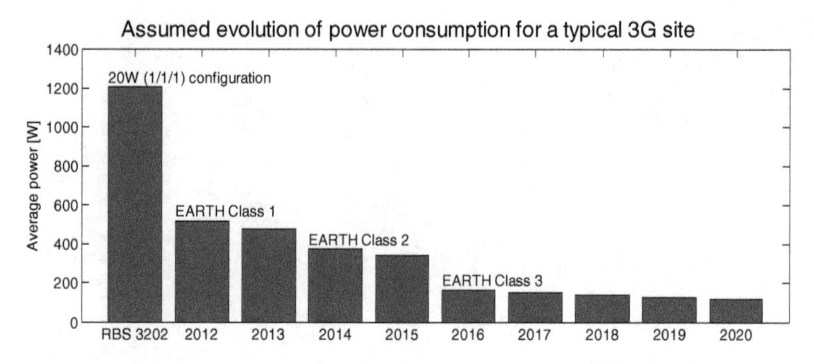

Fig. 3 Assumed evolution of power consumption for a typical future 3G (HSPA) site

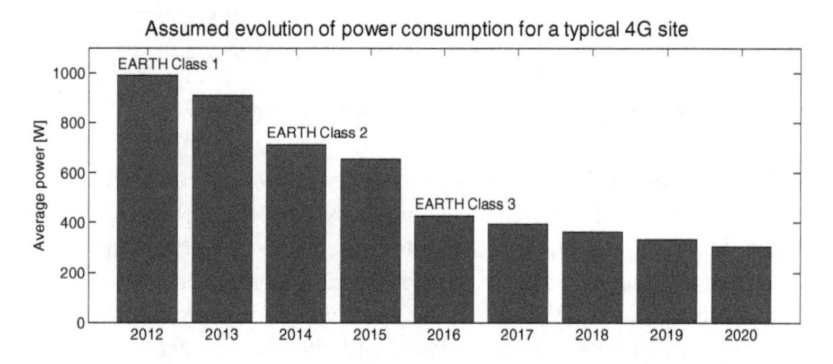

Fig. 4 Assumed evolution of power consumption for a typical future 4G (LTE) site

been assumed. The corresponding power consumed by a future LTE BS is depicted in Fig. 4. Note that for LTE power consumption in Fig. 4 higher RF output power and 2 transmitters per cell have been considered, which explains why the LTE power consumption is larger than the corresponding 3G one in Fig. 3.

3.3.3 BS Power Models for Different RATs and Different Years

The power models presented above are useful in order to estimate the average energy consumption in a typical radio BS that carries an average traffic volume. In order to derive models that can be used in load adaptive studies, slightly more detailed power models are needed. Using the same methodology as described above, BSs' power models as a linear function defined by a numerical triplet A, B and C can be derived and described, where A is the sleep power, B is the idle power and C is the maximum power of the BS.

For the considered RATs the following assumptions have been taken into account:

- LTE: 10 MHz BW, 2 TX, 40 W/sector, 3 sectors/site
- HSPA: 5 MHz BW, 1 TX, 40 W/sector, 3 sectors/site
- GSM: 1.4 MHz BW, 1 TX, 40 W/sector, 3 sectors/site

Furthermore, the "EARTH improvements" are assumed to be available from 2016, just as done above. From 2016 on, per-year 8 % improvement for the parameters A and B only is considered, as assumed in EARTH deliverable D2.1 [11]. For the maximum power of the BS (parameter C of the power model) any improvement is assumed to be made. This average annual improvement (taken as baseline continuous improvement) can be attributed to the technology scaling of semiconductors, as well as to improved RATs. With these assumptions, load dependent power models for LTE (Fig. 5), HSPA (Fig. 6) and GSM (Fig. 7) for the years 2012, 2016 and 2020 are derived.

The power consumption of a BS is determined by using the previously reported RAT-based power models; in principle, as can be seen in [3], the power model of a BS is a set of parameters describing the BS's power consumption P_{in} as a function

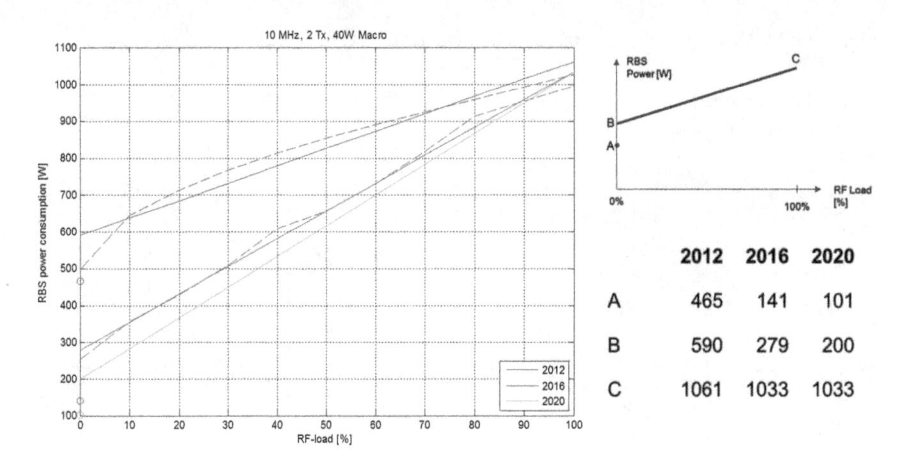

	2012	2016	2020
A	465	141	101
B	590	279	200
C	1061	1033	1033

Fig. 5 LTE BS's power models for 2012, 2016 and 2020

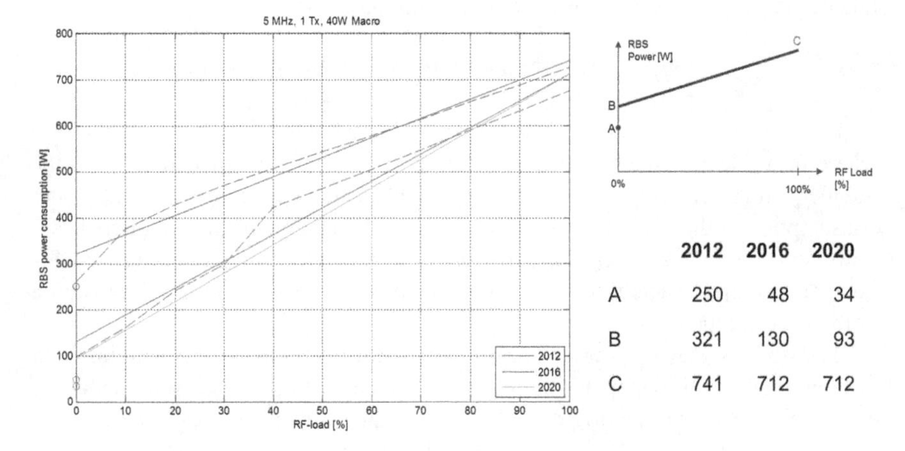

	2012	2016	2020
A	250	48	34
B	321	130	93
C	741	712	712

Fig. 6 HSPA BS's power models for 2012, 2016 and 2020

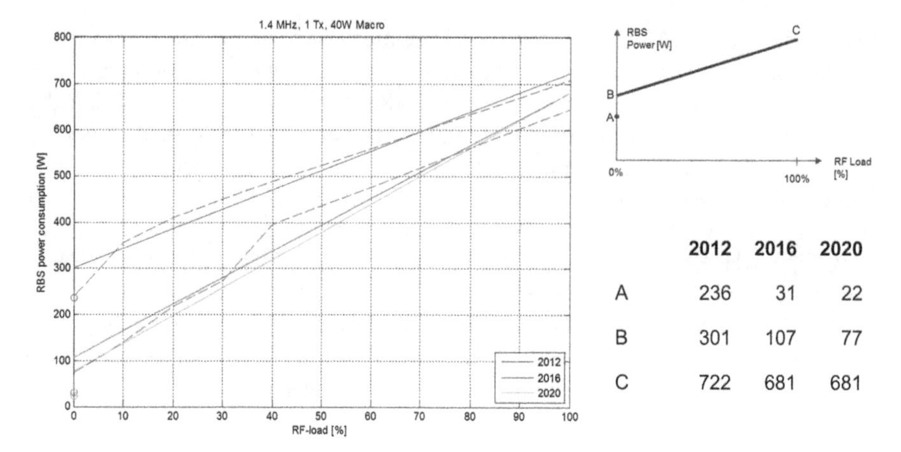

	2012	2016	2020
A	236	31	22
B	301	107	77
C	722	681	681

Fig. 7 GSM BS's power models for 2012, 2016 and 2020

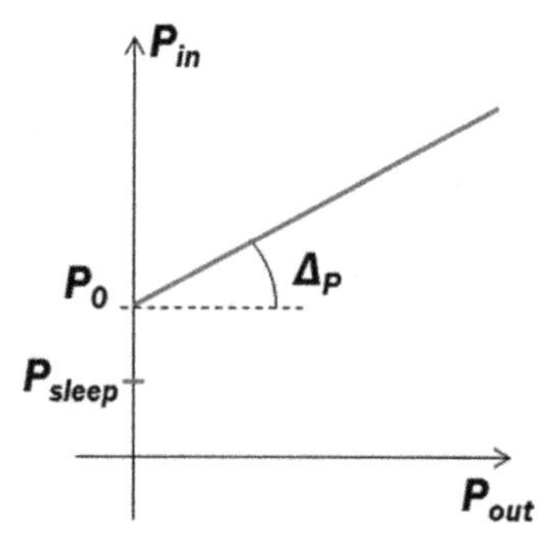

Fig. 8 Power model: parameters as explained in [3]

of the RF output power P_{out} (refer to Fig. 8). The mathematical expression which relates the power consumption P_{in} to the output power P_{out} is:

$$P_{in} = \begin{cases} N_{TRX} \cdot P_0 + N_{TRX} \cdot \Delta P \cdot P_{out} & 0 < P_{out} \leq P_{MAX} \\ N_{TRX} \cdot P_{sleep} & P_{out} = 0 \end{cases}$$

where P_{MAX} denotes the maximum RF output power at maximum load, N_{TRX} is the number of transceiver chains, P_0 is the linear model parameter representing the power consumption at the zero RF output power, ΔP is the slope of the load-dependent power consumption and P_{sleep} is the sleep mode power consumption (i.e., the BS's power consumption achieved when there is nothing to transmit, so fast deactivation of the BS's components is selected).

By assuming that P_{out} varies linearly with the throughput and by denoting with x the normalized throughput with respect to the capacity (i.e. peak throughput) of each considered RAT, the original EARTH formula is modified as follows:

$$P_{in} = \begin{cases} N_{TRX} \cdot P_0 + N_{TRX} \cdot \Delta P \cdot P_{MAX} \cdot x & 0 < x \leq 1 \\ N_{TRX} \cdot P_{sleep} & x = 0 \end{cases}$$

The above expression has been obtained by exploiting these equalities:

$$\begin{cases} A = N_{TRX} \cdot P_{sleep} \\ B = N_{TRX} \cdot P_0 \\ C = N_{TRX} (P_0 + \Delta P \cdot P_{MAX}) = N_{TRX} \cdot P_0 + N_{TRX} \cdot \Delta P \cdot P_{MAX} \\ \quad = B + N_{TRX} \cdot \Delta P \cdot P_{MAX} \end{cases}$$

Hence, the formula to be used in order to compute the BS's power consumption (according to the selected power model) as a function of the normalized traffic information is:

$$P_{in} = \begin{cases} B + (C - B) \cdot x & 0 < x \leq 1 \\ A & x = 0 \end{cases}$$

The latter formula has been used in all performed simulations, whose results will be shown later in Sect. 5.2.

A power model for future 5G BSs has also been derived for performance evaluation when considering 2G/3G phase-off and contextual 5G progressive introduction: to this end, the power model of a 2010 pico BS has been considered as a starting point for 5G RAT power models: in fact, the assumption is that 5G will be initially introduced as a small cell capacity layer. Then, it can be assumed an yearly improvement of 8 % for A and B power model's parameters only (as previously stated), the power model's parameters of a theoretical 5G BS (A, B and C values) are determined and reported in the following Table 1.

Table 1 Future (2020) theoretical 5G BS's power model

5G BS's power model			
Year	A [W]	B [W]	C [W]
2010[a]	8.60	13.60	14.64
2011	7.91	12.51	14.64
2012	7.28	11.51	14.64
2013	6.70	10.59	14.64
2014	6.16	9.74	14.64
2015	5.67	8.96	14.64
2016	5.21	8.25	14.64
2017	4.80	7.59	14.64
2018	4.41	6.98	14.64
2019	4.06	6.42	14.64
2020	3.74	5.91	14.64

[a] A, B and C obtained considering $N_{TRX} = 2$, $P_{MAX} = 0.13$ W, $P_0 = 6.8$ W, $\Delta P = 4$, $P_{sleep} = 4.3$ W

4 Energy Efficiency Features

Today's mobile networks are designed and deployed based on the peak traffic demand and kept active regardless of the low utilization during different times of the day. Even at the peak hours there are few cells that experience high load [8]. In order to scale the power consumption to the traffic dynamicity, adaptive network operation features can be considered, which are practical to apply and demonstrate even in today's mobile network operator sites. In this section, first of all different kind of Discontinuous Transmission (DTX) techniques are described and categorized in order to give an overview of the different possible features applicable to the mobile networks, then showing in more detail one of these techniques, namely micro DTX. Finally traffic steering techniques are described as energy efficiency enablers for operators, especially in multi-RAT environments and in the view of the future introduction of 5G systems.

DTX has been used for a long time in mobile terminals to achieve long battery life. The idea is to only transmit when there is a need and, otherwise, putting the transmitter in a low power state. At network side this technique is referred to as cell DTX and it is based on the hardware component de-activation feature which facilitates low power states. Two cell DTX versions can be distinguished: *fast cell DTX* and *long cell DTX*.

Fast cell DTX acts on slot/subframe level and exists in some different versions as well. Cell micro DTX (described in Sect. 4.1) is already possible in LTE Rel-8 and it means that when there is not any user data to transmit the radio is put into DTX (micro sleep) among transmissions of Cell-Specific Reference Symbols (CSRS). This technique is illustrated in Fig. 9 below, which reports the structure of an LTE downlink radio frame with 10 subframes showing CSRS for one antenna port, Physical Downlink Control CHannel (PDCCH) region with a size of one OFDM symbol, Primary and Secondary Synchronization Signals (PSS and SSS, respectively) and Broadcast CHannel (BCH).

In the other version, which is also referred to as MBSFN-based DTX, Multicast-Broadcast Single-Frequency Network (MBSFN) subframes are used to make room for longer sleep periods, since CSRS are not transmitted in these subframes, as depicted in Fig. 10, where a LTE downlink radio frame with 6 MBSFN subframes is shown. This is also possible in LTE Rel-8.

Finally, a cell short DTX is possible, in which the CSRS are assumed to be removed and replaced by DeModulation Reference Signal (DMRS), hence leaving room for even longer sleep periods, as shown in Fig. 11. This is not possible in the today's LTE standard, but may be in the future, see e.g. [12]. All these fast cell DTX versions are most effective in low load scenarios, but there is not any drawback at high load either.

On the other hand, **long cell DTX** acts on a slower time scale and in principle it refers to the cell being put into a low-activity mode [13]. As such it can be seen

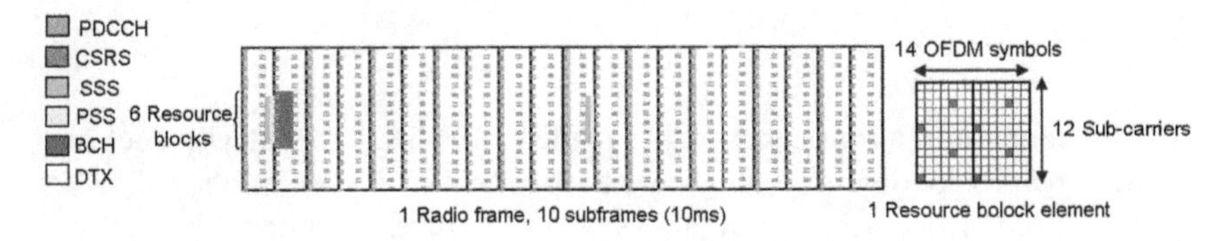

Fig. 9 LTE downlink radio frame. Only micro DTX is possible

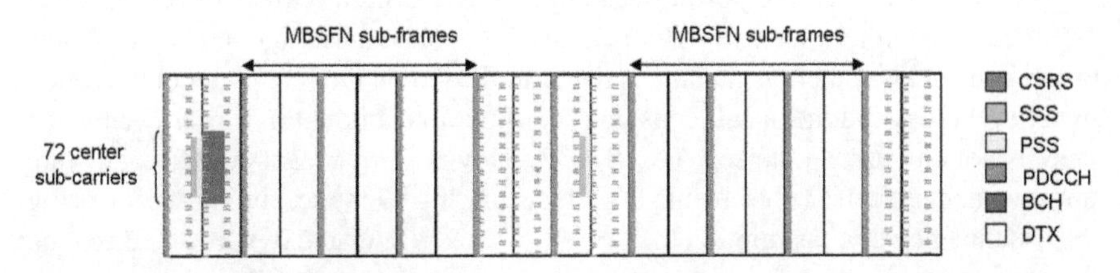

Fig. 10 LTE downlink radio frame configured with 6 MBSFN sub-frames

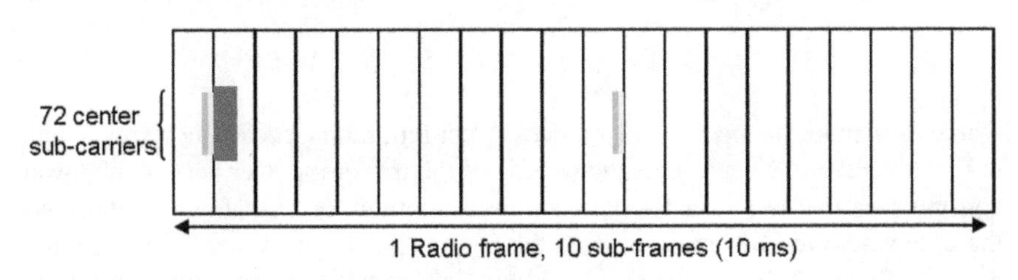

Fig. 11 Proposed LTE downlink radio frame without CSRS. Short DTX can be applied

as a cell sleep (on/off functionality) and it could be based on a deeper sleep state (lower power consumption) than the low power state considered above for the fast cell DTX versions. One use case or strategy for long cell DTX is in relatively densely deployed networks, where there is good coverage (dense urban, urban, suburban). Then capacity nodes can be put in long cell DTX during low-traffic periods. Hence, long cell DTX can be seen as an enabler/tool for many network management actions, such as small cell and multi-RAT management, but also macro cell management, as discussed in [13]. In principle, long cell DTX can be activated when there is not traffic demand or, alternatively, when, in periods of low traffic, cells are at low load and then traffic steering techniques are easily applicable without many "*ping-pong*" effects during the day. These techniques are considered in Sect. 4.2 as an enabler of great energy efficiency potential for operators, especially in multi-RAT network environments.

4.1 Micro DTX

In this section, short term sleep solutions (called micro DTX, or sometime referred as micro sleep[4]) are described; they have been proposed in [14] and they allow for dynamic sleep mode operations at the BS on a short time resolution, e.g., millisecond level.

In order to evaluate the power consumption of a BS under micro sleep, a cell is assumed to be either in active state, i.e., there is at least a user requesting a service, or in idle state, i.e., there is not any active user. Based on the linear model presented in the previous sections, a cell consumes considerable amount of power even when there is not any user in the cell, i.e., $B = P_0$. However, by means of BS's hardware improvements, a cell can be put into DTX mode during idle state, aiming at decreasing the baseline power consumption to $A = P_{sleep} = \delta \cdot P_0$, where $0 \leq \delta < 1$. Based on these assumptions, the average power consumption of the kth cell P_k with load $\eta_k \in \eta$ can be written as [15]:

$$P_k = \zeta \cdot \mathrm{P}_k \cdot \eta_k + (1 - \delta) \cdot P_0 \cdot \eta_k + \delta \cdot P_0$$

where P_k denotes the power spectral density per minimum scheduling resource unit in the kth cell and ζ represents the portion of the transmit power dependent power consumption due to feeder losses and power amplifier. Note that $\delta = 1$ represents the case where the BS does not have the DTX capability and therefore consumes $P_k = \zeta \cdot \mathrm{P}_k \cdot \eta_k + P_0$. In this case cell load only impacts the transmission-related power consumption, i.e., $\zeta \cdot \mathrm{P}_k \cdot \eta_k$.

Micro DTX enables node-level power consumption adaptation in accordance with traffic variation in a very short time scale (millisecond level) avoiding any network-level cooperation scheme. The energy saving potential of this feature has been discussed in the scientific literature [14, 16]. However, a quantitative analysis to evaluate the achievable savings has not been performed extensively. As an example, valid for all cell sleep modes, the interaction between network planning, cell load levels and actual deactivation time of the cells is an important aspect to be considered. In fact, energy saving obtained by means of sleep modes, in particular micro sleep, is closely related to the initial network deployment, essentially because network density determines the cell load levels in the network itself which, in turn, impact the deactivation time of each cell with micro sleep capability. Therefore, micro sleep should be taken into account at the planning stage, in order to maximize the achievable energy saving. These evaluation results will be showed in Sect. 5.1.

[4]In principle, the two terms could be distinguished (being the "micro sleep" the underlying HW functionality, and "micro DTX" the SW feature utilizing that HW functionality for energy saving). Anyway since they are used jointly and then relative to the same Energy Efficiency feature, for simplicity, both terms are used indifferently in the following.

4.2 Intra-sector Traffic Steering

Another energy efficiency feature is intra-sector traffic steering; basically the traffic carried by either 2G, 3G or both RATs' frequency layers within a sector is steered towards the 4G frequency layers of the same sector and, as a consequence, the emptied 2G and/or 3G layers can be deactivated. This is done in order to exploit the higher capacity of the 4G RAT and its more energy-efficient equipment compared to the 2G/3G one, in the view of achieving as much energy saving as possible.

Here below are described two possible *design conditions* (and related traffic thresholds) to be satisfied at the same time for the implementation of the intra-sector traffic steering:

Condition C_1:

- candidate 2G/3G layers for traffic steering are identified according to sector load: their traffic (normalized with respect to the peak throughput of each considered RAT) must be less than or equal to a predefined threshold th_1(e.g., 50 %).
- If steering is performed the empty layers are switched-off (i.e. the value of power consumption of those layers is the one that can be achieved when a sleep mode is activated).

Condition C_2:

- the candidate 4G layers able to handle the original 2G/3G traffic, are chosen in such a way that they will not saturate their capacity after traffic steering. This means that the 4G normalized traffic must remain less than or equal to a predefined threshold th_2 (e.g., 90 %).

where it must be highlighted that both the thresholds th_1 and th_2 can be set according to the operator preference and the values here considered are just exemplary ones.

In the following, a detailed explanation of the intra-sector traffic steering *working principle* is reported.

Let's consider all traffic data information of all RATs' frequency layers within a single sector (e.g., s_1) of a tri-sectorial site, indicated as $\underline{\mathbf{t}}_{0,s1}{}^{(1)}$ vector. This is a 6 elements row-vector in which the first two elements are the traffic information in charge of the 2G frequency layers, the third and fourth elements are the traffic information of the 3G frequency layers and the last two elements are related to the traffic on the 4G frequency layers. It must be specified that the above vector contains non-normalized traffic information whilst the superscript $^{(1)}$ refers to the considered time sample (i.e., 1st time sample).

$$\underline{\mathbf{t}}_{0,s1}{}^{(1)} = \left[t^{(1)}_{0,f2G\_1} \quad t^{(1)}_{0,f2G\_2} \quad t^{(1)}_{0,f3G\_1} \quad t^{(1)}_{0,f3G\_2} \quad t^{(1)}_{0,f4G\_1} \quad t^{(1)}_{0,f4G\_2} \right]$$

The vector indicating how the traffic of the considered sector is distributed after traffic steering is indicated as $\underline{\mathbf{t}}^{(1)}_{1,s1}$: this vector can be obtained by relating $\underline{\mathbf{t}}^{(1)}_{0,s1}$ to a 6-by-6 permutation matrix $\underline{\underline{\mathbf{M}}}^{(1)}_{s1}$ which represents the mathematical form of the energy efficiency feature for the considered sector (s_1) at the specified time sample (1st one). In formula:

$$\underline{\mathbf{t}}_{1,s1}{}^{(1)} = \underline{\mathbf{t}}_{0,s1}{}^{(1)} \cdot \underline{\underline{\mathbf{M}}}_{s1}^{(1)} = \left[t_{1,f2G\_1}^{(1)} \; t_{1,f2G\_2}^{(1)} \; t_{1,f3G\_1}^{(1)} \; t_{1,f3G\_2}^{(1)} \; t_{1,f4G\_1}^{(1)} \; t_{1,f4G\_2}^{(1)} \right]$$

The permutation matrix $\underline{\underline{\mathbf{M}}}_{s1}^{(1)}$ and, consequently, the post-steering traffic vector $\underline{\mathbf{t}}_{1,s1}{}^{(1)}$, is computed iteratively, by checking that the two conditions C_1 and C_2 are satisfied when steering the traffic from the 2G/3G (or both RATs) frequency layers towards the 4G ones.

At the beginning of the algorithm (i.e., 1st step of the iterative process) the permutation matrix $\underline{\underline{\mathbf{M}}}_{s1}^{(1)}$ is equal to the 6-by-6 identity matrix, i.e.:

$$\underline{\underline{\mathbf{M}}}_{s1}^{(1)} = \underline{\underline{\mathbf{M}}}_{1}^{(1)} = \underline{\underline{\mathbf{I}}}_{6} = \begin{bmatrix} 1\,0\,0\,0\,0\,0 \\ 0\,1\,0\,0\,0\,0 \\ 0\,0\,1\,0\,0\,0 \\ 0\,0\,0\,1\,0\,0 \\ 0\,0\,0\,0\,1\,0 \\ 0\,0\,0\,0\,0\,1 \end{bmatrix}$$

where the subscript "1" refers to the 1st step of the iterative algorithm.

Let us indicate with $x_{0,fXG\_i}^{(1)}$ the normalized pre-steering traffic data on the ith frequency layer of the XG RAT ($X = 2, 3, 4$) at the 1st time sample. Similarly, $x_{1,fXG\_i}^{(1)}$ indicates the normalized post-steering traffic data on the ith frequency layer of the XG RAT ($X = 2, 3, 4$) at the 1st time sample.

Let us now suppose that the traffic data originally handled by the 2G frequency layers can only be hypothetically steered and managed by the 4G ones, provided that conditions C_1 and C_2 are satisfied.

This means that, at the 2nd step of the algorithm, the normalized traffic data of the first 2G frequency layer (i.e., $x_{0,f2G\_1}^{(1)}$) must be less than or equal to th_1 and, at the same time, the normalized post-steering traffic data in charge of one between the two 4G frequency layers (i.e., $x_{1,f4G\_1}^{(1)}$ or $x_{1,f4G\_2}^{(1)}$, chosen according to post-steering saturation level) must be less than or equal to th_2. In formula:

$$\text{if } \left\{ x_{0,f2G\_1}^{(1)} \le th_1 \text{ AND } \left(x_{1,f4G\_1}^{(1)} \le th_2 \text{ OR } x_{1,f4G\_2}^{(1)} \le th_2 \right) \right\} \rightarrow t_{1,f4G\_i}^{(1)} = t_{0,f4G\_i}^{(1)} + t_{0,f2G\_1}^{(1)}$$

If the above statement is fulfilled the traffic data of the first 2G frequency layer is steered towards one (and only one) of the two available 4G frequency layers. Assuming that the first 4G frequency layer can accept the traffic originating from the first 2G layer, the permutation matrix at the 2nd step of the algorithm (i.e., $\underline{\underline{\mathbf{M}}}_{2}^{(1)}$) is expressed as:

$$\underline{\underline{\mathbf{M}}}_2^{(1)} = \begin{bmatrix} \boxed{0} & 0 & 0 & 0 & \boxed{1} & 0 \\ 0 & 1 & 0 & 0 & 0 & 0 \\ 0 & 0 & 1 & 0 & 0 & 0 \\ 0 & 0 & 0 & 1 & 0 & 0 \\ 0 & 0 & 0 & 0 & 1 & 0 \\ 0 & 0 & 0 & 0 & 0 & 1 \end{bmatrix}$$

and, consequently, the post-steering non-normalized traffic data vector of the considered sector at the 2nd step of the iterative process is:

$$\underline{\mathbf{t}}_{1,s1\_step2}{}^{(1)} = \underline{\mathbf{t}}_{0,s1}{}^{(1)} \cdot \underline{\underline{\mathbf{M}}}_2^{(1)} = \begin{bmatrix} 0 & t_{1,f2G\_2}^{(1)} & t_{1,f3G\_1}^{(1)} & t_{1,f3G\_2}^{(1)} & t_{1,f4G\_1}^{(1)} + t_{0,f2G\_1}^{(1)} & t_{1,f4G\_2}^{(1)} \end{bmatrix}$$

The possibility to steer the traffic data of the second 2G frequency layer towards one (and only one) of the available 4G frequency layers must now be analyzed. Again,

$$\text{if } \left\{x_{0,f2G\_2}^{(1)} \le th_1 \text{ AND } \left(x_{1,f4G\_1}^{(1)} \le th_2 \text{ OR } x_{1,f4G\_2}^{(1)} \le th_2\right)\right\} \to t_{1,f4G\_i}^{(1)} = t_{0,f4G\_i}^{(1)} + t_{0,f2G\_2}^{(1)}$$

Assuming that the first 4G frequency layer does not accept the traffic originating from the first 2G layer, but the second 4G layer does, the permutation matrix $\underline{\underline{\mathbf{M}}}_3^{(1)}$ at the 3rd step of the algorithm (which is the final step, since traffic steering for the 2G frequency layers only is performed) is expressed as:

$$\underline{\underline{\mathbf{M}}}_3^{(1)} = \underline{\underline{\mathbf{M}}}_{s1}^{(1)} = \begin{bmatrix} \boxed{0} & 0 & 0 & 0 & \boxed{1} & 0 \\ 0 & \boxed{0} & 0 & 0 & 0 & \boxed{1} \\ 0 & 0 & 1 & 0 & 0 & 0 \\ 0 & 0 & 0 & 1 & 0 & 0 \\ 0 & 0 & 0 & 0 & 1 & 0 \\ 0 & 0 & 0 & 0 & 0 & 1 \end{bmatrix}$$

and, consequently, the post-steering non-normalized traffic data vector of the considered sector at 1st time sample is:

$$\underline{\mathbf{t}}_{1,s1\_step3}{}^{(1)} = \underline{\mathbf{t}}_{1,s1}{}^{(1)} =$$
$$= \underline{\mathbf{t}}_{0,s1}{}^{(1)} \cdot \underline{\underline{\mathbf{M}}}_3^{(1)} = \begin{bmatrix} 0 & 0 & t_{1,f3G\_1}^{(1)} & t_{1,f3G\_2}^{(1)} & t_{1,f4G\_1}^{(1)} + t_{0,f2G\_1}^{(1)} & t_{1,f4G\_2}^{(1)} + t_{1,f2G\_2}^{(1)} \end{bmatrix}$$

Finally, the above expression shows that both the 2G frequency layers of the considered sector, for the 1st time sample, can be switched-off since their traffic is now jointly managed by the 4G layers.

The overall algorithm must be performed for the remaining sectors, leading to a set of permutation matrices $\underline{\underline{\mathbf{M}}}_{s\_i}^{(1)}$ (with $s\_i = 1, \ldots, N$, where N represents the number of sectors within the cluster of network sites) for the 1st time sample only; the same approach must be followed for the remaining $R - 1$ time samples, with R indicating the total number of time samples within the observation period.

5 Energy Efficiency Assessment

In the following subsections performance evaluations are presented, respectively, for micro sleep and intra-sector traffic steering.

5.1 *Micro Sleep*

In Sect. 4.1 it has been stated that micro sleep (here below also referred to as cell DTX) should be taken into account at the planning stage, in order to maximize the achievable energy saving. Here numerically demonstration of how the incorporation of this feature at the planning phase increases the achievable energy saving is reported; this has been accomplished by comparing the deployment for the lowest daily energy consumption R^* for:

i. Case 1: $\delta \in [0, 1) \rightarrow$ cell DTX is incorporated with clean-slate network deployment;
ii. Case 2: $\delta = 1 \rightarrow$ cells do not have DTX capability.

Figure 12 depicts the daily average area power consumption as a function of cell range for both cases (assuming $\delta = 0.1$ for Case 1). In this simulation, a mobile network with regular hexagonal layout, consisting of 19 sites equipped with a single omni-directional antenna with 15 dBi antenna gain and cell range varying between 100 and 800 m, has been considered. The simulated system operates at 2 GHz with 10 MHz BW. The Okumura-Hata pathloss model for an urban area based on 3GPP specifications [17] with 8 dB user noise figure has been utilized. Users are randomly distributed over the area with a density of $\rho = 1000$ users/km^2. To account for the traffic fluctuations $\alpha(t)$ during a day, an approximated daily pattern based on the downlink traffic measurements presented in [17] has been considered, by also assum-

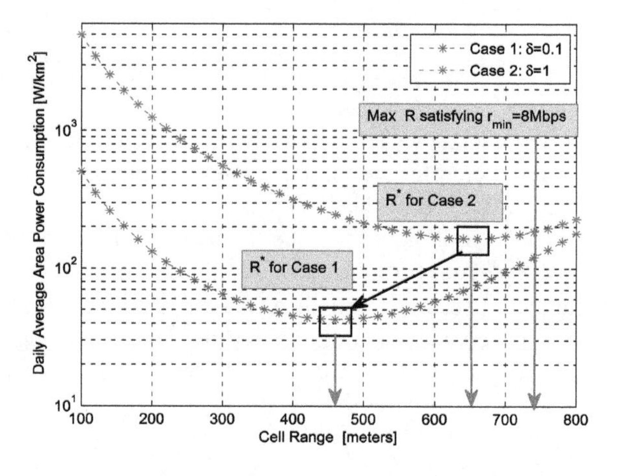

Fig. 12 Daily average area power consumption as a function of cell range for $\delta = 0.1$, $\delta = 1$ and $r_{min} = 8$ Mbps [15]

ing that user behaviour is unchanged. Here each user demanding $\Omega = 18\,\text{MB}$ within a duration of one hour is considered, which corresponds to a peak area traffic demand of $40\,\text{Mbps/km}^2$, a reasonable estimate for 2015 [14]. For the proposed algorithm $\zeta = 4.7$, $P_0 = 130\,\text{W}$ based on [17] and $\nu_{\max} = 6\,\text{bps/Hz}$ ($\nu_{\max}$ being the system's spectral efficiency) have been set. Moreover, Quality-of-Service (QoS) constraints are defined as $P_{\min} = -70\,\text{dBm}$, $r_{\min} = 8\,\text{Mbps}$ ($r_{\min}$ being the minimum data rate demand) [17].

From Fig. 12 it can be observed that when cell DTX is incorporated at the planning stage, higher densification tends to be preferred, which also brings significant energy savings. This is mainly due to the fact that lower cell load levels taken into account in Case 1, which create long deactivation periods originating from densification, can be efficiently exploited by cell DTX.

Normally, operators would design their networks for minimum network deployment cost, i.e., with as few BSs as possible. Assuming, however, that the main interest here is to obtain the most energy-efficient network deployment regardless of costs, the BSs' density that provides the lowest daily energy consumption in the considered cases has been found.

Finally, Fig. 13 displays how the new energy-efficient network incorporating cell DTX performs with respect to the daily area power consumption. To this end, three distinct cases have been considered:

(1) network deployment without cell DTX (blue bar, Case 2);
(2) no cell DTX considered in the network planning phase, but it is in operation (green bar, Case 2);
(3) network deployment with cell DTX (red bar, Case 1).

From Fig. 13 it can be noted that cell DTX brings striking energy savings (from blue to green bar) even when the network was not planned considering cell DTX. However, designing the network under the assumption that cells can be put into DTX mode during idle state results in additional 42 % saving, at the cost of deploying

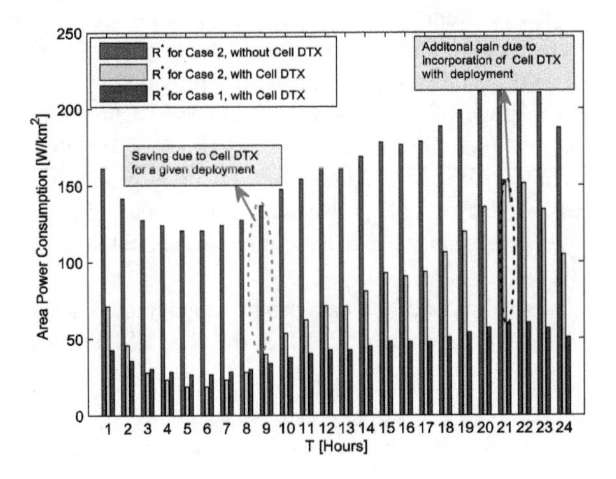

Fig. 13 Daily area power consumption variation for $\delta = 0.1$. [15]

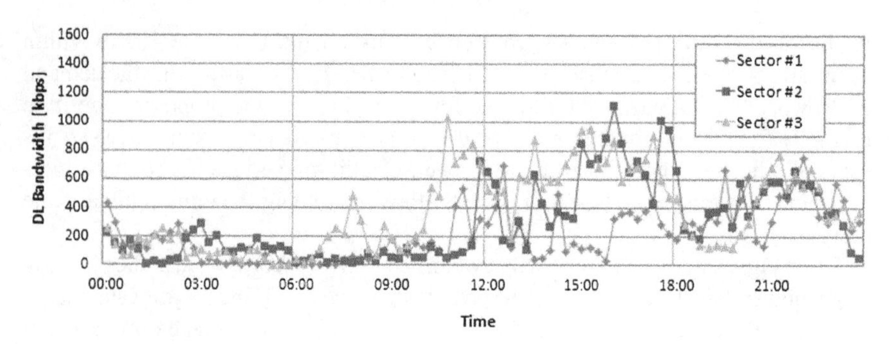

Fig. 14 Example of daily traffic profiles extracted from the live network (Downlink BW, 3 HSDPA sectors)

around 110 % more BSs. Furthermore, the resulting network deployment significantly improves the user QoS, due to the reduction of the overall network load. It can be concluded that, if the objective is to obtain the maximum energy savings, networks have to be designed taking into account that network deployment and operation are closely related. In other words, it can be stated that when modernizing and rolling out new technology, e.g. 5G, DTX capability should be taken into account in the network planning in order to maximize the energy savings.

5.2 *Intra-sector Traffic Steering*

In this section, simulation results related to the intra-sector traffic steering technique are shown and analyzed.[5] Energy efficiency performance have been assessed by considering a homogeneous cluster of mobile network sites and by using actual traffic profiles extracted from the live network, thus an energy consumption analysis has been done in realistic conditions. The set of data traffic profiles related to a BS currently running in the operator's mobile network was extracted by means of a network monitoring system with fixed time resolution. In particular, for each sector, a single data sample automatically provided by the monitoring system represents the average value of a specific Key Performance Indicator (KPI) over a 15 min time interval (e.g., for traffic profile in Fig. 14, a single sample represents the amount of data transferred in 15 min), hence a daily profile is represented by 96 consecutive values. Each sample is also averaged over the 5 working days of the week, in order to filter effects due to spurious data peaks and to increase the reliability of the provided profile.

An example of the data profiles used in the energy efficiency performance evaluations is reported in Fig. 14: in the plot, HSDPA downlink BW for the 3 sectors is

[5]These results have also been initially reported in [18], even if with less details on the considered methodology (which is described in this chapter with a more rigorous approach).

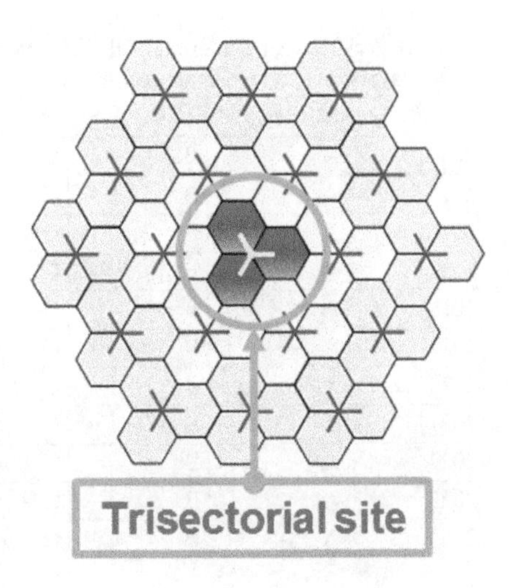

Fig. 15 System layout considered for network-level energy efficiency assessment

showed only, even if other traces can be extracted from the live network in order to analyze other KPIs.

The trend of extracted data profiles in Fig. 14 shows a typical daily oscillation of the traffic, consistent with average profiles available in literature [17, 18], but with the important difference that while literature curves are averaged over an entire network, data profiles used for the reported performance assessment are related to a single site, in order to better highlight particular burst effects of the traffic variation. The resulting curves are then suitable for a specific evaluation of the site's energy consumption, based on actual traffic load. Moreover, these profiles are also suitable for the evaluation of other energy efficiency features (e.g. like those proposed in [19] and ON-OFF schemes tested in [20] with commercial BSs).

Network-level energy efficiency assessment has been performed by evaluating the energy consumption of a homogeneous cluster of 19 tri-sectorial macro sites located in an urban environment, all equipped with two GSM frequency layers (900 and 1800 MHz), UMTS, HSDPA and two LTE frequency layers. This system layout is depicted in Fig. 15, where the traffic of the whole cluster has been generated by replicating the traffic information extracted from the site located in the actual operator's network through a set of *perturbation factors*. These factors are chosen in such a way that the resulting traffic distribution among all 19 sites in the cluster is, on average, as much homogeneous as possible. Traffic profiles of voice connections have been converted into "Equivalent Data Traffic" (i.e., from *Erl* to *kbps*), in order to jointly manage voice information with data traffic within the BS's power model formula.

In addition, aiming at assessing the yearly energy efficiency performance evaluations for traffic data evolution from 2014 to 2024, the November 2014s Ericsson Mobility Report [21] has been considered, by applying a Compound Annual Growth Rate (CAGR) of 40 % for mobile traffic growth.

Finally, to better exploit 4G capacity and its possible improvements (e.g., introduction of Carrier Aggregation, higher order MIMO), the amount of the total annual

Table 2 Year-by-year evolution of the non-normalized traffic data information on all RATs frequency layers related to a single sector of a site

Year	2G@ f_{2G_1}	2G@ f_{2G_2}	3G@ f_{3G_1}	3G@ f_{3G_2}	4G@ f_{4G_1}	4G@ f_{4G_2}
2014	A	B	C	D	E	F
2015	A	B	C	D	$E^{(1)}$	$F^{(1)}$
2016	A	B	C	D	$E^{(2)}$	$F^{(2)}$
2017	A	B	C	D	$E^{(3)}$	$F^{(3)}$
2018	A	B	C	D	$E^{(4)}$	$F^{(4)}$
2019	A	B	C	D	$E^{(5)}$	$F^{(5)}$
2020	A	B	C	D	$E^{(6)}$	$F^{(6)}$
2021	A	B	C	D	$E^{(7)}$	$F^{(7)}$
2022	A	B	C	D	$E^{(8)}$	$F^{(8)}$
2025	A	B	C	D	$E^{(9)}$	$F^{(9)}$
2024	A	B	C	D	$E^{(10)}$	$F^{(10)}$

Table 3 Peak throughput values for all considered RATs

RAT	Peak throughput (maximum capacity'/for traffic data normalization [Mbps]
2G	1.2[a]
3G Rel.'99	0.384
3G HSDPA	21.100
4G	150 (years: 2014 → 2018) 300 (years: 2019 → 2020) 600 (years: 2021 → 2022) 1200 (years: 2023 → 2024)

[a]Estimated in order to globally take into account Equivalent Voice Traffic, GPRS and EDGE traffic

traffic increase has been equally split between the two 4G frequency layers. Therefore, if the 2014 non-normalized traffic data information on all RATs frequency layers related to a single sector is represented by letters from A to F as reported in the first row in Table 2, the yearly evolution of the traffic information will be computed as:

$$
\begin{aligned}
X^{(n)} &= X^{(n-1)} + \Delta^{(n)} \\
\Delta^{(n)} &= \frac{0,4 \cdot \left(A+B+C+D+E^{(n-1)}+F^{(n-1)}\right)}{2} \\
X &= E, F
\end{aligned}
$$

As the power model formula requires the traffic information to be normalized with respect to the peak throughput value (i.e., maximum capacity) of each considered RAT, the 4G capacity must reflect the annual increase of the traffic demand within the network, thus simulations have been run by considering RATs' maximum capacity values as in Table 3.

Table 4 BSs' power models weighted combinations when evaluating the cluster's daily average energy consumption in the considered baseline scenarios

Year	BLS#1 Business as usual	BLS#2 Network renewal in 2017	BLS#3 Continuous network renewal
2014	100 % PM2012	100 % PM2012	100 % PM2012
2015	100 % PM2012	100 % PM2012	100 % PM2012
2016	100 % PM2012	100 % PM2012	100 % PM2012
2017	100 % PM2012	75 % PM2012 & 25 % PM2016	75 % PM2012 & 25 % PM201S
2018	100 % PM2012	50 % PM2012 & 50 % PM2016	50 % PM2012 & 50 % PM201S
2019	100 % PM2012	25 % PM2012 & 75 % PM2016	25 % PM2012 & 75 % PM2015
2020	100 % PM2012	100 % PM2016	100 % PM2016
2021	100 % PM2012	100 % PM2016	75 % PM2016 & 25 % PM2020
2022	100 % PM2012	100 % PM2016	50 % PM2016 & 50 % PM2020
2023	100 % PM2012	100 % PM2016	25 % PM2016 & 75 % PM2020
2024	100 % PM2012	100 % PM2016	100 % PM2020

4G capacity improvement and subsequent increase of the power model's parameters values (i.e., A, B and C) can be modelled as:

$$\begin{aligned} &Capacity_{4G} = 150 \cdot \tfrac{N}{2} \text{ [Mbps]} \\ &X_{4G}{}^{*} = X_{4G} \cdot \tfrac{N}{2} \text{ [W]} \\ &X = A, B, C \\ &N = 2, 4, 8, 16 \end{aligned}$$

where N is a parameter associated to the network feature allowing for the 4G capacity improvement and related power model's parameters increase (e.g. MIMO order or number of basebands to put together in Carrier Aggregation).

Performance of three baseline systems over the years from 2014 to 2024 have been assessed, in which different BSs replacement plans are considered and any kind of traffic steering option is taken into account (see Table 4):

1. *Business as Usual*: annual traffic increase from 2014 to 2024 and BSs' power models of the year 2012 (baseline power model).
2. *Network renewal in 2017*: annual traffic increase from 2014 to 2024 with network equipment replacement starting from 2017, by considering BSs' power models of the year 2016 and whose completion is due by 2020. The cluster's daily average energy consumption for traffic volumes from 2017 to 2019 is computed as

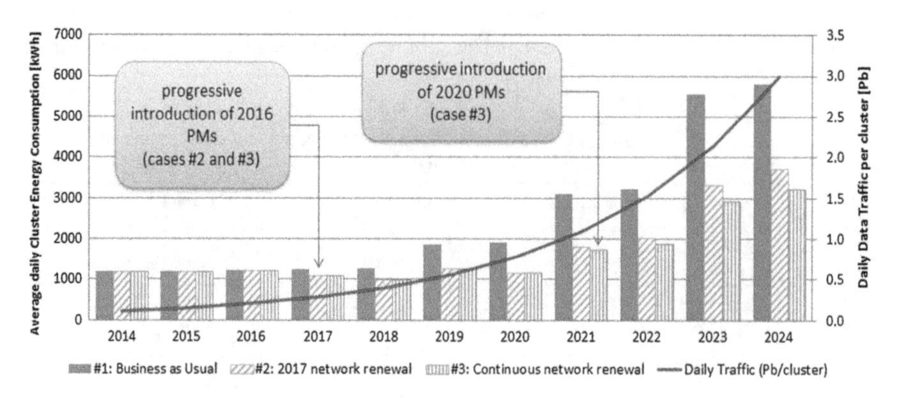

Fig. 16 Daily average energy consumption and data traffic within the considered cluster of mobile network sites

a weighted sum of energy consumptions obtained separately with both power models of 2012 and 2016, as a consequence of the stepwise BSs' replacement.

3. *Continuous network renewal*: annual traffic increase from 2014 to 2024 and continuous network equipment replacement to be carried out in two distinct phases:

    a. from 2017 to 2020, by considering progressive introduction of new BSs whose power consumption is computed as in baseline system labelled as "Network renewal in 2017". From 2021 to 2024, instead, the BSs' power model of the year 2016 is only considered;
    b. from 2021 to 2024, by considering progressive introduction of new BSs whose power consumption is computed by also considering the power model of the year 2020. In this case, the cluster's daily average energy consumption is computed as a weighted sum of energy consumptions obtained separately with both power models of 2016 and 2020, as a consequence of the stepwise BSs' replacement.

Figure 16 reports the daily average energy consumption (in kWh) related to the considered cluster of mobile network sites for the three baseline systems under evaluation. Daily data traffic (in Pb) is also shown.

As can be observed, main improvements in terms of energy savings, with respect to the "Business as Usual", can be achieved when considering network equipment replacement as supposed in "Network renewal in 2017" (i.e. by progressively replacing older BSs with new ones whose power consumption is modelled by means of the power model of the year 2016). In other words, a complete network equipment replacement in 2017 with BSs performing better in terms of power consumption will result in high savings also in subsequent years. Furthermore, a possible decision on a second round of network equipment replacement from 2020 can be postponed at a time when sufficiently high energy saving can be reached, thus motivating a second round of network investment. Besides, in 2020 or later on, the introduction of a new 5G RAT would represent another investment concurrent to the renewal of legacy equipment: that is the reason why it is not likely to foresee at this time baseline

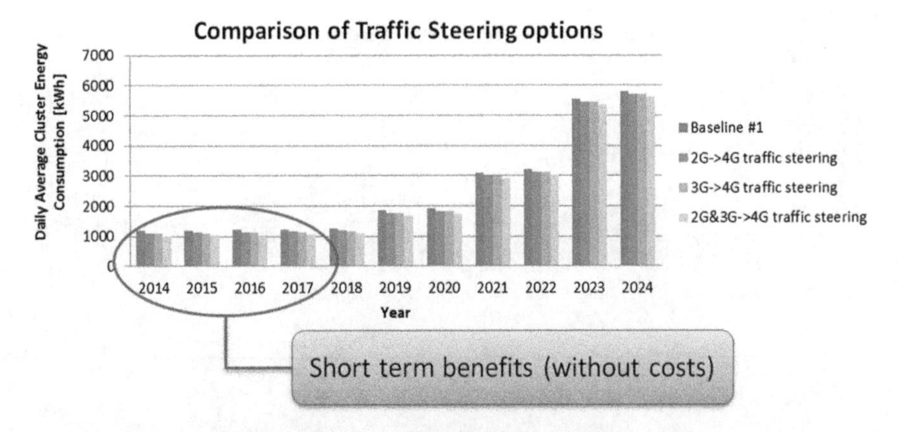

Fig. 17 Performance comparison of energy consumptions obtained by means of different traffic steering options with respect to the baseline system #1 ("Business as Usual")

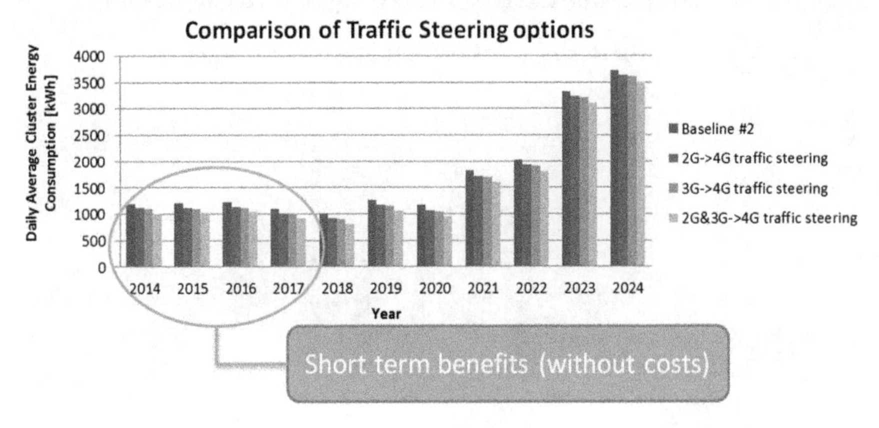

Fig. 18 Performance comparison of energy consumptions obtained by means of different traffic steering options with respect to the baseline system #2 ("Network renewal in 2017")

#3 as the most probable scenario, due to the possible presence of two concurrent investments, which are not necessarily affordable for the operator.

As a consequence, the "Network renewal in 2017" can be considered as the *best solution*, since it provides short term benefits with less costs when compared to the "Continuous network renewal" one; furthermore a 40 % energy saving with respect to the "Business as Usual" can be achieved in 2020, even if energy consumption of these baselines in 2020 is still higher than one in 2014.

Regarding frontline systems' evaluation, in the following Figs. 17, 18 and 19 evolutionary comparisons of different steering solutions with respect to the evaluated baseline systems ("Business as Usual", "Network renewal in 2017" and "Continuous network renewal") are reported, respectively.

As it can be observed, all the three pictures above show comparable trends and, in particular, the benefits of applying traffic steering from a single legacy RAT (2G

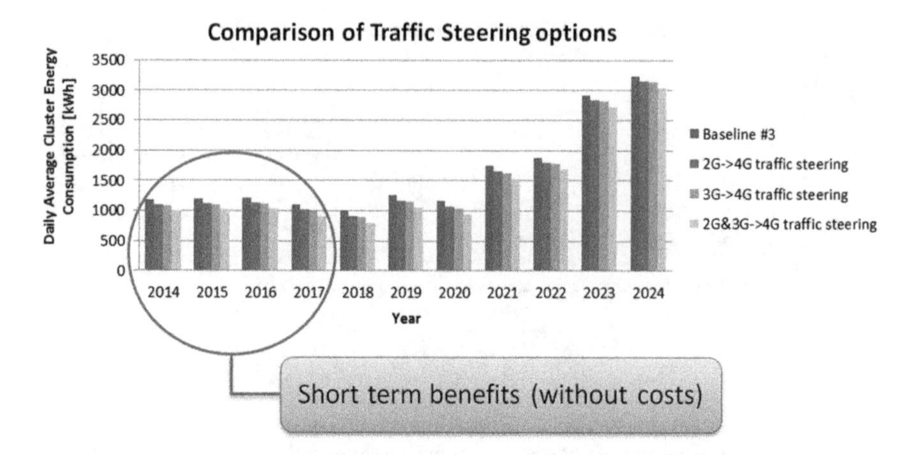

Fig. 19 Performance comparison of energy consumptions obtained by means of different traffic steering options with respect to the baseline system #3 ("Continuous network renewal")

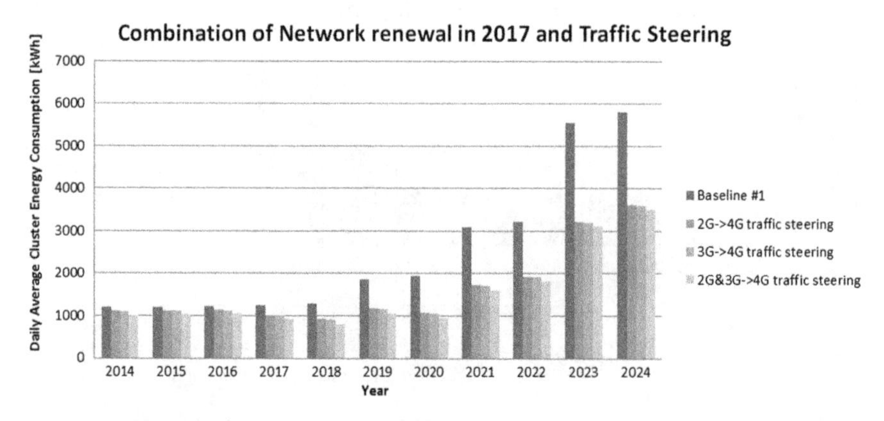

Fig. 20 Cluster's energy consumption performance obtained by jointly considering network renewal in 2017 and traffic steering options with respect to the baseline system #1 ("Business as Usual")

or 3G) to 4G are similar, due to similar performance of the legacy RAT (according to the considered power models). Furthermore, by jointly steering the traffic of both 2G and 3G RATs towards 4G, additional energy savings can be reached, with similar trends in terms of energy consumption decrease.

It is noteworthy to state that these traffic steering solutions do not require any investment for the network operator and they may be implemented since 2014, allowing for short term benefits' achievement.

In Fig. 20 cluster's energy consumption obtained by jointly considering network renewal in 2017 (as indicated in Table 4) and traffic steering options implemented since 2014, with respect to "Business as Usual" baseline system, is presented.

As expected, the combination of both traffic steering and stepwise network renewal permits to achieve short term benefits in terms of energy consumption by minimizing,

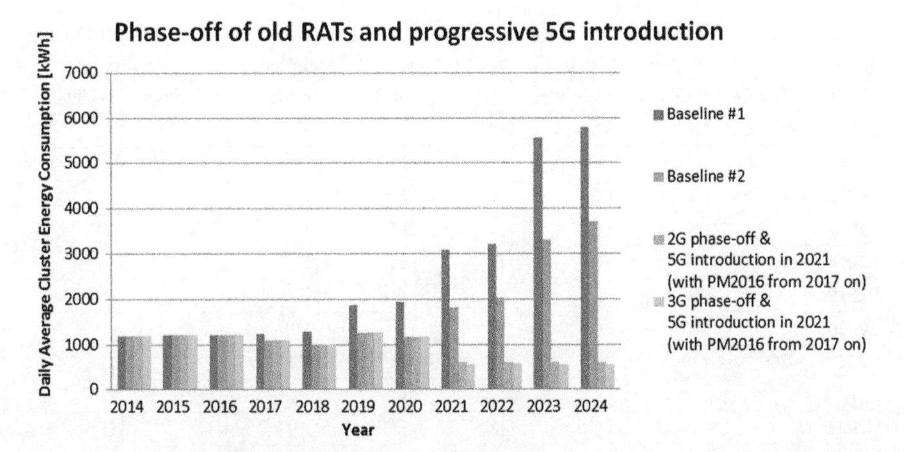

Fig. 21 Cluster's energy consumption performance obtained by jointly considering network renewal in 2017 and 5G introduction with contemporary phase-off of legacy RATs in 2021, with respect to both baseline systems #1 ("Business as Usual") and #2 ("Network renewal in 2017")

at the same time, the investment for the network operator. The holistic solution, of course, is still able to satisfy all the traffic demand in the network, allowing for huge energy savings with respect to the "Business as Usual" baseline system (38 % for 2G/3G traffic steering and 40 % for 2G&3G traffic steering, respectively, in 2024), even if power consumption in 2024 is still higher, in absolute terms, than the one in 2014. This can represent a motivation for 5G introduction, in order to further save energy by considering 4G offloading towards the new RAT, satisfying the additional traffic.

The last analyzed frontline scenario is the one in which the introduction in 2021 of the new 5G RAT with contextual phase-off of a legacy one, i.e. 2G or 3G, is jointly considered together with a BSs replacement plan as thought in "Network renewal in 2017" baseline system. Evolutionary performance of this scenario from 2014 to 2024 are shown in Fig. 21 and are compared with the ones related to both "Business as Usual" and "Network renewal in 2017" baseline systems.

Cluster's non-normalized traffic data evolution from 2014 to 2024 has been modeled as reported in Table 5, where the 2G phase-off is considered (the 3G phase-off case is similar); to this end, the availability of two 5G frequency layers is assumed.

Basically, a CAGR of 40 % for mobile traffic growth is still considered and the amount of the total annual traffic increase is now equally split between the two new 5G frequency layers, that is:
for 2021

$$
\begin{aligned}
G &= A + \Delta \\
H &= B + \Delta \\
\Delta &= \frac{0,4 \cdot \left(A + B + C + D + E^{(6)} + F^{(6)}\right)}{2}
\end{aligned}
$$

Table 5 Year-by-year evolution of the non-normalized traffic data information on all RATs frequency layers in a single sector of a site when introducing 5G and phasing-off 2G in 2021

Year	2G@f_{2G_1}	2G@f_{2G_2}	3G@f_{3G_1}	3G@f_{3G_2}	4G@f_{4G_1}	4G@f_{4G_2}	5G@f_{5G_1}	5G@f_{5G_2}
2014	A	B	C	D	E	F	–	–
2015	A	B	C	D	$E^{(1)}$	$F^{(1)}$	–	–
2016	A	B	C	D	$E^{(2)}$	$F^{(2)}$	–	–
2017	A	B	C	D	$E^{(3)}$	$F^{(3)}$	–	–
2018	A	B	C	D	$E^{(4)}$	$F^{(4)}$	–	–
2019	A	B	C	D	$E^{(5)}$	$F^{(5)}$	–	–
2020	A	B	C	D	$E^{(6)}$	$F^{(6)}$	–	–
2021	2G phase-off		C	D	$E^{(7)}$	$F^{(7)}$	G	H
2022			C	D	$E^{(8)}$	$F^{(8)}$	$G^{(1)}$	$H^{(1)}$
2023			C	D	$E^{(9)}$	$F^{(9)}$	$G^{(2)}$	$H^{(2)}$
2024			C	D	$E^{(10)}$	$F^{(10)}$	$G^{(3)}$	$H^{(3)}$

Table 6 Maximum throughput values for all considered RATs in phase-off scenarios

RAT	Peak throughput (maximum capacity) for traffic data normalization [Mbps]
2G	1.2[a]
3G Rel.'99	0.384
3G HSDPA	21.100
4G	300 (from 2021 on)
5G	1000 (from 2021 on)

[a]Estimated in order to gtobally take into account Equivalent Voice Traffic, GPRS and EDGE traffic

while, from 2022 to 2024

$$X^{(n)} = X^{(n-1)} + \Delta^{(n)}$$

$$\Delta^{(n)} = \frac{0,4 \cdot \left(C + D + E^{(6)} + F^{(6)} + G^{(n-1)} + H^{(n-1)}\right)}{2}$$

$$X = G, H$$

Regarding traffic data normalization, the considered peak throughput values (i.e., maximum capacity) for all considered RATs are reported in Table 6.

When 5G is deployed, the power consumption of remaining RATs (2G/3G and 4G) is still computed by considering BSs' power models of the year 2016 (according to the BSs replacement plan as in "Network renewal in 2017").

The introduction of the 5G RAT in substitution of a legacy one, i.e. 2G or 3G, allows for energy saving thanks to the elimination of a technologically obsolete

RAT. Furthermore, the partial 4G traffic offloading towards the 5G RAT provides additional energy savings since 2021, following the network renewal of the legacy network started in 2017. With 5G fully *on-field* in 2024 and 2G/3G RAT phased-off, a reduction of nearly 50 % of the cluster's energy consumption can be achieved with respect to the 2014 energy consumption.

6 Summary and Conclusions

In this chapter few selected techniques and actions that the network operator should consider for the mobile network's evolution towards 5G from a sustainability point of view have been presented and assessed. To this end, the energy consumption of the RAN when considering different load conditions (obtained by means of actual daily traffic profiles extracted from the live network) and different years (according to current traffic forecasts) has been taken into account. This study has been conducted in order to provide an overview of different "*what-if*" scenarios towards 2020 and beyond, as useful insights for operators aiming at analysing evolutionary steps of their mobile networks in the view of the introduction of future 5G systems.

In particular, in order to estimate the energy consumption of a future network, a model of the power consumption related to future BSs needed. Thus, following the methodology used in [3], results from the EARTH project [4] have been adopted, by considering different power models for different types of BSs. Evolutionary power models have also been derived for each RAT through the years, providing an essential framework for network-level energy efficiency assessments. Moreover, many different energy efficiency features to be possibly implemented within the mobile network have been analysed and, due to the different time scales of these features, their evaluation has been decoupled as follows:

- at shorter time scale (i.e., milliseconds) micro sleep has been considered and assessed by means of system-level simulations;
- at higher time scale (i.e., from minutes to years), progressive network renewal with traffic steering options and legacy RATs' phase-off policies has been evaluated by using a specific evaluation tool fed by actual traffic coming from the live network.

Regarding the first set of features, the maximum achievable energy saving when enabling cell micro sleep in mobile networks has been assessed, also considering the fact that the achievable energy saving is closely related to the initial network deployment. To this end, cell DTX has been incorporated within the planning stage of the network and the BSs' density that provides the lowest daily energy consumption satisfying certain coverage and QoS requirements has been determined. It has been shown that, if the DTX feature is taken into account in the planning stage of the network, the energy consumption can be reduced by means of denser deployment of lightly loaded cells. The drawback of such a solution is the additional deployment cost. In future works, the investigation regarding the optimum network deployment that minimizes the total network cost, also considering its energy consumption and incorporating cell DTX with sector sleep feature, are planned to be evaluated.

Finally, for the second set of energy efficiency features, different traffic steering strategies applied to different legacy RATs and RAT-based power models' evolution through the years have been considered. Therefore some "*what-if*" scenarios related to the introduction of a new 5G RAT have been evaluated, always by considering cluster-level energy efficiency performance (i.e., by taking into account the recent introduction of ETSI EE specifications, based on homogeneous clusters' evaluations that may be easily extrapolated at country level). An evolution of this study may include the full implementation of the methodology introduced in ETSI ES 203228 specification [5], allowing for a complete assessment of different clusters of sites, thus providing country-wide energy efficiency evaluations. In addition, a possible future step could take into account the introduction of the resource pooling in Cloud-RAN systems (and a complete power model including all elements in C-RAN architecture) as well as the Machine-to-Machine (M2M) infrastructure effects on the evolutionary energy consumption of the overall mobile network.

References

1. Cisco: Cisco Visual Networking Index: Forecast and Methodology, 2012–2017. [online document], White Paper, Nov 2014
2. Rapone, D., Sabella, D., Fodrini, M.: Energy efficiency solutions for the mobile network evolution towards 5G: an operator perspective. In: Sustainable Internet and ICT for Sustainability (SustainIT), 14–15 Apr 2015
3. EARTH project deliverable D2.3: Energy efficiency analysis of the reference systems, areas of improvements and target breakdown. https://bscw.ict-earth.eu/pub/bscw.cgi/d71252/EARTH_WP2_D2.3_v2.pdf
4. EARTH project deliverable D6.4: Final integrated concept. https://bscw.ict-earth.eu/pub/bscw.cgi/d49431/EARTH_WP6_D6.4.pdf
5. ETSI ES 203 228 V1.0.0 (2015-01): Assessment of mobile network energy efficiency. ETSI Environmental Engineering (EE), Jan 2015. http://www.etsi.org/deliver/etsi_es/203200_203299/203228/01.00.00_50/es_203228v010000m.pdf
6. EARTH project deliverable D2.4: Most suitable efficiency metrics and utility functions. https://bscw.ict-earth.eu/pub/bscw.cgi/d70454/EARTH_WP2_D2.4.pdf
7. Tombaz, S., Sung, K.W., Zander, J.: On metrics and models for energy efficiency in wireless access networks. IEEE Wirel. Commun. Lett. **3**(6), 649–652 (2014)
8. Frenger, P., Jading, Y., Turk, J.: A case study on estimating future radio network energy consumption and CO_2 emissions. In: Proceedings of PIMRC 2013
9. Desset, C., Debaillie, B., Louagie, F.: Towards a flexible and future-proof power model for cellular base stations. ccc (last access 2015), 23–25 Sept 2013, Green ICT—Tyrrhenian International Workshop on Digital Communications (TIWDC)
10. GrenTouch Consortium: Power model for today's and future base stations. Online power model webtool. http://www.imec.be/powermodel. Accessed Aug 2015
11. EARTH project deliverable D2.1: Economic and ecological impact of ICT. https://bscw.ict-earth.eu/pub/bscw.cgi/d38532/EARTH_WP2_D2.1_v2.pdf
12. Ericsson: Further details on design principle for a CRS-free additional carrier type. R1-121017, 3GPP TSG RAN WG1#68bis, 26–30 Mar 2012
13. EARTH project deliverable, D4.3: Final Report on Green Radio Technologies, June 2012

14. Frenger, P., Moberg, P., Malmodin, J., Jading, Y., Godor, I.: Reducing energy consumption in LTE with cell DTX. In: Proceedings of IEEE Vehicular Technology Conference (VTC Spring), Budapest, Hungary, May 2011
15. Tombaz, S., Han, S.W., Sung, K.W., Zander, J.: Energy efficient network deployment with cell DTX. IEEE Commun. Lett. **18**(6), 977–980 (2014)
16. Hiltunen, K.: Total power consumption of different network densification alternatives. In: Proceedings of IEEE Personal, Indoor and Mobile Radio Communication (PIMRC), Sydney, Australia, Sept 2012
17. Godor, I., et al.: Green communications: theoretical fundamentals, algorithms and applications. In: Green Wireless Access Networks, 1 edn. CRC Press, Cambridge (2012)
18. Auer, G., et al.: Cellular energy efficiency evaluation framework. In: Vehicular Technology Conference (VTC Spring), 2011 IEEE 73rd
19. Sabella, D., et al.: Evaluation of ON-OFF schemes and Linear Prediction methods for increasing energy efficiency in Mobile Broadband Networks. Networking 2012, Prague, Czech Republic, 25 May 2012
20. Tomaselli, W., Palestini, V., Sabella, D., Squizzato, V.: Energy Efficiency Performances of Selective Switch OFF Algorithm in LTE Mobile Networks. PIMRC, London (2013)
21. Ericsson: Ericsson Mobility Report—On the pulse of the networked society, Nov 2014. http://www.ericsson.com/res/docs/2014/ericsson-mobility-report-november-2014.pdf

Index Coding: A Greener Door for Wireless Networks

Mohammad Asad Rehman Chaudhry and Zakia Asad

Abstract Wireless networks are at the heart of telecommunication technology. Energy constrained wireless devices in the high-data-rate communication era are pushing energy efficiency to the top of the priority list for the network design. Improving the throughput without compromising energy efficiency as well as spectrum usage is pivotal for greener wireless networks. This chapter focuses on the Index Coding problem that presents an important class of practical problems in wireless networks. The Index Coding problem opens up a green door for the communication paradigm in wireless networks. This chapter not only provides a comprehensive overview of the Index coding problem abut also presents its mathematical formulation, taxonomy, as well as several solution techniques.

1 Introduction

The footprint of wireless networks is indispensable in modern day life. Wireless networks rely on energy constrained mobile devices including smartphones, tablets, laptops, and wearables to promote an "always-on" lifestyle. The constraint on energy is further topped up by the growth of high-data-rate applications for mobile devices. Increasing prevalence of energy-constrained wireless devices has fueled the efforts for development of energy-efficient and environmental friendly "green" networks.

A perceived nuisance associated with wireless networks is that the nodes that are not the intended receivers can also overhear the communication. This happens due to the broadcast nature of the wireless medium. The broadcast nature, however, can be used to improve energy efficiency and throughput in wireless networks by allowing the sender to code different packets together.

M.A.R. Chaudhry (✉)
Soptimizer, Toronto, Canada
e-mail: masadch@soptimizer.org

Z. Asad
University of Toronto, Toronto, Canada
e-mail: z.asad@mail.utoronto.ca

© Springer International Publishing Switzerland 2016

M.Z. Shakir et al. (eds.), *Energy Management in Wireless Cellular and Ad-hoc Networks*, Studies in Systems, Decision and Control 50,
DOI 10.1007/978-3-319-27568-0_18

Coding techniques—for example the Index Coding and the Network Coding—can optimize a network's resource-utilization by minimizing the volume of the communication, compared to the contemporary routing based approaches, without compromising on the rate of information exchange. The reduction in the volume of the communication can translate into an efficient utilization of the wireless spectrum as well as energy savings. For instance, authors in [1] show that the coding techniques offer a constant factor improvement in energy efficiency for fixed wireless networks. [1] also show that for dynamic wireless networks the advantage is bounded by a logarithmic factor depending on the number of nodes in the network. Similarly, the coding based techniques can achieve minimum energy consumption in wireless ad hoc networks [2]. Moreover, a coding based approach has been used to increase the bandwidth efficiency for reliable broadcast in wireless networks [3]. Similarly, the coding based techniques can construct minimum-energy multicast in wireless networks [4]. Furthermore, for multiple unicast networks, coding has been shown to save 25 % more power as compared to the pure routing [5]. Aside from the potential in better throughput, security, robustness, and energy efficiency, the coding based techniques have shown potential in recovering losses from half-duplex relays in 5G networks [6]. Hence, coding based techniques can play a vital role in achieving greener wireless networks by improving energy efficiency as well as overall resource-utilization.

Consider the following example to understand the benefits that the coding has to offer for wireless networks.

Example 1 (*Energy Efficiency Using Coding*) Consider a wireless network consisting of two nodes n_1, n_2 and a relay r as shown in the Fig. 1. The nodes n_1, n_2 can not communicate directly, and need the help of the relay to exchange data. Furthermore, the transmissions are done in rounds, one in each time slot. The traditional approach requires four transmissions in the system, one in each round. Specifically, two transmissions to deliver the packet a from node n_1 to node n_2, as firstly n_1 transmits the packet a to the relay, and then the relay transmits this packet to n_2. Similarly, two

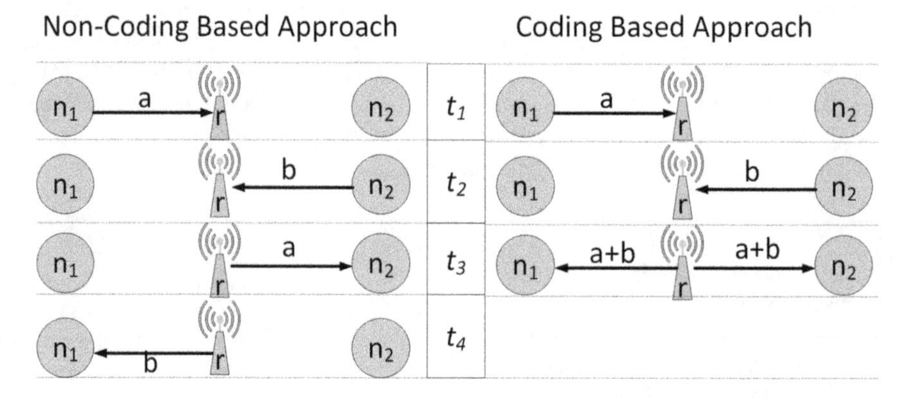

Fig. 1 Energy efficiency using coding in a wireless network of two nodes and one relay

transmissions are required to deliver the packet b from the node n_2 to the node n_1. However, by using coding the number of transmissions in the network can be reduced from four to three. More precisely, in case of the coding based approach, the relay r firstly receives the packet a from n_1 and waits until it receives the packet b from n_2. After receiving both the packets, the relay creates a combined packet $a + b$ and broadcasts it to both the nodes. Note, it results in a total of three transmissions in the network. Each node, after receiving the packet $a + b$, can decode its required packet. For example, n_1 already has packet a and it can use $(a + b) - a$ to decode the packet b, and similarly n_2 can decode the packet a. The reduction in the number of transmissions in the network translates into 25 % reduction in the energy consumption or 25 % less utilization of the bandwidth depending on the use-case, without any decrease in the rate of information exchange.

This chapter focuses on the Index Coding problem. The Index Coding problem can be considered as one of the basic problems in information theory [7–12]. The Index Coding problem can be seen as a multi-client network problem in which only one link in the communication network has finite capacity. In fact, the Index Coding and the Network Coding problems have been shown to be two equivalent problems. More precisely, it has been shown that, for each instance of the Index Coding problem, there is a corresponding instance of the Network Coding problem and vice versa [13, 14]. In addition to this, the Index Coding problem is also related to the Interference Alignment problem [15].

The Index Coding problem is defined for wireless networks comprising of a relay, and a set of clients. Each client is interested in a certain subset of the packets available at the relay, and might have a (different) subset of the packets available as the side information. The relay can broadcast the packets or encoding thereof. The goal is to find a scheme that requires the minimum number of transmissions to satisfy the requests of all the clients. By minimizing the number of transmissions, the Index Coding helps to achieve energy efficiency, as well as an optimal resource-utilization.

2 The Index Coding Problem

The Index Coding problem recently attracted significant interest from research community due to its theoretical as well practical significance. On theoretical side the Index Coding problem captures the limits of information exchange rates. On the practical side, its applications range from the satellite communications to the data centers. The Index Coding problem has been recognized to encompass many important applications like multi-way relay networks [16], sensor networks [17, 18], vehicular networks [19], big data processing [20], and interference management [21].

2.1 Model

An instance of the Index Coding problem is defined by a relay (coding server), a set of m clients $C = \{c_1, \ldots, c_m\}$, a set of n packets $P = \{p_1, \ldots, p_n\}$ that need to be delivered to the clients. Each client c_i requires a set of packet, $W_i \subseteq P$, known as its *want* set. Each client c_i might have access to some *side information*, $H_i \subseteq P$, known as its *has* set. The relay can transmit the packets in P, or their combinations (coded packets) to satisfy the demands of all the clients over an error-free broadcast channel. Each transmission x_i by the relay is specified by a function f over packets in P, and the coefficients from a finite field $\alpha_i = \{\alpha_i^j\} \in GF(q)$:

$$x_i = \overset{n}{\underset{j=1}{f}} (\alpha_i^j \cdot p_j). \tag{1}$$

The Index Coding problem is defined as:

Problem 1 (*The Index Coding Problem* (IC_q)) Find a scheme that minimizes the number of transmissions from the relay such that each client c_i can decode all the packets in its *want*. The optimal solution is denoted by $OPT(IC_q)$.

Remark 1 Without loss of generality, it can be assumed that for each packet $p_i \in P$ there exists at least one client $c_j \in C$ such that $p_i \in W_j$.

Remark 2 Without loss generality, it can be assumed that for each client $c_i \in C$ it holds that $H(c_i) \cap W(c_i) = \emptyset$.

Remark 3 If a client $c_i \in C$ requires more than one packets then without loss of generality it can be equivalently represented by multiple clients $c_i^1, c_i^2, \ldots,$ such that each client c_i^j requires only one packet and the "has" set $H(c_i^j)$ of each client $c_i^1, c_i^2, \ldots$ is identical to that of c_i. It easy to verify that the resulting instance has the same set of feasible solutions as the original problem.

Example 2 (*An Instance of the Index Coding Problem*) The Fig. 2 depicts an instance of the Index Coding problem. The instance consists of a relay node with four packets $p_1, \ldots, p_4$ destined for four clients $c_1, \ldots, c_4$. Each client c_i requires the packet p_i, and has access to some side information (shown as H_i in the Fig. 2). Using contemporary approach, the relay needs to broadcast all the four packets $p_1, \ldots, p_4$. However, if the relay deploys an Index Coding based solution then the demands of all the clients can be satisfied with only two transmissions. The transmissions are $p_1 \oplus p_2$, and $p_3 \oplus p_4$. Where $\alpha_1 = \{1\ 1\ 0\ 0\}$, and $\alpha_2 = \{0\ 0\ 1\ 1\}$ (both α_1 and α_2 $\in GF(2)$), and the function f is "addition over GF(2) (binary XOR)". Therefore, the Index Coding reduces the number of transmissions by from 4 to 2 resulting in 50 % lesser transmissions or 50 % energy savings compared to the contemporary approach. Note, each client can decode its required packet from the coded packets it

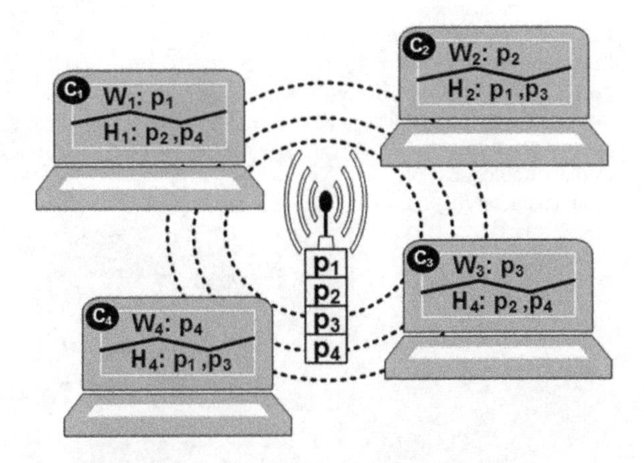

Fig. 2 An instance of the Index Coding problem

received using the side information available to it. For example c_1 can decode p_1 by performing the following operation:

$$(p_1 \oplus p_2) \oplus p_2 \tag{2}$$

2.2 Relation to the Field Size

It is interesting to note that the optimal solution to the Index Coding problem depends on the field size, i.e., by changing field size the minimum number of transmissions required to satisfy all clients can change [22]. This dependence on the field size is elaborated by the following examples.

Example 3 (*Effect of the Field Size on the Optimal Solution to the Index Coding Problem*) This example presents an instance of the Index Coding problem where increasing the field size decreases the number of transmissions by the relay. Consider an instance of the Index Coding problem with four packets p_1, p_2, p_3, p_4, and twelve clients $c_1, c_2, \ldots, c_{12}$. Corresponding *want* and *has* sets for each client are shown in the Fig. 3. It can be verified that the $|OPT(IC_2)| = 3$. Specifically, when the coding operations are performed over $GF(2)$ the relay needs to broadcast at least three packets which are $x_1 = p_1 \oplus p_3$, $x_2 = p_2 \oplus p_3$, and $x_3 = p_4$, where $\oplus$ refers to the addition over $GF(2)$. However, when the coding operation are performed over $GF(3)$ then the relay needs to broadcast only two packets which are $x_1 = p_1 + p_3 \oplus p_4$, and $x_2 = p_2 + p_3 \oplus 2p_4$, where $+$ refers to the addition over $GF(3)$.

As described in the Example 3 the optimal solution to the Index Coding problem depends on the field size but this is not a monotonic dependence, i.e., the larger field size does not necessarily means lesser number of transmissions by the relay. The following example elaborate this non-monotonic dependence on the field size.

Example 4 (*Non-monotonic Relationship Between the Field Size and the Optimal Solution to the Index Coding Problem*) Consider an instance of the Index Coding

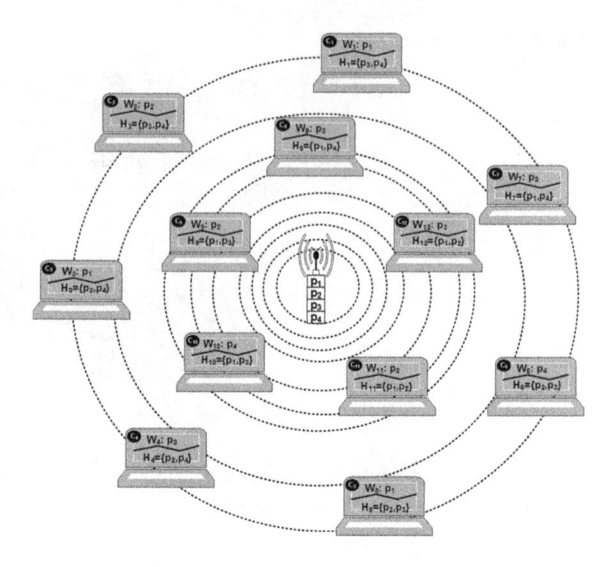

Fig. 3 An instance of the Index Coding problem for 4 packets and 12 clients. This instance showcase that the optimal solution to the Index Coding problem can be changed by varying field size

problem with seven packets $p_1, p_2, p_3, p_4, p_5, p_6, p_7$, and 18 clients $c_1, c_2, \cdots, c_{18}$. Corresponding *want* and *has* sets for each client are shown in the Fig. 4. It can be verified that $|OPT(IC_3)| = 3$, where the relay transmits the following three packets (+ refers to the addition over $GF(3)$):

$$x_1 = p_1 + p_4 + p_5 + p_7 \tag{3}$$

$$x_2 = p_2 + p_4 + p_6 + p_7 \tag{4}$$

$$x_3 = p_3 + p_5 + p_6 + p_7 \tag{5}$$

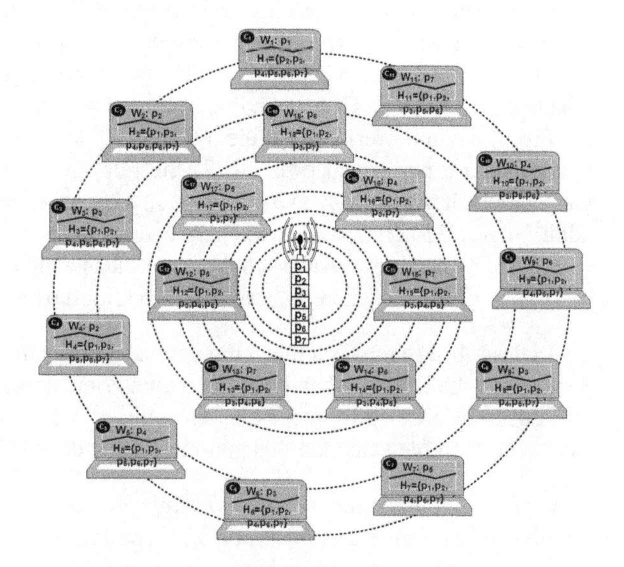

Fig. 4 An instance of the Index Coding problem for 7 packets and 20 clients. This instance showcases the fact that the optimal solution to the Index Coding problem is not monotonic with the field size. It can be verified that clients c_{16}, c_{17}, and c_{18} cannot decode the packets in their *want* set based on only three transmissions, i.e., x_1, x_2, and x_3 given by Eqs. 6, 7, and 8 respectively

However, for all $GF(2^k)$, i.e., the fields with even characteristic, the relay needs to transmit at least four packets i.e.,$|OPT(IC_{2^k})| = 4$ for all k. Specifically, the relay transmits the following four packets ($\oplus$ refers to the addition over $GF(q^k)$)

$$x_1 = p_1 \oplus p_4 \oplus p_5 \oplus p_7 \tag{6}$$

$$x_2 = p_2 \oplus p_4 \oplus p_6 \oplus p_7 \tag{7}$$

$$x_3 = p_3 \oplus p_5 \oplus p_6 \oplus p_7 \tag{8}$$

$$x_4 = p_4 \oplus p_5 \oplus p_6 \tag{9}$$

3 Taxonomy of the Index Coding Problem

This section describes the taxonomy of the Index Coding problem. The Index Coding problem can be classified based on following characteristics:

- Coding Methodology:
  - Linear Coding: A solution scheme to the Index Coding problem is said to be linear if for each transmission the function f (defined in the Sect. 2.1) defines a linear operation. An example of a linear coding scheme can be a scheme where each transmission $x_i = \sum_j^n \alpha_i^j \cdot p_j$.
  - Non-linear Coding: A non-linear coding scheme for the Index Coding problem is characterized by a solution using a non-linear function f (defined in the Sect. 2.1). An example of a non-linear coding scheme can be $x_i = \prod_j^n \alpha_i^j \cdot p_j$.
- Solution Granularity:
  - Scalar Solution: In a scalar solution to the Index Coding problem each packet is coded as a whole. For the instance of the Index Coding problem shown in the Fig. 5, the scalar-linear solution requires three transmissions i.e., $p1 \oplus p_2$, $p3 \oplus p_4, p_5$.
  - Vector Solution: In a vector solution for the Index Coding problem each packet p_i is subdivided into ℓ smaller size *subpackets* $p_i^1, \ldots, p_i^\ell$. Then, each transmitted packet, $x_i = \underset{j=1}{\overset{n}{f}} (\alpha_i^{j,k} \cdot p_j^k)$, is a combination of the subpackets $\{p_i^j| \ 1 \leq i \leq n, 1 \leq j \leq \ell\}$, rather than the original packets. With a vector coding scheme, the goal is to find an encoding schemes that minimize the ratio of $\frac{\mu}{\ell}$, where μ is the number of times a combination of subpackets is transmitted.
  It is interesting to note that compared to the scalar solution the vector solution can further reduce the volume of communication. For an instance of the Index Coding problem shown in the Fig. 5, the vector-linear solution requires

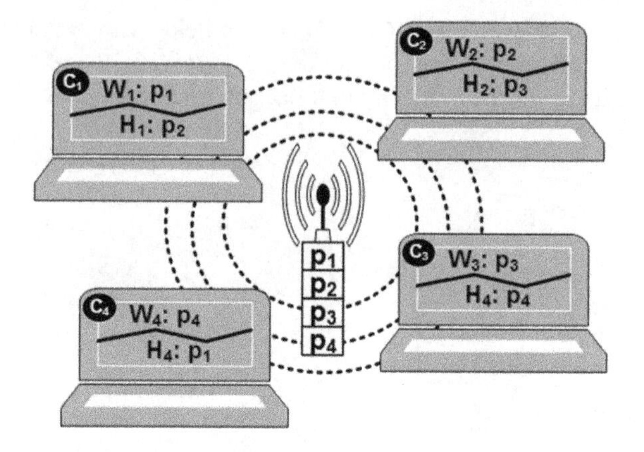

Fig. 5 An instance of the Index Coding problem for 4 packets and 4 clients. For this case the scalar solution would yield 3 transmissions, whereas a vector solution results in 2.5 transmissions

2.5 transmissions ($\frac{\mu}{\ell} = 2.5$) whereas the scalar-linear solution requires 3 transmissions. Specifically, each packet p_i is divided into two sub-packets p_i^1 and p_i^2 (i.e., $\ell = 2$). Then, the following five linear combinations of the subpackets are transmitted: $p_1^1 \oplus p_2^1, p_2^2 \oplus p_3^1, p_3^2 \oplus p_4^1, p_4^2 \oplus p_5^1$, and $p_5^2 \oplus p_1^2$.

- Code Density: The solution to the Index Coding problem can be classified based on the code density, i.e., the maximum number of packets that can be coded together in each transmission. Higher code density can be computationally expensive both in terms of encoding and decoding. Imposing a constraint on the code density can be helpful to achieve practically efficient solutions.
  - Sparse Solution: A solution to the Index Coding problem is referred to as $\gamma - sparse$ if the relay is restricted to code no more than γ packets in any transmission. For an instance of the Index Coding problem given in the Fig. 6, the optimal $2 - sparse$ solution results in two transmissions, i.e., $p_1 \oplus p_2$ and $p_3 \oplus p_4$.
  - Dense Solution: A solution to the Index Coding problem is referred to as dense if the code density is unrestricted. For an instance of the Index Coding problem given in the Fig. 6, the optimal dense solution results in one transmission, i.e., $p_1 \oplus p_2 \oplus p_3 \oplus p_4$.
- Objective Function: The Index Coding problem has been proven to be not only NP-hard but NP-hard to approximate as well [7, 11, 22, 23]. This motivates the need to investigate the Index Coding problem from a complementary perspective. Specifically, the objective of the Index Coding problem is to "minimize" the number of transmissions whereas the the objective of the Complementary Index Coding problem is to "maximize" the number of saved transmissions. It is interesting to note that there are several well-known NP-hard problems whose complementary problems exhibit significantly different behavior in terms of approximation. For instance, the problems of finding the minimum chromatic number of a graph

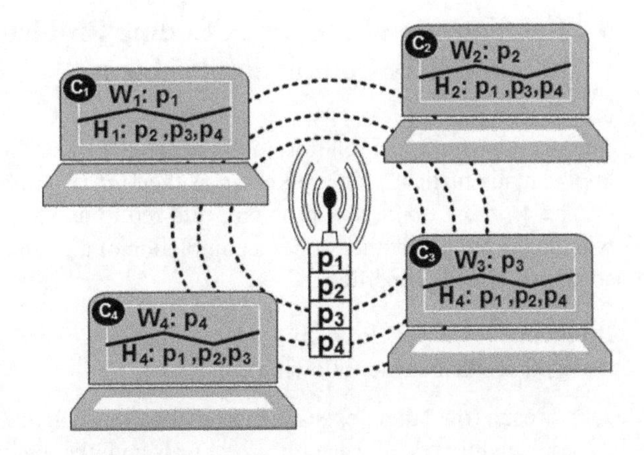

Fig. 6 An instance of the Index Coding problem for 4 packets and 4 clients. A 2-sparse solution to the Index coding problem yields 2 transmissions, whereas a dense solution results in only one transmission

(graph coloring) is in-approximable but its complementary problem of maximizing the number of colors saved (color saving problem) (see e.g., [24]) has good approximation algorithms.

- Minimization: Let μ be the number of transmissions, then the goal of the Index Coding problem is to minimize μ. For the instance of the Index Coding problem given in the Fig. 6, the optimal solution to Index Coding problem with the minimization objective needs only one transmission, i.e., $p_1 \oplus p_2 \oplus p_3 \oplus p_4$.
- Maximization: Let μ be the number of transmissions, then the goal of the Index Coding problem with the complementary objective function is to maximize $n - \mu$, i.e., to maximize the number of saved transmissions. The optimal solution for the instance of the Index Coding problem given in Fig. 6 with complementary objective saves 3 transmissions.

- Demand of the Clients:
  - Multiple Unicast: In a multiple unicast instance of the Index Coding problem each packet is required by just one client, i.e., $W_i \cap W_j = \phi \; \forall i, j$. Example 2 presents a multiple unicast instance of the Index Coding problem.
  - Multiple Multicast: An instance of the Index Coding problem where a packet can be required by more than one clients is referred to as a multiple multicast instance. Examples 3 and 4 both present multiple multicast instances of the Index Coding problem.
- Side-information available at the clients:
  - Uni-priori: An instance of the Index Coding problem is uni-priori if $|H_i| = 1 \; \forall i$ [25]. An instance of the Index Coding problem shown in the Fig. 6 is an example of a uni-priori Index Coding problem.
  - Multi-priori: An instance of the Index Coding problem is multi-priori if a client can have an arbitrary number of packets in its *Has* set. Example 2 is an instance of a multi-priori Index Coding problem.

4 Relationship of the Index Coding Problem to the Matrix Completion Problem

This section shows the relationship between the Index Coding problem and a well-known mathematical problem known as the *Matrix Completion* problem.

The *Matrix Completion* problem, referred to as the *MC* problem, is concerned with determining whether or not a completion of a partial matrix exists that satisfies some prescribed properties.

Problem 2 (*The MC Problem*) Let $A = \{a_{ij}\}$ be an $m \times n$ matrix over a finite field $GF(q)$ that satisfies the following conditions:

1. For each row i there exists only one entry a_{ij} such that $a_{ij} = 1$,
2. Each element $a_{ij} \neq 1$ is either zero or is an unknown x_{ij}.

Let $X = \{x_{ij}\}$ be the set of all unknowns. The objective is to assign values to X from a finite field that minimize the rank of A.

For an instance I for the *MC* problem, the minimum achievable rank of A is denote by $OPT(I)$.

The next two lemmas show that the IC_q problem and the *MC* problem are equivalent.

Lemma 1 *Let I be an instance of the IC_q problem. Then, an instance I' of the* MC *problem can be efficiently constructed in a polynomial time, such that $OPT(I) = OPT(I')$.*

Proof Given the instance I of the IC_q problem, the instance I' of the *MC* problem is constructed as follows. The matrix A has $|C|$ rows, and $|P|$ columns. For each client $c_i \in C$ there is a row in $r(c_i)$ in A, and for each packet $p_i \in P$ there is a column $\kappa(p_i)$ in A. Then,

- $a_{ij} = 1$, for row $r(c_i)$ and column $\kappa(p_j)$ if $p_j \in W_i$;
- $a_{ij} = X$, for row $r(c_i)$ and column $\kappa(p_j)$ if $p_j \in H_i$;
- $a_{ij} = 0$, otherwise.

The matrix A for the instance of the Index Coding problem from the Example 2 is given below:

$$M = \begin{bmatrix} 1 & X & 0 & X \\ X & 1 & X & 0 \\ 0 & X & 1 & X \\ X & 0 & X & 1 \end{bmatrix} \tag{10}$$

Next, in order to show that $OPT(I) \leq OPT(I')$, let us assume that $\Phi = \{\alpha_i\}$ be an optimum solution to I. Furthermore, the linear subspace generated by vectors in Φ is denoted by $\langle\Phi\rangle$. Note that for each client $c_i \in C$ there must be a vector $\alpha_i \in \langle\Phi\rangle$ that has the following properties:

1. Let p_j be a packet in W_i, then $\alpha_i^j = 1$
2. For each packet $p_j \in P \setminus \{W_i \cap_i)\}$ it holds that $\alpha_i^j = 0$

Then, for each client $c_i \in C$, the unknown values are set in a row of A that corresponds to c_i according to the vector $\alpha_i \in \langle \Phi \rangle$ that corresponds to c_i. Since each row vector in A belongs to the linear subspace spanned by ϕ, it follows that

$$OPT(I') = \text{rank(A)} \leq |\phi| = \text{OPT(I)}.$$

Finally, to show that $OPT(I') \leq OPT(I)$, assume that $X' = \{x'_{ij}\}$ be optimal assignment of unknown values in A that is an optimal solution to the instance I' of the MC problem. Also, let A' be a matrix formed from A by substituting the unknown values in A according to X'. The matrix A' has rank $OPT(I')$, thus there are $OPT(I')$ rows of A' which are linearly independent. Let Φ be the set of encoding vectors formed by $OPT(I')$ independent rows of A'. It is easy to verify that Φ is a feasible solution of to the instance I of the IC_q problem, hence $OPT \leq OPT(I')$.

Lemma 2 *Let I' be an instance of the* MC *problem. Then, an instance I of the IC_q problem can be efficiently constructed in a polynomial time, such that $OPT(I) = OPT(I')$.*

Proof Follows the same lines as the proof of Lemma 1.

5 Hardness Results for the Index Coding Problem

This section focus on NP-hardness of the Index coding problem in general setting. Later parts of this section present some insightful results related to the computational complexity of the Index Coding problem under different settings as presented in [7, 11, 12, 22, 23].

Theorem 1 *The Index Coding problem IC_q is NP-complete for any finite field $GF(q)$.*

Proof Let $G(V, E)$ be an undirected graph with n vertices $v_1, \ldots, v_n$, let $GF(q)$ be a finite filed, and let A be an $n \times n$ matrix over $GF(q)$ that satisfies the following requirements:

$$A_{ij} = \begin{cases} 1 \textbf{ if } v_i = v_j \\ 0 \textbf{ if } v_i \text{ is adjacent to } v_j \end{cases} \tag{11}$$

Then, *A fits G*.

For some field $GF(q)$ and an undirected graph $G(V, E)$, $|V| = n$, Let $\mathscr{A}(G, q)$ be a set of matrices over $GF(q)$ fitting G, i.e.,

$$\mathscr{A}(G, q) := \{A \in GF(q)^{n \times n} \mid \text{A fits G}\}.$$

In [26] it was proven, through the reduction from the problem of graph coloring (for definition of graph coloring please see Sect. 6.2.1), that for any finite field $GF(q)$, it is NP-hard to check whether or not $\mathscr{A}(G, q)$ contains a matrix of rank three. This implies, in turn, that it is NP-hard to determine whether the MC problem has a solution of size 3. It has been shown in Sect. 4 that the MC problem and the Index Coding problem are equivalent. Thus, by Lemmas 1 and 2 it holds, in turn, that the problem of determining whether there exists a feasible solution of size 3 or less to the IC_q problem is intractable.

Moreover, it has been shown that the Index Coding problem is not only NP-hard but *APX-hard* as well, i.e., a $(1 + \varepsilon)$ approximation—for $\varepsilon \geq 0$—of an optimal solution can not be found in the polynomial time unless $P = NP$ [7, 11, 12, 22, 23].

Furthermore, the authors in [27] have analyzed the computational complexity of the Index Coding problem with the complementary objective under different settings. Specifically, it has been shown that under the multiple unicast settings efficient scalar-linear, and vector-linear solutions can be constructed in the polynomial time with approximation ratios of $\Omega(\sqrt{n} \cdot \log n \cdot \log\log n)$, and $\Omega(\log n \cdot \log\log n)$ respectively. Moreover, constructing the scalar-linear solution under the multiple multicast settings is not only NP-hard, but NP-hard to approximate as well within a ratio of $n^{1-\varepsilon}$ for any constant $\varepsilon > 0$.

The computational complexity of finding a sparse solution to the Index Coding problem under different settings has been analyzed in [11]. Specifically, finding a sparse solution is NP-Complete. Furthermore, constructing a sparse solution with the complementary objective is quasi-NP-hard to approximate within a factor of $O(\log^{1-\varepsilon} n)$ for any constant $\varepsilon > 0$. Moreover, under the multiple unicast settings an efficient scalar-linear solution with an approximation ratio of $2 - \frac{1}{\sqrt{n}}$, and an optimal vector-linear solution can be constructed in the polynomial time. Additionally, a scalar-linear solution, having an approximation ratio of $\frac{1}{\sqrt{n}}$, with the complementary objective can be constructed in the polynomial time.

5.1 Bounds on the Green Gain

The *green gain* Γ is defined as the ratio between the minimum number of transmissions without coding and the minimum number of transmissions with coding, i.e.,

$$\Gamma = \frac{n}{OPT(IC_q)}$$

The lower and upper limits on the *green gain* are given below ([22]):

$$\frac{n}{n - \varphi} \leq \Gamma \leq \phi + 1 \tag{12}$$

where $\phi = \max_{c_i \in C} |H_i|$ and $\varphi = \min_{c_i \in C} |H_i|$.

6 Solution Techniques

In this section several solution techniques to the Index Coding problem are presented. Specifically, the Sect. 6.1 presents a SAT based technique that provides an optimal solution to the Index Coding problem. Moreover, the Sect. 6.2 presents several efficient solutions to the Index Coding problem.

6.1 An Optimal Solution to the Index Coding Problem Using Boolean Satisfiability Problem

This section provides an optimal solution by formulating the Index Coding problem as a Boolean Satisfiability (SAT) problem [10]. The presented SAT based formulation of the Index Coding problem for $GF(2)$ (referred to as the IC_2 problem) can be efficiently solved by SAT solvers such as *Chaff* or *Minisat* [28, 29]. Although the formulation given in this section is specific to $GF(2)$, but it is possible to extend it for the general finite fields.

In order to check whether it is possible to satisfy all the clients by μ transmissions, it is first required to check whether there exist μ encoding vectors $\alpha_1, \ldots, \alpha_\mu$ of size m, and m decoding vectors $\beta_1, \ldots, \beta_m$ of size μ that allow each client to decode the packet in its *want* set. Recall that α_i $0 \leq i \leq \mu$ is the vector of encoding coefficients for the packet x_i transmitted in round i, i.e., $x_i = \sum_{j=1}^{n} \alpha_i^j \cdot p_j$. For the decoding purposes each client $c_i \in C$ uses a corresponding decoding vector β_i to decode the packet in W_i. To be specific, each client c_i computes the following linear combination of the packets.

$$\begin{aligned}\sum_{i=1}^{\mu} x_j \cdot \beta_i^j &= \sum_{j=1}^{\mu} \beta_i^j \sum_{t=1}^{n} \alpha_j^t \cdot p_t \\ &= \sum_{t=1}^{n} p_t \sum_{j=1}^{\mu} \alpha_j^t \cdot \beta_i^j = \sum_{t=1}^{n} r_i^t p_t \end{aligned} \tag{13}$$

where $r_i = \sum_{j=1}^{\mu} \alpha_j^t \cdot \beta_i^j$ is a linear combination of the packets in P received by the client c_i. Note that a client c_i can decode the original packets under following conditions:

1. $r_i^j = 1$ if $p_j \in W_i$;
2. $r_i^j = 0$ if $p_j \notin \{H_i \cup W_i\}$, i.e., the packet p_j belongs to neither *has* set nor *want* set of c_i.

The above two conditions give rise to the following constraints on $\{\alpha_i\}$ and $\{\beta_i\}$:

$$\textstyle\sum_{j=1}^{\mu} \alpha_j^t \cdot \beta_i^j \equiv 1 \ \forall c_i \in C \text{ and } p_j \in W_i \tag{14}$$

$$\sum_{j=1}^{\mu} \alpha_j^t \cdot \beta_i^j \equiv 0 \; \forall c_i \in C \text{ and } p_j \notin \{H_i \cup W_i\} \tag{15}$$

The constraints given in the Eqs. 14 and 15 can be efficiently transformed into a SAT problem by substituting the summation and multiplication over $GF(2)$ by the XOR ($\oplus$) and AND ($\wedge$) operations respectively using the boolean variables $\{\alpha_i\}$ and $\{\beta_j\}$ as given below:

$$\oplus_{j=1}^{\mu} \left(\alpha_j^t \wedge \beta_i^j\right) \equiv 1 \; \forall c_i \in C \text{ and } p_j \in W_i; \tag{16}$$

$$\oplus_{j=1}^{\mu} \left(\alpha_j^t \wedge \beta_i^j\right) \equiv 0 \; \forall c_i \in C \text{ and } p_j \notin \{H_i \cup W_i\}. \tag{17}$$

The SAT problem given by formulae 16 and 17 is not in conjunctive normal form (CNF) required by most of the SAT solvers. A straightforward conversion of formulae 16 and 17 into equivalent CNF form may results in an exponential number of variables. Accordingly, in order to perform an efficient transformation the Tstein transformation [30] is used. Such a transformation guarantees that the CNF transformation is linear in the size of the original formulae.

For the Tseitin transformation an additional variable is defined for each non-literals subformula including the original formula. To understand the Tseitin transformation, consider the boolean formula given by 18.

$$(z_1 \wedge z_2) \oplus (z_3 \wedge z_4) \oplus (z_5 \wedge z_6) \tag{18}$$

Firstly, five additional variables—A, B, C, D, and E—are introduced for the formula 18 as given below:

$$A = \underbrace{(z_1 \wedge z_2)}_{B} \oplus \underbrace{\underbrace{(z_3 \wedge z_4)}_{C} \oplus \underbrace{(z_5 \wedge z_6)}_{D}}_{E} \tag{19}$$

Secondly, the formula 19 is transformed as below:

$$\begin{gathered} A \wedge (A \leftrightarrow (B \oplus E)) \wedge (E \leftrightarrow (C \oplus D)) \\ \wedge (B \leftrightarrow (z_1 \wedge z_2)) \wedge (C \leftrightarrow (z_3 \wedge z_4)) \\ \wedge (D \leftrightarrow (z_5 \wedge z_6)) \end{gathered} \tag{20}$$

Note that there are two types of subformulae in the formula 20, one represented by $A \leftrightarrow (B \oplus C)$, and the other represented by $A \leftrightarrow (B \wedge C)$. Each subformula of type $A \leftrightarrow (B \wedge C)$ is transformed into the CNF form as follows.

$$\begin{aligned} & A \leftrightarrow (B \wedge C) \\ &= (\bar{A} \vee B) \wedge (\bar{A} \vee C) \wedge (A \vee \bar{B} \vee \bar{C}) \end{aligned} \tag{21}$$

Similarly, each subformula of type $A \leftrightarrow (B \oplus C)$ is transformed into the CNF form as follows.

$$\begin{aligned} & A \leftrightarrow (B \oplus C) \\ & = (\bar{A} \vee B \vee C) \wedge (\bar{A} \vee \bar{B} \vee \bar{C}) \\ & \wedge (A \vee B \vee \bar{C}) \wedge (A \vee \bar{B} \vee C) \end{aligned} \tag{22}$$

The CNF transformation of the boolen formula 18 is given below:

$$\begin{aligned} & A \wedge (\bar{A} \vee B \vee E) \wedge (\bar{A} \vee \bar{B} \vee \bar{E}) \wedge (A \vee B \vee \bar{E}) \wedge (A \vee \bar{B} \vee E) \\ & \wedge (\bar{E} \vee C \vee D) \wedge (\bar{E} \vee \bar{C} \vee \bar{D}) \wedge (E \vee C \vee \bar{D}) \wedge (E \vee \bar{C} \vee D) \\ & \wedge (\bar{B} \vee z_1) \wedge (\bar{B} \vee z_2) \wedge (B \vee \bar{z}_1 \vee \bar{z}_2) \\ & \wedge (\bar{C} \vee z_3) \wedge (\bar{C} \vee z_4) \wedge (C \vee \bar{z}_3 \vee \bar{z}_4) \\ & \wedge (\bar{D} \vee z_5) \wedge (\bar{D} \vee z_6) \wedge (D \vee \bar{z}_5 \vee \bar{z}_6) \end{aligned} \tag{23}$$

6.2 *Computationally Efficient Solutions*

Both the IC_q problem and the boolean satisfiability (SAT) problem are NP-Hard problems, therefore using SAT based solution for finding an optimal solution for large instances is practically intractable. This opens the possibility of devising heuristic algorithms that can provide a near-optimal behavior. Accordingly, this section presents several solutions techniques and compare their performance.

6.2.1 Reduction to Graph Coloring

This section describes a graph coloring based solution to the Index Coding problem [10]. It has been shown in the Sect. 5 that the scalar-linear solution to the Index Coding problem under multiple unicast setting is related to problem of finding the minimum chromatic number of an undirected graph $G(V, E)$ also referred to as graph coloring problem. This section presents a scheme to find a solution to the Index Coding problem using graph coloring. This section assumes, without loss of generality (refer to Remark 3), that the *want* set of each client is of cardinality one.

Some of the terminologies and problem definitions used in the this section and the subsequent sections are given below.

Definition 1 (*Graph Coloring Problem*) Given a graph $G(V, E)$ with a vertex set V and an edge set E assign a color to each vertex $v \in V$ such that for any edge $(v, u) \in E$, the vertexes v and u are assigned a different color, and total number of the colors used is minimum.

Definition 2 (*Complementary Graph*) Given a graph $G(V, E)$, its complementary graph $\bar{G}(\bar{V}, \bar{E})$ is constructed as following:

- $\bar{V} = V$
- $\bar{E} = \{e(u, v) | e(u, v) \notin E\}$

Definition 3 (*Clique*) Given a graph $G(V, E)$, a clique is a set of vertices $V' \subseteq V$ such that $\forall (u, v) \in V'$, there exists an edge $e(u, v) \in E$.

Definition 4 (*Clique Partition Problem*) Given a graph $G(V, E)$, partition V into the minimum number of disjoint subsets $V_1, V_2, \ldots, V_k$, such that each V_i, $1 \leq i \leq k$, is a clique.

To solve the Index Coding problem under multicast setting using graph coloring problem, a dependency graph is defined as:

Definition 5 (*Dependency Graph*) Given an instance I of the IC_q problem, a dependency graph $G(V, E)$ is defined as follows.

- For each client $c_i \in C$ there is a corresponding vertex v_{c_i} in V
- Each two vertices v_{c_i} and v_{c_j} are connected by an edge if one of the following holds:
  - $W_i = W_j$, i.e., Clients c_i and c_j have identical *want* sets;
  - $W_i \subseteq H_j$ and $W_j \subseteq H_i$.

Lemma 3 *Let $\hat{V} \subseteq V$ be a clique in $G(V, E)$, then all clients that correspond to nodes in $\hat{V}$ can be satisfied by one transmission, which includes a linear combination of all packets in their* want *sets.*

Therefore, the number of transmissions for an instance of the Index Coding problem can be minimized by solving a *clique partition* problem for the corresponding dependency graph. The Graph coloring and the clique partitioning problems are interchangeable [31, 32] as follows.

Lemma 4 *Clique partitioning problem for graph $G(V, E)$ is equivalent to the graph coloring problem for the complementary graph $\bar{G}(\bar{V}, \bar{E})$.*

Hence, the number of transmissions for an instance of the Index Coding problem can be minimized by solving a *graph coloring* problem for the complementary graph of the corresponding dependency graph. The graph coloring based solution for the Index Coding problem is given in the Algorithm 1.

The Fig. 7 shows the relationship between the clique partitioning problem and the graph coloring problem for the instance of the Index Coding problem given in the Example 2. Specifically, the Fig. 7a shows the dependency graph, and its corresponding clique partitions are shown in the Fig. 7c. The Fig. 7b shows the complementary graph of the dependency graph shown in the Fig. 7a, and its corresponding graph coloring is given in the Fig. 7d. The packets corresponding to the same colored vertices are coded together, i.e., $p1_{+}p_4$ and $p_2 + p_3$. These two packets can satisfy the demands of all the four clients.

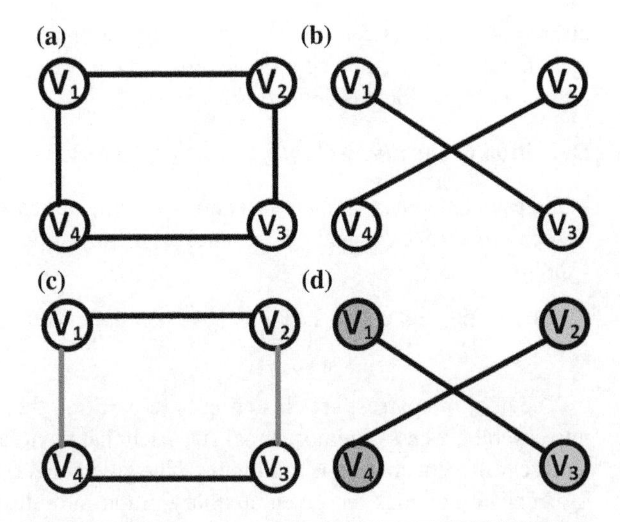

Fig. 7 **a** A dependency graph for the instance of the Index Coding problem given in Example 2. **b** Corresponding complementary graph. **c** Clique partitioning for the dependency graph shown in (**a**). **d** Graph Coloring for the instance in (**b**)

Algorithm 1 GC: A graph coloring based algorithm for the Index Coding problem

---

Require: An instance of the Index Coding problem
1: Construct an Dependency Graph $G(V, E)$
2: Create a complementary graph $\bar{G}(\bar{V}, \bar{E})$
3: Find graph coloring for G, let χ be the number of colors and let $\hat{V}_i$ be the set of vertices with same color i
4: **for** $i = 1 : \chi$ **do**
5: Construct a packet that satisfy all clients corresponding to vertices in $\hat{V}_i$ by XORing the packets in their *want* sets.
6: **end for**

6.2.2 Sparsest-Set Clustering Based Heuristic

Given the NP-hard nature of the Index Coding problem, the optimal solutions do not scale well with the number of clients. For example as shown in [10] that the optimal SAT-based solution can be used efficiently for only a limited number of clients. Similarly, the solution that uses the coloring heuristic can be used efficiently for a larger number of clients, but its time complexity grows considerably with the number of clients. This section presents a heuristic for solving the Index Coding problem for instances consisting of larger number of clients. The presented technique is based on the "divide-and-conquer" approach where the Index Coding problem is partitioned into smaller subproblems consisting of disjoint set of clients. The subproblem are solved separately. The solution for the original problem is the union of the solution obtained from individual subproblems.

An important decision in the partition-based approach is to perform an effective partitioning (clustering) of the clients such that each partition can be solved independently. Note that is it desirable to find a clustering in which the clients belonging to different groups have as few packets common in their *has* and *want* sets as possible to avoid compromising on the green gain. Such a partition is referred to as sparsest-set clustering, and has been shown to be computationally efficient as compared to the techniques given in the Sects. 6.1 and 6.2.1 [10]. The sparsest-set clustering works on an auxiliary graph as defined below:

Definition 6 (*Auxiliary Graph*) Construct a directed graph $G'(V', E')$ as follows:

- For each client $c_i \in C$ there is a corresponding vertex v_{c_i} in V'
- A pair of vertices v_{c_i} and v_{c_j} is connected by an arc (v_{c_i}, v_{c_j}) if one of the following holds:
  - $W_i = W_j$, i.e., clients c_i and c_j have identical *want* sets;
  - $W_i \subseteq H_j$.

The goal of sparsest-set clustering is to partition the auxiliary graph $G'(V', E')$ into several clusters of almost equal size such that the total number of the edges that connect different clusters is minimized. The greedy heuristic given in the Algorithm 2 recursively divides the given auxiliary graph into clusters until the clusters of a desired size are obtained. More specifically, the input vertex set V' is partitioned into two subsets V_1' and V_2' of equal size (in an arbitrary way). Then the proposed algorithm continue swapping the vertices in two subsets V_1' and V_2' until any vertex in V_1' (V_2') has no more neighbors in V_2' (V_1') than in V_1' (V_2'). Each partition thus obtained is then solved recursively.

Algorithm 2 Partition: A sparsest-set clustering based partition of the Index Coding problem

Require: V', desired cluster size θ
1: Partition V' in two sets, V_1' and V_2' arbitrarily
2: **if** $\exists\, v \in V_1'$ that has no more neighbors in V_2' than in V_1' **then**
3: $\quad V_2' = v \cup V_2'$, and $V_1' = V_1' \setminus v$
4: **end if**
5: **if** $\exists\, v \in V_2'$ that has no more neighbors in V_1' than in V_2' **then**
6: $\quad V_1' = v \cup V_1'$, and $V_2' = V_2' \setminus v$
7: **end if**
8: **if** $|V_1'| \leq \theta$ AND $|V_2'| \leq \theta$ **then**
9: $\quad$ Partition(V_1',$\theta = \frac{|V_1'|}{2}$)
10: $\quad$ Partition(V_2',$\theta = \frac{|V_2'|}{2}$)
11: **end if**

Time Versus Green Gain Tradeoff

It is interesting to note that the smaller clusters improve the time complexity at the cost of the green gain. The green gain using clustering based approach might be lesser than the green gain while solving the instance as a whole. To elaborate further, consider an instance of the Index Coding problem shown in the Fig. 6. Focus on the scenario when the instance is clustered in to two subproblems, namely s_1 and s_2. s_1 consists of the clients c_1, and c_2, and s_2 consists of the clients c_3, and c_4. It is interesting to note that solving each subproblem separately results in computational efficiency [10]. However, on the other end, solving each subproblem separately results in lower green green gain. Specifically, the sparsest-set clustering based solution requires two transmissions which are $p_1 + p_2$ corresponding to s_1, and $p_3 + p_4$ corresponding to s_2, resulting in a green gain of 2. However, solving the same instance as a whole can satisfy all the clients with one transmission $p_1 + p_2 + p_3 + p_4$ resulting in a higher green gain of 4.

6.2.3 Color Saving Based Heuristic for the Index Coding Problem with Complementary Objective

This section describes a heuristic solution for efficiently solving Index Coding problem with complementary objective. Recall that the complementary objective of the Index Coding problem is to maximize the number of unused colors. The scheme presented in this section that utilize the concept of color saving on a graph [10]. The color saving problem is defined as follows:

Definition 7 (*Color Saving*) Given a graph $G(V, E)$ the objective is to assign a color to each vertex such that for any edge $(v, u) \in E$ the vertices v and u are assigned different colors, and total number of the unused colors is maximized. Where the number of unused colors is the difference between the number of vertices in the graph and total number of the colors used for the coloring.

The steps of the color saving based scheme are given in the Algorithm 3. The algorithm greedily finds cliques of size 3, and then finds all the cliques of size 2 in the remaining graph (using maximum matching). Each clique corresponds to a packet that can satisfy the clients corresponding to that clique (Lemma 3). Let $\mathscr{C}_3$ be the number of clique of size 3 and $\mathscr{C}_2$ be the number of clique of size 2 identified by the Algorithm 3, then the algorithm saves $2 * \mathscr{C}_3 + \mathscr{C}_2$ transmissions compared to the contemporary routing based approaches. More precisely for each clique of size $\sigma, \sigma - 1$ transmissions are saved.

Algorithm 3 CSCIC: A color saving based algorithm for the Index Coding problem with complementary objective

Require: An instance of the Index Coding problem
1: Construct an Dependency Graph $G(V, E)$ as described in Sect. 6.2.1.
2: **while** $\exists$ a clique of size 3 in $G(V, E)$ **do**
3: Find a clique $\{v_i, v_j, v_k\}$ of size 3 in $G(V, E)$
4: Create a packet that satisfies all clients in $\{v_i, v_j, v_k\}$
5: $V := V \setminus \{v_i, v_j, v_k\}$
6: **end while**
7: Compute a maximum matching of $G(V, E)$
8: For each pair $\{v_i, v_j\}$ in the matching create a packet that satisfies all clients in $\{v_i, v_j\}$, i.e., $W_i + W_j$
9: Create a new packet for each one of the remaining vertices of V.

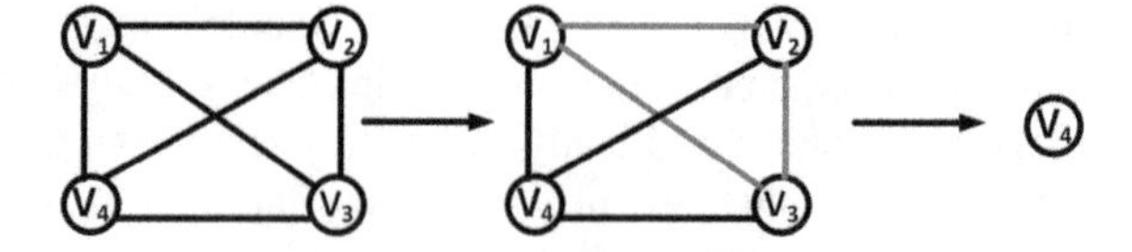

Fig. 8 Steps of the Algorithm 3 for the instance of the Index Coding problem given in the Fig. 6

For an instance of the Index Coding problem shown in the Fig. 6, the steps of the Algorithm 3 are shown in the Fig. 8. The algorithm results in two transmission $p_1 + p_2 + p_3$ and p_4. The number of transmissions saved is 2.

7 Conclusion

This chapter focuses on green optimization for an important class of problems in wireless networks by employing coding techniques. In particular, we address the Index Coding problem where the objective is to satisfy the demands of all the clients in a minimum number of transmissions by the relay without compromising on the rate of information exchange. Some potential applications of the Index Coding include the improved energy efficiency, and optimal spectral usage for the wireless cellular and ad hoc networks. We provide an in depth analysis of the Index Coding problem in terms of the relationship to the other problems, dependence on the field size, and performance bounds on green gain and computational complexity. Furthermore, we presents a number of technique to the Index Coding problem covering a wide spectrum of the solutions varying from the optimal to computationally efficient heuristics.

References

1. Fragouli, C., Widmer, J., Le Boudec, J.-Y.: Efficient broadcasting using network coding. IEEE/ACM Trans. Netw. **16**(2), 450–463 (2008)
2. Yunnan, W., Chou, P., Kung, S.-Y., et al.: Minimum-energymulticast inmobile ad hoc networks using network coding. IEEE Trans. Commun. **53**(11), 1906–1918 (2005)
3. Nguyen, D., Tran, T., Nguyen, T., Bose, B.: Wireless broadcast using network coding. Vehicular Technology, IEEE Transactions on **58**(2), 914–925 (2009)
4. Lun, D.S., Ratnakar, N., Koetter, R., Médard, M., Ahmed, E., Lee, H.: a decentralized approach based on network coding. In:INFOCOM 2005. Proceedings IEEE 24th Annual Joint Conference of the IEEE Computer and Communications Societies, vol. 3, pp. 1607–1617. IEEE (2005)
5. Cui, T., Chen, L., Ho, T.: Energy efficient opportunistic network coding for wireless networks. In: INFOCOM 2008. The 27th Conference on Computer Communications. IEEE (2008)
6. Thai, C.D.T., Popovski, P., Kaneko, M., De Carvalho, E.: Multi-flow scheduling for coordinated direct and relayed users in cellular systems. Communications, IEEE Transactions on **61**(2), 669–678 (2013)
7. Bar-Yossef, Z., Birk, Y., Jayram, T. S., Kol, T.: Index Coding with Side Information. In Proceedings of 47th Annual IEEE Symposium on Foundations of Computer Science, pp. 197–206 (2006)

8. Birk, Y., Kol, T.: Informed-Source Coding-on-Demand (ISCOD) over Broadcast Channels. In Proceedings of INFOCOM'98, 1998
9. Lubetzky, E., Stav, U.: Non-linear Index Coding Outperforming the Linear Optimum. In Proceedings of 48th Annual IEEE Symposium on Foundations of Computer Science, pp. 161–168, 2007
10. Chaudhry, M.A.R., Sprintson, A.: Efficient algorithms for index coding. In Proceedings of INFOCOM Workshop, 2008
11. Chaudhry, M.A.R., Asad, Z., Sprintson, A., Langberg, M.: Finding sparse solutions for the index coding problem. In 2011 IEEE Global Telecommunications Conference (GLOBECOM 2011), pp. 1–5. IEEE, 2011
12. Chaudhry, M.A.R., Asad, Z., Sprintson, A., Langberg, M.: On the complementary index coding problem. In *Information Theory Proceedings (ISIT), 2011 IEEE International Symposium on*, pp. 244–248, July 2011
13. El Rouayheb, S., Sprintson, A., Georghiades, C.: On the index coding problem and its relation to network coding and matroid theory. Information Theory, IEEE Transactions on **56**(7), 3187–3195 (2010)
14. Effros, M., El Rouayheb, S., Langberg, M.: An equivalence between network coding and index coding. Information Theory, IEEE Transactions on (2015)
15. Maleki, H., Cadambe, V.R., Jafar, S., et al.: Index coding-an interference alignment perspective. Information Theory, IEEE Transactions on **60**(9), 5402–5432 (2014)
16. Yazdi, S.M., Savari, S., Kramer, G., et al.: Network coding in node-constrained line and star networks. Information Theory, IEEE Transactions on **57**(7), 4452–4468 (2011)
17. Amdouni, I., Adjih, S., Plesse, T.: Network coding in military wireless ad hoc and sensor networks: Experimentation with gardinet. In Military Communications and Information Systems (ICMCIS), 2015 International Conference on, pp. 1–9. IEEE, 2015
18. Liu, H., Zhang, B., Mouftah, B.T., Shen, X., Ma, J.: Opportunistic routing for wireless ad hoc and sensor networks: Present and future directions. Communications Magazine, IEEE, **47**(12), 103–109 (2009)
19. Li, M., Yang, Z., Lou, W.: Codeon: Cooperative popular content distribution for vehicular networks using symbol level network coding. Selected Areas in Communications, IEEE Journal on **29**(1), 223–235 (2011)
20. Asad, Z., Chaudhry, M.A.R., Malone, D.: Codhoop: A system for optimizing big data processing. In Systems Conference (SysCon), 2015 9th Annual IEEE International, pp. 295–300. IEEE, 2015
21. Jafar, S., et al.: Topological interference management through index coding. Information Theory, IEEE Transactions on **60**(1), 529–568 (2014)
22. El Rouayheb, S.Y., Chaudhry, M.A.R., Sprintson, A.: On the Minimum Number of Transmissions in Single-Hop Wireless Coding Networks. In Proceedings of IEEE Information Theory Workshop, Lake Tahoe, CA (2007)
23. Langberg, M., Sprintson, A.: On the Hardness of Approximating the Network Coding Capacity. In the Proceedings of International Symposium on Information Theory (ISIT) 2008
24. Duh, R., Fürer, M.: Approximation of k-set cover by semi-local optimization. In Proceedings of the twenty-ninth annual ACM symposium on Theory of computing, STOC'97, May 1997
25. Ong, L., Ho, C.K.: Optimal index codes for a class of multicast networks with receiver side information. In Communications (ICC), 2012 IEEE International Conference on, pp. 2213–2218. IEEE, 2012
26. Peters, R.: Orthogonal Representations over Finite Fields and the Chromatic Number of Graphs. Combinatorica **16**(3), 417–431 (1996)
27. Chaudhry, M.A.R., Asad, Z., Sprintson, A., Langberg, M.: On the complementary index coding problem. In IEEE International Symposium on Information Theory Proceedings (ISIT), pp. 244–248. IEEE, 2011
28. Moskewicz, M., Madigan, C., Zhao, Y., Zhang, L., Malik, S.: Chaff: Engineering an Efficient SAT Solver. Las Vegas, NV (2001)

29. Een, N., Srensson, N.: An Extensible SAT-solver. In Proceedings of 6th International Conference on Theory and Applications of Satisfiability Testing, 2003
30. Tseitin, G.S.: On the Complexity of Derivation in Propositional Calculus. In: Slisenko, A.O. (ed.) Studies in Constructive Mathematics and Mathematical Logic. Part 2, pp. 115–125. Consultants Bureau, New York (1970)
31. Karp, R.M.: Reducibility among combinatorial problems. Springer, 1972
32. Michael, R.G., David, S.J.: Computers and intractability: a guide to the theory of np-completeness., : San Francisco, p. 1979. Freeman, LA (1979)

Erratum to: RF Energy Harvesting Communications: Recent Advances and Research Issues

M. Majid Butt, Ioannis Krikidis, Amr Mohamed and Mohsen Guizani

Erratum to: Chapter 'RF Energy Harvesting Communications: Recent Advances and Research Issues' in: M.Z. Shakir et al. (eds.), *Energy Management in Wireless Cellularand Ad-hoc Networks*, Studies in Systems, Decision and Control 50, DOI 10.1007/978-3-319-27568-0_15

The original version of Chapter 15 was inadvertently published with incorrect affiliation for Ioannis Krikidis. The correct affiliation is given below.

Department of Electrical and Computer Engineering, University of Cyprus, Nicosia, Cyprus.

The online version of the original chapter can be found under
DOI 10.1007/978-3-319-27568-0_15

M.M. Butt (✉) · A. Mohamed · M. Guizani
Department of computer science and engineering, Qatar University, Doha, Qatar
e-mail: majid.butt@ieee.org

I. Krikidis
Department of Electrical and Computer Engineering, University of Cyprus, Nicosia, Cyprus

© Springer International Publishing Switzerland 2016

M.Z. Shakir et al. (eds.), *Energy Management in Wireless Cellular and Ad-hoc Networks*, Studies in Systems, Decision and Control 50,
DOI 10.1007/978-3-319-27568-0_19

CPSIA information can be obtained
at www.ICGtesting.com
Printed in the USA
LVHW020055311218
602233LV00003B/71/P